COLLECTIVE DYNAMICS: TOPICS ON COMPETITION AND COOPERATION IN THE BIOSCIENCES

To learn more about AIP Conference Proceedings, including the Conference Proceedings Series, please visit the webpage
http://proceedings.aip.org/proceedings

COLLECTIVE DYNAMICS: TOPICS ON COMPETITION AND COOPERATION IN THE BIOSCIENCES

A selection of papers in the

Proceedings of the BIOCOMP2007 International Conference

Vietri sul Mare, Italy 24 – 28 September 2007

EDITORS

Luigi M. Ricciardi
Aniello Buonocore
Enrica Pirozzi
University of Naples Federico II, Italy

All papers have been peer reviewed

SPONSORING ORGANIZATIONS

IBM Italy, S.p.A.
Kay Systems Italia, s.r.l.
National Group for Scientific Computation (GNCS) of the National Institute for Higher Mathematics (INdAM)
Neatec S.p.A.
Sanpaolo Banco di Napoli

Melville, New York, 2008
AIP CONFERENCE PROCEEDINGS ■ VOLUME 1028

Editors:

Luigi M. Ricciardi
Aniello Buonocore
Enrica Pirozzi

Dipartimento di Matematica e Applicazioni
Università di Napoli Federico II
Via Cintia
I-80126 Napoli
Italy

E-mail: luigi.ricciardi@unina.it
aniello.buonocore@unina.it
enrica.pirozzi@unina.it

L.C. Catalog Card No. 2008929302
ISBN 978-0-7354-0552-3
ISSN 0094-243X
Printed in the United States of America

CONTENTS

FOREWORD

The international conference *BIOCOMP2007 - Collective Dynamics: Topics on Competition and Cooperation in the Biosciences* was held 24–28 September 2007 in Vietri, Italy, on the Amalfi Coast. This latest conference in the *BIOCOMP* series was dedicated to the memory of Professor Ei Teramoto on the tenth anniversary of his death. Professor Teramoto's scientific interests and his leadership in establishing theoretical biology as a new research field in Japan and in promoting cooperation among scientists around the world are too well-known to require repetition here. What may be less known to those who did not have the opportunity to become acquainted with him are certain unique aspects of his personality, as well as the essential role he played in establishing a long and fruitful collaboration between Japanese and Italian research groups. These new insights appear in the last paper of this volume.[1]

Like its predecessors, *BIOCOMP2002 - Topics in Biomathematics and Related Computational Problems at the Beginning of the Third Millennium,*[2] and *BIOCOMP2005 - Diffusion Processes in Neurobiology and Subcellular Biology,*[3] *BIOCOMP2007* was conceived as a program of invited lectures and contributed papers designed to bring together a group of mathematicians, physicists, and biologists for an in-depth discussion of model-building and computational strategies. At this meeting, topics focused on information processing and coding in neuronal systems, molecular motors, and quantitative approaches to ecology and population dynamics. Invited talks also addressed current problems within the areas of stochastic processes and applications of quantitative paradigms to the life sciences. In addition, there were overviews of the heritage of Charles Darwin, and of the mathematical contributions of Paul Levy and Norbert Wiener related to the discovery of the Brownian motion. Indeed, most papers presented at this conference acknoledged their roots in the seminal works of Robert Brown, Charles Darwin, Paul Levy and Norbert Wiener.

This volume contains twenty-five articles chosen from the more than ninety papers submitted for the conference proceedings. These articles were selected with a careful eye to their scientific quality and relevance to the specific topics of the conference, or to the general area of methods and tools for biomathematics. The articles explore a variety of topics, including: critical issues in the approach to understanding brains along certain physical, neurophysiologic and psychiatric lines; cooperative and competitive computation; mathematical, physical and computer-based studies of neuronal models and of

---

[1] A complementary information, mostly in Japanese, can be found in the volume "In Memory of Professor Ei Teramoto", compiled by M. Higashi, Y. Iwasa, K. Kawasaki, H. Matsuda, H. Nakajima, N. Shigesada, S. Takeno, N. Yamamura. Yoshioka Publishing Company (Kyoto), pp. 138, 1997.

[2] Proceedings in Special Issues of (i) *Sci. Math. Jap.*, vol. 58, no. 2, 221–494, 2003. ISSN 1346-0862. International Society for Mathematical Sciences (Osaka, Japan); (ii) *BioSystems*, vol. 71, nos. 1-2, 2003, 1–248. ISSN 0303-2647. Amsterdam: Elsevier (The Netherlands); (iii) *Math. Biosciences*, vol. 188, 2004, 1–238. ISSN 0025-5564. New York: Elsevier Science Inc. (USA).

[3] Proceedings in Special Issues of (i) *Sci. Math. Jap.*, vol. 64, no. 2, 173–507, 2006. ISSN 1346-0862. International Society for Mathematical Sciences (Osaka, Japan); (ii) *BioSystems*, vol. 88, no.3, 2005, 175–356. ISSN 0303-2647. Amsterdam: Elsevier (The Netherlands); (iii) *Math. Biosciences*, vol. 207, no.2, 161–401, 2007. ISSN 0025-5564. New York: Elsevier Science Inc. (USA).

related information processing and transmission; stochastic and statistical descriptions of single neuron activities; analytic and numerical models for three-dimensional vessel growth; and various topics in population dynamics and genetics, with a special focus on issues on computation, decision and optimization. Hida Calculus based on the white noise approach to time or space-time dependent random functions is used to model the randomness and variability systematically observed in living beings.

BIOCOMP2007 was developed under the high patronage of the President of the Italian Republic and the patronages of the Ministry of the Cultural Heritage and Activities, the Ministry of Environment, Land and Sea, the Ministry of Foreign Affairs, the Ministry of Technology Innovation, the Ministry of University and Research, the Campania Region, the Federico II Naples University (High Patronage), the Naples Academy of Physical and Mathematical Sciences of the National Society of Science, Literature and Art, and the University of Salerno.

BIOCOMP2007 was co-organized, under the chairmanship of one of us (LMR), by the BIOCOMP ASSOCIATION, the Department of Mathematics and its Applications of Federico II Naples University, the Department of Mathematics and Informatics of University of Salerno, the International Institute for Advanced Scientific Studies "Eduardo R. Caianiello" (IIASS) and the Italian Institute for Philosophical Studies. Participating institutions included the Center for Information Services (CSI) of Federico II Naples University, the CNR Institute for Cybernetics "Eduardo Caianiello", the Institute for Sciences and Cybernetic Technology of the University of Gran Canaria at Las Palmas, the International Society for Mathematical Sciences (ISMS) and the Japan Science and Technology Agency (JST).

Sponsors of the conference included the IBM Italy S.p.A., the Kay Systems Italy s.r.l., the National Group for Scientific Computation (GNCS) of the National Institute for Higher Mathematics (INdAM), the Naples Azienda di Soggiorno e Turismo, the Neatec S.p.A., the Salerno EPT and the Sanpaolo Banco di Napoli.

We are especially grateful to Avv. Gerardo Marotta, the President of the Istituto Italiano per gli Studi Filosofici (the Italian Institute for Philosophical Studies), for co-organizing this conference, and to Professor Antonio Gargano, the Secretary General, for his contributing to its success.

Numerous friends and colleagues have generously and efficiently helped us during all phases of the organization and development of the conference, in particular W. Balzano, G. Colaps, M.R. Del Sorbo, A. Di Crescenzo, V. Giorno, M. Longobardi, A.G. Nobile, F. Palmieri, and L. Sacerdote. To them we extend our most sincere thanks. We are particularly grateful to L. Caputo and to P. Festa for their dedication to the organization of *BIOCOMP2007* and to the preparation of the Proceedings.

Luigi M. Ricciardi
(on the behalf of the Organizing Committee and of the Editors)

Scientific Works by Wiener and Lévy After Brown

Takeyuki Hida

Nagoya University, Japan, E-mail: takeyuki@math.nagoya-u.ac.jp

Abstract. Brownian motion was recognized as a scientific research object by R. Brown in 1827, and then many scientists studied the motion within science. Two famous mathematicians, N. Wiener and P. Lévy, studied Browniam motion mainly in mathematics. We appreciate their works and would like to note that their results play very important role in the study of biological phenomena.

Keywords: Brownian motion, cybernetics, innovation, white noise.
PACS: 01.30.Rr, 02.50.Cw, 02.50.Fz

1. INTRODUCTION

It was 1827 when Robert Brown started scientific study of Brownian motion. Since then, the theory of Brownian motion has extensively developed, and now everybody agrees with the assertion that Brownian motion provides the most basic objects in the modern sciences.

During the last 180 years after Brown, particularly during the earlier half of the last century, significant contributions to the study of Brownian motion were made by three persons; A. Einstein, N. Wiener and P. Lévy.

Brown's observation of grains of pollen suspended in the water was made in 1827 by using a simple microscope with one and the same lens, the focal length of which is about $\frac{1}{32}$nd of an inch. The point that his study is appreciated is that, quite different from what were done before, he did try to study the structure of the pollen and to inquire into its mode of action. The detailed results were reported in his paper [1]. Thus, it is often said that Brownian motion was discovered by R. Brown.

A. Einstein made a Brownian motion to be a research object of mathematics, from which physical application came out. The basic paper [2] describing this result appeared in 1905; the year is called a miracle year for him as is well-known. Right after this paper, followed [3]. Since then, there have been many scientists who succeeded to his study.

Significant mathematical and/or applied mathematical development of the theory of Brownian motion had been made by many scientists; among others we should like to mention two names: Norbert Wiener and Paul Lévy.

Needless to say, N. Wiener and P. Lévy are great mathematicians, specifically in the fields of probability and its applications. N. Wiener is known as the founder of *cybernetics*, so that one can expect that some of his works are closely related to biological sciences related to Brownian motion. This is true, and we shall never be disappointed whenever we follow his works in search of any results related to biological sciences.

The present report aims, for the first place, at a short introduction of Wiener's works

CP1028, *Collective Dynamics: Topics on Competition and Cooperation in the Biosciences*
edited by L. M. Ricciardi, A. Buonocore, and E. Pirozzi

that have close connection with the field of BIOCOMP. We shall emphasize the fact that many of his works had stimulated research in probability. This will be illustrated in what follows successively.

On the other hand, P. Lévy is usually viewed as a pure mathematician; this is true. People therefore may not expect that any of his works would have connections with biology; we can say so in some sense. However, what is claimed here is that some of Lévy's works have been giving deep insight into random phenomena that may be observed in the living creature. Discovering such phenomena is a good homework for us all. While we are doing so, we may find new significant mathematical theories and the effort will never be brought to naught.

The probabilistic works of the two mathematicians are *similar* in the sense that those works have the same source, although realizations and appearance are different, if we understand correctly. The source that we mean is nothing but a Brownian motion, or equivalently a white noise, which is the time derivative of a Brownian motion. This is not an accident, but there is a reason why rich mathematical characteristics are inherent in Brownian motion and they are going to be discovered in step with the development of analysis and other fields of mathematics. This fact can be illustrated as their works are reviewed step by step.

One of the motivations of their earlier works are related to infinite dimensional analysis, in particular the functional analysis. Since the number of variables of functions are infinite, it is naturally requested to have the average of functions in those variables or integration with respect to those variables. Hence, there is required to introduce a measure on infinite dimensional space. Eventually, it became to be known that the most suitable measure has to be Gaussian; in other words the differential space in the Wiener's sense is necessary as a background. See [4]. It is noted that there is referred to the Lévy's work on functional analysis. In fact, this is not a surprise. Thus, we may say that Brownian motion has been a bridge that connects Wiener and Lévy as is seen in their researches in the earlier time. Or, we may say that a Brownian motion is a beautiful realization of their ideas in mathematics.

In short, R, Brown recognized that a Brownian motion is a subject of scientific research. A. Einstein made it a mathematical object. Then, many mathematicians, involving Wiener and Lévy, have kept a Brownian motion as a basic notion of mathematical sciences. Physicists say it is within physical phenomena, biologists claim that it is investigated in biology, and engineers regard it a typical noise and fluctuation that might disturb the communication as well as a useful tool in engineering. All are correct!

Let us now focus our attention to the contribution of Brownian motion to biological sciences.

2. NORBERT WIENER, 1894-1964

Allow me to start with my personal experiences related to Wiener.

I first met him in 1953 at Nagoya, where he delivered a public lecture on cybernetics and where a panel discussion was also organized. To be lucky, I was given an opportunity to participate in the discussion. There he emphasized the importance of the study of brain; brain waves and others. He was delighted to know that "Brain" is "No", which is in Japanese. Later I came back to this problem when I read his book "Nonlinear problems in random theory". This will be mentioned later.

Giacomo Della Riccia, who succeeded Wiener's idea on quantum theory, spent one year with me in US 1964-5 academic year. He explained a lot about the Wiener's opinion on the study of sciences, in particular quantum dynamics. He finished the joint paper [11] with Wiener in 1966.

Also, Pesi Masani, who was the last professor working with Wiener. He edited the Collected works of Norbert Wiener. I was invited by Professor Masani to Indiana University for the academic year 1964-5, so that we talked about Wiener's papers a lot.

There is one more story related to Wiener. He passed away on March 18th, 1964. Exactly four years later, on March 18th, 1968, I was in the midst of teaching graduate course on Probability at Princeton University, so I took this opportunity to talk on Wiener's theory for nonlinear networks and to pay our respect to his memory. See [24] Section 10.

We are now ready to have a quick review of the Wiener's works related to Brownian motion.

2.1. Differential space [4]

An important fact to be mentioned on this work is that Wiener was motivated by integration in infinite dimensional space, not by actual Brownian motion. As is noted in the Introduction, this paper owed its inception to a conversation with P. Lévy on integration of functionals (see [12] and [13]). Having mentioned the Brownian movement briefly, Wiener introduced the differential space, where he took not a Brownian motion itself, but its increments that are independent Gaussian random variables. Those increments determine an infinite direct product of standard Gaussian measures, so that the average of a functional as well as Fourier coefficients can be computed by integration. Thus an infinite dimensional calculus launched actually.

2.2. Fourier transforms in the complex domain [9]

The Fourier series approach to defining a Brownian motion appeared in a concrete formula in 1934. In fact, a complex Brownian motion can be constructed as follows. Let $X_k, Y_k, k = \cdots - 1, 0, 1, 2, \cdots$, be a sequence of independent standard Gaussian random variables. Define complex Gaussian random variables Z_k by

$$Z_k = \frac{1}{\sqrt{2}}(X_k + iY_k).$$

Let $Z_1(t)$ be given by

$$Z_1(t) = tZ_0 + \sum_{n\geq 1} \frac{Z_n(e^{int} - 1)}{in} + \sum_{n\geq 1} \frac{Z_{-n}(e^{-int} - 1)}{-in}.$$

The above sums strongly converges in L^2-sense.

The complex Gaussian process $Z(t) = \frac{1}{\sqrt{2\pi}} Z_1(t)$ is nothing but a complex Brownian motion. Indeed,

$$E(Z(t)) = 0, \quad E(Z(t)Z(s)^c) = t \wedge s,$$

where Z^c denotes the complex conjugate. It is proved that there is a decomposition

$$Z(t) = \frac{1}{\sqrt{2}}(X(t) + iY(t)),$$

such that $X(t)$ and $Y(t)$ are real valued, mutually independent standard Brownian motions.

Thus, a construction of a Brownian motion is given through the random Fourier series. Such a method is interesting for a construction, done in the spirit of Fourier analysis. In addition, the Fourier series expansion is often useful for analysis of nonlinear functionals of Brownian motion $B(t), t \in R$, since the variables of functionals can be expressed in terms of ordinary (real valued) Gaussian random variables as many as *countably infinite*. The partial derivatives can therefore be reduced to the Gâteaux derivatives. Later we shall compare with another significant method of construction due to Lévy (see 4) after the comments on Lévy's book [16]).

These results [4], [9] as well as [5] below, by Wiener provided a background of his approaches to the cybernetics, nonlinear problems in random theory and others.

2.3. The homogeneous chaos [5]

From the viewpoint of probability theory, this paper is one of the most important papers on stochastic analysis. After [4] Wiener discussed nonlinear functions of a Brownian motion. The collection of those functionals with finite variance forms a Hilbert space (in our notation (L^2)) which is the basic space, where the analysis of functionals of a Brownian motion is carried on. An interesting tool is the so-called Fock space which admits a direct sum decomposition into the space H_n of homogeneous chaos of degree n:

$$(L^2) = \oplus_0^\infty H_n.$$

The space H_n is spanned by the multiple Wiener integrals of degree n due to K. Itô. The Fock space plays a role of a *milestone* of stochastic analysis.

2.4. Cybernetics [6]

As is indicated by the subtitle of this book by such words as "Control and communication in the animal and the machine", he was interested in the behavior of animals. As for the theoretical part, Wiener wished to form stochastic processes as functions of a Brownian motion as wide as possible and to discuss the analysis of them. And he tried to apply the established analysis to the theory of stochastic processes and time series, where the time development is always taken into account. Naturally, we did find good applications to the information and communication theories. Wiener further proceeded to discuss brain waves and self-organizing systems. It is noted that in the important places much contribution is made by a Brownian motion.

Any complex systems that admit Brownian input can be identified by observing the output. There Lee-Wiener's networks are essential tools (cf. [26]). Incidentally, mathematical theory behind those mechanism of the instruments is the Hellinger-Hahn theorem for determination of the spectrum of unitary operators. See e.g. [25]. The details are also described in the following book [8].

One more topic has to be mentioned related to cybernetics. Wiener introduced the notion of entropy, maybe independently of Shannon. Having been motivated by this fact Kiho et al. (including the present author) have discussed a behavior of protein.

2.5. Nonlinear problems in random theory [8]

Throughout this book, Brownian motion plays dominant roles. Starting with a construction of a Brownian motion, denoted by $x(t)$ (instead of our notation $\dot{B}(t)$) by successive approximation, follow discussions on homogeneous polynomial functionals in $x(t)$'s. Stationary stochastic processes with Brownian input are discussed. Those processes share the time shift (the flow) with Brownian motion, and the spectra of the processes are computed. Together with the ergodic property of the flow of Brownian motion, many significant applications are discussed. Study of brain waves would be a good proposal towards physiological study. Identification of black boxes is a useful technology that comes from significant properties of the flow of Brownian motion. In reality, countably infinite spectral type of the flow implies the technique in question (see [25] for details). Now, we should like to mention a very interesting result obtained by using this method of identification of unknown organ. It is Ken Naka's result [22], where the reaction of retina (actually, catfish retina) to a white noise (a Brownian) input is studied. Note that, in addition to the spectral type, Brownian input is the best choice among the possible input signals to a black box. We have to skip the interpretation by the reason to save pages, to our regret.

This book shows how profound and useful properties are found out of Brownian motion. We shall be encouraged to be in search of some other basic properties. This fact will be briefly mentioned at the end of the present report.

Some additional notes are now in order.

In [10] the authors (Wiener and A. Rosenblueth) have discussed formulation of the actions of cardiac muscle mathematically. It is interesting to note that they use Poisson noise which occupies a similar position to that of Brownian motion (or white noise). Wiener and Wintner published a paper in 1943 under the title "Discrete chaos", which gives a background of the research in [10]. Incidentally, probabilistic interpretation on Poisson noise has been given by [23], where optimality and symmetry of the noise are emphasized.

The title of his 1956 autobiography is "I am a Mathematician". Surely he is so, but we should like to say that Wiener is an outstanding "Scientists". We see that his way of writing is neither like ε and δ type, nor "Satz und Beweiss" style like Landau. He is hasty to express his idea even in a naive stage. Anyhow, we learn from him a lot, in particular about Brownian motion.

3. PAUL LÉVY, 1886-1971

Everybody agrees, I am sure, with the opinion that Lévy's most important contribution to biological sciences is that he has appreciated the *significance* of Brownian motion and has discovered its profound properties that have influence over the study of biological phenomena.

P. Lévy is a man of probability theory. The main research object is *stochastic processes* with special emphasis on Brownian motion. We know that Brownian motion is like a king of the circle of stochastic processes sitting in the center. Having studied his works, we really recognize that he did make a frontal attack on Brownian motion.

He passed away on December 15, 1971. Around mid December of the year, we have often organized the Lévy Seminar at Nagoya in honor of him. Last year the 5th Lévy Seminar was held. At each Lévy seminar we have realized again and again the importance of his legacy and the significance of his works.

Allow me to start with personal affairs concerning Lévy.

1) I started to communicate with him by mail early 1950's. He used to send his reprints and to give me valuable comments on my papers.

2) I met him first in the winter of 1968 at his house in Paris 16e. This was the only opportunity to meet him face to face.

On this occasion he kindly recommended me to work multi-dimensional parameter Brownian motion (see [19]); now it is called Lévy's Brownian motion (not a Brownian sheet). There is an episode to know how he put the importance on Lévy's Brownian motion. It is this: When he was appointed as a member of the Academy 1964, he was given some budget to prepare a formal gown. In reality he did not buy it, but the money was contributed to the Prize (the Lévy prize) given to the author who publish excellent paper on Lévy's Brownian motion.

Let me continue the story about my visiting him. We discussed his formula the so-called *infinitesimal equation* for a stochastic process $X(t)$:

$$\delta X(t) = \Phi(X(s), s \leq t; Y(t), t, dt).$$

This variational equation for $X(t)$ appeared in his 1953 Lecture Notes [18].

The $Y(t)$ is the *innovation.* I have been deeply impressed at this formal equation; although it has only an intuitive significance, but tells us deep meaning. In fact it has directed me to the idea of "Reduction" and eventually to the innovation theory.

3) Some months after we met, he kindly communicated my paper to C. R. Acad. Sc. Paris.

4) There were more private communications (including non-mathematical matters).

It was 1969 when he kindly sent me a letter, where his additional autobiography was written, together with the introduction of his family, at my request. Also, on the occasion when his autobiography published 1970, he kindly sent me a copy. I tried hard to translate it into Japanese with my friend Yamamoto, so that Lévy's works and ideas would become popular in Japan. He hoped so with pleasure. To be extremely sad, he passed away in Dec. 15, 1971, before the Japanese translation appeared in 1973. However, L. Schwartz sent us a forward to the translation; it is our great pleasure and we express our deep thanks to him.

We now turn our eyes to the monographs that were written by him.

Leçons d'analyse fonctionnelle [12].

Wiener was influenced by this book, as was explained before (see the Section Differential space). In order to discuss calculus of functionals, Lévy wished to have a measure on function space. Start with the volume of the n-sphere for large n. We recognize the volume getting to be concentrated on the surface as n is getting larger. To be interested, if the radius is taken to be $\sqrt{n}$, then the projection of uniform probability measure on the surface of the volume down to a one-dimensional space is close to the standard Gaussian measure. Each direction determines the same measure and different directions imply independence. Intuitively speaking we shall be given a countably infinite dimensional standard Gaussian measure. This is nothing but the measure (probability distribution) μ of Gaussian white noise, a realization of which is obtained by taking the time-derivative $\dot{B}(t)$ of a Brownian motion $B(t)$. Thus, the measure μ is thought of as the probability distribution of the generalized stochastic process $\dot{B}(t)$. The increments of $B(t)$ that are dealt with in Wiener's differential space are close to the method of Lévy just explained above.

Incidentally, his original paper on this fact was published in 1919.

After this book appeared, the contents, in particular Part III, had extensively developed and the second edition appeared in 1951 under the title

Problèmes concrets d'analyse fonctionnell [17].

Much of what are written can be rephrased in terms of stochastic calculus, so that this book may be said to be a fountain, as it were, of ideas for white noise analysis.

Théorie de l'addition des variables aléatoires [15].

As is written in the Preface of this book (cf. [13]), he did not quite accept the classical probability theory. The title of this book says the sum of random variables. And Brownian motion appears. This is not surprising; in fact, Brownian motion has independent increments, so that it is a continuous analogue of sum of independent random variables. If an additive process satisfies some conditions on sample functions and the increment is stationary in time, then the additive process is decomposed into independent atomic additive processes: loosely speaking, it is a sum of a Brownian motion up to constant and compound Poisson process which consists of independent Poisson processes with various values of jump. All components are mutually independent. This is the Lévy decomposition, or called Lévy-Itô decomposition, of an additive process. Because of the *independent* increment property, additive processes are basic, and among those basic processes Brownian motion occupies a central position. We understand the reason why a great deal of this book is devoted to Brownian motion.

Next, we see the most important monograph on Brownian motion by Lévy.

Processus stochastiques et mouvement brownien [16].

The main part of this book is devoted to the theory of Brownian motion including the cases where the values are two-dimensional and where the parameter space is multi-dimensional. We should like to note that an effective determination of a Brownian motion is in Chapter 1. It shows his method of constructing a Brownian motion, through which its profound properties including latent traits can be clarified. Before explain this fact, we shall see his way of construction.

Prepare a system of independent standard Gaussian random variables $\{Y_n\}$. Start from $X_1(t), t \in [0,1]$, given by

$$X_1(t) = tY_1.$$

The sequence of processes $\{X_n(t), t \in [0,1]\}$ is formed by induction. Let T_n be the set of binary numbers $k/2^{n-1}, k = 0,1,2,\cdots,2^{n-1}$, and set $T_0 = \cup_{n\geq 1} T_n$. Assume that $X_j(t) = X_j(t,\omega), j \leq n$, are defined. Then, we set

$$X_{n+1}(t) = \begin{cases} X_n(t),\ t \in T_n, \\ \dfrac{X_n(t+2^{-n}) + X_n(t-2^{-n})}{2} + 2^{-\frac{n}{2}} Y_k,\ t \in T_{n+1} - T_n, \\ \qquad\qquad k = k(t) = 2^{n-1} + \frac{2^n t+1}{2}, \\ (k+1-2^n t) X_{n+1}(k2^{-n}) + (2^n t - k) X_{n+1}((k+1)2^{-n}), \\ \qquad\qquad t \in [k2^{-n}, (k+1)2^{-n}]. \end{cases} \tag{1}$$

It is easy to see that the sequence $X_n(t), n \geq 1$, is consistent in n and that the uniform L^2-limit of the $X_n(t)$ exists. The limit is denoted by $\tilde{X}(t)$. It is proved that it has independent increments and

$$\begin{aligned} E(|\tilde{X}(t) - \tilde{X}(s)|^2) &\leq |t-s|, \\ \Gamma(\tilde{X}(t), \tilde{X}(s)) &= t \wedge s, \end{aligned}$$

where Γ denotes the covariance.

Furthermore it can be proved that $X_n(t)$ is an almost sure convergent sequence and we denote the limit by $X(t)$. It is a Brownian motion.

This construction is quite unique and significant. It is different from what is proposed by Wiener. The reasons why we say that Lévy's method of construction (i.e. approximation) is significant are now in order.

1) The time parameter t always appears explicitly in every step of approximation. Hence, evolutional phenomena that are expressed as functionals of Brownian motion can be approximated without losing the time parameter, so that way of time-developing can also be approximated.

2) The time derivatives, either from right or from left, of $X_n(t)$ can approximate a white noise $\dot{X}(t)$.

Even more, a quadratic form of the renormalized square of white noise can well be approximated by this method. This property is a characteristic of quadratic forms. If a fluctuation of a trajectory of some dynamics is represented by a Brownian notion, then the kinetic energy may be dealt with by using this fact.

3) Since any real number t has the binary expansion, so the system $\dot{X}(t), t \in [0,1]$, may behave as if it were a base of the space of Brownian functionals, however they span

a separable space denoted by $H_1^{(-1)}$, which is an extension of H_1 defined as the space of linear functionals of $X(t)$ with finite variance. The collection $\{\dot{X}(t)\}$ is not quite a base, but it is *total* in the space $H_1^{(-1)}$.
Remark. The statement above can be illustrated rigorously. The proof needs, however, the profound results on white noise analysis.

4) Since white noise $\dot{X}(t)$'s (in our earlier terminology $\dot{B}(t)$'s) are taken to be the variables of functionals, we may think of partial derivatives $\partial_t = \frac{\partial}{\partial \dot{X}(t)}$. This corresponds to the Fréchet derivative that comes from the Volterra form of variation of functionals. We can further define the adjoint ∂_t^* of ∂_t and Laplacians, too.

And so forth.

To follow what are written in this book [16] is not easy. However, while we are trying to understand, we often discover essential meaning behind the author's description. In particular, we realize this fact in later chapters of this book.

As he recommended (private communication and [19]), it seems reasonable that one chapter, Chapter 8, is provided to the discussions on multi-dimensional parameter Brownian motion, i.e. the Lévy Brownian motion. There we learn a lot, and are given interesting problems to be investigated.

Quelques aspects de la penseé d'un mathématicien [21].

This is the Lévy's autobiography, where we can see some of his idea which can not be found from his mathematical papers and books (also see [20]. Since he is not a person who is willing to write his ideas, this autobiography is helpful to understand his mathematics. See [14].

There is a quotation from the introduction of [21]:

J'ai bâti un eíage de cet édifice; que d'autres continuent!

4. CONCLUDING REMARKS

We have so far quickly reviewed the works by N. Wiener and P. Lévy on Brownian motion, noticing that they are mainly discussed from theoretical side. Some results are immediately applied to actual problems, while some are just theoretical. However, theoretical results often describes latent traits of living cells and some other biological movements. We are suggested to be in search of implicit properties of actual random phenomena, in addition to the study of probabilistic properties of Brownian motion.

4.1. Open problems

1. Duality between Brownian motion and Poisson process.
   Both Wiener and Lévy have often discussed Poisson process, so they appreciate significance of Poisson process (noise). There are lots of similarity in analysis of their functionals between the two, and dissimilarity should be more interesting (cf. [23]). Duality may help the dissimilarity.

2. Invariance of Brownian motion.
   Conformal invariance of Brownian motion will serve to describe the characteristics of random complex systems.
3. Optimalities of trajectory for Brownian motion (cf. [7]) and Poisson process should be discussed in details.
4. Random fields, including Lévy's Brownian motion, have rich favorite properties and are fitting for applications to biology, quantum dynamics and geometry, as we see in the recent topics.

REFERENCES

1. R. Brown, *The Philisophical magazine and annals of philosophy* New series, **4** (21), 161–178 (1928).
2. A. Einstein, *Ann. der Physik*, **17**, 549–560 (1905).
3. A. Einstein, *Ann. der Physik*, **19**, 371–381 (1906).
4. N. Wiener, *J. Math. and Physics*, **2**, 131–174 (1923).
5. N. Wiener, *American JU. of Mathematics*, **60**, 897–936 (1938).
6. N. Wiener, *Cybernetics: or control and communication in the animal and the machine*, The MIT Press, 1948 and 1961.
7. N. Wiener, *Extraporation, interpolation and smoothing of stationary time series*, The MIT Press, 1949.
8. N. Wiener, *Nonlinear problems in random theory*, The MIT Press, 1958.
9. N. Wiener, R.E.A.C. Paley, *Fourier transforms in the complex domain*, American Math. Soc, 1934.
10. N. Wiener, A. Rosenblueth, *Arch. Inst. Cardiol. Méxicana*, **16**, 205–265 (1946).
11. N. Wiener, G. Della Riccia, *J. Math. Phys*, **7**, 1372–1383 (1966).
12. P. Lévy, *Leçons d'analyse fonctionnelle*, Gauthier-Villars, 1922.
13. P. Lévy, *Calcul des probabilités*, Gauthier-Villars, 1925.
14. P. Lévy, *Notice sur les travaux scientifiques*, Hermann, 1935.
15. P. Lévy, *Théorie de l'addition des variables aléatoires*, Gauthier-Villars, 1937 [2éme ed. 1954].
16. P. Lévy, *Processus stochastiques et mouvement brownien*, Gauthier-Villars, 1948 [2éme éd. revue et augmentée. 1965].
17. P. Lévy, *Problèmes concrets d'analyse fonctionnelle*, Gauthier-Villars, 1951.
18. P. Lévy, *Random functions: General theory with special reference to Laplacian random functions*, Univ. Calif. Pub. in Statistics, **I**, 331–358 (1953).
19. P. Lévy, *Le mouvement brownien fonction d'un ou de plusieurs paramètres*, Rendiconti di Matematica, **22**, 24–101 (1963).
20. P. Lévy, *Nouvelle Notice sur les travaux scientifiques*, 1964.
21. P. Lévy, *Quelques aspects de la penseé d'un mathématicien*, Blanchard, 1970.
22. K. Naka and V. Bhanot, "White-Noise analysis in retinal physiology", in *Advanced Mathematical Approach to Biology*, edited by T. Hida, World Scientific, 1997, pp. 109–267.
23. S. Si, *IDAQP*, **6**, 609–617 (2003).
24. T. Hida, *Stationary stochastic processes*, Blanchard, 1970.
25. T. Hida, *Brownian motion*, Springer-Verlag, 1980.
26. M. Schetzen, *The Volterra and Wiener theories of nonlinear systems*, Krieger Pub. Co., 2006.

Philosophical Approaches towards Sciences of Life in Early Cybernetics

Leone Montagnini

Dipartimento di Matematica e Applicazioni, Università di Napoli Federico II, and Istituto di Cibernetica "Eduardo Caianiello", Napoli, Italy

Abstract. The article focuses on the different conceptual and philosophical approaches towards the sciences of life operating in the backstage of Early Cybernetics. After a short reconstruction of the main steps characterizing the origins of Cybernetics, from 1940 until 1948, the paper examines the complementary conceptual views between Norbert Wiener and John von Neumann, as a "fuzzy thinking" versus a "logical thinking", and the marked difference between the "methodological individualism" shared by both of them versus the "methodological collectivism" of most of the numerous scientists of life and society attending the Macy Conferences on Cybernetics. The main thesis sustained here is that these different approaches, quite invisible to the participants, were different, maybe even opposite, but they could provoke clashes, as well as cooperate in a synergic way.

Keywords: Cybernetics, history of Computer Science, individualism, organicism, Macy conferences.
PACS: 01.70.+w

1. INTRODUCTION

Given the theme of Biocomp2007 on "Competition and Cooperation in the Biosciences", I considered useful to draw an historical outline of the early developments of Cybernetics in the USA, during World War Two and its aftermath, focusing in particular on the different conceptual and philosophical approaches towards the sciences of life operating in the backstage. It is my belief that this history contains indeed lots of aspects requiring an in-depth study, dealing with the always hard-to-define epistemological approaches and also with records at times sunk into oblivion if not unknown at all.

2. THE GENESIS OF CYBERNETICS IN THE USA

During the war - according to the Norbert Wiener's recollections [19] - the main steps of the dawning Cybernetics were represented by Wiener's *Memorandum* of 1940 on digital computers, by his work with the engineer Julian Bigelow on anti-aircraft systems, condensed in the book nicknamed *Yellow Peril* [20], containing his fundamental mathematical "prediction theory", and by the article *Behavior, Purpose and Teleology,* written with Bigelow and the physiologist Arturo Rosenblueth [21], in which they proposed to consider the intentional (or "purposive") human or animal behaviors in parallel with the self-controlled machines, based on feedback systems. A more obscure period (1943-45) followed, in which Wiener went back to computing studies, became familiar with the

CP1028, *Collective Dynamics: Topics on Competition and Cooperation in the Biosciences*
edited by L. M. Ricciardi, A. Buonocore, and E. Pirozzi

1943's McCulloch-Pitts work on neural nets [7], and carried out a lesser known activity as an advisor to help von Neumann in computing researches.

A very rare evidence of their collaboration on computer science during the war is a letter from von Neumann to Wiener, of November 29, 1946. Von Neumann states:

> Our thoughts - I mean yours and Pitts' and mine - were so far mainly focused on the subject of neurology, and more specifically on the human nervous system, and there primarily on the central nervous system. Thus, in trying to understand the function of automata and the general principles governing them, we selected for prompt action the most complicated object under the sun - literally. In spite of its formidable complexity this subject has yielded very interesting information under the pressure of the efforts of Pitts and McCulloch, Pitts, Wiener and Rosenblueth. Our thinking - or at any rate mine - on the entire subject of automata would be much more muddled than it is, if these extremely bold efforts - with which I would like to put on one par the very un-neurological thesis of R. Turing - had not been made. (Von Neumann to Wiener, in [18], pp. 278-282)

Roughly speaking, von Neumann took the schemes of McCulloch and Pitts, substituting their theoretical neurons with relays or with electronic valves. After that, using the "very un-neurological thesis" of Turing about computability, he managed to design a general purpose computer, the machine that he had begun to build at the IAS just in 1946.

Such an itinerary comes to an early synthesis with the meeting organized by Wiener and von Neumann at Princeton, on January 6 and 7, 1945 [19]. As Wiener reported to Rosenblueth:

> The first day von Neumann spoke on computing machines and I spoke on communication engineering. The second day Lorente de Nó and McCulloch joined forces for a very convincing presentation of the present status of the problem of the organization of the brain. In the end we were all convinced that the subject embracing both the engineering and neurology aspects is essentially one, and we should go ahead with plans to embody these ideas in a permanent program of research (Wiener to Rosenblueth, January 24, 1945; quoted from [3], pp. 157-158)

After that, a series of ten conferences organized by Macy Foundation of New York (1946-50) appeared to be the ideal occasion where the topics could be dealt with. Here the group of the 1945 Princeton meeting gathered with another group of scientists including Gestalt psychologists, social scientists interested in relational phenomena, and George E. Hutchinson, one of the fathers of the modern scientific ecology. This other group hoped to clarify mutual causal phenomena in society and in biological systems, and therefore it was interested in the Cybernetics ideas and, in particular, in the concept of feedback [2].

In fact, we could consider the constitutive path towards Cybernetics concluded with the publication of Wiener's book [19], in 1948, which gave the new field its name.

In Wiener's opinion, Cybernetics was essentially - as Aldo De Luca have recently stressed - a "research program" [1], whose his wartime researches were paradigmatic to draw the guidelines. Those had been researches parallelizing technical devices and biological systems. On one hand the control theory used to project self-correcting and goal-seeking devices and to describe the nervous system in the animal; on the other hand the computer science applied to describe brains and computers or "electronic brains". There was in addition a unifying idea: Cybernetics would be a sort of generalized "communication engineering", focused on the production, elaboration, transmission and use of information, applied to describe the machines, as well as the biological organisms.

3. THE SCIENTIFIC APPROACHES OF WIENER AND VON NEUMANN

Steve J. Heims, one of the major historians of Early Cybernetics, stressed the marked differences between the approaches of Wiener and von Neumann, speaking of a theoretical opposition, as "chaos" versus "logic" [3]. Pesi R. Masani, another of the major biographer of Wiener and his former collaborator, pointed out on the contrary the analogies between the two men [6]. As I have shown in my book on the Norbert Wiener's intellectual itinerary, *Le Armonie del Disordine*, Wiener's way of making science was clearly influenced by the Pragmatism of his old teachers at Harvard, where he got his Ph.D. in philosophy. From Pragmatism he inherited the idea of an "essential irregularity of the universe" and of the superiority of the statistical and stochastic methods [10]. On the other hand, John von Neumann - a pupil of Hilbert and partly influenced by Logical Positivism - was in various aspects of his scientific work more inclined to create very rigorous and "closed" logical models, from the set theory to the quantum mechanics, searching for the basic components of the complex systems.

In any case, they were able to work very well in a complementary way. As far as Cybernetics goes, they both believed that the key to understanding the world of life and mind was "information"; and that could be possible to extend the technological ideas to living organisms when considered as systems elaborating information. In fact a sort of division of tasks arose quite naturally between them. Computer science became for von Neumann his natural field, while Wiener went on working with Rosenblueth and Pitts to apply control theory to the nervous system but also looking for an extension of the neural nets in a stochastic sense.

4. VON NEUMANN'S DOUBTS

However as far back as the end of 1946 von Neumann began to have serious doubts about the possibility to go on with his part of the cybernetic program. His letter to Wiener, quoted above, went on:

> What seems worth emphasizing to me is, however, that after the great positive contribution of Turing - cum - Pitts - and-McCulloch is assimilated, the situation is rather worse than better than before. Indeed, these authors have demonstrated in absolute and hopeless generality, that anything and everything Brouwerian can be done by an appropriate mechanism, and specifically

> by a neural mechanism -and that even one, definite mechanism can be "universal". Inverting the argument: Nothing that we may know or learn about the functioning of the organism can give, without "microscopic", cytological work any clues regarding the further details of the neural mechanism. (Von Neumann to Wiener, in [18], pp. 278-282)

All right, von Neumann seems to say, we now have a machine that could - at least theoretically - deal with every possible Brouwerian concept, that is, with every notion considered in a finitary sense. But this great result is - in von Neumann's opinion - also the symptom of a failure. In fact, the general purpose computer is the living proof that one can build a general purpose machine not only with a the simplified natural neuron used by McCulloch and Pitts, but also by means of relays, flip-flops and so on.

Therefore the McCulloch-Pitts' neural nets could not have a guide function anymore for the new discoveries about the real brains. With their simplification of neurons, McCulloch and Pitts had put the real neuron in a black box, but now one had to go back into the black box, that is to study it again at the "cytological" level. Therefore the letter proposed to go back to the "standard component" of the nervous system and every other systems of a living organism, that is the cell, and prospected to Wiener a wide program including X-ray analysis of the cellular structure. He assured him he was able to formulate rigourously the problem of self-reproduction: "I can formulate – von Neumann wrote – the problem rigourously, in about the style in which Turing did it for his mechanisms." (Von Neumann to Wiener, in [18], pp. 278-282)

Von Neumann went on working on the parallel between computer and brain until his death, but this letter shows how precociously the "computational" side of Cybernetics was tempted to take an autonomous path. However with this new point of view von Neumann widened the Cybernetics field including molecular biology. And as Lily Kay pointed out, von Neumann - and Wiener as well - will have also an active role in that history [4].

5. METHODOLOGICAL INDIVIDUALISM OR METHODOLOGICAL COLLECTIVISM

The Macy Conferences on Cybernetics became the place where this and other questions should have been discussed. In particular the earliest conferences, in 1946 and 1947, were very exciting but also rather tumultuous. The officials of Macy Foundations spoke of "problems of communication" [17]. In general the differences in training between "hard" and socio-human scientists seemed to explain these difficulties. But the stumbling-block was more likely represented by the different epistemologies which were operating in the "backstage": "methodological individualism" and "methodological collectivism", if we follow Popper's terminology [12].

Von Neumann, but also Wiener, and in general every participants to the 1945 Princeton meeting were all, more or less, "methodological individualists", and this was true notwithstanding the presence in that group of physiologists, a type of scientists traditionally keener to welcome organicist ideas. In addition they wanted to continue studying the concrete questions of Cybernetics, in particular the relation between central and peripheral nervous system, with the newly born computers and self-controlled systems [8].

On the contrary the group of socio-human scientists and a man like George E. Hutchinson, desired to know how to study in a better way, maybe in a mathematical way, the phenomena characterized by circular causality, that is when, for example, "variable A affects B, B affects C, and C affects A" again, to quote the anonymous (possibly Bateson himself) that wrote the "Introductory statement" for the Conference on Teleological Mechanisms [13] [9].

The difference becomes just clearly perceivable if we compare the title of Wiener's book (*Cybernetics, or control and communication in the animal and the machine*) [19] to the one chosen by the curators of the official proceedings of Macy Conferences: *Cybernetics, Circular Causal, and Feedback Mechanisms in Biological and Social Systems* [17].

6. THE 1946'S CONFERENCE ON TELEOLOGICAL MECHANISMS

In order to understand the problems arising from the encounter between the two epistemologies it's useful to pay attention in particular to the meeting on 'Teleological Mechanisms', which took place on October 21 and 22, 1946, under the aegis of The New York Academy of Sciences. A meeting collateral to the ordinary Macy Conferences which had been thought on purpose to study circular causality. We need to integrate the official proceedings [11] with other materials such as the manuscript of the "Preprinted abstracts" [13], including never published papers (the Bateson's, Lazarsfeld's and von Neumann's ones) or papers completely rewritten as in the case of Wiener's "Self-correcting and goal-seeking devices and their breakdown" [13], whose original version has been published only in 1985. For the official publication Wiener in fact preferred to substitute it with a long paper on his idea about Cybernetics as a generalized communication engineering. Actually in the feedback itself, Wiener saw essentially a message without exchange of matter or energy, and in particular a message which comes back, rather than a loop.

In the official proceeding the real interest of the "methodological collectivists" is clear in the paper of the neuro-surgeon William Livingston that spoke on "The vicious circle in Causalgia," [11, 13] a disease characterized by burning pain, supposed to be caused by a reverberating circuit provoked by an injury in efferent nerves. The same point of view was evident in the paper of Hutchinson on "Circular causal mechanisms in ecology" [11, 13], where he spoke about the self-regulatory mechanisms exemplified by the Carbon Cycle in Biosphere, the Phosphorus Cycle in Inland Lakes, the prey-predator models etc.

During the Sixties, M. Maruyama [5] sustained that early Cybernetics had neglected positive feedbacks and mutual causality. That's true in some way. But this meeting showed how several scientists agreed with the importance of those notions. McColl himself, the author of the most influential handbook on automatic controls of that period, didn't object at all to the application of the control theory outside of the technical context [11]. Wiener on the contrary didn't appear then very enthusiastic about the application of Cybernetics outside of the nervous system. Why? Partly he feared the advent of an illiberal social-engineering. But he didn't appear keen on the application of Cybernetics to ecology either. This position seems to have depended on his choice to privilege informational feedbacks compared to those feedbacks in which matter and energy are most important. And Hutchinson's paper had clearly pointed out the importance of matter and energy in circular causal phenomena in ecology. Probably just a clearer reflexion about the differences between the two kinds of feedbacks would have helped towards a more definite comprehension of cybernetic problems in general.

6.1. 7. Almost a Conclusion

Notwithstanding the difficulties, the different conceptual and philosophical views operating in the backstage, more or less evident to the characters involved, went on to cohabit in Cybernetics and also in the disciplines in which it will fragment itself (see, e. g., Settimo Termini's considerations in [16]), sometimes opposing and clashing, but other times constructively cooperating, as shown, in the case of Italy, by many papers of the book *Imagination and Rigor* [15], which deals with Eduardo R. Caianiello and his school. A school whose ideas still continue to be a useful guide for present day research as witnessed, for instance, by the challenging paper by Luigi M. Ricciardi [14].

REFERENCES

1. A. De Luca, *Scientiae Mathematicae Japonicae*, **64**: 2, 243–253 (2006).
2. S. J. Heims, *The Cybernetics Group*, MIT Press, Cambridge (MA) 1991.
3. S. J. Heims, *John von Neumann and Norbert Wiener. >From Mathematics to the Technologies of Life and Death*, MIT Press, Cambridge (MA), 1980; 2nd ed. 1984.
4. L. E. Kay, *Who wrote the book of life? A history of the genetic code*, Stanford University Press, Stanford, 2000.
5. M. Maruyama, *American Scientist*, **51**: 2, 164–179 (1963).
6. P. R. Masani, *Norbert Wiener, 1894-1964*, Birkhäuser Verlag, Basel - Boston - Berlin, 1990.
7. W. S. McCulloch and W. Pitts, '*Bulletin of Mathematical Biophysics*, **5**, 115–133 (1943).
8. W. S. McCulloch, "To the members of the conference on teleological mechanisms - Oct. 23 & 24, 1947," Typescript, in McCulloch's Archive, Library of the APS, Philadelphia. Box no. 19.
9. L. Montagnini, *Kybernetes*, **36**: 7/8, 1012–1021 (2007).
10. L. Montagnini, *Le Armonie del Disordine. Norbert Wiener Matematico-Filosofo del Novecento*, Istituto Veneto di Scienze Lettere ed Arti, Venezia, 2005.
11. New York Academy of Sciences, "Conference on Teleological Mechanisms," October 21st and 22nd, 1946, in *Annals of the New York Academy of Sciences*, **50**, pp. 185-278 (1948).
12. K. Popper, *The poverty of historicism*, 1st ed. 1944, Beacon Press, Boston, 1957.
13. "Preprinted abstracts of Conference on Teleological Mechanisms", October 21st and 22nd , 1946. Typescript, in Box 0-14 of M. Mead's Archive, Library of Congress, Washington.
14. L. M. Ricciardi, in [15], 133–145 (2006).
15. S. Termini (ed), *Imagination and Rigor. Essays on Eduardo R. Caianiello's Scientific Heritage*, Springer, Milan - Berlin - New York, 2006.
16. S. Termini, *Scientiae Mathematicae Japonicae*, **64**: 2, 461-468 (2006).
17. H. von Förster (ed.), *Cybernetics : Circular Causal, and Feedback Mechanisms in Biological and Social Systems*, Josiah Macy Foundation, New York. Conferences from 6th to 10th (1949-1953).
18. J. von Neumann, *Selected letters*, edited by Miklos Redei, Providence: American Mathematical Society; London: London Mathematical Society, 2005.

19. N. Wiener, *Cybernetics, or Control and Communication in the Animal and the Machine*, MIT Press, Cambridge (MA), 1948.
20. N. Wiener, *Extrapolation, interpolation, and smoothing of stationary time series, with engineering applications*, New York and London: Wiley & Sons; Cambridge, MA: MIT Press (1949). First published during the war as a classified report to Section D2, National Defense Research Committee and nicknamed "Yellow Peril".
21. N. Wiener, A. Rosenblueth and J. Bigelow, *Philosophy of Science*, **10**, 18–24 (1943).

Darwin and his Mathematical Inspirations

Andrea Pugliese

Department of Mathematics, University of Trento, Povo (TN), Italy, E-mail: andrea.pugliese@unitn.it

Abstract. I have been kindly asked by the organizers of the BIOCOMP2007 conference to provide a short sketch of Charles Darwin's contribution to science, and of the role mathematics has played in his discoveries and in subsequent developments. I felt flattered by the invitation but rather unfit to it, since I have no particular expertise in evolutionary theory, and even less in its history; eventually, I decided to accept the invitation, appreciating the opportunity to read some more about Darwin, and the importance of making his contribution better known, at a time where teaching at school the theory of evolution is coming under attack also in Italy (perhaps under American influence). I hope to be able here to give a glimpse of the history of Darwinian thought, and of some current research areas, that will lead some readers towards further reading. There are many excellent books available now about Darwin and Darwinian theory, and my presentation is based on many of them, listed in the Bibliography; I found especially illuminating the book by Gayon *Darwinism's Struggle for Survival*, a history of theoretical Darwinism illustrating the scientific content, and the philosophical implications, of the debates on evolutionary theory at Darwin's time and up to the "modern synthesis".

Keywords: Darwin, natural selection, theory of evolution, mathematical models of evolution.
PACS: 01.60.+q, 01.65.+g

1. SYSTEMATIC BIOLOGY AT (AND BEFORE) DARWIN'S TIME

The century before Darwin had seen the rise of systematic biology, of which the famous Linnean classification is a permanent legacy. While the concept of higher taxa was considered a fiction by some (indeed, now I see it difficult to give them any meaning outside an evolutionary interpretation), the undisputed central entity was that of species; the general view at the time (by great naturalists that will afterwards be known as "fixists") was that species are represented by a type (the specimen conserved in the museum); variation exists in species around the type but only within a "sphere of variation", and species cannot be turned into other species.

Geographical discoveries, and especially the fossils, caused the development of evolutionary ideas: Buffon conceded that species could "transmutate" (the word used at the time for "evolve") within genera, but certainly not further; other thinkers (among them Erasmus Darwin, Charles's grandfather) had a larger view of evolution.

Jean-Baptiste Lamarck was the first to present a general theory of evolution in "Philosophie zoologique" (1809). He clearly recognized that time is an essential ingredient for the evolution of life as we know it:

> *Time and favourable conditions are the two principal means which nature has employed in giving existence to all her productions. We know that for her time has no limit, and that consequently she always has it at her disposal.*

CP1028, *Collective Dynamics: Topics on Competition and Cooperation in the Biosciences*
edited by L. M. Ricciardi, A. Buonocore, and E. Pirozzi

He maintained that the main mechanism of evolution was the transmission of acquired traits. He formulated this in two laws: "First Law": use or disuse causes structures to enlarge or shrink. "Second Law": all such changes are heritable.

Lamarckian's theory thus differs from modern views, and also from Darwin's (though later editions of "The origin of species" may leave some space for that) in the reliance on the heritability of acquired traits as the mechanism of evolution. Even more importantly, especially from the philosophical point of view, is that Lamarck sees evolution as a process of increasing complexity and "perfection", not driven by chance:

> *Nature, in producing in succession every species of animal, and beginning with the least perfect or simplest to end her work with the most perfect, has gradually complicated their structure.*

The idea of an inherent direction in evolution is clearly foreign to Darwin's view, and to any of the modern theories.

Lamarckian's theory was at time closely linked with revolutionary politics, and materialist philosophy. As such, it provoked heated debates, a particular famous one occurring in 1829-30 between Etienne Geoffroy Saint-Hilaire and Georges Cuvier, both professors at the French National Museum of Natural History (as also Lamarck had been); the debate started about technical aspects of comparative anatomy, but then touched "transmutation", common descent and materialism. It is generally acknowledged that Cuvier (probably the most influential biologist of the time, often considered Aristotle's worthy successor) had won the debate, and "transformist" views came to be generally discredited in 1830–50.

2. A SKETCH OF DARWIN'S BIOGRAPHY

- 1809 Charles Darwin born in Shrewsbury, England, son of Robert, a physician, and of Susan Wedgwood, daughter of the founder of the famous pottery industry.
- 1825-27 Darwin studies Medicine in Edinburgh, getting bored by all courses except for chemistry; he starts a friendship with several naturalists, among which Robert Grant, an expert in marine biology and also a supporter of Lamarckian views; eventually, he goes back home without a degree.
- 1828-31 studies Theology in Cambridge; in that time follows the botany courses by John Henslow, who becomes a good friend. In January 1831 obtains the Bachelor of Arts. On advice by Henslow, he then starts the study of geology, and takes part to a geological expedition in Wales.
- 1831-36 on suggestion by Henslow, he is chosen as naturalist for a scientific survey of South America and South Pacific on "The Beagle". The 5 years provide a wealth of observations about the changes of characters of species in space and time, most famous examples being Darwin's finches in Galapagos Islands. In Figure 1, a picture of four species, reproduced from "The Voyage of the Beagle", drawn by John Gould, taken from Wikimedia Commons, http://commons.wikimedia.org/wiki/
- 1837-42 in London; publishes many scientific papers on the results of the Beagle voyage; becomes secretary of the Geological Society, and member of the Royal

Society; in 1839 gets married with his cousin Emma Wedgwood; in 1842 writes an unpublished Sketch of the Theory of Evolution by Natural Selection; the Darwin family moves to Downe, a small village.

- 1843-58 publishes several scientific papers, and continues the elaboration of the theory of evolution.
- 1858 Communication by Darwin and Wallace to the Linnean Society.
- 1859 First edition of the "Origin of species".
- 1860-72 Further editions of the "Origin of species", "The variation of animals and plants under domestication", "The descent of man".
- 1873-82 writes several books on botany, especially on plant reproductive system; writes the Autobiography; in 1882 Charles Darwin dies in Downe, and is then buried at Westminster Abbey.

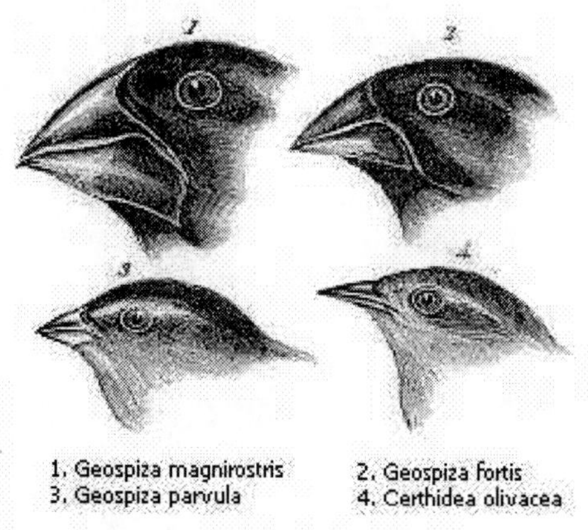

FIGURE 1. Finches of Galapagos Archipelago

3. THE THEORY OF EVOLUTION BY NATURAL SELECTION

As seen in the list above, Darwin had started thinking about the evolution of species shortly after the return from the voyage in the "Beagle", and had already written a sketch of the theory of evolution by natural selection in 1842 and an essay in 1844, both unpublished, but discussed with several colleagues, especially Charles Lyell, the most famous geologist of the time, and the botanist Hooker. Presumably, he refrained from publishing anything, because he thought he needed a very convincing evidence for a theory that at the time had a subversive connotation. Indeed, in 1856 he had started writing a manuscript, described as a "big book on the theory of descent with modification through natural selection", of which in 1858 he had written 11 chapters.

At that time, Darwin received a letter from Alfred Russel Wallace containing a short essay, exposing a conception of the formation of species extremely similar to Darwin's

FIGURE 2. On the left Darwin aged 59. Photographed by Julia Margaret Cameron in July 1869, reproduced with permission from John van Wyhe ed., The Complete Work of Charles Darwin Online (http://darwin-online.org.uk/). On the right a photograph of Alfred Russel Wallace. Reproduced with permission from John van Wyhe ed., The Complete Work of Charles Darwin Online (http://darwin-online.org.uk/).

theory of natural selection. Wallace was a young self-learnt naturalist who maintained himself around the world by selling embalmed birds and butterflies, and had been for some time in correspondence with Darwin. Wallace asked Darwin to show the work to Lyell, if he thought it deserved that. Darwin, shocked by the coincidence and the possibility to lose all claims to his prior work, passed the essay to Lyell and Hooker, suggesting to publish it wherever appropriate but expressing his dismay for being beaten to publication. Lyell and Hooker found a solution that gave credit to both Darwin and Wallace: on 1 July 1858 they made an oral presentation at the Linnean Society, under the title "*On the Tendency of Species to form Varieties; and on the Perpetuation of Varieties and Species by Natural Means of Selection. By CHARLES DARWIN, Esq., F.R.S., F.L.S., & F.G.S., and ALFRED WALLACE, Esq.*". The formally joint communication by Darwin and Wallace contained three papers: an 'extract' of Darwin's 1844 Essay, a summary of Darwin's 1857 letter to the American naturalist Asa Gray which gave a schematic presentation of his theory, and Wallace's essay. Wallace accepted this arrangement, said himself gratified of having stimulated Darwin's publication, and became a strong advocate of Darwinian theory, taking for himself a second-place position.

After this publication, Darwin worked hastily at preparing an abstract of the "big book", abstract that became "On the origin of species by means of natural selection", certainly one of the most influential books in the history of science and quite readable with pleasure and interest also nowadays.

Before briefly discussing the theory exposed in it, it is worth presenting its Table of Contents:

1. Variation under Domestication

2. Variation under Nature
3. Struggle for Existence
4. Natural Selection; [or the Survival of the Fittest, from the 5th Edition]
5. Laws of Variation
6. Difficulties of the theory
7. Miscellaneous objections to the theory of Natural Selection
8. Instinct
9. Hybridism
10. On the imperfection of the geological record
11. On the geological succession of organic beings
12. Geographical distribution
13. Geographical distribution (continued)
14. Mutual affinities of organic beings: morphology: embryology: rudimentary organs

In Figure 3, the front page of the first edition, published on 24 November 1859, reproduced with permission from John van Wyhe ed., The Complete Work of Charles Darwin Online (http://darwin-online.org.uk/).

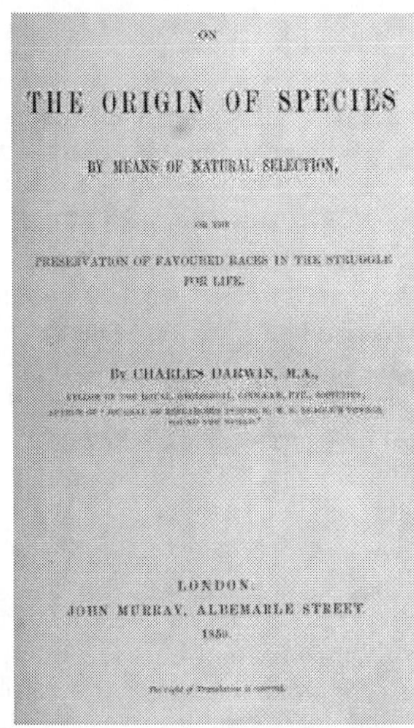

ON

THE ORIGIN OF SPECIES

BY MEANS OF NATURAL SELECTION,

[illegible]

PRESERVATION OF FAVOURED RACES IN THE STRUGGLE FOR LIFE.

BY CHARLES DARWIN, M.A.,

[illegible]

LONDON:
JOHN MURRAY, ALBEMARLE STREET.
1859.

[illegible]

FIGURE 3. Front page of the first edition. Reproduced with permission from John van Wyhe ed., The Complete Work of Charles Darwin Online (http://darwin-online.org.uk/).

Does mathematics have a role in "The origin of species", and more generally in Darwin's theory? I would answer yes, although it is a very simple mathematics. A crucial point of Chapter 3, “The struggle for existence”, is Malthus’s principle of the “Geometrical Ratio of Increase”, used also by Wallace; quoting Darwin

> *Every being must suffer destruction during some period of its life; otherwise, on the principle of geometrical increase, its numbers would quickly become so inordinately great that no country could support the product. There must in every case be a struggle for existence. It is the doctrine of Malthus applied with manifold force to the whole animal and vegetable kingdoms*

Later he goes to a numerical illustration

> *The elephant is reckoned the slowest breeder of all known animals, and I have taken some pains to estimate its probable minimum rate of natural increase; it will be safest to assume that it begins breeding when 30 years old, and goes on breeding till 90 years old, bringing forth 6 young in the interval, and surviving till 100 years old; if this be so, after a period of from 740 to 750 years there would be nearly 19 million elephants alive, descended from the first pair.*

The birth, because of the geometrical ratio of increase, of many more individuals than can survive and reproduce, is at the core of what Darwin calls “natural selection”.

> *How will the struggle for existence, act in regard to variation? Can the principle of selection, which we have seen is so potent in the hands of man, apply in nature? [...] Can we doubt (**remembering that many more individuals are born than can possibly survive**) that **individuals having any advantage, however slight**, over others, would have the **best chance of surviving and of procreating** their kind? On the other hand, we may feel sure that any **variation** in the least degree **injurious would be** rigidly **destroyed**. This preservation of favourable variations and the rejection of injurious variations, I call Natural Selection. [my emphasis]*

Some authors maintain that natural selection is not a scientific theory, but a tautology or an unfalsifiable statement. Indeed, Spencer, who coined the phrase “survival of the fittest”, considered that an *a priori* truth that did not require any empirical proof, and argued that the only serious question was what facts this principle can explain. In a sense, modern theories about replicators follow in that scheme. Replicators are entities endowed with 4 properties: replication; transmission; innovation (= mutation); competition (= selection). They will then evolve through mutation-selection (or competition-innovation) processes. The idea can be ascribed to Dawkins, but it has been formalized by several authors (see [1], [2]). An interesting account of how this idea can be used similarly in evolutionary biology, or in the evolution of technology can be found in a recent book by Dercole and Rinaldi (see [3]).

Darwin had a rather different view of this. As clearly illustrated in the Introduction to “The variation of animals and plants under domestication”, he distinguishes between the “mere hypothesis” of natural selection and the theory of evolution by natural selection. The whole “On the origin of species” is a “long argument” for the theory. The hypothesis of natural selection is presented in the first part of the book, and rests on some generalizations:

- All organisms tend to increase geometrically
- Hence, individuals compete for survival and reproduction
- There is heritable variation for many characters
- Some variation affects the chances of survival and reproduction
- Moreover, domestic races have been formed by artificial selection

In the second part of the book, natural selection is a principle that

> *explains several large and independent classes of facts, such as the geological succession of organic beings, their distribution in past and present lines, and their mutual affinities and homologies". A theory with the "power of absorbing new facts...of interpreting phenomena previously looked upon as anomalies.*

Indeed, the "long argument" appeared quite convincing, and the theory made a lot of proselytes among naturalists and scientists, and also raised strong objections, especially concerning the descent of man.

4. THE NEED FOR A THEORY OF HEREDITY

A particularly acute criticism of Darwin's theory was raised in 1867 by Fleeming Jenkin, an engineering professor, in a review of the "Origin of Species". Darwin, in a letter to Hooker, mentions "*Fleeming Jenkin has given me much trouble, but has been of more real use than any other essay or review*". Jenkin raises three many objections:

1. The length of time required for natural selection to produce the variety of living things is incompatible with Lord Kelvin's estimate of the time since the Earth had solidified (at most 20-40 M years).
2. Species cannot indefinitely vary in a given direction.
3. Under blending inheritance (the mainly accepted theory of heredity at the time), natural selection is probably unable to modify the 'mean species type'.

About the first objection, based on thermo-dynamical estimates of the cooling rate of Earth, Darwin could say little. Eventually, by the discovery of radioactivity, Lord Kelvin was proved wrong, and geologists' estimates correct.

The second objection mainly repeats the "fixists"' objection, although in an updated version.

The third objection instead forced Darwin to make clearer how exactly selection and heredity acted. The main idea about heredity in the "Origin of Species" is that "true breeds true"; hence, if some individual is slightly more successful than the others and produces more offspring, these will on average bear similar traits and so the species average will change.

Jenkin examines more closely the statement and shows how the consequences will depend on many details. We can state his objection through a numerical example:
Suppose a given variation is present just in a few individuals. Assume only in 1. Take a

population where 10^6 are born, and 10^4 reproduce. An individual bearing a variant trait is born with twice the chances to reproduce. Still, it will have only 1 chance in 50 to survive, so that it is very likely that the very favourable variant will disappear. Even if it survives, if we accept blending inheritance, its 100 offspring will have an advantage equal only to 1.5, since their other parent will have carried the normal trait. Their surviving offspring will generally have an advantage 1.25, again because the other parent will have been a normal individual. Even if the variant does not disappear altogether, the variation is swamped. The other possibility is that variants do not occur in single individuals, but that there always is a continuous variation in any population. One must then understand the nature of this variation, and how selection operates on a distribution.

In "The variation of animals and plants under domestication" (1868) Darwin presented a theory of "pangenesis", an old hypothesis intended to explain heredity and variation, rephrased in terms of cells. It is difficult for me now to exactly understand the hypothesis, but definitely it left some room for Lamarckian transmission of acquired traits; this could respond to Jenkin's objections, since evolution would have been speeded up, and variation would have been continuously created, but no clear statement in this sense exists in later editions of the "Origin of species".

In retrospect, an adequate answer to Jenkin's objection required a proper theory of heredity, and also a mathematical toolbox, to analyse its theoretical consequences. Both were absent at Darwin's time, but we can say that, this notwithstanding, the core of his theory was essentially correct. Shortly after the publication of "The origin of species", Galton started a theory of heredity, based on the idea of a continuum of variation, giving rise to the biometric school; Galton had an anti-Darwinian attitude, but then biometric methods were used by Weldon and Pearson, in efforts to demonstrate natural selection in action. Quantitative population genetics can be considered heir to this school of thought.

The idea that variation is generated through unique new variants gained instead a strong support after the rediscovery of Mendel's law (1900). Initially, it seemed that the new science of genetics (as it was first named in 1905) had given the last blow to Darwinian theory, and instead favoured a "mutational view" of evolution. It became clear, however, that genetics provided the necessary basis to Darwinian theory. Hardy-Weinberg laws, discovered independently (1909) by the famous mathematician Thomas Hardy and by the physician Wilhelm Weinberg, provided a sort of law of inertia for genetic composition.

5. THE MODERN SYNTHESIS AND AFTER

On the basis of genetics, and its consequences at the level of populations ("population genetics"), the so-called modern synthesis could view evolution as a change in the genetic composition of populations. A theoretical basis was provided by the mathematical elaborations provided by the famous trinity: Ronald Fisher, Sewall Wright and J.B.S. Haldane. Their work is definitely too complex to be summarized: Fisher is famous for the fundamental theorem of natural selection, on which he based a physically inspired statistical view of evolution; Wright coined the metaphor of adaptive landscape and emphasized the role of random genetic drift in small population; Kimura's (see [4]) neutral molecular evolution theory can be considered an heir to that tradition. Haldane has not provided any such generalizations, but instead contributed in several ways to the understanding of the evolutionary process; among other things, he fitted a selection model to the data on the substitution, in the course of the industrial revolution, of the melanic to the clear forms of *Bison betularia* , a moth living in central England; he showed that data were consistent with a reasonable selection pressure, and later his analysis was confirmed experimentally by Kettlewell.

Mathematical methods are continuing to contribute to the development of evolutionary theory, discussing also features that could seem problematic within the classical evolutionary framework. Among this, I can quote the evolution of sexual reproduction (see [5]– [8]), the evolution of altruism and cooperation (see [9], [10]), the mechanisms of speciation (see [11], [12]) (it has been said that the "Origin of the species" does not really deal with the subject of its title; also in the "modern synthesis", speciation was generally left to outside accidents, such as geographical isolation).

We can probably distinguish two main schools of thought in current theories: one that emphasizes the role of "fitness", adequately revisited, and has been recently formalized as adaptive dynamics (see [13], [14]); the other one, in which the role of selection is joint to and limited by the self-organizing properties of complex systems (see [15]).

Going back to the title, I would say that, although the use of mathematics by Darwin himself was quite limited, there has been a great cross-fertilization between mathematics and evolutionary theory in the century and a half since the publication of the Darwin-Wallace memoir.

Recent scientific publications have been referenced in the above text. For less recent scientific publications, I refer to the bibliography of the books cited below. I wish only to say that all publications by Darwin (including most letters, notebooks...) are available on the Web at http://darwin-online.org.uk. The classical source for the history of evolutionary biology is [16]. The so-called modern synthesis is presented extensively in [17]. Among the many books dealing with Darwin, the history and philosophy of Darwinism, I have consulted [18]– [22].

REFERENCES

1. R. Dawkins, *The Extended Phenotype*, Oxford Univ. Press, Oxford, 2006.
2. J. Hofbauer, and K. Sigmund,, *The Theory of Evolution and Dynamical Systems*, Cambridge Univ. Press, 2006.
3. F. Dercole and S. Rinaldi, *Analysis of Evolutionary Processes: The Adaptive Dynamics Approach and Its Applications*, Princeton Univ. Press, 2008.
4. M. Kimura, *The neutral theory of molecular evolution*, Cambridge Univ. Press, 1985.
5. G. C. Williams, *Sex and Evolution*, Princeton Univ. Press, 1975.
6. J. Maynard Smith, *The Evolution of Sex*, Cambridge Univ. Press, 1978.
7. A. S. Kondrashov, *Nature* **336**, 435–440 (1988).
8. W. D. Hamilton, R. Axelrod and R. Tanese, *Proc. Natl. Acad. Sci. USA* **87**, 3566–3573 (1990).
9. R. Axelrod and W. D. Hamilton, *Science* **211**, 1390–1396 (1981).
10. R. Axelrod and W. D. Hamilton, *Science* **314**, 1560–1563 (2006).
11. U. Dieckmann and M. Doebeli, *Nature* **400**, 354–357 (1999).
12. R. Bürger, *Evolution* **60**, 2185–2206 (2006).
13. J. A. Metz, S. A. H. Geritz, G. Meszena, F. J. A. Jacobs and J. S. van Heerwaarden, "Adaptive dynamics, a geometrical study of the consequences of nearly faithful reproduction", in *Stochastic and spatial structures of dynamical systems*, edited by S.J. van Strien and S.M. Verduyn Lunel,

North Holland, 1996.
14. U. Dieckmann, *Trends Ecol. Evol.* **12**, 128–131 (1997).
15. S. A. Kaufmann, *The origins of order*, Oxford Univ. Press, New York 1993.
16. E. Mayr, *The growth of biological thought*, Harvard Univ. Press, Cambridge, Ma. 1982.
17. W. B. Provine, *The origins of theoretical population genetics*, Univ. of Chicago Press, 1971.
18. D. C. Dennett, *Darwin's dangerous idea*, Simon and Schuster, New York 1995.
19. D. J. Depew and B.H. Weber, *Darwinism evolving: systems dynamics and the genealogy of natural selection*, MIT Press, 1995.
20. A. Desmond and J. Moore, *Darwin*, Michael Joseph Ltd, London 1991.
21. J. Gayon, *DarwinismŠs struggle for survival*, Cambridge Univ. Press, Cambridge 1998.
22. M. Ruse, *Darwinism and its discontents*, Cambridge Univ. Press, Cambridge-New York 2006.

The Brain: What is Critical About It?

Dante R. Chialvo*, Pablo Balenzuela† and Daniel Fraiman**

*Department of Physiology, Feinberg Medical School, Northwestern University, 303 East Chicago Ave. Chicago, IL 60611, USA

†Departmento de Física, Facultad de Ciencias Exactas y Naturales, Universidad de Buenos Aires, Pabellón 1, Ciudad Universitaria (1428), Buenos Aires, Argentina

**Departamento de Matemática y Ciencias, Universidad de San Andrés, Vito Dumas 284 (1644), Buenos Aires, Argentina

Abstract. We review the recent proposal that the most fascinating brain properties are related to the fact that it always stays close to a second order phase transition. In such conditions, the collective of neuronal groups can reliably generate robust and flexible behavior, because it is known that at the critical point there is the largest abundance of metastable states to choose from. Here we review the motivation, arguments and recent results, as well as further implications of this view of the functioning brain.

Keywords: Brain, critical phenomena, complex networks.
PACS: 87.19.L-, 89.75.-k, 87.85.Xd

1. INTRODUCTION

The brain is a complex adaptive nonlinear system that can be studied along with other problems in nonlinear physics from a dynamical standpoint. With this perspective here we discuss a proposal [5, 15, 16, 17, 18, 19] claiming that the brain is spontaneously posed at the border of a second order phase transition. The claim is that the most fascinating properties of the brain are -simply- generic properties found at this dynamical state, suggesting a different angle to study how the brain works. From this viewpoint, all human behaviors, including thoughts, undirected or goal oriented actions, or simply any state of mind, are the outcome of a dynamical system -the brain- at or near a critical state.

The starting point for this conjecture is that it is only at the critical point that the largest behavioral repertoire can be attained with the smallest number of degrees of freedom. *Behavioral repertoire* refers to the set of actions useful for the survival of the brain and *degrees of freedom* are the number of (loosely defined) specialized brain areas engaged in generating such actions. A number of ideas from statistical physics can be used to understand how the brain works by looking at the problem from this angle.

This article is dedicated to discussing the basis and specifics of this proposition along with its implications. The paper is organized as follows: The second section begins by reviewing the problem. Basic features of the physics of critical phenomena are introduced and used to support the Darwinian notion that brains are needed to survive in a critical world. The third section addresses predictable observations, and the fourth section reviews recent results that support the idea of a critical state in brain function. The paper closes with a short discussion of its implications.

CP1028, *Collective Dynamics: Topics on Competition and Cooperation in the Biosciences*
edited by L. M. Ricciardi, A. Buonocore, and E. Pirozzi

2. WHAT IS THE PROBLEM?

New fascinating discoveries about brain physiology are reported every week , each one uncovering a relatively isolated aspect of brain dynamics. Yet the reverse process - how these isolated pieces can be integrated to explain how the brain works- is rarely discussed. Large-scale knowledge of the nervous system is generally only casted in psychological terms, with little discussion of underlying mechanisms. The goal should be, as it is in physics, to explain all macroscopic phenomena -regardless of their nature- on the basis of their underlying microscopic dynamics.

The problem discussed here concerns which underlying properties allow the brain to work as a collective of neuronal groups. How chief brain abilities work in concert, how perception and action are engaged, and how the conscious mind emerges out of electrical impulses and neurochemistry is what we wish to understand, to name a few. This is essentially equivalent, for instance, to understanding how culture (or any other community emergent property) emerges from each individual's intellectual capital. It is clear that the solution of these questions, as for other complex systems, requires more than the mere enumeration of all the knowledge about the individual components.

The task of understanding how a collective works together is challenging, but even more is in the case of the brain. As a whole the brain has some notoriously conflictive demands. In some cases it needs to stay "integrated" and in others must be able to work "segregated", as discussed extensively by Tononi and colleagues [48, 49].[1] This is a non trivial constraint, nevertheless mastered by the brain as it is illustrated with plenty of neurobiological phenomenology. Any conscious experience always comprises a single undecomposable scene [48], i.e., an integrated state. This integration is such that once a cognitive event is committed, there is a refractory period in which nothing else can be thought of. At the same time segregation properties allow for a large number of conscious states to be accessed over a short time interval. As an analogy, the integration property we are referring to could be understood as the capacity to act (and react) on an all-or-nothing basis, similar to an action potential or a travelling wave in a excitable system. The segregation property could be then visualized as the capacity to evoke equal or different all-or-nothing events using different elements of the system. In fact, this metaphor may be more than applicable.

It will be discussed below that the segregation-integration conflict shares many similarities with the dynamics seen in nonlinear systems near a order disorder phase transition.

2.1. What is special about being critical

Work in the last two decades has shown that complexity in nature often originates from the tendency of non-equilibrium extended nonlinear systems to drift towards a critical point. There are many examples in which this connection was made more

[1] Perhaps the same conflict can be identified also in other complex systems.

or less rigorously including problems in physics, economy, biology, macroeconomics, cosmology and so on [32, 41, 50, 5, 10]. It had been argued already [5, 15, 16, 18, 19] that the same approach should be used to understand large scale brain dynamics.

To review such a proposal we will briefly discuss which features of the critical state are pertinent to the conjecture that the brain is critical. As an example we will use the well known Ising model of the magnetization in ferromagnetic materials, but it should be noted that the important point are the universal features of the phase transition and not the model itself.

We can describe the Ising model by considering a relatively small square lattice containing $N = LxL$ sites, with each i site associated with a variable s_i, where $s_i = +1$ represents an "up" spin and $s_i = -1$ a "down" spin. Then any particular configuration of the lattice is specified by the set of variables $s_1, s_2, ..., s_N$. The energy in absence of external magnetic field is given by

$$E = -J \sum_{i,j=nn(i)}^{N} s_i s_j \tag{1}$$

where J is the coupling constant and the sum of j runs over the nearest neighbors of a given site i ($nn(i)$). The simulation is usually implemented with the Metropolis Monte Carlo algorithm [30, 46] solving for a given heat bath temperature T.

Collectively, spins will show different degrees of order and magnetization values depending on the temperature, as seen in the ferromagnetic-paramagnetic phase transition illustrated in Figure 1. A material is ferromagnetic if it displays a spontaneous magnetization in absence of any external magnetic field. If we increase the temperature the magnetization gets smaller and finally reaches zero. At low temperature the system is very ordered with only very large domains of equally oriented spins, a state almost invariant in time. At very high temperatures, spin orientation changes constantly and become correlated only at very short distances resulting in vanishing magnetization. Only in between these two homogeneous states, at the critical temperature, does the system exhibit peculiar fluctuations both in time and space. The temporal fluctuations of the magnetization is scale invariant. Similarly, the spatial distribution of spins clusters show long range (power law) correlations and scale invariance reflected in a fractal structure of clusters of aligned spins. It is important to realize two points: 1) these large structures only emerge at the critical point, and 2) they extend up to the system size despite the fact that the interactions between the systems elements are only *short-range* (i.e., between the nearest neighbors). Thus, at the critical temperature, the system is able to maintain correlation between far away sites (up to the size of the system) staying long periods of time in a given meta-stable state but also exploring a large diversity of such states. This behavior is reflected in the maximization of the fluctuations of magnetization, a typical signature of a second order phase transition.

We propose that this dynamical scenario -generic for any second order phase transition- is strikingly similar to the integrated-segregated dilemma discussed above, and is necessary for the brain to operate as a conscious device. It is important to note that there are no other conceivable dynamical scenarios or robust attractors known to exhibit these two properties simultaneously. Of course, any system could trivially

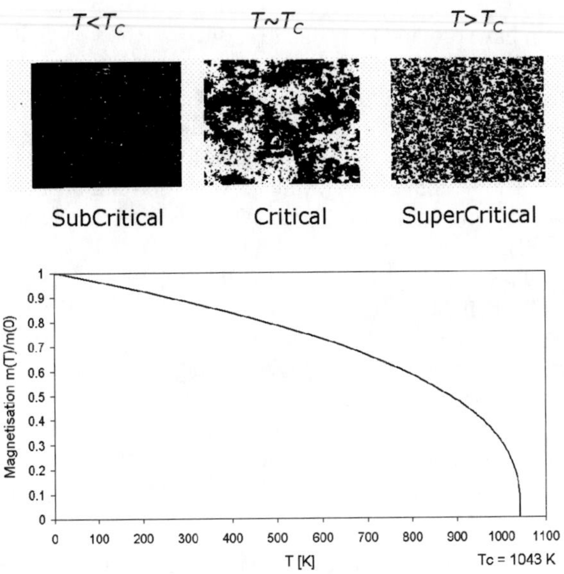

FIGURE 1. Ferromagnetic-paramagnetic phase transition. Bottom: Temperature dependence of magnetization m(T) for Fe. Top three panels are snapshots of the spins configuration at one moment in time for three temperatures: subcritical, critical, and supercritical.

achieve integration and long range correlations in space by increasing links' strength among faraway sites, but these strong bonds prevent any segregated state.

2.2. Why do we need a brain?

This question may sound frivolous but it is not at all, because in Darwinian terms it is necessary to consider the brain embedded in the rest of nature, and co-evolving according with the constraints of natural selection. Although some views could advocate for computational properties in specific neural circuits and find mathematical justification for it existence, we simply think that the brains we see today are the ones that -for whatever means- got an edge and survived. How consistent is our view of the brain near a critical point will be answered by considering these Darwinian constraints. We propose that the brains we see today are critical because the world in which they have to survive is up to some degree critical as well. Let us look at the other possibilities. If the world were sub-critical then everything around will be simple and uniform (as in the left top panel of Figure 1); there would be nothing to learn, a brain will be superfluous. In a supercritical world, everything would be changing all the time (as in the right top panel of Figure 1); in these conditions it would be impossible to learn. Thus in neither extreme could a brain have provided an edge to survive- in the very uniform world there is nothing to learn and in the wildly fluctuating one there is no use for learning. The brain, therefore, must only be necessary to navigation in a complex, critical world.[2] In a critical world, things are most often the same, but there is always room for surprises. To us, this is -intuitively speaking- how the dynamics with power law correlations look like, there is always a very unlikely event that always surprises us, i.e., some novelty on a background of well known usual things. We "need" a brain *because* the world is critical [5, 6, 7, 15, 32].

Furthermore, a brain not only needs to learn and remember, but also has to be able to forget and adapt. If the brain were sub-critical then all brain states would be strongly correlated with the consequence that brain memories would be frozen. On the other extreme, a supercritical brain would have patterns changing all the time, resulting in the inability to hold any long term memory. One must conclude therefore that in order to be highly susceptible the brain itself has to be near the critical state.

Of course these ideas are not entirely new, indeed almost the same intuition prompted Turing half a century ago to speculate about learning machines using similar terms [51].

3. WHAT SHOULD BE SEEN?

In previous writings we have advanced a tentative list of features of the critical point that should be observed in brain experiments. These included:

1. At large scale:
   Cortical long range correlations in space and time. Large scale anti-correlated cortical states.
2. At small scale:
   "Neuronal avalanches", as the normal homeostatic state for most neocortical circuits. "Cortical-quakes" continuously shaping the large scale synaptic landscape providing "stability" to the cortex.
3. At behavioral level:
   All adaptive behavior should be "bursty" and apparently unstable, always at the "edge of failing". Life-long learning should be critical due to the effect of continuously "rising the bar".

In addition one should be able to demonstrate that a brain behaving in a critical world performs optimally at some critical point, thus confirming the intuition that the problem can be better understood by considering the environment from which brains evolved.

In the list above, the first item concerns the most elemental facts about critical phenomena: despite the well known short range connectivity of the cortical columns, long range structures appear and disappear continuously. The presence of inhibition as well as excitation together with elementary stability constraints determine that cortical dynamics should exhibit large scale anti-correlated structures as well [22]. The features at smaller scales could have been anticipated from theoretical considerations, but

[2] It has been already argued elsewhere [5, 32] that the world at large is critical.

avalanches were first observed empirically in cortical cultures and slices by Plenz and colleagues [11]. An important point that is left to understand is how these quakes of activity shape the neuronal synaptic profile during development. At the next level this proposal suggests that human (and animal [13, 37]) behavior itself should show evidence of criticality and learning also should be included. For example, in teaching any skill one chooses increasing challenging levels that are easy enough to engage the pupils but difficult enough not to bore them. This "raising the bar" effect continues through life, pushing the learner continuously to the edge of failure! It would be interesting to measure some order parameter for sport performance to see if shows some of these features for the most efficient teaching strategies.

4. RECENT RESULTS

4.1. Neuronal avalanches in cortical networks

The first demonstration that neuronal populations can exhibit critical dynamics were the experiments reported by Plenz' lab [11]. What they uncovered was a novel type of electrical activity for the brain cortex. This type of population activity, which they termed "neuronal avalanches", sits half way in between two well known patterns: the oscillatory or wave-like highly coherent activity on one side and the asynchronic and incoherent spiking on the other. In each neuronal avalanche it is typical of a large probability to engage only few neurons and a very low probability to spread and activate the whole cortical tissue. In very elegant experiments Plenz and colleagues estimated a number of properties indicative of critical behavior including a power law with an exponent $\sim 3/2$ for the density of avalanche sizes (see Figure 2). This agrees exactly with the theoretical expectation for a critical branching process [57]. Further experiments in other settings, including monkey and rat in vivo recordings, have already confirmed and expanded upon these initial estimations [12, 34, 45, 33].

An unsolved problem here is to elucidate the precise neuronal mechanisms leading to this behavior. Avalanches of activity such as the one observed by Plenz could be the reflection of completely different scenarios. It could be that the power law distribution of avalanches sizes reflect several non- homogeneous Poisson processes that when added together look like a scale free process. This is unlikely, and scaling analysis should show that this is not the case. It could also reflect a structural (i.e., anatomical) substrate over which travelling waves in the peculiar form of avalanches occur. This would imply that the long range correlations detected are trivially due to long range connections. If that is the case, as was discussed above, this would have nothing to do with criticality, and furthermore would imply that segregation is impossible. Based of what is known about the connectivity, it is reasonable to think of a dynamical mechanism responsible for this type of activity. One can assume that the neuronal avalanches occur over a population of locally connected neurons. Their ongoing collective history will permanently keep them near the border of avalanching and each collective event will only excite enough neurons to dissipate the excess of activity. This is the most likely scenario, following the ideas put forward by Bak and colleagues [5, 6, 7, 32]; however, there is no theoretical formalization of these results as of yet.

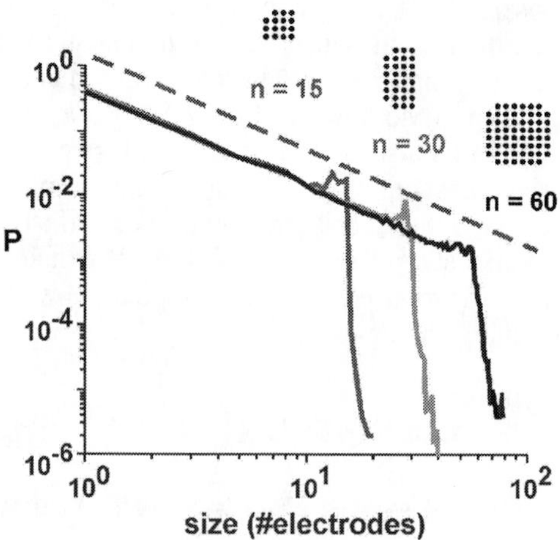

FIGURE 2. Scaling in neuronal avalanches of mature cortical cultured networks. The distribution of sizes follows a power law with an exponent $\sim 3/2$ (dashed line) up to a cutoff which depends on the grid size. The data, re-plotted from Figure 4 of [11], shows the probability of observing an avalanche covering a given number of electrodes for three sets of grid sizes shown in the insets with n=15, 30 or 60 sensing electrodes (equally spaced at $200\mu m$). The statistics is taken from data collected from 7 cultures in recordings lasting a total of 70 hours and accumulating 58000 ($+-$ 55000) avalanches per hour (mean $+-$ SD).

The most significant theoretical effort to elucidate the mechanisms underlying neuronal avalanches was reported recently by Levina and colleagues [28]. They considered a network model of excitable elements with random connectivity in which the coupling is activity dependent, such that, as in reality, too much activity exhausts the synaptic resources. This induces a decreasing in coupling strength which in turn decreases the propagation of activity. The interaction between activity and coupling results in a self-organized drifting of the dynamic towards a critical avalanching activity with the statistics reported in Plenz' experiments. Further work is needed to see other spatiotemporal properties of neuronal avalanches to check if they follows the mechanism suggested by Levina et al. [28].

4.2. Functional brain networks are complex

Functional magnetic resonance imaging (fMRI) allows us to non-invasively monitor spatio-temporal brain activity under various cognitive conditions. Recent work using this imaging technique demonstrated complex functional networks of correlated dynamics responding to the traffic between regions, during behavior or even at rest (see methods in [21]. The data was analyzed in the context of complex networks (for a review see [42]). During any given task the networks were constructed first by calculating linear correlations between the time series of the blood oxygenated level dependent (BOLD) signal in each of $36 \times 64 \times 64$ brain sites called voxels. After that, links were defined

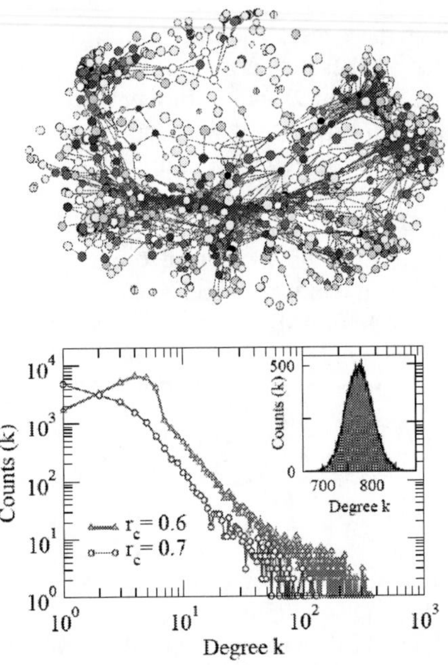

FIGURE 3. A typical brain network extracted from the correlations of functional magnetic resonance images. Top panel shows a pictorial representation of the network. The bottom panel shows the degree distribution for two correlation thresholds r_c. The inset depicts the degree distribution for an equivalent randomly connected network. Data re-plotted from [21].

between those brain sites whose BOLD temporal evolutions were correlated beyond a pre-established value r_c.

Figure 3, show a typical brain functional network extracted with this technique. The top panel illustrates the interconnected networks' nodes and the bottom panel shows the statistics of the number of links (i.e., the degree) per node. There are a few very well connected nodes in one extreme and a great number of nodes with a single connection. The typical degree distribution approaches a power law distribution with an exponent around 2. Other measures revealed that the number of links as a function of -physical-distance between brain sites also decays as a power law, something already confirmed by others [39] using different techniques. Two statistical properties of these networks, path length and clustering, were computed as well. The path length (L) between two voxels is the minimum number of links necessary to connect both voxels. Clustering (C) is the fraction of connections between the topological neighbors of a voxel with respect to the maximum possible. Measurements of L and C were also made in a randomized version of the brain network. L remained relatively constant in both cases while C in the random case were much smaller, implying that brain networks are "small world" nets, a property with several implications in terms of cortical connectivity, as discussed further in [43, 42, 4, 38]. In summary, the work in [21] shows that functional brain networks exhibit highly heterogeneous scale free functional connectivity with small world properties. Although these results admit a few other interpretations, the long range correlations demonstrated in these experiments are consistent with the picture of the brain operating near a critical point, as will be further discussed below. Of course, further experiments are needed to specifically define and measure some order parameter to clarify the precise nature of these correlations. Furthermore, as more detailed knowledge of the properties of these networks is achieved, the need to integrate this data in a cohesive picture grows [44].

To gain insight into the possible dynamical origins of Eguiluz [21] findings we simulated the Ising model on a relatively small square lattice at the critical temperature. Then, as was done with the brain fMRI data, the linear correlations between the time series of each one the lattice points ($s_i = \pm 1$) were calculated:

$$r(i,j) = \frac{\langle s_i(t)s_j(t)\rangle - \langle s_i(t)\rangle\langle s_j(t)\rangle}{\sigma(s_i(t)\sigma(s_j(t)))}, \qquad (2)$$

where $\sigma^2(s(t) = \langle s^2(t)\rangle - \langle s(t)\rangle^2)$.

Figure 4 illustrates typical results for the critical temperature. The distribution of correlations is approximately Gaussian, encompassing both positive as well as negative correlations (see the left panel of Figure 4). This is related to the large domains of equally oriented spins found at the critical temperature, which are positively correlated amongst themselves and negatively correlated to domains with opposite spin orientation. These counterbalanced correlations are only present at the critical temperature, since for supercritical temperatures all correlations vanish and for subcritical values only a large domain of a given orientation survive.

In analogy with Eguiluz et al. methods, a correlation network was constructed by defining links between those lattice points whose fluctuations correlated beyond a a given r_c value. The degree distribution for $r_c = 0.4$ is depicted in Figure 4 and 5, where it can be seen that there is a mode centered around four (i.e., the number of neighbors in the simulation) and then a long tail which resembles very much the experimental results shown previously. Further details can be appreciated more clearly in Figure 5. The top right panel shows the degree for each lattice point, and the top left a correlation map. Notice that the tail of the degree distribution in the previous figure corresponds here to the points in the two clusters with highest degree (colored yellow-red). In the left panel, the origin of these clusters is clarified by selecting one of them as a seed (labelled S) and plotting its correlation values with the rest of the lattice points. Typical time series of two nodes placed far away from the seed: one positively correlated (P) and the other negatively correlated (N) are also plotted in Fig. 5. Note that the two large anti-correlated domains correspond to the two hubs in the degree map.

Of course, these numerical experiments are very far from representing anything close to the details of brain physiology. Nevertheless, they serve the purpose of showing that key features of the correlations seen in the fMRI experiments are also observed in a paradigmatic critical system. The main point of these results is to demonstrate that a correlation network with scale free degree distribution as reported by Eguiluz et al. [21] can be extracted from a dynamical system, providing is at a critical point, regardless of the underlying connectivity. The example shown here uses the worst case scenario of a lattice with only *local* connectivity, but we expect the main conclusions to remain the

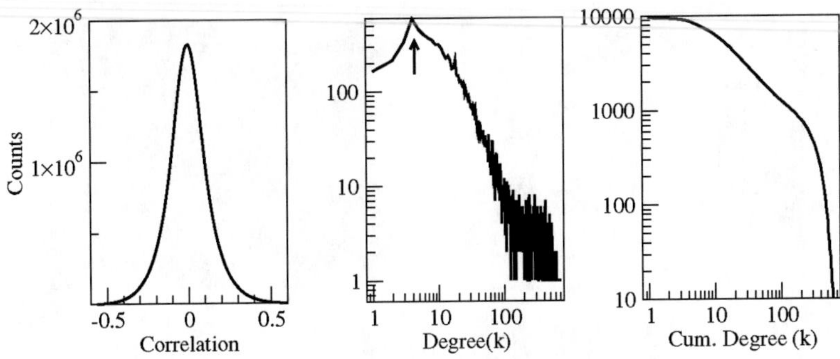

FIGURE 4. Ising model at the critical temperature: Left plot shows the distribution of correlation values. Middle and right plots depict the degree distribution of the network extracted. Arrow point at degree=nn=4. Correlation network constructed as in [21] using a threshold $r_c > 0.4$. Simulation of Eq. 1 with $k = 1$ and $J = 1$, discarding a transient $N_{equil} = 10^8$ steps, we chose $N_{time} = 1000$ configurations every $N_{sample} = LxL = 10^4$ steps. Each time step corresponds to a single spin flip. In all cases the system is at the critical temperature ($T_c \simeq 2.3J/k_B$)

same using other less ordered topologies.

It is important to remark that the dynamics described arises in the Ising model with ferromagnetic interactions, i.e., there is only positive correlations between neighbor sites (analogous to have only "excitatory synapsis"). Despite its absence in Eq. 1, negative correlations emerge as a collective property of the critical dynamics. Accordingly, these negative correlations manifest at relatively long time scales (reflecting the collective movements of spins) and not at short time scales. This agrees well both with observations made from fMRI experiments and with those extracted from a detailed model of the cortex [26]. Finally another aspect to note is that the ratio between the area covered by positive and negative correlations equal to one (see Fig. 5), just as it observed in the brain of healthy people [8] as discussed in the next section.

4.3. What state is the brain "resting state"?

Over a decade ago [9] BOLD low-frequency fluctuations were shown to be correlated across widely spatially separated but functionally related brain regions (between left and right sensorimotor cortices) in subjects at rest. Brain "rest" can be defined - more or less unsuccessfully- as the state in which there is no explicit brain input or output.[3]

Various groups have suggested that these fluctuations are of neuronal origin and correspond to the neuronal baseline or idle activity of the brain. These fluctuations exhibit long-range correlations with the power of the spectrum decaying as $1/f^{\beta}$, with

[3] Readers familiar with Italian traditions advantageously can specify brain rest as the brain state resulting from *"dolce fare niente"*. Translated literally it reads "sweet do nothing" or also the "sweet act of doing nothing".

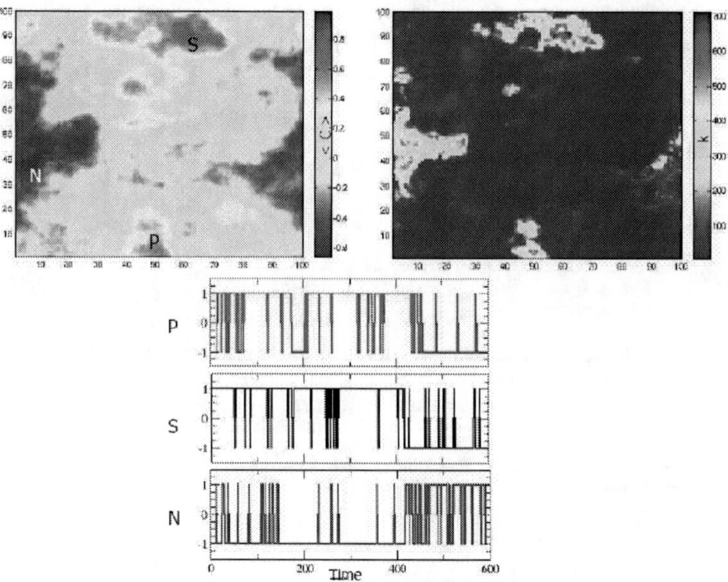

FIGURE 5. Ising model at the critical temperature: Top left: Correlations between the seed (S) and the rest of the lattice. Top right: Degree (k) of each lattice point. Bottom: Time series of three selected places, one for the seed (S) and one for a negatively (N) or positively (P) correlated point. Each time step here corresponds to 10^4 single flip spins.

$\beta \sim 1$. Up until recently these observations were considered a nuisance in the majority of neuroimaging studies and disregarded as unwanted noise, despite the fact that they are the baseline against which other task-related conditions are usually compared.

The notion of a specific network of brain regions active in rest states was reinforced by the observation of a consistent pattern of deactivations seen across many goal-oriented tasks [40]. This observation coupled, with studies of cerebral blood flow led Raichle and colleagues [35] to propose a theory for the so called brain "default mode networks". This view sees BOLD signal decreases during cognitive tasks as one way to identify how the brain is active during rest. In other words, what part of the brain was more active during rest is inferred by identifying what is being deactivated during a given task.

One simple way used to study this network is to look at the linear correlations between the time series of BOLD activity of different regions of the brain [22]. Figure 6 shows a typical result from experiments in with the subjects were ask to track the height of a moving bar varying in time during fMRI data collection [8]. The depicted correlation maps were constructed by first extracting time series for the seeds (small green circles in Fig. 6, obtaining averaging a cube of 3x3x3 voxels) and then computing its correlation coefficient with the time series of all the other brain voxels. This is equivalent to the correlation map shown previously for the Ising model (Fig. 5 top left panel).

Figure 6 shows correlation maps associated with six predefined seed regions. Based on previous results, these seeds are known to be sensitive to the task being conducted.

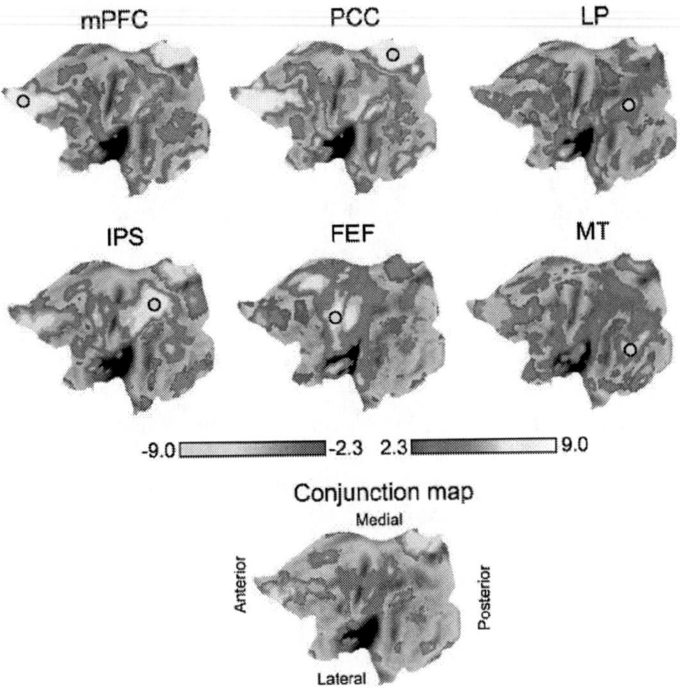

FIGURE 6. Typical balanced correlated-anticorrelated spatial domains of brain fMRI recorded from human volunteers during a simple attention task (replotted from Baliki et al. [8]. These patterns are typical of healthy individuals, where the total area covered by positive correlations is approximately the same as that covered by negatively correlations. The data shows averaged z-score maps (for a group of 15 volunteers) showing regions with significant correlations with the six seed regions (small circles). The results shown correspond to three task-negative seed regions: mPFC, PCC, and LP, as well as to three task-positive seed regions: IPS, FEF, and MT. Colors indicate regions with positive correlations (red-yellow) and negative correlations (blue-green); both have z-scores > 2.3 ($p<0.01$). The group z-score conjunction map below shows voxels significantly correlated or anti-correlated with at least five of the six seed regions.

Three regions, referred to as task-positive regions, exhibit activity increases during the task, and three regions, referred to as task-negative regions, exhibit activity decreases (de-activation) during the attention task [22, 14]. Task-positive regions were centered in the intraparietal sulcus (IPS), the frontal eye field region (FEF), and the middle temporal region (MT). Task-negative regions were centered in the medial prefrontal (mPFC), posterior cingulate/precuneus (PCC), and lateral parietal cortex (LP).

The correlation maps of Figure 6 summarize the functional co-activation between a given seed region and the rest of the cortex. These maps replicate very closely the ones described at rest [22, 29, 35, 36], since it is known that in minimally demanding tasks brain functional connectivity approximates the functional connectivity seen during rest [22, 24, 25]. It displays brain regions that are positively correlated (red-yellow colors) and regions that are negatively correlated (blue-green colors) with any of the chosen seeds. An important experimental finding was that the ratio between the area covered by positive correlations and those with negative correlations was always very close to one [8] (see Fig. 5 and Fig. 6). This was consistently found in all healthy volunteers analyzed up to now. However, the same analysis carried out in patients that have suffered chronic pain for many years, revealed a ratio up to forty times larger [8]. This suggests a healthy dynamic balance of the resting state network, which deserves to be explored further.

The brain is clearly not a lattice and the connectivity is not homogeneous. Moreover the "small world" features revealed by fMRI described earlier are also found in the anatomical connectivity [43]. Thus, finding in any given complex spatiotemporal patterns what is due to the dynamics and what is induced by the underlying structure is still a difficult problem.

In an attempt to gain insight into the brain resting state fluctuations, Honey et al. [26], simulated the cerebral cortex using neuronal dynamics under the real structural connections given by known large scale connectivity. According with their results, coupled excitable elements embedded in this type of anatomical architecture, favors the emergence of spatio-temporal patterns such as those observed in the brain at resting condition. For instance, they found that the functional connectivity seen in the BOLD signal are present at low frequency as a result of fluctuations in the aggregate number of transients couplings and decoupling occurring at a more rapid scale ($\approx 10Hz$). At the slow time scale they identify two major anti-correlated functional clusters which, in their interpretation, are coordinated via anatomical connection patterns. Nevertheless, the results shown in Fig. 5 suggest that these anticorrelated clusters can be originated solely by the critical dynamics.

The possibility which we favor is that the correlations seen during resting state are very similar to those described for the Ising model at the critical temperature. Of course, this similarity is not in the details, but in the fundamental aspects of the dynamics. In this view, spins are represented by entire regions of coherent neuronal groups, say for instance any of the seeds we choose in Figure 6. Thus, at each moment in time, each cortical region competes or cooperates according with the connectivity and the dynamics at that moment. The experimental observation that at any given time positive and negative correlations are equal is awaiting to be explained, and its implications for disease further explored. We claim that the brain is always near criticality such that the spatiotemporal patterns illustrated above should be scale-invariant, and some other temporal variables describing its evolution power law distributed. If that is the case, then the resting state dynamical equivalent is criticality as in other extended non-linear systems near the edge of a phase transition.

4.4. Epileptic seizures as brain quakes?

In a recent paper, Osorio et al. [31] shows an interesting analysis of the temporal organization of epileptic seizures. They studied very large catalogues of seismic activity and epileptic seizures with special attention to the statistical distribution of event sizes, and waiting time between these events. Their analysis reveals an striking analogy between the dynamics of seizures and well known power laws governing earthquakes such as the Gutemberg-Ritcher and Omori laws.

Some counterintuitive conclusions are worth to mention, already noted for earthquakes [5], such as the meaningless use of intensity and duration to characterize a given seizure. This is analogous to the scale invariance noticed already in the analysis of earthquakes. In earthquakes, (as now seems in seizures) it is known that to establish the probability of any event one must specify a time window, a spatial grid size, and a given intensity of that event.

Osorio et al.'s approach also elegantly answers the classical question of why a seizure stops. An earthquake stops spontaneously whenever it has released the excess energy accumulated. In geology terminology an earthquake "starts without knowing how big is going to be or how long it is going to last". Neuronal avalanches, according to Plenz' work described earlier, also obey the same laws. According to the findings of Osorio et al. the mechanism by which seizures stop is related with the same critical process that triggers them. The authors comment that "scale invariance in seizures may be conceptualized as the hallmark of certain complex systems (the brain in this case) in which, at or near the critical point, its component elements (neurons) are correlated over all existing spatial (minicolum, column, macrocolum, etc.) and temporal scales (microseconds, seconds, tens of seconds, etc.)"

The similarities uncovered by Osorio et al. suggest that the researchers' intuition regarding the statistical laws governing epileptic seizures need to be adjusted accordingly.

4.5. Senses are critical

Of course brains are useful to escape from predators, move around, choose a mate or find food, and in these respects the sensory apparatus is critical for any animal survival. Recent results indicate that senses are also critical in the thermodynamic sense of the word. Consider first the fact that the density distribution of the various form of energy around us is clearly inhomogeneous, at any level of biological reality, from the sound loudness any animal have to adapt, to the amount of rain a vegetable have to take advantage. From the extreme darkness of a deep cave to the brightest flash of light there are several order of magnitude changes; nevertheless our sensory apparatus is able to inform the brain of such changes.

It is well known that isolated neurons are unable to do that because of their limited dynamic range, which spans only a single order of magnitude. This is the oldest unsolved problem in the field of psychophysics, tackled very recently by Kinouchi and Copelli [27] by showing that the dynamics emerging from the *interaction of coupled excitable elements*, is the key to solving the problem. Their results show that a network of excitable elements set precisely at the edge of a phase transition - or, at criticality - can be both, extremely sensitive to small perturbations and still able to detect large inputs without saturation. This is generic for any network regardless of the neurons' individual sophistication. The key aspect in the model is a local parameter controlling the amplification of any initial firing activity. Whenever the average amplification is

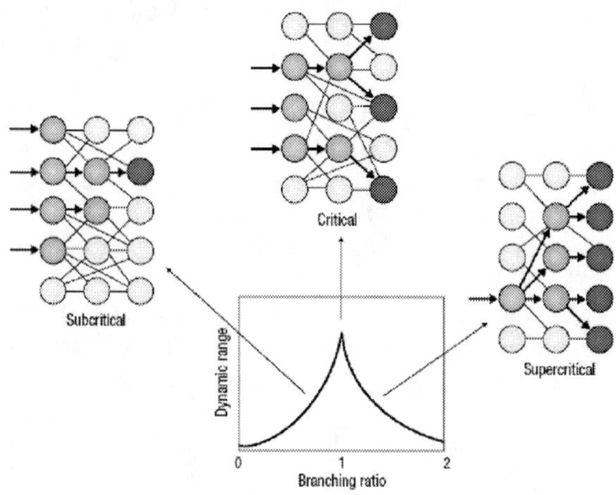

FIGURE 7. Sensory networks constructed with branching ratios close to one maintain, on the average, the input activity (green, followed by yellow and red), thus optimizing the dynamic range. In supercritical networks, however, activity explodes, while in subcritical ones are unable to sustain any input pattern. Redrawn from [17].

very small activity dies out, as it can be seen in the cartoon of Figure 7. In this case the system is subcritical and not sensitive to small inputs. On the other hand, choosing an amplification very large one sets up the conditions for a supercritical reaction in which for any - even very small - inputs the entire network fires. It is only in between these two extremes that the networks have the largest dynamic range. Thus, amplification around unity, i.e., at criticality, seems to be the optimum condition for detecting large energy changes as an animal encounters in the real world [17]. Of course, in a critical world energy is dissipated as a fractal in space and time with the characteristic highly inhomogeneous fluctuations. As long as the world around is critical, it seems that the evolving organisms embedded in it have no better choice than to be critical as well.

5. OUTLOOK

The study of collective phenomena is at the center of statistical physics. It is not surprising then, to see the recent outburst of physics and mathematics publications studying this type of phenomena in the context of computer science as well as social and economic settings. While in all these fields there is a clear transfer of methods and ideas from statistical physics, identical flow has yet to start in brain science.

This lack of communication is even more intriguing if one considers that most see brain science precisely as the study of collective patterns of neuronal activity. Nonetheless, this acknowledgment had not yet been translated into useful approaches. To the contrary, the literature contains numerous old and new promises to understand brain function by way of very large (and in some cases very detailed) numerical simulations of millions of neurons, completely orphan of considerations related with the statistical physics of collective phenomena in large systems.

To make the point above relevance of these ideas, let us recall once more the results presented in Figure 5 and the connectivity between the sites revealed in Figure 4. As it was discussed, on one side the connectivity says that the system is a lattice with only four nearest neighbors, but the correlations reveal a network with scale-free degree distribution. It seems to be a gross contradiction, but the apparent divorce between the patterns dictated by the coupling equations and those found by the analysis of the spatial correlations will not surprise those already familiar with emergent phenomena at the critical state. Again, lets remind ourselves that the divorce between "anatomy" (i.e., the coupling) and dynamics disappears both in the supercritical and subcritical state (as correlations vanish). Now, let suppose that the time series data in Figure 5 were to be from a typical brain experiment. Classical approaches of brain connectivity, based either in the analysis of correlations (i.e., so-called "functional" connectivity or in the anatomical (i.e., "effective") connections could never reach to the right conclusion and solve the puzzle. It is only by knowing about the features of critical phenomena that the apparent puzzle can be solved. As far as we know, there is no report in the literature suggesting changes in the character of the functional connectivity due to the dynamics at the critical point as we suggest here.

In summary, according with the proposal reviewed here, several relevant aspects of brain dynamics *can* be only understood using the theoretical framework as for any nonequilibrium thermodynamic system near the critical point of a second order phase transition. That include the understanding of neuronal dynamics at small scale, the cooperative-competitive equilibrium seen at rest in the healthy cortex, the burst of brain quakes during seizures and the optimization of the dynamic range at the sensory periphery. We have mentioned but left out the discussion of behavior, which understanding we submit should also benefit from this approach.

Some of the ideas here are novel, but the motivation is not, since Ashby was probably the first to indicate how fundamental is to understand the way self-organization shapes brain function [1]. Nevertheless, these views are gaining momentum, and is refreshing to read recent reviews [53, 54, 55] advocating the further study of phase transitions, metastability and criticality in cognitive models and experiments. This enlightening perspective is even more meaningful coming from those that first introduced information theory to the study of sensation in neuroscience,... forty three years ago [56].

ACKNOWLEDGMENTS

Work supported by NIH NINDS of USA. Thanks to Drs. D. Plenz and J.P. Segundo for stimulating discussions, and to E. Parks for proofreading the manuscript. PB and DF are researchers supported by CONICET, Argentina.

REFERENCES

1. W. Ross Ashby. (1962). Principles of the self-organizing system. In *Principles of Self-Organization* Transactions of the University of Illinois Symposium, edited by H. Von Foerster and G. W. Zopf. Jr,

Pergamon Press, London, UK , pp. 255–278.
2. S. Achard, R. Salvador, B. Whitcher, J. Suckling, E. Bullmore. Resilient, low-frequency, small-world human brain functional network with highly connected association cortical hubs. *J Neurosci* **26**, 63–72 (2006).
3. D.S. Bassett, A. Meyer-Lindenberg, S. Achard, T. Duke, E. Bullmore. Adaptive reconfiguration of fractal small-world human brain functional networks. *Proc. Natl. Acad. Sci. USA* **103**, 19518 – 19523 (2006).
4. D.S. Bassett & E. Bullmore. Small-World Brain Networks. *The Neuroscientist* **12**, 512 – 523 (2006).
5. P. Bak (1997). *How Nature works.* Oxford University Press, Oxford UK, 1–212.
6. P. Bak, C. Tang, K. Wiesenfeld. Self-organized criticality: an explanation of the 1/f noise. *Phys. Rev. Lett.* **59**, 381 (1987).
7. P. Bak & D.R. Chialvo. Adaptive learning by extremal dynamics and negative feedback. *Phys. Rev. E* **63**, 031912 (2001).
8. M.N. Baliki, P.Y. Geha, A.V. Apkarian, D.R. Chialvo. Beyond feeling: Chronic pain hurts the brain disrupting the default mode networks. *J. Neuroscience* **28**, 1398 – 1403 (2008).
9. B. Biswal, F. Yetkin, V. Haughton, J. Hyde. Functional connectivity in the motor cortex of resting human brain using echo-planar MRI. *Magn. Res. Med.* **34**, 537–541 (1995).
10. M. Buchanan . *Ubiquity.* Weidenfeld and Nicolson. London, UK, (2000).
11. J.M. Beggs & D. Plenz. Neuronal avalanches in neocortical circuits. *J. Neuroscience* **23**, 11167–11177 (2003).
12. J.M. Beggs, & D. Plenz. Neuronal avalanches are diverse and precise activity patterns that are stable for many hours in cortical slice cultures. *J. Neuroscience* **24** 5216–5229, (2004).
13. D. Boyer, G. Ramos-Fernández, O. Miramontes, J. L. Mateos, G. Cocho, H. Larralde, H. Ramos, F. Rojas. Scale-free foraging by primates emerges from their interaction with a complex environment. *Proceedings of the Royal Society of London 717 Series B: Biological Sciences* **273**, 1743–1750 (2006). Also as http://xxx.lanl.gov/abs/q-bio.PE/0601024.
14. M. Corbetta, & G.L. Shulman. Control of goal-directed and stimulus driven attention in the brain. *Nat Rev Neurosci* **3**, 201–215 (2002).
15. D.R. Chialvo & P. Bak. Learning from mistakes. *Neuroscience* **90**, 1137–1148 (1999).
16. D.R. Chialvo. Critical brain networks. *Physica A* **340**, 756–765 (2004).
17. D.R. Chialvo. Are our senses critical? *Nature Physics* **2**, 301–302 (2006).
18. D.R. Chialvo. The brain near the edge. In: *Cooperative behavior in neural systems: Ninth Granada Lectures.* AIP Conference Proceedings, 887, pp.1–12 (2007).
19. D.R. Chialvo. Emergent complexity: What uphill analysis or downhill invention cannot do. *New Ideas in Psychology* (in press), (2008).
20. S. Dehaene & L. Nagache. Towards a cognitive neuroscience of consciousness: basic evidence and a workspace framework. *Cognition* **79**,1–37 (2001).
21. V.M. Eguiluz, D.R. Chialvo, G. Cecchi, M. Baliki, A.V. Apkarian. Scale free brain functional networks. *Phys Rev Lett.* **94**, 018102 (2005).
22. M.D. Fox, A.Z. Snyder, J.L. Vincent, M. Corbetta, D.C. van Essen, M. E. Raichle. The human brain is intrinsically organized into dynamic, anticorrelated functional networks. *Proc. Natl. Acad. Sci. U.S.A.* **102**, 9673–9678 (2005).
23. M.D. Fox & M.E. Raichle. Spontaneous fluctuations in brain activity observed with functional magnetic resonance imaging. *Nat. Rev. Neurosci.* **8**,700–711 (2007).
24. M.D. Greicius, B. Krasnow, A.L. Reiss, V. Menon. Functional connectivity in the resting brain: a network analysis of the default mode hypothesis. *Proc Natl Acad Sci U S A* **100**, 253–258 (2003).
25. M.D. Greicius, G. Srivastava, A.L. Reiss, V. Menon. Default-mode network activity distinguishes Alzheimer's disease from healthy aging: Evidence from functional MRI. *Proc Natl Acad Sci U S A* **101**, 4637–4642 (2004).
26. C.J. Honey, R. Kotter, M. Breakspear, O. Sporns. Network structure of cerebral cortex shapes functional connectivity on multiple time scales. *Proc Natl Acad Sci U S A* **104**, 10240–10245 (2007).
27. O. Kinouchi & M. Copelli. Optimal dynamical range of excitable networks at criticality. *Nature Physics* **2**, 348–352 (2006).
28. A. Levina, J.M. Herrmann, T. Geisel. Dynamical synapses causing self-organized criticality in neural networks. *Nature Physics* **3**, 857–860 (2007).

29. M.F. Mason, M.I. Norton, J.D Van Horn, D.M. Wegner, S.T. Grafton, C.N. Macrae. Wandering minds: the default network and stimulus-independent thought. *Science* **315**, 393–395 (2007).
30. N. Metropolis, A.W. Rosenbluth, M.N. Rosenbluth, A.H. Teller, E. Teller. Equations of State Calculations by Fast Computing Machines. *Journal of Chemical Physics* **21**, 1087–1092 (1953).
31. I. Osorio, M.G. Frei, D. Sornette, J. Milton, Y-C. Lai. Epileptic Seizures: Quakes of the brain? http://arxiv.org/abs/0712.3929, (2008).
32. M. Paczuski & P. Bak. Self organization of complex systems. In Proceedings of 12th Chris Engelbrecht Summer School, (1999). Also as http://www.arxiv.org/abs/cond-mat/9906077.
33. T. Petermann, M.A. Lebedev, M. Nicolelis, D. Plenz. Neuronal avalanches in vivo. *Society for Neuroscience Abstracts* 531.1, (2006).
34. D. Plenz & T.C. Thiagarajan. The organizing principles of neuronal avalanche activity: cell assemblies in the cortex? *Trends in Neuroscience* **30**,101–110 (2007).
35. M.E. Raichle, A.M. MacLeod,A.Z. Snyder, W.J. Powers, D.A. Gusnard, G.L. Shulman. A default mode of brain function. *Proc Natl Acad Sci U S A* **98**, 676–682 (2001).
36. M.E. Raichle. Neuroscience. The brain's dark energy. *Science* **314**, 1249–1250 (2006).
37. G. Ramos-Fernández, D. Boyer, V.P. Gómez. A complex social structure with fission-fusion properties can emerge from a simple foraging model. *Behav Ecol Sociobiol* **60**, 536–549 (2006).
38. J.C. Reijneveld, S.C. Ponten, H.W. Berendse, C.J. Stam. The application of graph theoretical analysis to complex networks in the brain. *Clinical Neurophysiology* **118**, 2317–2331 (2007).
39. R. Salvador, J. Suckling, M. R. Coleman, J. D. Pickard, D. Menon, E. Bullmore. Neurophysiological architecture of functional magnetic resonance images of human brain. *Cerebral Cortex* **15**, 1332–1342 (2005).
40. G.L. Shulman, J. Fiez , M. Corbetta, R. Buckner, F.M. Miezin, M.E., Raichle, S. Petersen. Common blood flow changes across visual task: II decreases in cerebral cortex. *J Cognitive Neuroscience* **9**, 648–663 (1997).
41. R. Sole & B.C. Goodwin. *Signs of Life: How Complexity Pervades Biology* Basic Books, (2000).
42. O. Sporns, D.R. Chialvo, M. Kaiser, C.C. Hilgetag. Organization, development and function of complex brain networks. *Trends Cog. Sci.* **8**, 418–425 (2004).
43. O. Sporns & J.D. Zwi. The small world of the cerebral cortex. *Neuroinformatics* **2**, 145–162 (2004).
44. O. Sporns, G. Tononi, R. Kotter. The human connectome: a structural description of the human brain. *PLoS Comput Biol* **1**, 245–251 (2006).
45. C.V. Stewart & D. Plenz. Inverted-U profile of dopamine-NMDA-mediated spontaneous avalanche recurrence in superficial layers of rat prefrontal cortex. *J Neurosci.* **26**, 8148–8159 (2006).
46. H. Gould & J. Tobochnik. An introduction to computer simulations methods, Addison Wesley, (1996).
47. G. Tononi, G.M. Edelman, O. Sporns. Complexity and coherency: integrating information in the brain. *Trends Cog. Sci.* **2**, 474–484 (1998).
48. G. Tononi & G.M. Edelman. Consciousness and complexity. Science **282**, 1846–1851 (1998).
49. G. Tononi. An information integration theory of consciousness. *BMC Neurosci.* **5**, 42 (2004).
50. D.L. Turcotte. Self-Organized Criticality. *Reports on Progress in Physics* **62**, 1377–1429 (1999).
51. A. Turing. Computing machinery and intelligence. Mind, **59**, 433–460. (1950/1963). I am quoting from E. A. Feigenbaum and J. Feldman (eds.), *Computers and thought*. New York: McGraw-Hill.
52. J.L. Vincent, G.H. Patel, M.D. Fox, A.Z. Snyder, J.T. Baker, D.C. Van Essen, J.M. Zempel, L.H. Snyder, M. Corbetta, M.E. Raichle. Intrinsic functional architecture in the anesthetized monkey brain. *Nature* **447**, 83–86 (2007).
53. G. Werner. Perspectives on the neuroscience and consciousness. *BioSystems* doi:10.1016/j.biosystems.2006.03.007, (2006).
54. G. Werner. Metastability, criticality and phase transitions in brain and its models. *BioSystems* doi:10.1016/j.biosystems.2006.12.001, (2006).
55. G. Werner. Dynamics across levels of organization. *Journal of Physiology - Paris* (2008, in press).
56. G. Werner & V.B. Mountcastle. Neural activity in mechanoreceptive cutaneous afferents stimulus response functions, Weber functions and information transfer. *J. of Neurophysiology* **28**, 359–397 (1965).
57. S. Zapperi, L. K. Baekgaard, H.E. Stanley. Self-organized branching processes: mean-field theory for avalanches. *Phys Rev Lett* **75**, 4071–4074 (1995).

Mathematical Modeling of Spreading Cortical Depression: Spiral and Reverberating Waves

Henry C. Tuckwell

Max Planck Institute for Mathematics in the Sciences, Inselstr. 22, Leipzig, D-04103 Germany

Abstract. Mathematical models of spreading depression are considered in the form of reaction-diffusion systems in two space dimensions. The systems are solved numerically. In the two component model with potassium and calcium ion concentrations, we demonstrate, using updated parameter values, travelling solitary waves of increased potassium and decreased calcium. These have circular wavefronts emanating from a region of application of potassium chloride. The collision of two such waves does not, as in one space dimension, result in annihilation but the formation of a unified wave with a large wavefront. For the first time we show that the mathematical model reproduces the actual properties of spreading depression waves in cortical structures. With attention to geometry, timing and location of stimuli we have succeeded in finding reverberating waves matching experiment. By simulating the technique of anodal block, spiral waves have also been demonstrated which parallel those found experimentally. The six-component model, which contains additionally sodium, chloride, glutamate and GABA, is also investigated in 2 space dimensions, including an experimentally based exchange pump for sodium and potassium. Solutions are obtained without (amplitude 29 mM external K^+) and with action potentials (amplitude 44 mM external K^+) with speeds of propagation, allowing for tortuosity, of 1.4 mm/minute and 2.7 mm/minute, respectively. When action potentials are included a somewhat higher pump strength is required to ensure the return to resting state.

Keywords: Spreading depression, brain dynamics, ion and transmitter movements, migraine.
PACS: 84.35.+i, 87.10.+e, 87.16.Ac

1. INTRODUCTION

Spreading depression (SD) is a complex wave of transient depolarization of neurons and glia that propagates across cortical and subcortical gray matter at speeds of 2-5 mm/min. It arises mainly as a response to brain injury or pathology. By itself SD does not usually damage brain tissue, but during stroke and head trauma SD can arise repeatedly near the site of injury and may promote neuronal damage. One of the characteristics of SD is a large increase in extracellular potassium and a dramatic fall in extracellular calcium and other ions [1,2]. For reviews, see [3,4].

There has been much evidence for the idea that SD is a concomitant or cause of migraine [5-7]. A strong link exists between glutamate and migraine [8] and glutamate has been found to be important in the propagation of SD [9]. SD is associated with focal ischemia, traumatic brain injury [10,11], seizure activity [12] and spinal cord injury [13]. New effects and roles for SD have been discovered in the last several years. It has been demonstrated in human neocortical slices [14], has been found to suppress γ-activity in rabbit cortex [15] and it has been elicited in the brainstem [16], thought previously not to be capable of supporting SD. SD has also been recently shown to release ATP into the extracellular compartment [17] and sustained elevated levels of

CP1028, *Collective Dynamics: Topics on Competition and Cooperation in the Biosciences*
edited by L. M. Ricciardi, A. Buonocore, and E. Pirozzi

potassium ion concentration have been found to lead to significant amounts of neuronal death [18]. Another finding, revealed by MRI, is that SD has components called primary and secondary events, the former having greater range and greater speed [19].

SD involves neuronal, including synaptic, and glial elements and there are a very large number of physiological and anatomical details which are relevant to its propagation, all of which are difficult to include in a mathematical model. These details include those of the dynamics of many neuronal and glial ion channels, pumps and other clearance mechanisms, blood supply, gap junctions as well as diffusion in the extracellular space. Just the aspects of presynaptic terminals and the dynamics of transmitter release are extremely complex as there are hundreds of different channel types in these specialized parts of the nervous system [20]. Glia and neurons are connected by gap junctions which are ubiquitous in brain circuits [21]. Some gap-junction blockers have been found to impede SD, resulting in reduced amplitude and duration [22]. The NMDA blocker MK-801 was found to prevent SD in mouse neocortical slices but not an accompanying astrocytic calcium wave, though its speed was reduced [23]. Further, the specific gap junction blocker carbenoxolon did not prevent SD but also reduced the speed of the calcium wave. It is noteworthy that increased potassium reduces the efficacy of NMDA receptor antagonists to block SD [11].

There appears to be a reciprocal interaction between neurons and glia through neuro-transmitter release, uptake and calcium fluxes [24, 25]. Astrocytes have been found to play a role in the modulation of synaptic inhibition [26] and glia have been suggested as playing an important role in SD [27]. However, the effects of gap junction blockers may be more than just blocking gap junctions, which clouds the role of gap junctions in SD. For example, some alcohols have been shown to affect both receptor-activated ion channels and voltage-gated ion channels [28]. These effects include inhibition of sub-types of NMDA-glutamate receptor ion channels and potentiation of certain subtypes of GABA-A receptor ion channels. Altered properties of these and other ion channels may contribute to the difficulty of eliciting SD in the presence of gap-junction blockers. Furthermore, alcohols may change the probabilities of opening of sodium channels [29, 30].

Mathematical models of SD have usually taken the form of reaction-diffusion systems - that is systems of parabolic partial differential equations involving spatial diffusion, with one or several equations for each neurochemical variable. Solutions of the equations are obtained by numerical methods, although some useful information about solution properties can be obtained analytically for the simpler models. Theoretical and experimental aspects of reaction-diffusion systems in the brain have been comprehensively reviewed by Nicholson [31].

Properties of reaction-diffusion systems in other areas such as the Belousov-Zhabotinsky reaction [32] may extend to neural phenomena including SD. Computational models of SD have sometimes adopted the cellular automata approach which is a useful first approximation [33,34]. Properties of SD, such as topographical constraints, have also been usefully investigated with mathematical models of general excitable systems, typified by the Fitzhugh-Nagumo equations [33, 35-37] and metabolic models have been related to migraine aura [38]. Shapiro's [39] model included a great amount of physiological detail including gap-junctional mechanisms and volume changes. Continuum neural models have mathematical properties of interest, such as travelling wave solutions [40] and pattern formation [41]. Theoretical models of SD include those which incorporaete a single-cell approach [42,43], which is useful for delineating local effects, and electrodiffusion models [44]. In this article we consider some properties of reaction-diffusion models with 2 and 6 components in two space dimensions in order to investigate their properties and to demonstrate for the first time the experimental phenomena of spiral waves and reverberating waves for SD.

2. THE SIMPLIFIED TWO-COMPONENT MODEL IN TWO SPACE DIMENSIONS

Diffusion through brain extracellular space proceeds quite readily even for larger molecules [48,49] so there is doubtless a substantial contribution to the propagation of SD by diffusion. This is supported by the fact that models which allow for diffusion and tortuosity give the correct speed for SD. Significant changes in extracellular space occur during SD [50,51] but these changes are neglected here because we concentrate mainly on low-amplitude waves and moderate changes in external K^+ have little such effect [52,53]. In the simplified model in two space dimensions, with extracellular potassium and calcium ion concentrations, denoted by $K^o(x,y,t)$ and $Ca^o(x,y,t)$, playing key roles, the model equations are as follows, slightly modified from [46],

$$\frac{\partial K^o}{\partial t} = D_K \nabla^2 K^0 + f(K^o, Ca^o)$$

$$\frac{\partial Ca^o}{\partial t} = D_{Ca} \nabla^2 Ca^o + g(K^o, Ca^o)$$

where $\nabla^2 = \frac{\partial^2}{\partial x^2} + \frac{\partial^2}{\partial y^2}$, D_K and D_{Ca} are diffusion coefficients and f and g are the reaction terms to be described below. To reduce the number of differential equations with little loss of accuracy the corresponding internal concentrations are given by

$$K^i(x,y,t) = K^{i,R} - \frac{\alpha}{\beta}\left[K^o(x,y,t) - K^{o,R}\right]$$

$$Ca^i(x,y,t) = Ca^{i,R} - \frac{\alpha}{\gamma}\left[Ca^o(x,y,t) - Ca^{o,R}\right]$$

where the superscript R denotes resting level and where α/β and α/γ are ratios of the volume of extracellular space to those of the appropriate intracellular compartments. For potassium, contributions to f come from flux into postsynaptic compartments and a pump which acts to restore resting ionic concentrations. We may put $f = f_{source} - f_{pump}$ where

$$f_{source} = k_1(V_K - V)(V - V_{Ca})g_{Ca}(V)$$

and

$$f_{pump} = k_2\left[1 - \exp\left\{-k_3(K^o(x,y,t) - K^{o,R})\right\}\right].$$

Here V_K and V_{Ca} are the Nernst potentials for potassium and calcium

$$V_K = \frac{RT}{F}\ln(K^o/K^i)$$

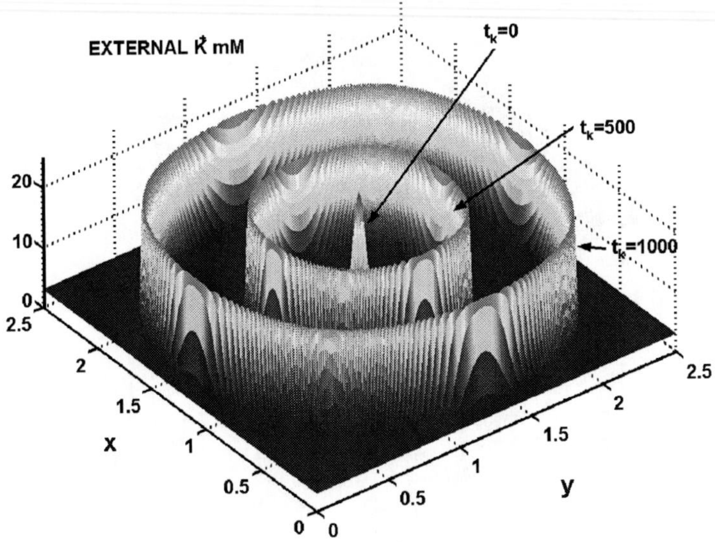

FIGURE 1. Showing the extracellular potassium ion concentration at t=2.5 and t=5.0 for the two-component model with standard parameters.

$$V_{Ca} = \frac{RT}{2F} \ln(Ca^o/Ca^i)$$

and V is the equilibrium membrane potential

$$V = \frac{RT}{F} \ln \left[\frac{K^o + p_{Na}Na^o + p_{Cl}Cl^i}{K^i + p_{Na}Na^i + p_{Cl}Cl^o} \right]$$

where $Na^{i,o}$, $Cl^{i,o}$ are the appropriate internal and external concentrations of sodium and chloride ions and p_{Na}, p_{Cl} are the corresponding relative permeabilities.

$g_{Ca}(V)$ is the calcium conductance of presynaptic membrane, given by

$$g_{Ca}(V) = [1 + \tanh[k_7(V + V_T)] - k_8]H(V - V_c)$$

where V_c is a cut off voltage and putting $k_8 = 1 + \tanh(k_7(V_c + VT))$ ensures that the conductance rises smoothly from zero. $H(x - x_o)$ is a unit step function at x_o. Similarly we put

$$g = g_{pump} - g_{sink}$$

where

$$g_{sink} = k_4(V_{Ca} - V)g_{Ca}(V)$$

$$g_{pump} = k_5\left[1 - \exp\left\{-k_6(Ca^o - Ca^{o,R})\right\}\right].$$

The value of $\frac{RT}{F}$ is chosen appropriately for 37 degrees Celsius. All concentrations are in mM.

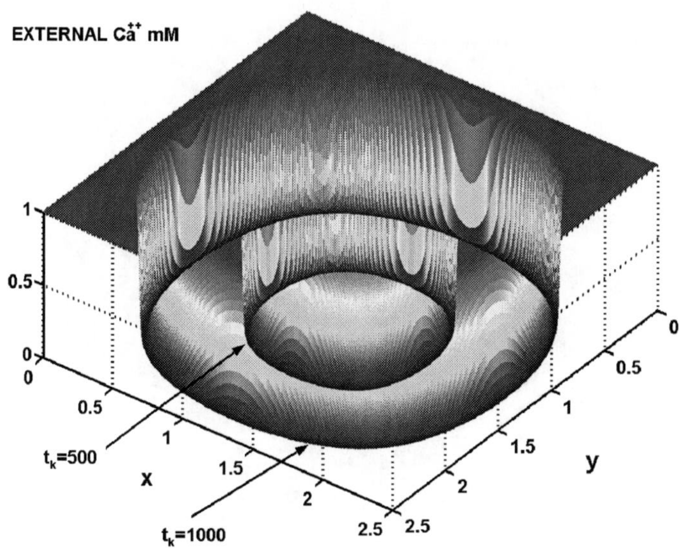

FIGURE 2. Showing the extracellular calcium ion concentration (looking from below) at t=2.5 and t=5.0 for the two-component model with standard parameters.

What we will call the standard parameter set consists of the following: $k_1 = 3.3$, $k_2 = 208$, $k_3 = 10$, $k_4 = 0.3$, $k_5 = 2.38$, $k_6 = 40$, $k_7 = 0.11$, $K^{i,R} = 140$, $K^{o,R} = 3$, $Ca^{i,R} = 0.0001$, $Ca^{o,R} = 1$, $V_T = 45$ mV, $V_c = -70$ mV, $\alpha/\beta = 0.53$, $\alpha/\gamma = 0.207$. The accepted value for the fraction of the brain which is extracellular space is 0.2 [49] so the values of α/β and α/γ differ considerably from those used previously, being based on [54]. Note that the actual intracellular volumes are not necessarily those available for occupation by ions and molecules because of various intracellular organelles such as mitochondria, filaments and other elements. The model was tested with large ranges for these parameters. Since here sodium and chloride movements are ignored, in contrast with the 6-component model, the quantities

$$\gamma = p_{Na}Na^o + p_{Cl}Cl^i$$

$$\delta = p_{Na}Na^i + p_{Cl}Cl^o$$

are held fixed at 9 mM and 40 mM, respectively. The quantity RT/F is set at 60.09. Numerical integration is performed on $x \in [0, 2.5]$, $y \in [0, 2, 5]$ with an explicit method and the results were checked against those for an implicit method. For the numerical integration $\Delta x = \Delta y = 2.5/300$ and $\Delta t = .005$. (Distances are scaled so that the diffusion coefficients are 0.0025 and 0.00125 for potassium and calcium.) With the standard set of parameters, a stable solitary wave with amplitude $K^o_{max} = 17.14$ mM and $Ca^o_{min} = 0.04$ mM formed from supra-threshold local elevations of external potassium

$$K^o(x, y, 0) = K^{o,R} + 20\exp\left[-\left\{\left(\frac{x-1.25}{0.05}\right)^2 + \left(\frac{y-1.25}{0.05}\right)^2\right\}\right],$$

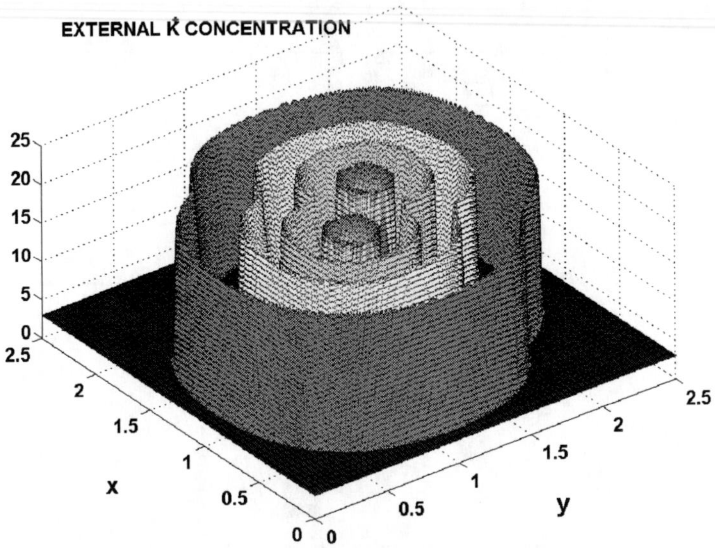

FIGURE 3. Showing the extracellular potassium ion concentration during a collision in the two component model with standard parameter set. The times indicated are $t_k = 200, 400, 600$ & 900.

with $Ca^o(x,y,0) = Ca^{o,R}$. Throughout the paper, a unit of distance corresponds to about 5.2 mm and a unit of time is about 26 seconds [47]. The discrete time points are denoted by $t_k, k = 0, 1, ..., T/\Delta t$. Figures 1 and 2 show potassium and calcium waves for standard parameters when action potentials are ignored.

2.1. Collision

In Figure 3 is shown the result for a collision between two SD waves, one starting at $(x,y) = (1.05, 1.05)$ and the other at $(x,y) = (1.45, 1.45)$. It can be seen that after colliding the waves merge to form a wave of a large wavefront, unlike the case in 1 spatial dimension where colliding waves annihilate.

2.2. Reverberating SD waves

One very interesting aspect of waves of spreading depression was the demonstration of reverberating waves; that is, waves which keep circling an obstacle for a very long time. In the experiments [55] in rat cortex, spreading depresssion waves could circulate for several cycles, with more cycles when lesions were made in some regions than others. On occasion, up to 27 cycles were observed. The principal method of instigation of reverberating SD was with the topographical set up of Figure 4. As depicted in A, a lesion or obstacle Q of large enough dimension is made in the cortex. A suitable

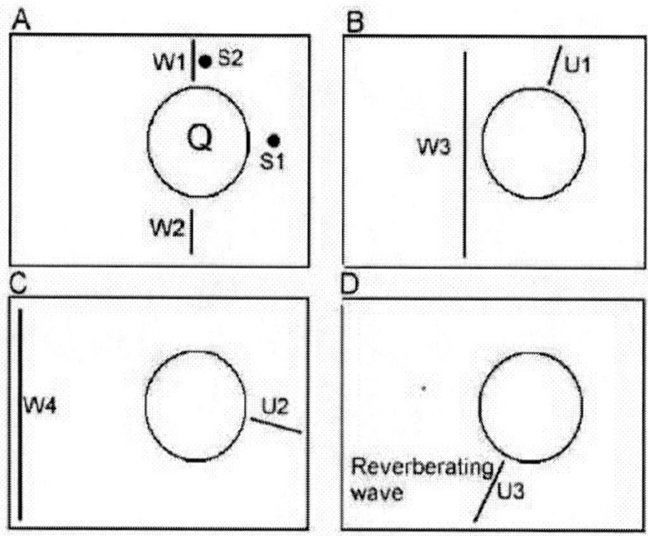

FIGURE 4. The topographical setup for obtaining reverberating waves of spreading depression. In A is shown a brain region containing a lesion or obstacle Q which blocks the passage of SD. S1 is the first stimulus which gives rise to waves W1 and W2 which pass around the obstacle. At a suitable time and location, a second stimulus S2 is applied in the back of the first wave W1. As shown in B, this wave can only move to the right which it does into a non-refractory zone. Meanwhile the first waves advance firstly to W3 and then as seen in C to the boundary of the region where they die, leaving the sole SD wave to circulate unimpeded as in D.

stimulus, such as a local increase of potassium chloride concentration sufficient to elicit a wave, is applied at S1. After a certain time interval, the wavefront has split into two components W1 and W2 on either side of Q. At this point a second stimulus S2 is applied so that its emitted waves move on one side into the refractory zone of W1 and on the other side into the recovered zone in the wake of W1. If the timing and location of the second stimulus are appropriate, then as shown in B the first waves have passed the obstacle as wavefront W3 and meanwhile the surviving part of the secondary wave is at U1. In C, the first wave has reached the boundary as W4 where it will die and the secondary wave has advanced to U2. Hereafter, as in D, the secondary wave is able to circulate unimpeded around the obstacle. In experimental reverberating SD, the wave motion starts to slow after several cycles and eventually disappears due to exhaustion of local metabolic resources after successive recoveries from SD. This latter aspect is not addressed with the present model but it could possibly be incorporated in a model such as that of [56, 57] where metabolic variables were included in a model of stroke which embraced SD.

The computational details for a reverberating wave are as follows. The model equations were as in the beginning of Section 2 with what we have called the standard parameter set. The rectangle $x \in [0, 0.8]$ and $y \in [0, 0.9]$ is used with an obstacle centered at $x = 0.4$, $y = 0.45$ with radius 0.2. The increments in Δx, Δy and Δt are 0.8/60, 0.9/70

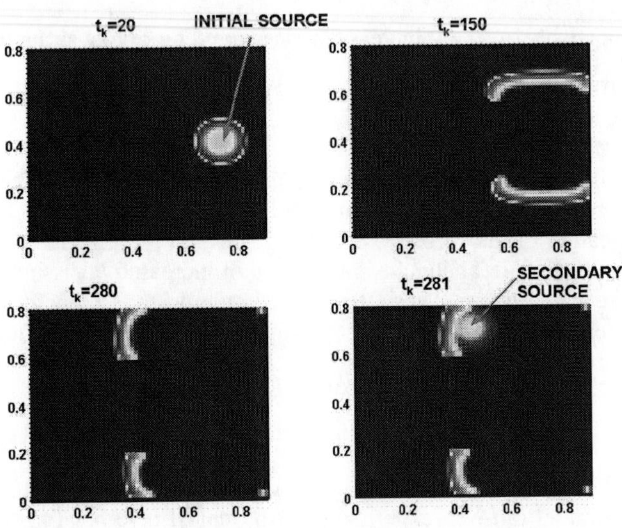

FIGURE 5. Showing the first phases of the development of a reverberating SD wave as seen in experiments on rat cortex. For parameter values, see text.

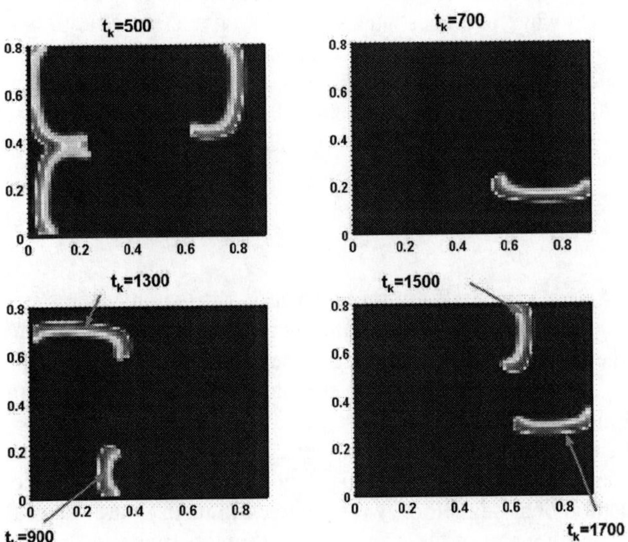

FIGURE 6. Showing the established reverberating SD wave.

and 0.01, respectively. At $t = 0$ increased potassium (chloride) was applied as

$$K^o(x,y,0) = K^{o,R} + 20\exp\left[-\left(((x-0.4)/0.05)^2 + ((y-0.725)/(0.05))^2\right)\right],$$

with $Ca^o(x,y,0) = Ca^{o,R}$.

The resulting wave pattern was observed and it was noted when one branch of the primary wave was at W1 as indicated in Figure 4A. There was not much freedom in the timing and location of the second stimulus S2. Several placements led to a non-clean secondary wave which left behind a small patch of increased potassium that developed into a complex wave pattern. Thus after several attempts the following second stimulus was found to lead to a clean SD wave emanating from S2 in the back direction

$$K^o(x,y,281\Delta t) = K^o(x,y,280\Delta t)$$

$$+10\exp\left[-\left(((x-0.7)/0.06)^2 + ((y-0.44)/(0.06))^2\right)\right],$$

and $Ca^o(x,y,281\Delta t) = Ca^o(x,y,280\Delta t)$. The results are shown in plan view in Figures 5 and 6 and parallel those in experimentally produced reverberating waves in retinal SD. At the 20-th time point ($t_k = 20$), the initial Gaussian distribution has spread a small amount and, by $t_k = 150$, two SD wavefronts are observed travelling in opposite directions away from the source. At $t_k = 280$ the wavefronts are at about the narrowest part of the region between obstacle and boundary and this is when the second stimulus is applied at $t_k = 281$ - see the bottom right part of Figure 5. At $t_k = 500$ (see Figure 6), the initial two wave branches waves are merged and about to pass into the left-hand boundary at $y = 0$ where they are absorbed whereas the secondary wave is seen travelling at top right on its first circuit of the obstacle. At $t_k = 700$ the sole secondary wave can be seen travelling down past the right side of the obstacle and it continues indefinitely, in the absence of metabolic constraints, in a clockwise motion about the obstacle.

2.3. Spiral SD waves

Spiral waves are frequently found in reaction-diffusion systems involving excitable and recovering elements [41, 35, 58] and there have been many theoretical attempts to understand them [59-61]. Experimentally they have been observed for SD in chicken retina [62, 63]. However they have never been obtained in *bona fide* models of spreading depression so we decided to investigate their possible occurrence in the improved two-component model outlined above. Time did not allow their investigation in more complex models. In one set of spreading depression experiments carried out on chicken retina, the method of anodal block was used to extinguish a part of an SD wavefront [62]. The end of remaining part of the wave curled around behind the almost plane-wave front to give a spiral. This could be done by dissolving one end of a wave or a middle segment. In the latter case two spirals lurched towards each other.

To study this phenomenon it is necessary to effect the mathematical equivalent of an anodal block. There are doubtless several ways to do this, but the following was adopted. A wave is started at point P in Figure 7 in the lower left quadrant. A block of all entry of an SD wave into the lower right (shaded) quadrant is effected by holding the ionic

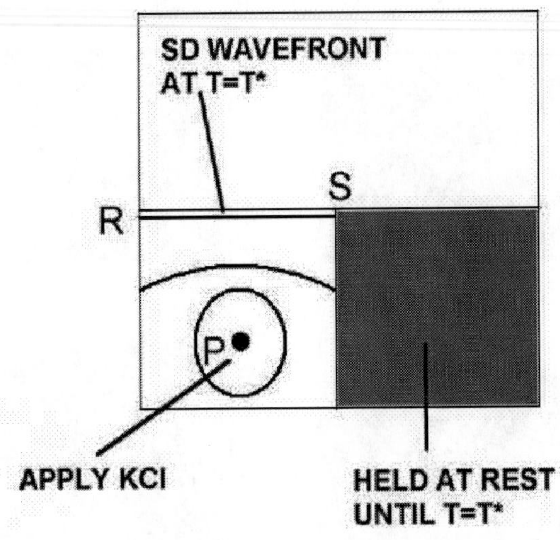

FIGURE 7. Showing the set up for development of a spiral wave of SD. For full explanation, see text.

concentrations at their resting levels until the time when the wavefront from the source at P is just at the upper edge RS of the lower left quadrant. At this instant the clamp on the lower right quadrant is removed, leaving a wavefront along RS with a completely unrefractory region to its right and a refractory region immediately behind it. At the time of the removal of the clamp, the set up is equivalent to having extinguished the right branch of a travelling wavefront.

The results of the computations are shown in Figure 8. The region considered is $x \in [0,2]$, $y \in [0,2]$ with 151 space points in both x and y directions. The time step is 0.01. The lower right quadrant was held at the resting level until $t_k = 400$ and a wave started with the initial distribution

$$K^o(x,y,0) = K^{o,R} + 20\exp\left[-\left(((x-0.3)/0.05)^2 + ((y-0.5)/(0.05))^2\right)\right]$$

concentrated in the center of the lower part of the bottom left quadrant. The wave front becomes almost a plane wave front at $t_k = 450$ ($t = 4.5$) occupying approximately the upper boundary of the lower left quadrant as seen in the Figure. At $t_k = 600$ the right hand end of the wave has begun to curl around behind the plane wave and this spiralling continues for several turns until the simulation is terminated at $t_k = 900$. Note that these spirals are actually circular but the graphic limitations have made them appear rather squarish. The spiralling is exactly analogous to that found in experiments with the anodal block technique.

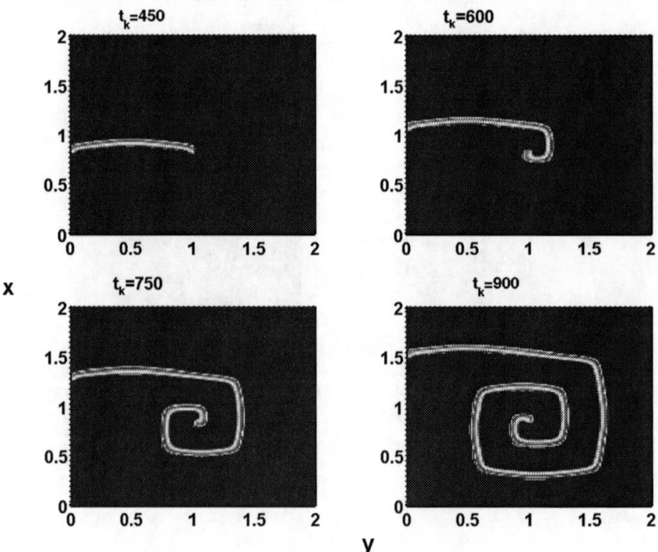

FIGURE 8. Showing the development of a spiral wave as seen in experiments on retinal spreading depression. The graphics has distorted the wavefronts which are actually circular. For remaining parameter values, see text.

3. THE 6-COMPONENT MODEL WITH K^+, Ca^{++}, Na^+, Cl^- AND EXCITATORY AND INHIBITORY TRANSMITTERS

The above two-component model is useful for studying certain phenomena associated with spreading depression. However, a more extensive model [47] considers the 4 ions K^+, Ca^{++}, Na^+, Cl^- and an excitatory transmitter, denoted by T_E, which is expected to be mainly glutamate, and an inhibitory transmitter, denoted by T_I, mostly GABA. Letting the vector of external ion and transmitter concentrations be $\mathbf{u}(x,y,t)$ with $u_1, ..., u_6$, as those of K^+, Ca^{++}, Na^+, Cl^-, T_E and T_I, respectively, then quite generally

$$\frac{\partial \mathbf{u}}{\partial t} = \nabla^2 \mathbf{u} + \mathbf{F}(\mathbf{u})$$

with an initital condition

$$\mathbf{u}(x,y,0) = \mathbf{u}_0(x,y),$$

and suitable conditions at the boundary of the region under consideration. SD is actually a phenomenon in 3 space dimensions, but we consider only two for economy of computation.

We consider two intracellular compartments, one pertaining to synapses and the other to nonsynaptic processes which may include contributions from glia. These are assigned possibly different ratios of extracellular to intracellular volumes, denoted by α_1 and α_2, respectively. The internal ion concentrations, denoted by $u_i^{int}, i = 1,2,3,4$, are assumed to be given by the local conservation equations, which for potassium, sodium and chloride are, with R denoting a resting equilibrium value,

$$u_i^{int}(x,y,t) = u_i^{int,R} + \alpha_1[u_i^R - u_i(x,y,t)], i = 1,3,4,$$

for potassium, sodium and chloride whereas for calcium

$$u_2^{int}(x,y,t) = u_2^{int,R} + \alpha_2[u_2^R - u_2(x,y,t)].$$

It is more transparent to use $K^{o,i}$, $Ca^{o.i}$, $Na^{o,i}$, $Cl^{o,i}$, $T_E^{o,i}$ and $T_I^{o,i}$ for the ion and transmitter concentrations and it is expeditious to omit the space-time coordinates (x,y,t). The membrane potential is assumed given by the Goldman formula

$$V_M = \frac{RT}{F}\ln\left[\frac{K^o + p_{Na}Na^o + p_{Cl}Cl^i}{K^i + p_{Na}Na^i + p_{Cl}Cl^o}\right]$$

and the Nernst potentials are as given in Section 2 for K and Ca and by similar formulas for Na and Cl. The very complex dynamics of calcium at presynaptic terminals have been the subject of many experimental and theoretical studies mainly with a view to quantitatively understanding transmitter release [64-66]. Other works have concentrated on calcium dynamics in neurons during action potentials [67]. We include a major component of calcium fluxes, that associated with the activation of synapses because of its relevance to transmitter release. Although flows through other membranes are doubtless significant they are for the most part neglected in the present model. Their inclusion is no more difficult and their quantitative aspects just as uncertain but to maintain a degree of simplicity they are omitted.

The source and sink terms are slightly modified from those given in [46]. For potassium,

$$f_K = k_1(V_M - V_K)\left[\frac{T_E^o}{T_E^o + k_2} + \frac{k_3 T_I^o}{T_I^o + k_4}\right] - P_{K,Na} + f_{K,p} + k_5,$$

where $P_{K,Na}$ is the pump term and $f_{K,p}$ is a passive flux term given by

$$f_{K,p} = k_6(V_M - V_{M,R})(V_M - V_K)H(V_M - V_{M,R})$$

where $V_{M,R}$ is resting membrane potential. The constant k_5 ensures that $f_K = 0$ at resting levels and it is assumed that the transmitter induced conductance changes are zero unless T_E^o, T_I^o are positive. Although ion pumps have a complicated dependence on concentrations of several ion species [68,69], we have adopted a model with an explicit and relatively simple form for the sodium-potassium exchange pump [70],

$$P_{K,Na} = k_{17}\left(1 + \frac{k_{18}}{Na^i}\right)^{-3}\left(1 + \frac{k_{19}}{K^o}\right)^{-2},$$

where it is assumed that $Na^i > 0$ and $K^o > 0$.

For calcium,

$$f_{Ca} = k_7(V_M - V_{Ca})g_{Ca} + P_{Ca} - k_8$$

where the calcium conductance is

$$g_{Ca} = (1+\tanh[k_{31}(V_M+V_M^*)]-k_{32})H(V_M-V_M^T),$$

V_M^T being a cut-off potential with

$$k_{32} = 1+\tanh[k_{31}(V_M^T+V_M^*)]$$

to ensure g_{Ca} rises smoothly up from zero as V_M increases through V_M^T. The calcium pump is simply

$$P_{Ca} = \frac{k_{20}Ca^iH(Ca^i)}{Ca^i+k_{21}}.$$

The sodium and chloride terms contain transmitter-induced conductance changes and pumps

$$f_{Na} = k_9(V_M-V_{Na})\left[\frac{T_E^o}{T_E^o+k_2}+\frac{k_{10}T_I^o}{T_I^o+k_4}\right]-k_{22}P_{K,Na}-k_{11}$$

$$f_{Cl} = k_{12}(V_M-V_{Cl})\left[\frac{k_{13}T_E^o}{T_E^o+k_2}+\frac{T_I^o}{T_I^o+k_4}\right]+P_{Cl}-k_{14}$$

where

$$P_{Cl} = \frac{k_{25}Cl^iH(Cl^i)}{Cl^i+k_{26}}.$$

Glutamate NMDA receptors have been strongly implicated in SD as known blockers of them prevent SD [71,72]. Rates of transmitter release are assumed proportional to calcium flux so

$$f_{T_E} = k_{15}(V_M-V_{Ca})g_{Ca}-P_E$$

$$f_{T_I} = k_{16}(V_M-V_{Ca})g_{Ca}-P_I,$$

where

$$P_E = \frac{k_{27}T_E^oH(T_E^o)}{T_E^o+k_{28}}$$

$$P_I = \frac{k_{29}T_I^oH(T_I^o)}{T_I^o+k_{30}}.$$

Glutamate may also be released from glia during SD [73,74], but this contribution is not explicitly taken into account here. The pump terms for glutamate and GABA represents the clearance of these transmitters, for example into glial cells [75].

3.1. Results with the standard parameter set

The above system of six reaction-diffusion equations was integrated using an explicit method. In particular, the first results are obtained with the following set of constants - called the standard set.

Ratios of extracellular to intracellular volumes
$\alpha_1 = 0.25$, $\alpha_2 = 2.0$.

Diffusion coefficients in units of 10^{-5} cm^2 sec^{-1}
$D_K = 2.5, D_{Ca} = 1.0, D_{Na} = 1.7, D_{Cl} = 2.5, D_{T_E} = D_{T_I} = 1.3.$

Resting concentrations in mM
$K^{o,R} = 3, K^{i,R} = 140, Ca^{o,R} = 1, Ca^{i,R} = 0.0001, Na^{o,R} = 120$
$Na^{i,R} = 15, Cl^{o,R} = 136.25, Cl^{i,R} = 6.$

Permeabilities
$p_{Na} = 0.05, p_{Cl} = 0.4.$

Calcium conductance parameters in mV
$V_M^* = 45, V_M^T = -60.$

Dynamical constants
$k_1 = 78.091, k_2 = 1.5, k_3 = 0, k_4 = 1.5, k_5 = 0, k_6 = 0.00015$
$k_7 = 0.2, k_8 = 0.0003998, k_9 = 1.6, k_{10} = 0, k_{11} = 39.8140, k_{12} = -104.05$
$k_{13} = 0, k_{14} = 104.064, k_{15} = -3.47, k_{16} = -3.15, k_{17} = 577.895, k_{18} = 2.5$
$k_{19} = 2.5, k_{20} = 0.8, k_{21} = 0.2, k_{22} = 0.3677, k_{23} = 0.11, k_{24} = 0.0711$
$k_{25} = 260.16, k_{26} = 9.0, k_{27} = 47.124, k_{28} = 1.0, k_{29} = 47.124, k_{30} = 1.00.$

Figures 9 and 10 show some results for the standard set of parameters. In Figure 9 can be seen the waves of increasing or decreasing ion and transmitter concentrations spreading out from a local source in which both potassium and chloride ion concentrations are increased. The amplitudes of the components in mM at their maxima or minima are as follows: $K^o = 29.15, Ca^o = 0.076, Na^o = 107.9, Cl^o = 81.68, T_E = 8.38, T_I = 7.11.$ The details of the profiles of the waves in space are shown in Figure 10. The transmitter concentrations are not known accurately and should only be considered as within multiplicative constants. However, the ion concentrations are all feasible in comparison with experiment, both in magnitudes and time courses. More detailed investigations will be reported later.

3.2. Action potentials

Action potential contributions to potassium flux were included by a time coarse-graining technique in [46], and the procedure is vindicated by the results of the single cell model [42]. However an alternative and similar procedure is as follows. Kainic acid application in hippocampus showed that rapid neuronal firing occurred roughly at depolarizations between 15 and 41.7 mV [76] which suggests the following approximate but realistic term for the contribution to potassium sources from action potentials,

$$f_{K,AP} = c[H(V - v_1) - H(v - v_2)](V - v_1)(V - v_2)(V_{Na} - V_K)(V - V_{Ca})g_{Ca}(V)$$

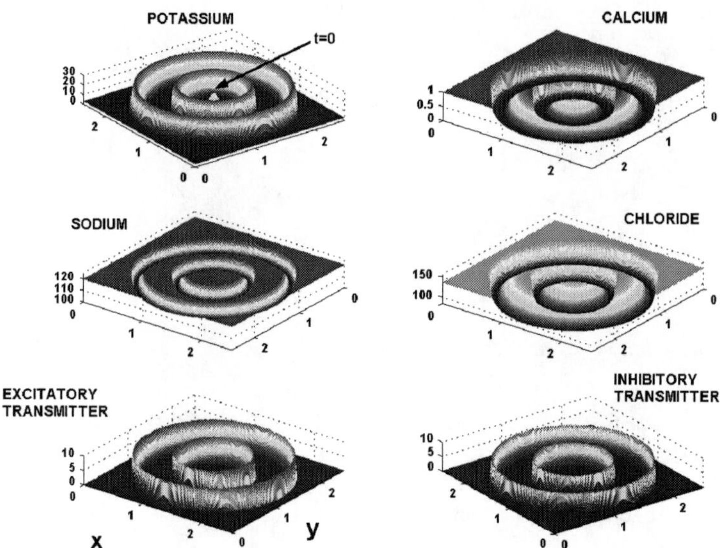

FIGURE 9. The response when potassium chloride is added at the center of the square for the 6-component model with the standard parameter set. Shown spreading from the center are solitary waves of increased extracellular potassium and transmitters and decreased calcium, sodium and chloride. The times shown are $t_k = 0$, $t_k = 300$ and $t_k = 600$ ($\Delta t = 0.01$).

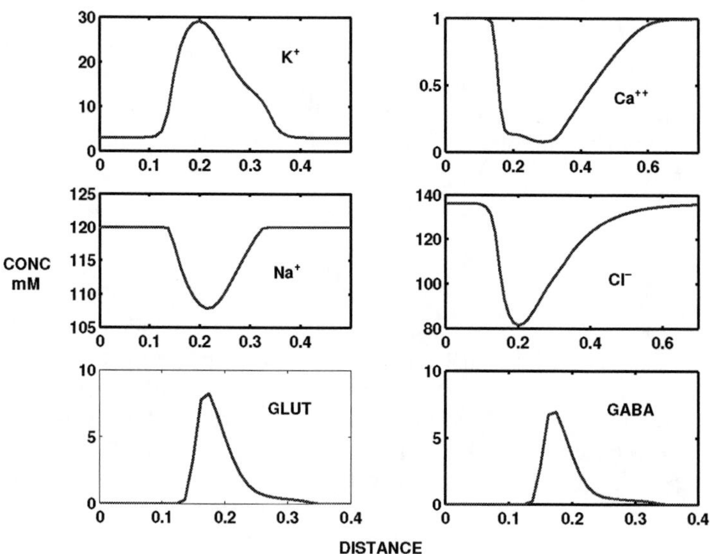

FIGURE 10. Showing the profiles of the external ion and transmitter concentrations at $t_k = 600$ corresponding to the waves shown in the previous figure.

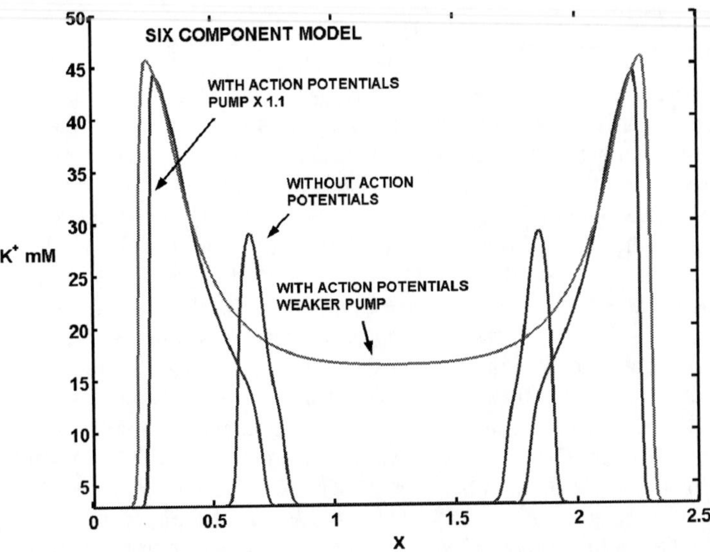

FIGURE 11. The external potassium ion concentration at $t_k = 350$ as a function of distance in the 6-component model both with and without the inclusion of action potentials. These results are for two-space dimensions of which only one is shown here. With action potentials results for two values of the sodium-potassium exchange pump strength are shown - the same as without action potentials and a 10% stronger pump which results in large amplitude faster solitary waves. The initial distribution in all cases was a Gaussian suprathreshold application of KCl at the center of the space interval.

where c is a constant and V_{Na} is the sodium Nernst potential. The membrane potentials between which action potentials are emitted are v_1 and v_2, with $v_1 < v_2$. We now have

$$f_K = k_1(V_M - V_K)\left[\frac{T_E^o}{T_E^o + k_2} + \frac{k_3 T_I^o}{T_I^o + k_4}\right] - P_{K,Na} + f_{K,p} + k_5 + f_{K,AP}.$$

In the calculations we set $v_1 = -55$ mV and $v_2 = -20$ mV.

The distribution of potassium ion concentration with the inclusion of action potentials is shown in Figure 11. Three distributions are shown at $t_k = 350$ ($\Delta t = 0.01$). The smaller wave is the solution with the standard parameter set and no action potential contribution. The red curve shows the effect of including the action potential term with $c = 0.0015$ and no change in the strength of the sodium-potassium exchange pump. It can be seen that the ion concentrations do not return to rest, although they may do so eventually. Increasing the pump strength parameter by 10% so that $k_{17} = 635.6845$ gave a properly formed homoclinic orbit with a return of ion concentrations to resting levels. In this particular calculation the external sodium ion concentration did not fall very much because the exchange pump returns sodium to the extracellular compartment more rapidly. No attempt was made to increase the inward sodium or calcium fluxes due to action potentials, this aspect being left for later work.

4. DISCUSSION

There has been much interest in both experimental and theoretical aspects of SD in recent years, mainly because it has become increasingly apparent that SD plays a significant role in many pathologies of the nervous system. The most prominent example and that which has attracted the most attention is migraine headache. However, there has been much interest also in SD as a concomitant of stroke and of seizures. Furthermore, the implication that SD occurs in spinal cord injury is remarkable. It has also been hypothesized that waves similar to SD might accompany orgasm [77].

The phenomenon of SD has been investigated for over 60 years and comprehensive modeling (as opposed to an *ad hoc* approach where elements are either excited or not and may excite their neighbours) of the kind discussed in the present article was commenced 30 years ago. Since that time there have been a few attempts to address the physiological and anatomical substrates of SD. It is clear that the number of neural, glial, synaptic, metabolic and neurochemical variables which are involved in some way with the formation and passage of an SD wave is very large. A knowledge of all the relevant factors is hard to acquire, not only because of the size of the task, but also because of the uncertainties in the numerical values which one should ascribe to many parameter values. As an example, the ratio, let us call it α, of extracellular to intracellular volumes is an important variable. The usually quoted value for this parameter is 0.2. However, what does this mean? When we consider the space surrounding a synaptic terminal and the space within a synaptic terminal it is not at all clear what α is because the portion of the synaptic terminal that is available for the free motion of ions is known to be much less than its actual physical volume. Calcium is not able to diffuse freely for more than a very short distance in terminals [78] due to buffering. If the volume of a terminal is about 0.5 μ^3 [79] and one takes the rough density of synapses to be 1 per square micron of cortical area [80], then the value of α becomes much larger than 0.2. Despite the limitations on the accurate knowledge of many parameters and the neglect of certain variables, the present computations have been successful in predicting or agreeing with many of the observed properties of SD waves.

ACKNOWLEDGMENTS

The author thanks Prof. Dr Juergen Jost for his hospitality and Prof. Dr Luigi Ricciardi for the opportunity to present this material in Vietri.

REFERENCES

1. Nicholson, C. et al., PNAS USA, **74**, 1287-1290 (1977).
2. Somjen, G. G. & Giacchino, J.L., J. Neurophysiol. **53**, 1098-1108 (1985).
3. Martins-Ferreira,H., Nedergaard, M. & Nicholson, C., Brain Res. Rev. **32**, 215-234 (2000).
4. Somjen, G.G., Physiol. Rev. **81**, 1066-1096 (2001).
5. Goadsby, P.J., Trends Mol. Med. **13**, 39-44 (2006).
6. Bigal, M.E. & Lipton, R.B., Headache: J Head & Face Pain **48**, 7-15 (2008).
7. Cucchiara,B. & Detre, J., Med. Hypotheses **70**, 860-865 (2008).
8. Lauritzen, M. Brain **117**, 199-210 (1994).

9. Van Harreveld,A. & Fifková E., J. Neurobiol. **2**, 13?29 (1970).
10. Revett, K., Ruppin, E., Goodall, S. & Reggia, J.A., J. Cereb. Blood Flow & Metab. **18**, 998-1007 (1998).
11. Petzold, G.C. et al., J. Cereb. Blood Flow & Metab. **25**, s470-476 (2005).
12. Olsson, T. et al., Neuroscience **140**, 505-515 (2006).
13. Gorji, A. et al., Neurobiol. Disease **15**, 70-79(2004).
14. Gorji, A. et al., Brain Res. **906**, 74-83 (2001).
15. Koroleva, V.I., Davydov, V.I. & Roschia, G. Ya., Neurosci. Behav. Physiol. **36**, 625-630 (2006).
16. Richter, F., et al., J. Cereb. Blood Flow & Metab., Adv. Online Pub., 1-11 (2007).
17. Schock, S.C. et al., Brain Res. **1168**, 129-138 (2007).
18. Leis, J.A., Bekar, L.K. & Walz, W., GLIA **50**, 407-416 (2005).
19. Bockhurst, K.H.J. et al., J Magn. Res. Imag. **12**, 722-733 (2000).
20. Meir, A. et al., Physiol. Rev. **79**, 1019-1088 (1999).
21. Connors, B.W. & Long, M.A., Annu. Rev. Neurosci. **27**, 393-418 (2004).
22. Margineanu, D.G. & Klitgaard, H., Brain Res. Bull. **71**, 23-28 (2006).
23. Peters, O. et al., J. Neurosci. **23**, 9888-9896 (2003).
24. Nedergaard, M.. Science **263**, 1768-1771 (1994).
25. Fellin,T.& Carmignoto, G., J. Physiol. **559**, 3-15 (2004).
26. Kang, J. et al., Nature Neurosci. **1**, 683-692 (1998).
27. Leibowitz, D. H., Proc. Roy. Soc. **250**, 287-295 (1992).
28. Crews, F.T. et al., Int. Rev. Neurobiol. **39**, 283-367 (1996).
29. Elliot, J.R. & Elliot, A.A., Prog. Neurobiol. **42**, 611-683 (1994).
30. Klein,G. et al., Forens. Sci. Int. **171**, 131-135 (2007).
31. Nicholson, C., Rep. Prog. Phys. **64**, 815-884 (2001).
32. Dolzmann, K. et al., Int. J. Bifurc. Chaos, **17**, 1329-1335 (2007).
33. Reggia, J.A. & Montgomery, D., Comput. Biol. Med. **26**, 133-141 (1996).
34. Monteira, L.H.A., Paiva, D.C. & Piqueira, J.R.C., J. Biol. Syst. **14**, 617-629 (2006).
35. Dahlem, M.A. & Chronicle, E.P., Prog. Neurobiol. **74**, 351-361 (2004)
36. Grenier, E. et al., Prog. Biophys. Mol. Biol., in press (2008).
38. Dahlem, M.A., Schneider, F.M. & Scholl, E. , J. Theor. Biol. **251**, 202-209 (2008).
38. Ruppin, E. & Reggia, J.A., Neurol. Res. **23**, 447-456 (2001).
39. Shapiro, B., J. Comput. Neurosci. **10**, 99-120 (2001).
40. Ruktamatakul, S., Bell, J. & Lenbury, Y., IMA J. Appl. Math. **71**, 544-564 (2006).
41. Riaz, S.S. & Ray, D.S., J. Chem. Phys. **123**, 174506 (2005).
42. Kager, H., Wadman, W.J. & Somjen, G.G., J. Neurophysiol. **84**, 495-512 (2000).
43. Makaraova, J. et al., Biophys. J. 92, 4216-4232 (2007).
44. Almeida, A.C.G. et al., IEEE Trans. Biomed. Eng. **51**, 450-458 (2003).
45. Tuckwell, H.C. & Miura, R.M., Biophys. J. **23**, 257-276 (1978).
46. Tuckwell, H.C. Int. J. Neurosci. **10**, 145-165 (1980).
47. Tuckwell, H.C. & Hermansen,C.L., Int. J. Neurosci. **12**, 109-135 (1981).
48. Lehmenkuhler, A. et al., Neurosci. **55**, 339-351 (1993).
49. Nicholson,C.& Sykova, E., TINS **21**, 207-215 (1998).
50. Ying, J., Aitken, P.G. & Somjen, G.G., J. Neurophysiol. **71**, 2548-2551 (1994).
51. Mazel, T. et al., Physiol. Res. **51**, Supp. 1, S85-93 (2002).
52. Dietzel, I. et al., Exptl. Brain Res. **40**, 432-439 (1980).
53. Yan, G-X. et al., J. Physiol. **490**, 215-228 (1996).
54. Ren, J. Q. et al., Exp. Brain Res. **92**, 1-14 (1992).
55. Shibata,M. & Bures,J., J. Neurobiol. **5**, 107-118 (1975).
56. Chapuisat,G. et al., Prog. Biophys. Molec. Biol., in press (2008).
57. Chapuisat,G., ESAIM: Proc. **18**, 87-98 (2007).
58. Beaumont, J. et al., Biophys.J. **75**, 1-14 (1998).
59. Mikhailov, A.S. & Zykov, V.S., Physica D **52**, 379-397 (1991).
60. Hess, B., Naturwissenschaften **87**, 199-211 (2000).
61. Lindner, B. et al., Phys. Rep. **392**, 321-424 (2004).
62. Gorelova,N.A. & Bures,J., J. Neurobiol. **14**, 353-363 (1983).
63. Dahlem, M.A. & Muller, S.C., Exp. Brain. Res. **115**, 319-324 (1997).

64. Heidelberger,R, et al., Nature **371**, 513-515 (1994).
65. Fossier, P., Tauc, L. & Baux, G., TINS **22**, 161-166 (1999).
66. Koester, H.J. & Sakmann,B., J. Physiol. **529**, 625-646 (2000).
67. Schilller, J., Helmchen, F. & Sakmann, B. J. Physiol. **487**, 583-600 (1995).
68. Yingst, D.R., Davis, J. & Schiebinger, R., Am. J. Physiol. Cell Physiol. **280**, C119-C125 (2001).
69. Torok, T.L., Prog. Neurobiol. **82**. 287-347 (2007).
70. Garay, R,P. & Garrahan, P.J., J. Physiol. **231**, 297-325.
71. Obrenovitch, T.P. & Zilkha, E., Br J. Pharmacol. **117**, 931-937 (1996).
72. Anderson, T.R, & Andrew, R.D., J. Neurophysiol. **88**, 2713-2725 (2002).
73. Basarsky, T.A., Feighan, D. & MacVicar, B.A., J. Neurosci. **15**, 6439-6445.
74. Larrosa, B. et al., Neuroscience **141**, 1057-1068 (2006).
75. Zoremba, N. et al., Exp. Neurol. **203**, 34-41 (2007).
76. Le Duigou, C. et al., J. Physiol **569**, 833-847 (2005).
77. Tuckwell, H.C., Int. J. Neurosci. **44**, 143-148 (1989).
78. Burrone, J. et al., Neuron **33**, 101-112 (2002).
79. Egelman, D.M. & Montague, P.R., Biophys.J. **76**, 1856-1867 (1999).
80. Koch, C., *Biophysics of Computation: Information Processing in Single Neurons*, Oxford Univ. Press, Oxford (1998).

Computational Approach to Schizophrenia: Disconnection Syndrome and Dynamical Pharmacology

Péter Érdi*, Brad Flaugher†, Trevor Jones†, Balázs Ujfalussy**, László Zalányi** and Vaibhav A. Diwadkar‡

**Center for Complex Systems Studies, Kalamazoo College, Kalamazoo, Michigan, USA and Department of Biophysics, KFKI Research Institute for Particle and Nuclear Physics of the Hungarian Academy of Sciences, Budapest, Hungary*
†Center for Complex Systems Studies, Kalamazoo College, Kalamazoo, Michigan, USA
***Department of Biophysics, KFKI Research Institute for Particle and Nuclear Physics of the Hungarian Academy of Sciences, Budapest, Hungary*
‡Psychiatry and Behavioral Neurosciences, Wayne State Univ. School of Medicine, Detroit, MI, USA

Abstract. Schizophrenia may be best understood in terms of abnormal interactions between different brain regions. Tasks such as associative learning that engage different brain regions may be ideal for studying altered brain function in the illness. Preliminary data suggest that the hippocampus is involved in the encoding (learning) and the prefrontal cortex in the retrieval of associative memories. Specific changes in the fMRI activities have also been observed based on comparative studies between stable schizophrenia patients and healthy control subjects. Disconnectivity, observed between brain regions in schizophrenic patients could result from abnormal modulation of N-methyl-D-aspartate (NMDA)-dependent plasticity implicated in schizophrenia...

Keywords: Schizophrenia models, associative learning, dynamical diseases.
PACS: 87.19.lv, 87.19.x, 87.85.dq

1. INTRODUCTION

Schizophrenia is one of the most debilitating mental illnesses in the world. Global prevalence rates are estimated at between $1-2\%$ and the illness has profound personal costs for patients and their families, and widespread societal costs. Despite the illness being widely accepted as biological following decades of biological research, still we have serious challenges toward the understanding of schizophrenia. Experimental studies have proliferated the literature in several in vivo imaging modalities including functional and structural MRI and MR spectroscopy . In conjunction with post-mortem studies of brain tissue , these studies have provided innumerable examples of specific and general deficits in function and structure in the living and deceased schizophrenia brain. Yet very significant shortcomings in understanding remain. Few studies offer theoretical frameworks to provide formal computational models of brain function to interpret experimental data. Notwithstanding a handful of efforts , such an absence is glaring because the diverse biological findings are rarely reconciled within a formal framework that

CP1028, *Collective Dynamics: Topics on Competition and Cooperation in the Biosciences*
edited by L. M. Ricciardi, A. Buonocore, and E. Pirozzi

can be provided by such models.

2. GENERAL FRAMEWORK: DYNAMICAL DISEASES

Neurological and psychiatric disorders can be interpreted as dynamical diseases. The concept emerged about ten years ago [1], and seems to be fruitful to explain a variety of disorders. For example, the emergence of seizures was explained by computational models to interpret transitions between the two states of a bistable system, namely between normal and epileptic activity [2]. The prediction and control of epileptic seizures became a hot, and at this moment, controversial topic [3, 4]. It has become clear that dynamical models can predict seizure development and the administration of drugs could be designed accordingly providing novel therapeutic procedures for epileptic patients. Although, statistical analysis helps to predict the emergence of seizures, we still need to be cautious regarding its potential clinical applications [5, 6]. There are many other diseases where the dynamical characteristics seem to be relevant and the concept of "dynamical disease" can be applied. For example, ten years ago the question was raised whether Parkinson's disease is a dynamical disease [7]. Along this line, more recently, a method was suggested for detecting preclinical tremor in Parkinson's disease [8]. Using nonlinear dynamical theories and calculations [9] it has been shown that patients with Parkinson's disease had EEG series' with higher complexity than normal persons during the performance of complicated motor tasks. The explanatory hypothesis for this increased complexity states that additional superfluous cortical networks are recruited due to impaired inhibition. Depression is also thought to be dynamical disease [10], and now it seems to be clear that there is a correspondence between clinical and electro-physiological dimensions [11], i.e. clinical remission and brain dynamics reorganization.

Nonlinear dynamical methods gave new insights to study the neurodynamics in Alzheimer's disease [14, 15]. The analysis suggested that there is a reduced complexity and level of synchronization in the EEG patterns presumably due to impaired connectivity among different cortical regions. Dynamically evolving processes also can be captured in studies of migraine, i.e. migraine aura lasts less than an hour, and precedes headache, and it is characterized by visual and other distortions. Hallucinatory patterns have been modeled by reaction-diffusion models leading to waves as has been observed in other excitable media (for a review see [16]).

Applications to Schizophrenia are discussed in Sec. 3. Dynamical disease occurs due to the impairment of the control system and the steps of a procedure of a computational neuropharmacology [17] is the following:

- Develop realistic mathematical models and study effects of parameter changes
- Neurobiological interpretation
- Integration of molecular, cellular and system neuroscience
- Therapeutic strategies

A summary of applying the concept of dynamical diseases for a number of disorders is illustrated on Fig 1.

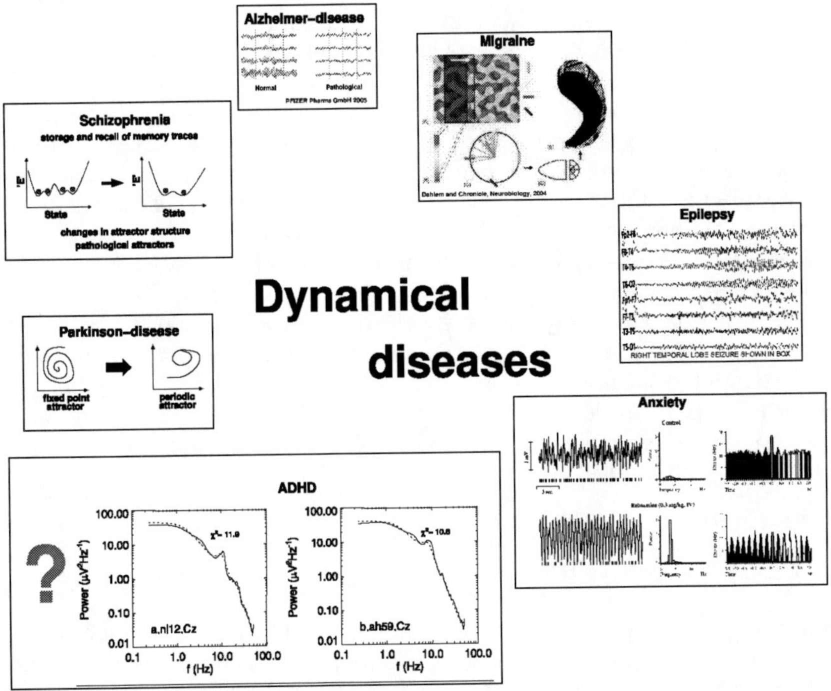

FIGURE 1. Dynamical diseases. Illustrative panels of using dynamical diseases to model neurological and psychiatric disorders.

3. DYNAMIC APPROACH TO SCHIZOPHRENIA

3.1. Concepts and models

3.1.1. Cortical pruning hypothesis

It is generally accepted that schizophrenia is related to excessive pruning of cortical connections, and simple network studies [18] have shown that that cortical pruning may lead to formation of "pathological attractors".

"Computation with attractors"became a paradigm which suggests that dynamic system theory is a proper conceptual framework for understanding computational mechanisms in self-organizing systems such as certain complex physical structures, computing devices, and neural networks. Its standard form is characterized by a few properties. Some of them are listed here: (i) the attractors are fixed points; (ii) a separate learning stage precedes the recall process whereby the former is described by a static 'one-shot' rule; (ii) the time-dependent inputs are neglected; (iv) the mathematical objects to be classified are the static initial values: those of them which are allocated in the same basin evolve towards the same attractor, and can recall the memory trace stored there. In an extended form, not only fixed points but also limit cycles and strange attractors can be involved. A continuous learning rule may be adopted but, in this case, the basins of the attractors can be distorted which may even lead to qualitative changes in the nature of the attractors. Some family of models of the cortex can be interpreted as special cases of attractor neural networks. More realistic models, which take explicitly into account the continuous interaction with the environment, however, are non-autonomous in mathematical sense [20, 21]. Such systems do not have attractors in the general case. Attractor neural network models cannot be considered as general frameworks of cortical models, but give some insight to memory storage and recall.

Pathological attractors may implement the dynamic generation of positive symptoms in schizophrenia as these symptoms, including delusions and hallucinations can be activated in the absence of external cues. Related to the modifiability of the attractor-basin portrait, a model based on the NMDA receptor delayed maturation was also suggested as a possible mechanism of the pathogenesis of schizophrenic psychotic symptoms [19].

3.1.2. Nonlinear Dynamics and Schizophrenia

Dynamic system theory offers conceptual and mathematical tools for describing the performance of neural systems at very different levels of the organization [22].

Non-invasive brain-imaging techniques will have a major role in relating neural structures to function. Present techniques fall into two groups, depending on the physical quantities on which the imaging is based. First, electric and magnetic signals generated by the neural tissue are used. These methods such as EEG and MEG, typically recorded outside from the scalp, so they generally provide a spatial resolution around 1cm, with a temporal resolution in the region of milliseconds. Second, neural activity is estimated based on measurements of haemodynamic and metabolic events. Positron emission tomography (PET) and functional magnetic resonance imaging (fMRI) produce data with several millimeter resolution, and a time resolution on the order of few seconds for fMRI and to tens of seconds for PET. The electrophysiological methods serve better dynamic temporal information but only at a few brain centers, while the PET/fMRI methods give information about the brain regions and the strengths of interconnections among them during the performance of cognitive tasks. The combination of techniques is useful: first the brain regions to be involved in some cognitive task can be determined, following which the detailed dynamics of this site can be measured by EEG.

Dynamical systems hypotheses are based on the assumption that pathological symptoms are related to changes in the geometry of the attractor basin portrait [23]. A network model of excitatory and inhibitory neurons built by Leaky integrate-and-fire models was used to design several simulation experiments to study the effects of changes in synaptic conductances on overall network performance. Reduction in synaptic conductances connected to glutamatergic NMDA receptors implied flatter attractor basins, and consequently less stable memory storage. Combined reduction of NMDA and GABA receptors imply such changes in the attractor structure, that may implement such positive symptoms, as hallucinations and delusion.

More analyses are need to relate impairment of global (interregional) and local (intraregional) connections to the emergence of schizophrenia. Nonlinear theories of schizophrenia have been suggested [24] based on EEG recordings, but whereas EEG analysis is very extensively used, the theoretical bases remain unclear.

On the one hand, classical signal analysis considers EEG records as the realization of (often stationary) stochastic processes, and spectral (and later also wavelet) analysis has been the conventional method to extract the dominant frequencies and other parameters of the rhythms.

On the other hand, the occurrence of chaotic temporal patterns has been reported at different hierarchical levels of neural organization. Chaotic patterns can be generated at the single neuron level, due to the nonlinearity of voltage-dependent channel kinetics of the ionic currents, at the multicellular network level, due to the interactions among neurons, and at the global level in consequence of spatiotemporal integration.

Dynamic systems theory offers a conceptual approach to EEG signal processing, different from the classical analysis. Time series, even irregular ones, are considered as deterministic phenomena generated by nonlinear differential equations. Though the methodological difficulties of interpreting the calculated quantities (Lyapunov exponents, fractal dimensions, entropies etc.) to characterise neurological categories are now well-known [25], the application of dynamic systems theory brought a breath of fresh air to the methodology of processing of neural signals. Schizophrenic symptoms may occur due to impairment in coupling of processes taking place at different temporal and spatial scale, and the structural basis of pathological changes in dynamics and behavior can be unrevealed by using dynamical models.

3.1.3. Disconnection syndrome

The old idea that the schizophrenia is caused by pathological connection between brain regions [27, 28], has re-emerged as the disconnectivity hypothesis. Dynamical analysis of scalp EEG data have been used to address the question of whether schizophrenia originates from the reduction of functional connectivity among brain regions [12]. Evidence for reduced functional connectivities are derived from brain imaging experiments. These reduced connectivities are due to impairments in synaptic transmission and plasticity [26]. Disconnectivity, observed between brain regions in schizophrenia patients could result from abnormal modulation of NMDA-dependent plasticity implicated in schizophrenia. Our own ongoing project is to build a dynamical causal model for the five interconnected brain regions to estimate functional loss in schizophrenia patients, see 6.5.1.

A functional computational model [13] suggests that schizophrenia might be the results of massive pruning of local connections in association cortex.

4. A DYNAMICAL APPROACH BASED ON COMBINED BEHAVIORAL AND FMRI DATA

4.1. The experimental paradigm

During the task subjects alternated between blocks of consolidation, rest/rehearsal and retrieval. During consolidation, nine equi-familiar objects with monosyllabic object names were presented in sequential random order (3s/object) in grid locations for naming "bed" and "boo" are depicted). Following a brief rest/rehearsal interval, memory for object-location associations was tested by cuing grid locations for retrieving objects associated with them (3s/cue). Object names were monosyllabic to minimize head motion. Eight blocks (each cycling between consolidation, rest and retrieval) were employed.

Healthy Controls (n=11; mean age=22 yrs, sd=5; 5 females) and stable early course schizophrenia patients (n=11; mean age=26 yrs; sd=5; 3 females) gave informed consent. Groups did not differ in terms of age (p>.10). Patients were diagnosed using DSM-IV, SCID and consensus diagnosis. All were on a regimen of atypical antipsychotics (Risperidone, Olanzapine or Aripiprazole).

4.2. Basic Behavioral Data

Subjects participated in an associative learning/memory task adapted from previous studies [29]). During the course of the task, subjects learned the associations between nine equi-familiar objects drawn from a standardized battery ([30]) over a series of encoding/consolidation and retrieval epochs. During each epoch, objects were presented in their associated location in space (3 s/object) for naming. All nine objects were shown in random order. Following a brief rest interval, the nine locations were cued (with a square) in random order and subjects were required to recall the object associated with the location. Eight blocks (each cycling between encoding, rest and retrieval) were employed. A schematic of the task is depicted in Figure 2. The temporal profile of performance in terms of the number of trials is expressed by a learning curve. A learning curve of healthy controls and schizophrenia patients is visualized in Fig 3.

4.3. Basic fMRI data and some analysis

Data from preliminary fMRI studies demonstrate the following: a) feasibility to conduct complex fMRI paradigms in a 4T environment and b) using a complex tasks such as associative learning to assess trends in the data that indicate altered learning dynamics in BOLD in schizophrenia patients. The fMRI paradigm is depicted in Figure 4.

Our data analysis focused on finding regions, where the activity is correlated with block of learning or recall. Figure 5 shows the active regions during encoding (black)

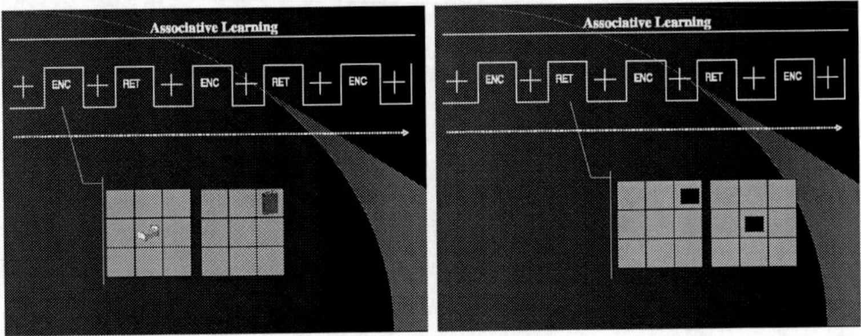

FIGURE 2. Structure of the experimental paradigm is depicted with two examples of associations presented during encoding/consolidation ("bed" and "book") and examples of those locations cued during recall/retrieval.

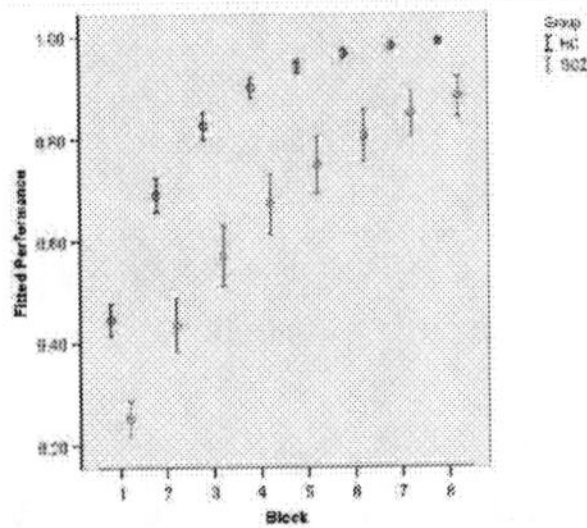

FIGURE 3. Learning dynamics in controls and schizophrenia patients over time are plotted. The data provide evidence of generally asymptotic learning in both groups, with reduced learning rates in patients compared to controls.

and recall (red) in healthy control (open symbols) and schizophrenia subjects (filled symbols). The row data were de-trended and normalized, and correlated with a simple signal that is one during the block learning and zero otherwise. Similar analysis was performed with the recall signal. The data presented here are averaged over subjects. The results show high activation in the visual areas V1 and SP both

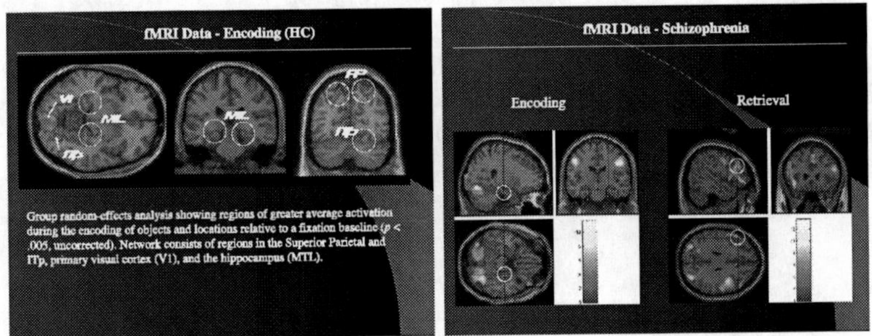

FIGURE 4.

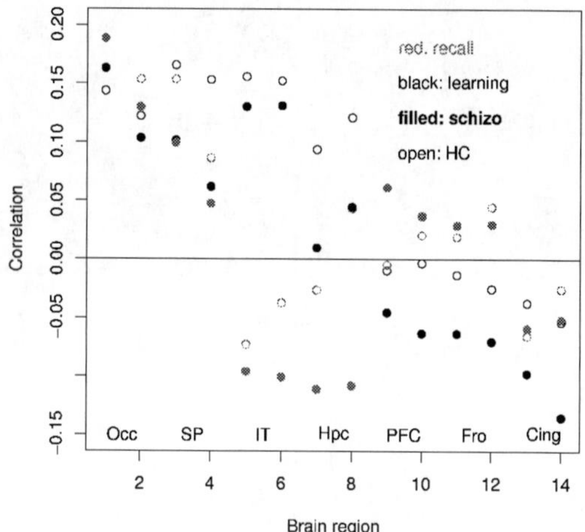

FIGURE 5. Correlations with blocks. Vertical axis: different regions: Occ: occipital, SP: superior parietal, IT inferio-temporal, Hpc Hippocampal, PFC prefrontal, Fro: Frontal, CNG: Cingulum. Data from left and right hemispheres are plotted separately.

during learning and recall, but only during encoding in the inferior temporal cortex, the part of the ventral stream engaged in object recognition. Interestingly, the the hippocampus was mostly active during encoding, whereas the prefrontal regions (PFC, Fro) during the recall of information. Differences between healthy controls and schizophrenic patients are remarkable in the hippocampus and the prefrontal regions.

4.4. The functional macro-network for associative memory

Brain areas involved
Based on the available data on the activity of five interconnected regions (superior parietal cortex, inferio-temporal cortex, prefrontal cortex, primary visual cortex and the hippocampus) are supposed to form the macro network. In accordance with the spirit of the "disconnection syndrome" a question to be answered is which connections are impaired during schizophrenia, and what is the measure of functional reduction of the information flow?

4.5. Associative learning

Associative learning relies on the consolidation and retrieval of associations between diverse memoranda, sensory inputs and streams of neural activity, particularly by hippocampal and medial temporal lobe neurons . This detection and

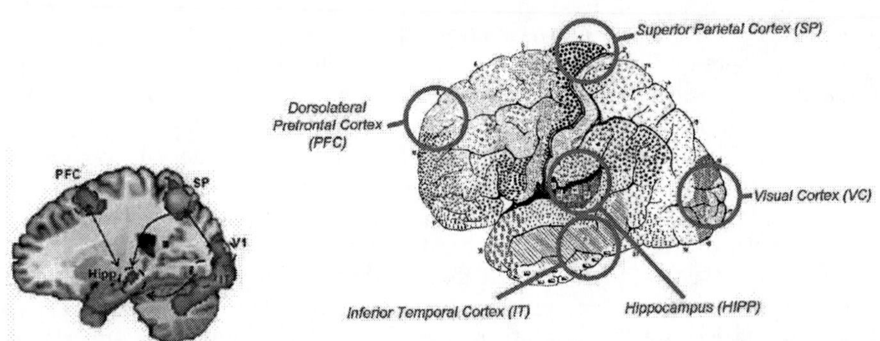

FIGURE 6. Left: Information flow during object-location associative memory (based on [31].). The cortical pathway is overlaid on a medial slice depicting brain activity (p<.001) during memory consolidation during object-location associative learning (see Preliminary data for further details). Regions labeled are: V1: Primary Visual Cortex; IT: Inferior Temporal Cortex; SP: Superior Parietal Cortex; Hipp: Hippocampus; PFC: Dorso-lateral prefrontal cortex. Right: An enlarged view of the brain areas involved.

consolidation of correlated spatio-temporal patterns of neuronal activity was proposed in classic neuroscience as a centerpiece of learning and memory (Post-hebbian learning mechanisms and algorithms are reviewed in [32]). In neuroimaging, the collection of large scale in vivo and surface imaging modalities has allowed the estimation of coherence or connectivity using a variety of statistical and clustering techniques including coherence analysis, principal components analyzes and analyzes of functional and effective connectivity between brain regions . What do we know about the macro-network for associative memory?

The weight of experimental evidence clearly indicates that the medial temporal lobe, including the hippocampus and its sub-units such as the cornu ammonis (CA), the dentate gyrus (DG) and the subiculum, and adjacent structures such as the entorhinal cortex are central to the formation of long term memories. These regions occupy a unique anatomical place within the realm of cortical and subcortical connections receiving inputs from the sensory areas in unimodal association cortex and from heteromodal areas such as the dorsal and ventral prefrontal cortex via the entorhinal cortex. This region is therefore uniquely positioned in a "hierarchy of associativity" tointegrate multi-modal inputs from unimodal areas before redistribution of potentiated associations into the neo-cortex . This general framework provides a good explanation for the patterns of anterograde amnesia in classic neuropsychological studies of patients with hippocampal lesions in whom the retention of memories before the lesion is preserved but the formation of new long term memories is impaired . It also is consistent with experimental work in animals: lesions that are applied to the hippocampus early during learning devastate trace conditioning preventing eventual consolidation of traces in long term memory . When the function of the medial temporal lobe is impaired during learning of associations, memorial representations that rely on this hippocampal activity are either not formed or are formed to inadequate strength . Thus, memory is inadequately established and is unavailable at the fidelity needed when recall is required. Therefore in a disorder such as schizophrenia, impaired hippocampal activity during critical periods of learning and memory may form a central basis of impaired memory formation.

Why are the hippocampus and its sub-units crucial to memory formation? In Figure 6 we provide a schematic model for object-location association, and of the unique pattern of uni- and poly-modal inputs to the hippocampus, where these inputs are subsequently bound into a-modal associations. Network locations are depicted on a statistical map of fMRI-measured brain activity in a single subject during memory consolidation in our object-location association task. As seen, location and object information are relayed from dorsal (Superior Parietal) and ventral (Inferior Temporal) inputs respectively to the hippocampus. The CA and DG sub-regions in the hippocampus process information via uni- or bi-directional connections from the entorhinal cortex. CA regions (primarily CA3) and DG form a mutual excitatory network for encoding amodal information (associations). Entorhinal connectivity with prefrontal cortex provides access to intermediately encoded associations (short time scales) and for the eventual disbursal of memories into the neo- cortex (longer time scales). Through this funneling of information, the degree of abstraction of the information increases through this pathway before assuming its most abstract form in the encoding within the CA and DG units within the hippocampus . In the context of schizophrenia, it is plausible that abnormal prefrontal-hippocampal and glutamate-dopamine interactions lead to associative memory deficits. As the figure indicates, multiple neural regions can be targeted with fMRI to identify impaired function in a network disorder such as schizophrenia.

5. A SIMPLE MODEL

5.1. Model description

Our initial modeling efforts were directed toward a simple model to simulate the behavioral associative learning task, with model output as learning curves depicting performance over each iteration of recall [33]. As will be evident, the model incorporates the separation between encoding/consolidation and cued recall while also retaining biological plausible relationships between model architecture and neural systems, as well as known learning parameters in the brain. In particular, the model accounts for (i) the separation between "where" and "what" regions (ii) reduced synaptic plasticity in schizophrenia and reduced cognitive capacity in schizophrenia .

Separate neural systems are represented by two separate nine (nine objects and nine locations) dimensional binary vector inputs supplied to the model representing the object shown to the subject and the location of the object named a_L and a_O respectively. Nine unique object-location vector pairs for each trial represent the nine unique object-location relationships in the task. Background neural activity is simulated by the addition of a normally distributed noise term (mean= .5) to each element. Each vector pair is dyadically multiplied to form a single object-location association matrix A, which has elements that fall into three categories. The element that results from multiplying the active signal of a_L and $a-O$ contains the strongest association. Every other element in the row and column that it occupies is the product of active signal and a noise term. The remaining elements are the product of only noise terms and contain no meaningful signal.

The learning rule adopted was motivated by the Rescorla-Wagner rule [34]. Each A is added to form $W(t)$ with nine elements that hold correct multiplicative associations with noise, and 72 elements that hold incorrect additive associations. $W(t)$ is multiplied by a learning rate $r(t)$ modulating the strength of associations on a trial-wise basis to form $W(t+1)$ as in 1. Synaptic plasticity can be represented by the rate parameter rmax, modulating the learning rate $r(t)$:

$$r(t) = r_{max}(S_{max} - S(t-1)) \tag{1}$$

Cognitive capacity is indexed by S_{max} and $S(t-1)$ is performance on the previous iteration of the task. During encoding in, the learning rate $r(t)$ functions as a supervisory parameter by modulating the strength of the encoding matrix W at any instant during learning, depending on its own parametrization. Crucially, at any given time "t" during learning the association matrix $W(t)$ represents the strengths between associations to be learned during the task. During recall, the model is given a noiseless input a_L which represents the location cue. a_L is multiplied with the encoding matrix $W(t+1)$ to select the column of $W(t)$ that contains the information of which object is associated with the chosen a_L. This recalled column is a vector y:

$$y = a_L W(t) \tag{2}$$

Each element in y is tested by a threshold function (τ to determine if it is an active recall or a noise induced value:

$$\tau := 1/81 \sum_{i=1}^{9} \sum_{j=1}^{9} W_{ij}(t+1) + \pi[maxW(t+1))] \tag{3}$$

where i is a column, and j is a row of $W(t)$ and π is a chosen multiplier between 0 and 1. The threshold is determined by averaging the elements of $W(t)$ and adding the largest element in $W(t+1)$ multiplied by π. This premium controls the sensitivity of the model to noise. As seen in Figure 7, model performance for "controls" ($S_m ax = 1$ and $r_m ax = 0.4$) and "patient" behavior ($S_m ax = 0.7$ and $r_m ax = 0.2$) provide reasonable simulations of control and patient data (see Figure 3).

5.2. Simulation results

Healthy Control performance: For healthy controls, the plausible parameter range for $r_m ax$ was 0.2 to 0.55. The representative healthy control parameters were

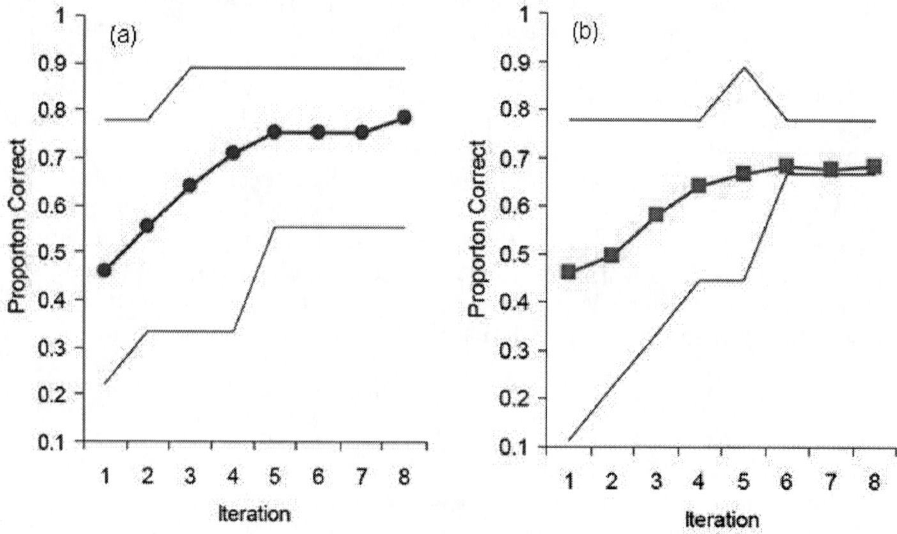

FIGURE 7. (a) Maximum, minimum, and average values for healthy control patients. Produced with parameter values $S_m ax = 1$ and $r_m ax = 0.4$. Maximum and minimum curves are the basis for finding the parameter ranges. (b) Maximum, minimum, and average values for "schizophrenia" patients. Produced with parameter values $S_m ax = 0.7$ and $r_m ax = 0.2$.

$r_m ax = 0.4$ and $S_m ax = 1$, yielding a behavior curve similar to achieved control data. The plausible parameter range can be estimated by looking at the maximum and minimum values given by the ideal parameters and determining the parameter values that will give, on average, the maximum and minimum values for an average healthy control subject. Figure 7A shows the performance of the model based on ideal parameter values.

"Schizophrenia" performance: For "schizophrenia" like performance, the plausible parameter range for $r_m ax$ was .1 to .25 with representative performance achieved for $r_m ax = 0.2$ and $S_m ax = .7$. Figure 7B depicts average and range of performance for schizophrenia behavior.

These results indicate how limitations/changes in synaptic plasticity and memory capacity can predict control or schizophrenia-like behavior in learning and memory dynamics.

6. A FUNCTIONAL NEURAL MODEL OF NORMAL AND PATHOLOGICAL ASSOCIATIVE LEARNING

A model has been built order to compare the (1) activities with the fMRI data; (ii) the performance with the behavioral data. Five brain regions were involved and their more-or-less known functions were exploited:

- VC: visual signal processing (receptive fields)
- IT: object recognition
- SP: location recognition
- HIPP: associative memory
- PFC: motivation, attention, context

6.1. Model Outline

The model has two parts: A simple visual system, and a more detailed model of the hippocampal formation.

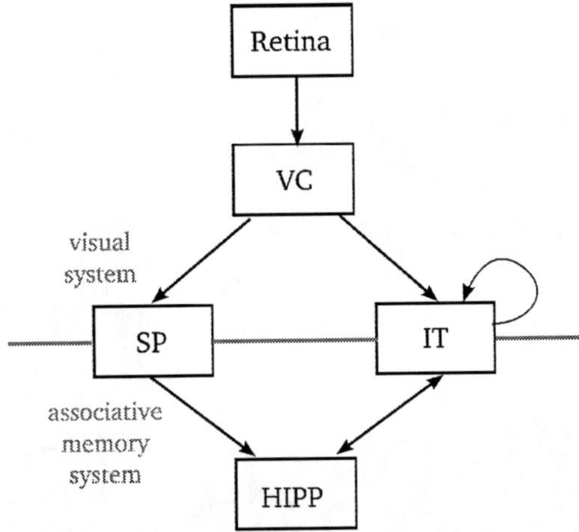

FIGURE 8. The connectivity between areas participated in the model. Horizontal red line separates the visual and the associative memory part of the system. Higher order visual areas (SP and IT) take part in both systems.

6.2. Modeling the visual system

We did not intended to model the visual signal processing system in its details, because these mainly sensory areas are not affected by the illness. However, we implemented a feed-forward network to analyse the retinal image, and to create the representation of the object (the model of the area IT in the ventral stream), and its location (as the area SP in the dorsal stream). The proposed role of the hippocampus is to bind these two representation together [36] so that when cued by the location, the correct object could be recalled.

The retina

The retinal images we use are 8x8 pixel sized, random images placed on a 16x16 pixel arena. The 9 different positions are overlapping.

FIGURE 9. Left: The original screen used in the associative memory experiments. Right: Retinal images used in the model were composed of 8x8 randomly coloured pixels on 16x16 uniform background.

The visual cortex

- Recieves input from retina
- Use receptive fields to process images. ($2^4 - 1$ possible patterns can be detected at each location of the retina.)
- Sends processed data to the IT and SP

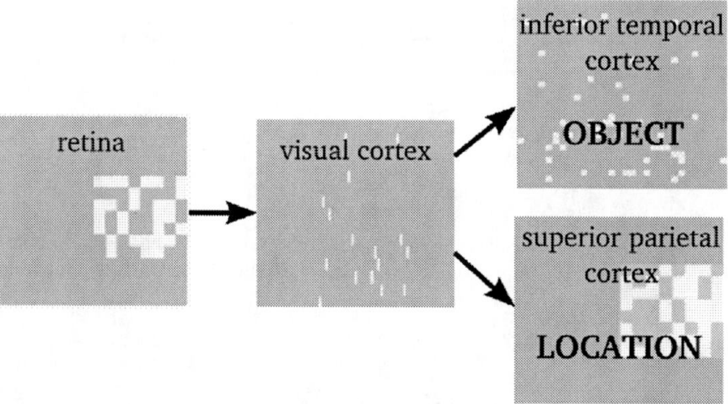

FIGURE 10. The model of different visual areas. In the retina, combined encoding of object and its location is modeled by units with different shaped receptive fields at different positions. The superior parietal cortex implements object invariant location encoding by reading the positional information from the V1, while position invariant representation of objects is modeled in the inferior temporal cortex.

6.3. Modeling the hippocampus

The highly precessed sensory input enters the hippocampus through the mossy fiber pathway, originating in the entorhinal cortex which is not explicitly modelled here. The EC itself has reciprocal connections with both the hippocampus and various neocortical, including visual areas and considered as a relay for information coming from multimodal association areas.. Mossy fibers terminate on the dentate granule cells and hippocampal pyramidal neurons. Two regions of the hippocampal formation was modeled: the dentate gyrus and the CA3 region. We used firing rate models, where the activation of each unit was calculated by the linear sum of its input. Synaptic connections in the IT, DG and CA3 was modified by simple Hebbian plasticity. We used large number of neurons in the simulations (typically 500 in one layer) in order to be able to implement distributed encoding in a realistic range of sparsity (0.1 in the hippocampus).

Our hippocampal model was built according to the following key assumptions [35]:

1. The DG performs pattern separation by competitive learning: it removes redundancies from the input and produce a sparse representation for learning in the CA3 region. This process can be considered as a translation from the neocortical to the hippocampal language.

2. The granule cells in the DG innervates CA3 pyramidal cells with particularly large and efficient synapses (the mossy fiber pathway) that makes postsynaptic neurons fire. Hebbian plasticity between active CA3 neurons and the perforant path axons associates the activity pattern in the CA3 to its incoming input (hetero-associative plasticity). After the encoding, the same CA3 assembly can be activated by the presentation of the partial or noisy version of the original input (e.g., only the object or the location).

3. Next, connections between CA3 cells and IT cells are modified, to translate back the hippocampal to the neocortical code.

4. Finally, objects are stored in a long term memory system in the inferio-temporal cortex forming an attractor network. During recall, the activity of this subsystem converges to one of the stored items (objects).

The performance of the hippocampal model on the associative learning task is shown on Figure 13. We note, that this is not the ideal performance of the model: The capacity of the system with 500 units and 0.1 sparsity is around a few hundreds of associations. However, with random initial synaptic weights and small learning rate, it requires some repetitions to learn new associations appropriately.

As an other bottle-neck of the system is the domain of attraction of the attractor network in the IT. If the attraction basin is smaller, than the recall cue should be more precise. Our results with the hippocampal model indicate, that the poorer performance of schizophrenic patients on the associative memory task is mainly due to the shallower attractor basin and not necessary to a lower learning rate in the hippocampus.

Learning

Before the experiment, we initialize our network by

- storing different number of objects in the recurrent network of IT;
- random synaptic matrices, modelling associations not relevant for the current context.

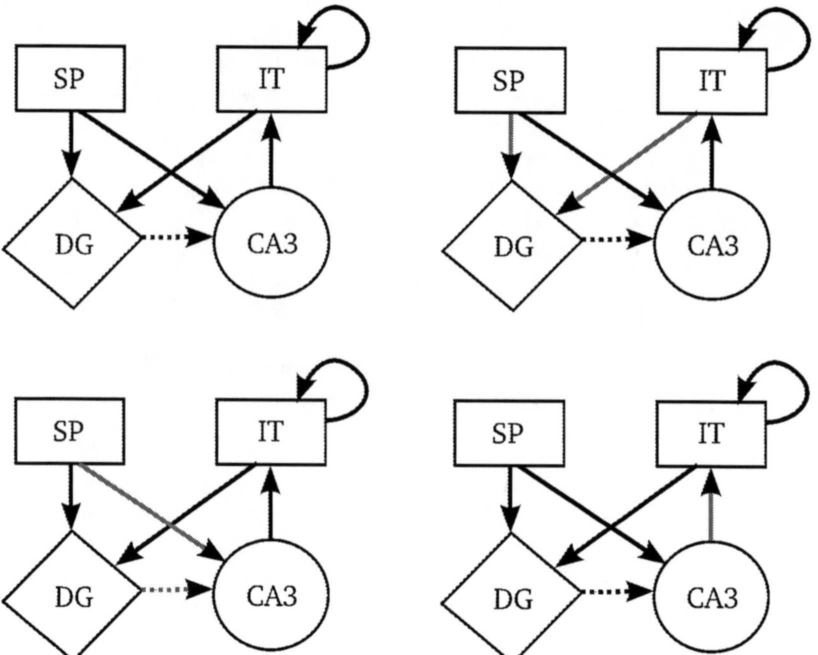

FIGURE 11. A, Connectivity within the associative memory system (A) and the operation of the circuitry (B-C). Arrows pointing towards the DG represent indirect connections through the entorhinal cortex. Dotted arrow from DG to CA3 represent unmodifiable synapses. (B) Visual input activate the representation of object and its location in the IT and SP. Granule cells in the Dentate Gyrus (DG) gradually learn orthogonal representation of object-location pairs through competitive learning. (C) Granule cells activites a portion of CA3 pyramidals, and the location of the current object is associated to these hippocampal neurons. (D) The hippocampal output is associated to the current neocortical representation.

DG - competitive network

$$a^i_{dg} = \sum^j w^{ji}_{sp2dg} r^j_{sp} + \sum^k w^{ki}_{it2dg} r^k_{it}$$
$$r^i_{dg} = F(a_{dg}, sp_{dg})$$
$$\Delta w^{ij}_{sp2dg} = \alpha r^j_{dg}(r^i_{sp} - w^{ij}_{sp2dg})$$
$$\Delta w^{ij}_{it2dg} = \alpha r^j_{dg}(r^i_{it} - w^{ij}_{it2dg})$$

Competitive network in the Dentate Gyrus, to make unique, orthogonal representation for each object-location pair. Cortical signals (from SP and IT) arrive to the hippocampus through the *entorhinal cortex.*

CA3 - heteroassociative stage

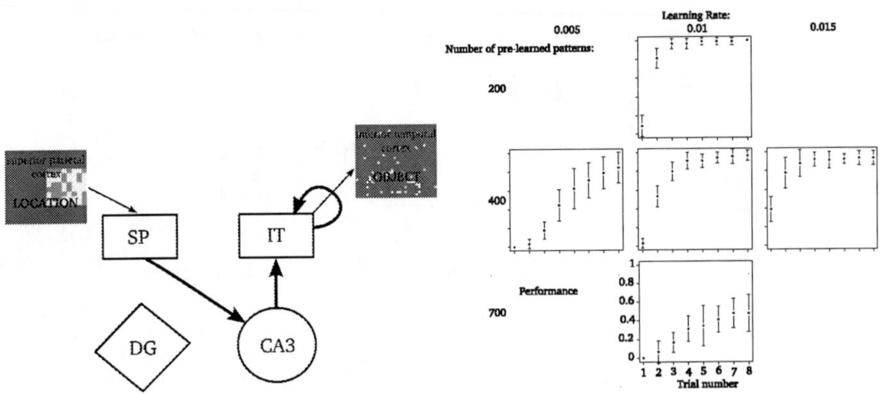

FIGURE 12. Illustration of the operation of the model during recall. Left: Given a location as input, the model recalls one of the 9 object. Right: performance of the model at different parameter values.

$$a^i_{ca} = \sum^j w^{ji}_{dg2ca} r^j_{dg}$$
$$r^i_{ca} = F(a_{ca}, sp_{ca})$$
$$\Delta w^{ij}_{sp2ca} = \alpha r^j_{ca}(r^i_{sp} - w^{ij}_{sp2ca})$$

The strong mossy fiber synapses act as teacher signal for CA3 pyramidal neurons. Perforant path synapses are modified via Hebbian learning.

Back to the cortex - heteroassociative stage

$$\Delta w^{ij}_{ca2it} = \alpha r^j_{it}(r^i_{ca} - w^{ij}_{ca2it})$$

The hippocampal representations are associated to the representation of the original object in the IT.

Recall

During the recall connections are not modified. The attractor network in the IT help the recall by converging to one of the learned objects.

6.4. Results

Model performance is shown on Fig. 13. We note that this is not the ideal performance of the model: The capacity of the system with 500 units and 0.1 sparseness is around a few hundred associations. However, with random initial synaptic weights and small learning rate, it requires some repetitions to learn new associations appropriately.

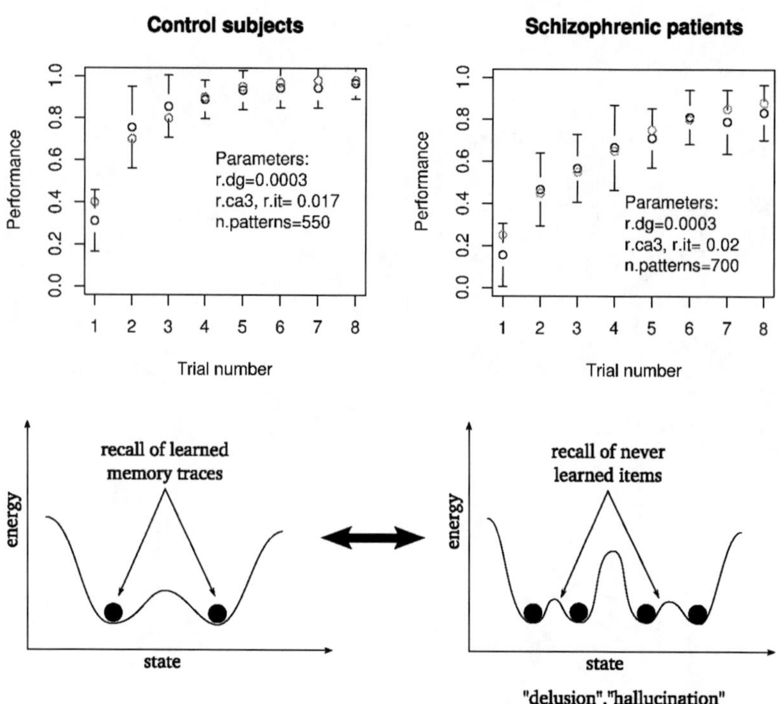

FIGURE 13. Upper row: Comparison of the performance of the model with the experimental data. Red and green circles: the same experimental data as on Fig. 3. Black: performance of the model. Error bars show the standard deviation. Learning rate and the number of pre-learned objects are higher in schizophrenic subjects. Lower row: Illustration shows that learning more objects swallows the basin of attraction of each individual items, and results in the recall of not learned items.

6.5. Much left to be done

Our own research project is scheduled for the stages:

1. fMRI analysis:
2. the inclusion of prefrontal cortex into the model
3. the role of the reduced NMDA-related plasticity. Kinetic model of the drug-altered glutamate-NMDA receptor interaction: test and design of drugs

6.5.1. fMRI analysis and dynamical causal modeling

Based on available data on the activity of five interconnected regions (superior parietal cortex, inferio-temporal cortex, prefrontal cortex, primary visual cortex and the hippocampus) a computational model will be established to understand

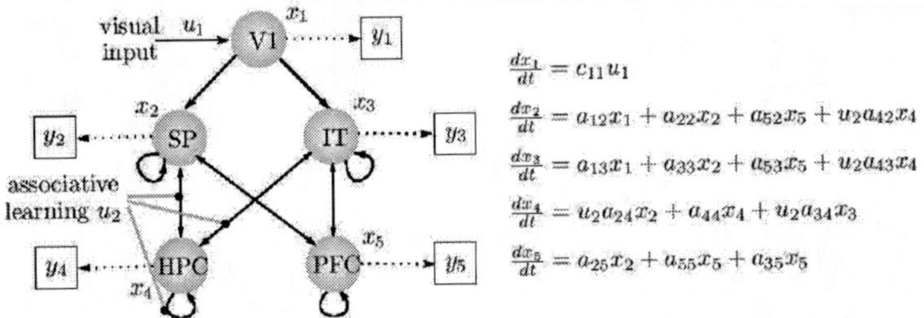

FIGURE 14. DCM desribeing the dynamics in the hierarchical system involved in associative learning. Each area is represented by a single state variable (x). Black arrows represent connections, grey arrows represent external inputs into the system, and thin dotted arrows indicate the transformation from neural states to haemodynamic observations (y). In this example visual stimuli drive the activity in V1 which is propagated through the dorsal and ventral stream to the hippocampus and the PFC. The higher order connections are allowed to change between different blocks (learning, retrieval).

the generation of normal and pathological temporal patterns. A dynamical casual model will be established and used it to solve the "inverse problem", i.e. to estimate the effective connectivity parameters. This model framework Fig. 14 would address impaired connections in schizophrenia, and the measure of functional reduction of the information flow.

6.5.2. The inclusion of prefrontal cortex into the model: a structurally plausible functional model

Our present research aims to study the cooperation between the prefrontal working memory and the hippocampal associative memory systems in both normal and pathological subjects. Our current model of interacting brain regions, as it was described in Sect. 6, neglects the role of the prefrontal cortex. To provide a more biologically inclusive framework, we will construct and test more plausible models that include mechanisms of neural synchrony for understanding the interaction between different memory modules, and the relevance of these interactions for schizophrenia.

In the hippocampus, memories are stored by the synaptic connection between pyramidal neurons. The modification of these synapses is a relatively slow process, and requires multiple presentations of the same pattern, but the capacity of the hippocampal system is high. The prefrontal working memory system stores memories in the persistent activity of cell-assemblies. The storage is fast, single presentation of a specific pattern leads to the formation of the memory but this system has a very limited capacity.

Visual short term memory (VSTM) can be readily distinguished from verbal short term memory and can be sub-divided into spatial and object sub-systems. Our preliminary hippocampal model will be completed by working memory mod-

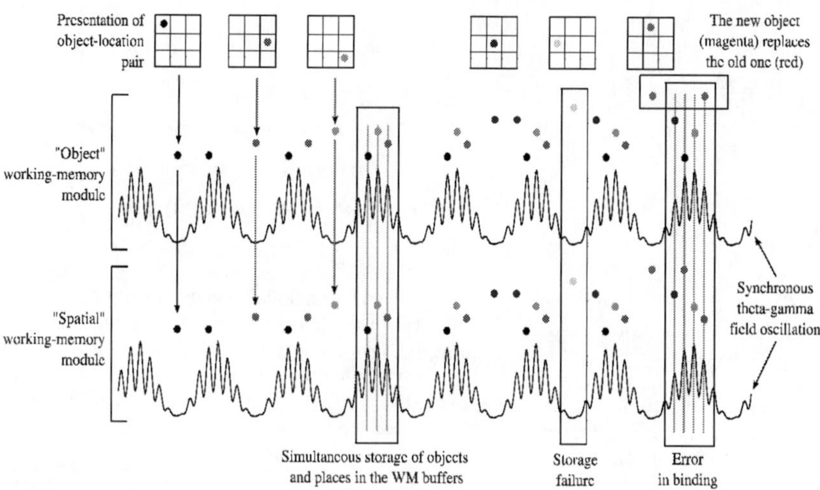

FIGURE 15. Schematic illustration of the involvement of working memory systems in making short term associations. The different modules stores different features of the same input by the persistent activation of neural assemblies. Oscillations allow the network to store multiple items at the same time. However, the capacity of the system is limited, which results in storage failures, or in the overwriting of previously stored items. The binding between these features can be realized by synchronization.

ules. We will adapt the model [37] which is capable of storing multiple items in an oscillatory network. A memory is encoded by a subset of principal neurons that fire synchronously in a particular gamma sub cycle. Firing is maintained by a membrane process intrinsic to each cell Fig. 15. We propose two similar working memory modules involved in the short term storage of locations and object, respectively. These different features are connected by gamma frequency synchronization among cortical regions, similarly to the mechanism proposed for sensory binding . The involvement of the prefrontal regions in associative learning has dual function: (a) The WM system modulates the hippocampal associative memory system by the repeated presentation of the information. (b) The WM memory buffer itself can store the memories through the delay period (independently from the hippocampus), and can increase performance by recalling elements not currently stored in the hippocampus.

6.6. Multi-scale models of drug design

Technically it is a control-theoretical problem to shift the system from a pathological dynamics regime to the normal one. The incorporation of a detailed kinetic model of the modulatory effects of the drugs on the transmitter-receptor kinetics. into a neural network model, as a paradigm of multi-scale modeling, was recently suggested in context of mood regulation [38]).

Our working hypothesis is that (i) for given putative c drugs we can test its effect for the fMRI data curve by (i1) starting from pharmacodynamic data (i.e from dose-response plots) (i2) setting a the kinetic scheme and a set of rate constants, (i3) simulating the drug-modulated glutamate -NMDA recepetor interaction (i4) calculating the generated postsynaptic current, and (i5) integrating the result into the network model of the cortico-hippocampal system (i6) simulating the emergent global BOLD activity (i7) evaluating the result and to decide whether the drug tested has a desirable effect (ii) we can find a kinetic scheme and set of rate constants which gives the best fit to a prescribed "desired" activity curve.

7. CONCLUDING REMARKS

What might be an appropriate framework to understand some aspects of schizophrenia pathology? We propose the following translational and integrative approach:

- The use of combined EEG and fMRI studies as the tool of studying the neural basis. As an imaging technique, fMRI is noteworthy for its ability to measure blood-flow related surrogates of neuronal activity , thereby providing the most current in vivo measures of function and dysfunction available .
- The use of functional paradigms that have at least two attributes: That they involve functions that may be attributable to pharmacological systems that are of relevance to schizophrenia and that they target specific cortical systems and engage inter-regional connectivity.
- The use of computational macro-network models of inter-regional interaction to provide detailed specifications of how task behavior arises in normal and pathological conditions.
- The use of theoretical models of network behavior that will explain the experimental fMRI data in both normal and pathological groups.

The goal is to integrate findings across several domains of basic research, and translate them into clinical and pharmacological practice.

8. ACKNOWLEDGMENTS

This work was supported by MH68680 (VAD), the Children's Research Center of Michigan (CRCM) and the Joe Young Sr. Fund to the Dept of Psychiatry & Behavioral Neuroscience. VAD thanks Matcheri S. Keshavan, R. Marciano and C. Zajac-Benitez for assistance in patient recruitment and assessment, E. Murphy, S. Chakraborty, M. Benton and D. Khatib for assistance in conducting the experiments, N. Seraji-Borzorgzad and S. Fedorov for programming assistance, and J. Stanley for helpful discussions. PÉ thanks the Henry Luce Foundation for general support.

REFERENCES

1. Belair, J. et al. (1995) Dynamical disease: Identification, temporal aspects and treatment strategies of human illness. Chaos 5, 1-7
2. Lopes da Silva, F. et al. (2003) Epilepsies as dynamical diseases of brain systems: basic models of the transition between normal and epileptic activity. Epilepsia 44 Suppl. 12, 72-83
3. Iasemidis, L.D. et al. (2005) Long-term prospective on-line real-time seizure prediction. Clin. Neurophysiol. 116, 32-44
4. Ebersole, J.S. (2005) In search of seizure prediction: a critique. Clin Neurophysiol. 116, 489-92
5. Mormann, F. et al. (2005) On the predictability of epileptic seizures. Clin. Neurophysiol. 116, 569-587
6. Schelter B., Winterhalder M., Feldwisch H., Wohlmuth J., Nawrath J., Brandt A., Schulze-Bonhage A., Timmer J. (2007): Seizure prediction: The impact of long prediction horizons. Epilepsy Research 73, 213-217
7. Beuter, A. et al. (1995) Is Parkinson's disease a dynamical disease? Chaos 5, 35-42.
8. Beuter, A. et al. (2005) Characterization of subclinical tremor in Parkinson's disease. Mov Disord. 20, 945-950
9. Muller, V. et al. (2001) Investigation of brain dynamics in Parkinson's disease by methods derived from nonlinear dynamics. Exp Brain Res. 137, 103-110
10. Pezard, L. et al. (1996) Depression as a Dynamical Disease. Biological Psychiatry 39, 991-999
11. Nandrino, J.L. et al. (2004) Emotional information processing in first and recurrent major depressive episodes. J. Psychiatr. Res. 38, 475-84
12. Breakspear, M. et al. (2003) A disturbance of nonlinear interdependence in scalp EEG of subjects with first episode schizophrenia. NeuroImage 20, 466 - 478
13. Siekmeier, P.J. and Hoffman, R.E. (2002) Enhanced semantic priming in schizophrenia: a computer model based on excessive pruning of local connections in association cortex. British Journal of Psychiatry 180, 345-350
14. Jeong, J. (2004) EEG dynamics in patients with Alzheimer's disease. Clin. Neurophysiol. 115, 1490-1505
15. Stam, C.J. et al. (2005) Disturbed fluctuations of resting state EEG synchronization in Alzheimer's disease. Clin Neurophysiol. 116, 708-715
16. Dahlem, M.A. and Chronicle, E.P (2004) Computational perspective on migraine aura. Prog. Neurobiol. 74, 351-361
17. Aradi, P. Érdi (2006). Computational neuropharmacology: dynamical approaches in drug discovery. Trends in Pharmacological Sciences 27(5) 240-243
18. Hoffman RE, McGlashan TH. (2001) Neural network models of schizophrenia. Neuroscientist. 7(:441-54).
19. Ruppin E (2000) NMDA receptor delayed maturation and schizophrenia. Medical Hypotheses, 54, 693-697.
20. Érdi P, Grőbler T, Tóth J., 1992: On the classification of some classification problems. Int. Symp. on Information Physics, Kyushu Inst. Technol., Iizuka, pp. 110-117.
21. Aradi I, Barna G, Érdi P, Grőbler T: 1995. Chaos and learning in the olfactory bulb. Int. J. Intel. Syst. 10:89-117.
22. Érdi P (2000) On the 'Dynamic Brain' Metaphor. Brain and Mind 1(119-145).
23. Loh, M, Rolls ET and Deco G (2007): A Dynamical Systems Hypothesis of Schizophrenia. PLoS Comput Biol. 3(2255-2265)
24. Breakspear M (2006): The nonlinear theory of schizophrenia. Australian and New Zeeland J. Psychiatry 40(0-35).
25. Palus, M. 1999: Nonlinear dynamics in the EEG analysis: Disappointments and perspectives. In: Nonlinear Dynamics and Brain Functioning., Rapp, P.E., Pradhan, N., Sreenivasan, R. (eds.) Nova Science Publishers.
26. Stephan KE, Baldeweg T, and Friston KJ (2006): Synaptic Plasticity and Dysconnection in Schizophrenia. Biol. Psychiatry. 59(929-939).
27. Friston, K.J. and Frith, C.D., 1995. Schizophrenia: A disconnection syndrome? Clin. Neurosci. 3(88-97).

28. Weinberger DR (1999) Cell biology of the hippocampal formation in schizophrenia. Biol. Psychiatry, 45:395-402;
29. Buchel C, Coull JT, Friston KJ (1999) The predictive value of changes in effective connectivity for human learning. Science 283:1538-1541.
30. Snodgrass JG, Vanderwart M (1980) A standardized set of 260 pictures: norms for name agreement, image agreement, familiarity, and visual complexity. J Exp Psychol [Hum Learn] 6, 174-215.
31. Lavenex & Amaral, 2000: Lavenex P, Amaral DG (2000) Hippocampal-neocortical interaction: A hierarchy of associativity. Hippocampus 10:420-430.
32. Érdi P, Somogyvári Z (2002) Post-Hebbian Learning Algorithms. In: Arbib M (ed) The Handbook of Brain Theory and Neural Networks, Second edition, The MIT Press, Cambridge, MA. p 533-539.
33. Vaibhav et al. Impaired Associative Learning in Schizophrenia: Behavioral and Computational Studies (submitted)
34. Rescorla RA, Wagner AR (1972): A theory of Pavlovian conditioning: Variations in the effectiveness of reinforcement and nonreinforcement. In: Black A, Prokasy W (eds). Classical Conditioning II: Current Research and Theory New York: Appleton Century Crofts, p 64.
35. :Treves A, Rolls ET. (1994): Computational analysis of the role of the hippocampus in memory. Hippocampus. 4:374-91.
36. Milner B, Johnsrude I, Crane J (1997): Right medial temporal-lobe contribution to object-location memory. Philos Trans R Soc Lond B Biol Sci 352:1469-1474.
37. Jensen O, Lisman JE. (1996): Theta/gamma networks with slow NMDA channels learn sequences and encode episodic memory: role of NMDA channels in recall. Learning and Memory. 3:264-78.
38. Érdi P, Kiss T, Tóth J, Ujfalussy B and Zalányi L: From systems biology to dynamical neuropharmacology: Proposal for a new methodology. IEE Proceedings in Systems Biology 153(4) (2006) 299-308

A Framework for Modeling Competitive and Cooperative Computation in Retinal Processing

Roberto Moreno-Díaz*, Gabriel de Blasio* and Arminda Moreno-Díaz†

*Institute of Cybernetics, Universidad de las Palmas de Gran Canaria, Spain
†School of Computer Science, Technical University of Madrid, Spain

Abstract. The structure of the retina suggests that it should be treated (at least from the computational point of view), as a layered computer. Different retinal cells contribute to the coding of the signals down to ganglion cells. Also, because of the nature of the specialization of some ganglion cells, the structure suggests that all these specialization processes should take place at the inner plexiform layer and they should be of a local character, prior to a global integration and frequency-spike coding by the ganglion cells.

The framework we propose consists of a layered computational structure, where outer layers provide essentially with band-pass space-time filtered signals which are progressively delayed, at least for their formal treatment. Specialization is supposed to take place at the inner plexiform layer by the action of spatio-temporal microkernels (acting very locally), and having a center-periphery space-time structure. The resulting signals are then integrated by the ganglion cells through macrokernels structures.

Practically all types of specialization found in different vertebrate retinas, as well as the quasi-linear behavior in some higher vertebrates, can be modeled and simulated within this framework.

Finally, possible feedback from central structures is considered. Though their relevance to retinal processing is not definitive, it is included here for the sake of completeness, since it is a formal requisite for recursiveness.

Keywords: Cooperative computation, layered computation, lateral interaction.
PACS: 87.19.lj, 87.19.lq

1. INTRODUCTION

Cooperative and competitive computation in neuron-like systems takes place in a layered computational structure. This is formed by layers of similar computing units, which communicate laterally and with themselves (local feedback). Their outputs are passed on to the next layer.

The retina, as well as the cortex, presents this type of anatomical and, at least partially, functional architecture. At the photoreceptor level, lateral interaction, which can take place in the inputs to bipolars, may consist simply of a lateral spreading of information by a "competitive" process that produces lateral defacilitation or inhibition, already found in lower vertebrates [1, 2, 3].

It is not clear what computational role the horizontal cells, whose axons carry local information far from its origin, play [1]. It has been argued that they are involved in directional selectivity [4]. But this is a difficult claim to maintain, as it would require special directionally selective bipolars that would then contact exclusively directionally selective ganglion cells [5].

The retinal structure and the coexistence of specialized and non-specialized cells

CP1028, *Collective Dynamics: Topics on Competition and Cooperation in the Biosciences*
edited by L. M. Ricciardi, A. Buonocore, and E. Pirozzi

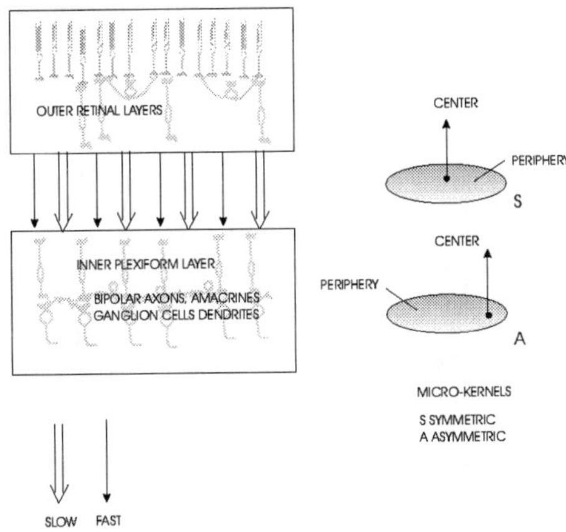

FIGURE 1. Generation of signals by outer retina layers

"looking" out through the same sets of bipolars and photoreceptors suggests that the inner plexiform layer is the site where very local signals from bipolar axons and amacrines interact with dendrites of ganglion cells, thus producing the resulting specializations. Local interactions in the inner plexiform layer may occur through what we call "microkernels". Their action is typically non-linear; they have a center-periphery organization, and they are different for each type of ganglion cell. They determine the local properties of ganglion receptive fields.

Finally, signals are integrated in the ganglion cells by means of facilitatory or defacilitatory processes, which often have a center-periphery organization [6]. This interaction is performed by what we call "macrokernels". Their action is non-local, and they have a more or less wide projection from the photoreceptors, the so-called ganglion cell receptive field.

2. LOCAL MICROKERNELS

Many local contrast detectors - temporal ON-OFF, directionality and color coding, linear and non-linear - can be explained by models in which there is a distribution of local microkernels acting on two versions of the signals generated in the photoreceptor: a fast signal and a delayed or slow signal. These two signals are understood to be generated by the outer layers of the retina, and they interact locally within the inner plexiform layer, as illustrated in Figure 1.

Figure 1 also shows two examples of center-periphery microkernels - symmetric and asymmetric - where the center area has been reduced to a delta function. The center and periphery of each microkernel operate on slow and fast signals and interact [7].

Figure 2 shows the symbolic representation that will later be generalized to the continuum. In this diagram, outer retinal layers generate two new signal planes from the image falling on the retina (the image in the input plane). One plane contains the fast version of the input signal, and the other contains the delayed version.

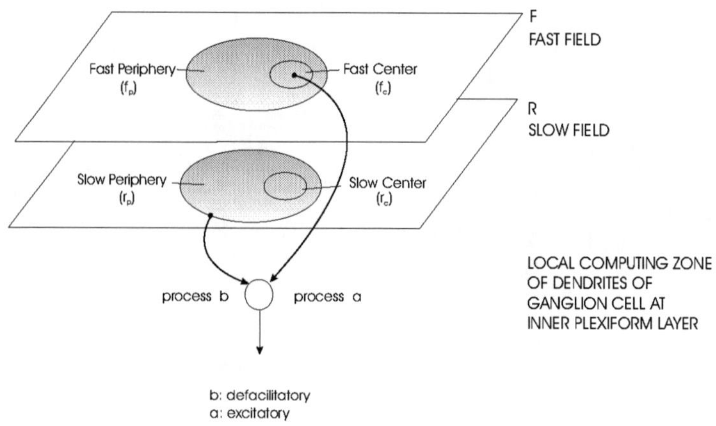

FIGURE 2. Symbolic representation

A microkernel having the center-periphery structure is defined in each field. For a two-input local computing mechanism, there are four possibilities of interaction: $f_c \leftrightarrow r_p$, $f_c \leftrightarrow r_c$, $f_p \leftrightarrow r_c$ and $f_p \leftrightarrow f_c$. In each type of interaction, each interacting signal can be either facilitatory (excitatory) or defacilitatory (inhibitory). In the diagram, when f_c is facilitatory and r_p is competitive or inhibitory, the resulting microkernel computes local stationary spatial and asymmetrical contrast detection; it has an overall ON behavior and provokes directional selectivity for bright spots moving from right to left. The simplest ON operation is an $f_c\,(+) \leftrightarrow r_c\,(-)$ interaction, where the sign $(+)$ indicates facilitation and $(-)$, competition. The simplest contrast detector is the symmetric $f_c\,(+) \leftrightarrow f_p\,(-)$.

A local OFF response is yielded by the interaction $f_c\,(-) \leftrightarrow r_c\,(+)$. Notice that ON-OFF responses require the concurrency of two-input interacting mechanisms, each followed by a rectifying non-linear function (or transfer function of the type $exp\,[f_c\,(+) \leftrightarrow r_c\,(-)]$-1) such that the total non-linear ON-OFF behavior is given by the additional interaction

$$[exp(\,([f_c\,(+) \leftrightarrow r_c\,(-)) - 1](+) \leftrightarrow [exp(\,([f_c\,(-) \leftrightarrow r_c\,(+)) - 1](+)$$

Notice that the equivalent pure spatial interaction

$$[exp(\,([f_c\,(+) \leftrightarrow f_p\,(-)) - 1](+) \leftrightarrow [exp(\,([f_c\,(-) \leftrightarrow f_p\,(+)) - 1](+)$$

is a spatial contrast detector that is independent of the contrast polarity.

A basic two-input competitive computing unit is obtained by generalizing a formal McCulloch neuron that competes with itself and with an external input (see Figure 3(a)). The neuron output facilitates itself (process α) and defacilitates the input (process β), and the external input does the same. Therefore they compete to trigger an output. McCulloch often interpreted processes α and β at a much higher level than the simplest logical and analytical interpretations of McCulloch-Pitts' neural nets [8].

If only positive or even Boolean signals are allowed, the α process is excitatory and process β is totally inhibitory (inspired by presynaptic inhibition). The resulting formal neuron (Figure 3(b)) computes the exclusive OR of the two inputs x_1 and x_2. If α is a exponential or positive rectification process, the unit computes a process similar to the two-input interacting mechanism for symmetric kernels. If $x_1 = f_c$ and $x_2 = r_c$, the unit has nonlinear ON-OFF behavior. For $x_1 = f_c$, $x_2 = r_p$, the unit is a locally nonlinear ON-OFF contrast detector. This case is shown for a fast and slow signals given by the solution of:

$$\dot{x}_1 = -0.5x_1 + u$$
$$\dot{x}_2 = -0.2x_2 + u$$

where the input u is a double step of light. Figure 3(c) shows the analog ON-OFF response and Figure 3(d) the spike-frequency coded output.

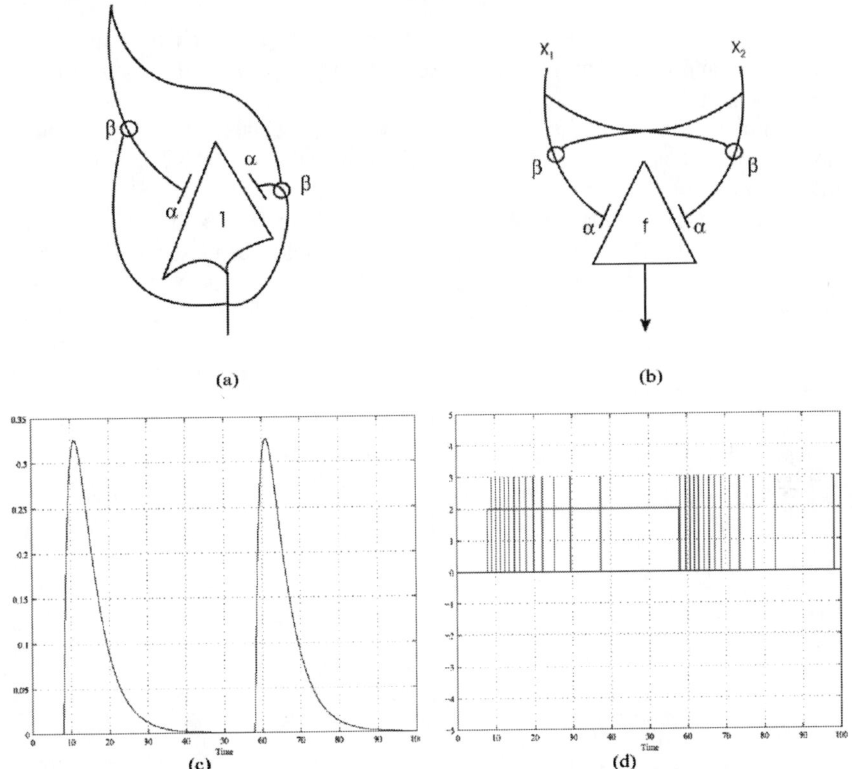

FIGURE 3. Generalization of a formal McCulloch neuron

An evident conclusion is that linear ON-OFF operation is performed by a "center-periphery" microkernel which extends also in time. This is illustrated in Figure 4. Now, there are three planes that are to contain, at least operationally, one fast and two delayed versions of the signals. That is, signals can be now symbolically represented in three input planes and by microkernels with a double center-periphery structure. The center of this space-time microkernel is in the first slow plane S_1 and the "periphery" extends

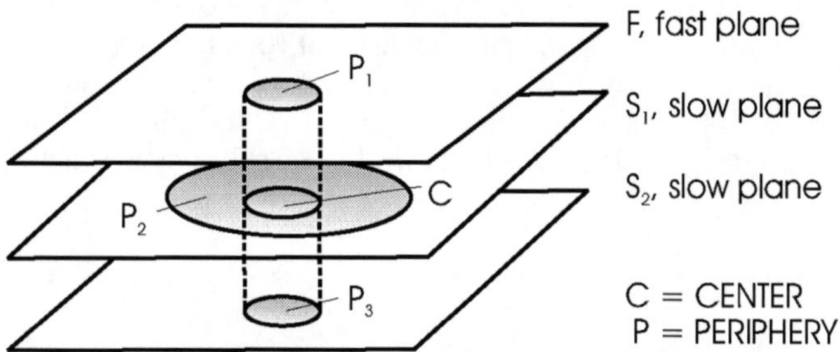

FIGURE 4. Temporal microkernel

to all three planes. When the center signal c is facilitatory or excitatory and periphery signals p_1, p_2 and p_3 are competitive or defacilitatory, it behaves like a linear ON-OFF contrast detector.

For example, consider a three time-delay system in which x_1 is a fast signal given by the solution of $\dot{x}_1 = x_1 + u$ where u is the local input (an ON-OFF double step of light); x_2 and x_3 are the two delayed signals given by $x_2 = x_1(t + \Delta)$ and $x_3 = x_2(t + \Delta)$. The temporal kernel is simply $Pos(-x_1 + 2x_2 - x_3)$, where Pos is the positive part (half wave rectification). Figure 5(a) shows the analog ON-OFF response and figure 5(b) the corresponding spike-frequency coded output. Notice the delay between the ON front of the input and the starting of the ON output, which is typical for linear ON-OFF behavior.

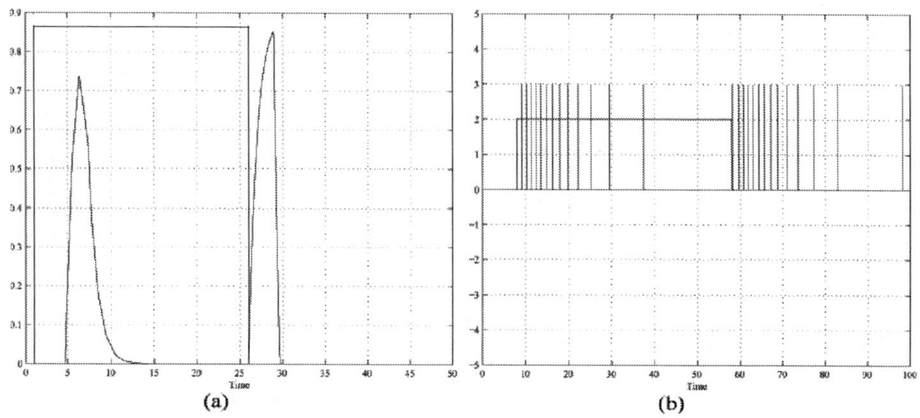

FIGURE 5. A time-delay system

3. MACRO-KERNELS GENERATED BY LAYERS OF MICROPROCESSORS

Macrokernels for a ganglion cell of a type provide for space-time operations to be performed on the results of the local or microkernel layer. In many cases, macrokernels

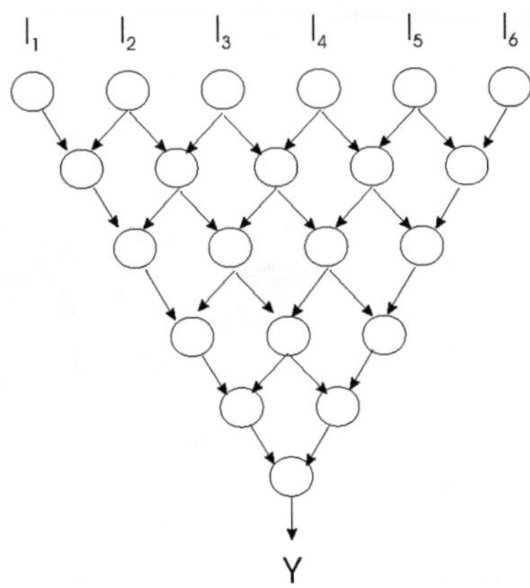

FIGURE 6. A layered structure with two-input computing units

have a center-periphery, or a multiple-ring organization across the retinal receptive field of the ganglion cell.

An interesting particular case is when macrokernels are the result of many computational layers that converge to a final or ganglion unit. This is illustrated in Figure 6 for a one dimensional spatial case where the operations in each layer are performed by computing units with two inputs.

In this case the macrokernel represents the overall rules for handling input signals $I_1, I_2, \ldots I_n$, providing output Y. In general, this has to be computed by giving the rules of each of the two-input computing units. We consider two interesting and simple cases. First, units compete with the fuzzy rule $Sup[x_1, x_2]$; $Coord(Sup)$, which outputs the highest of the two inputs and its coordinates. At output Y, we find the highest of all inputs and its coordinates, which is a behavior similar to what is found in the frog's tectum sameness cells, that is, it is a maximum selector network [9].

Second, computing units perform linear or nonlinear operations on the two inputs, which may be cooperative (the result is greater than any of the two inputs) or competitive (the sign of the right-left difference is carried into the output). For the simplest case of addition/subtraction, if all layers are additive, a Gaussian type macrokernel is obtained. If there are, for example, two layers of competing units, the macrokernel has a center-periphery organization. For more layers of competing units, the receptive field has a "multiple rings" structure.

Figure 7(a) shows the macrokernel resulting for the six input lines of figure 6 with two rows of competitive (subtractive) interaction. The center-periphery structure is already evident.

It can be shown that, in the limit, linear cooperative or competitive operations performed by the units of the computing layers generate Hermite-type of macrokernel structures, having multiple ring excitatory inhibitory zones [10]. Figure 7(b) shows the corresponding kernel for two competitive "rows", corresponding to the Hermite function of order 2, which provides for the classical "Mexican hat" shape.

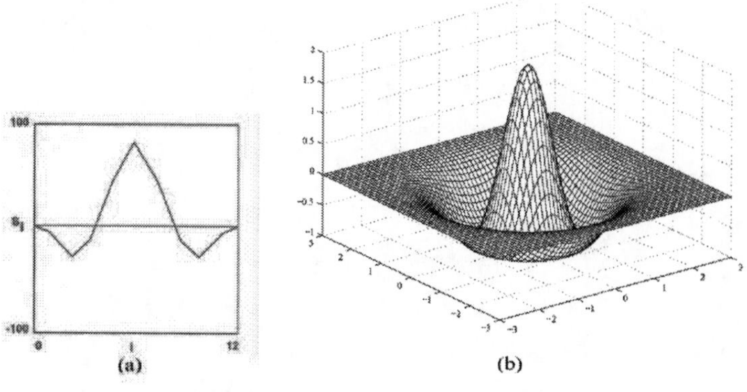

FIGURE 7. Macrokernel structures

4. GENERATING A SYMBOLIC SPACE-TIME INPUT SPACE

The representation of the microkernels discussed in section 2, mainly as illustrated by the linear ON-OFF temporal kernel, suggests that local operations can be considered as a function of the spatio-temporal data present in some space-time input space. Microkernels are now space-time "volumes", which, in the typical center-periphery organization, have a "center volume" and a "peripheral volume". Now, there is only one type of interaction, say $f_c \leftrightarrow f_p$, where f_c stands for the operations performed on the central volume data, and f_p represents the peripheral volume operations. The directional selectivity in time (OR on OFF) or the time symmetric "contrast detection" (ON-OFF) depends only upon the relative positions of the center and periphery along the temporal axis. Similarly, directional selectivity in space and symmetric or asymmetric contrast detection depend on the relative positions of the center and periphery in the $X-Y$ plane. This is illustrated in Figure 8.

Notice that, because of the finite spatial extension and the finite memory of both input space and microkernel volumes, operations are limited to a parallelepiped with a depth of memory T. It is formed by a stack of "slides", the last at time t. Microkernels perform a kind of generalized space-time convolution. Figure 8 also shows kernels for the space-time symmetric contrast detector 8(a), ON-space contrast detector 8(b), ON-OFF directionally selective unit 8(c) and OFF directionally selective unit 8(d).

Figure 9 illustrates functions within the inner plexiform layer where a new input space is generated from the microkernel outputs, all of which are of a similar type for a given ganglion cell (directionally selective, locally ON-OFF cell). Notice that if the central and peripheral volumes are randomly situated, the global directionality is lost on the ganglion cell. Macrokernels for a ganglion cell are then defined in this new input space, by, say, a center-periphery structure.

Again, the nature of the $f_c \leftrightarrow f_p$ interactions can be formulated by means of different mathematical or formal tools, as in the case of microkernels described in section 2.

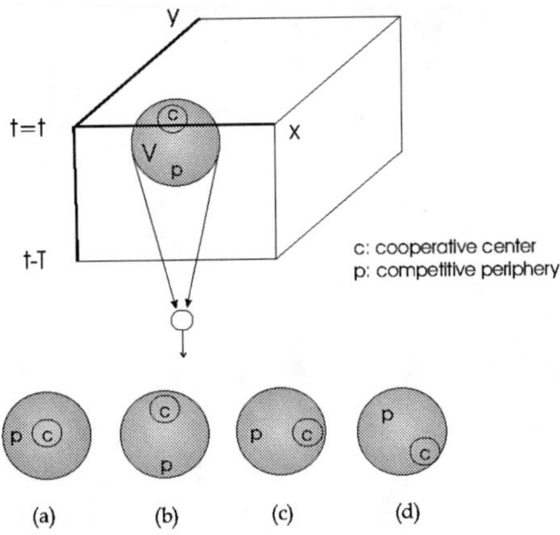

FIGURE 8. Symbolic space-time input space

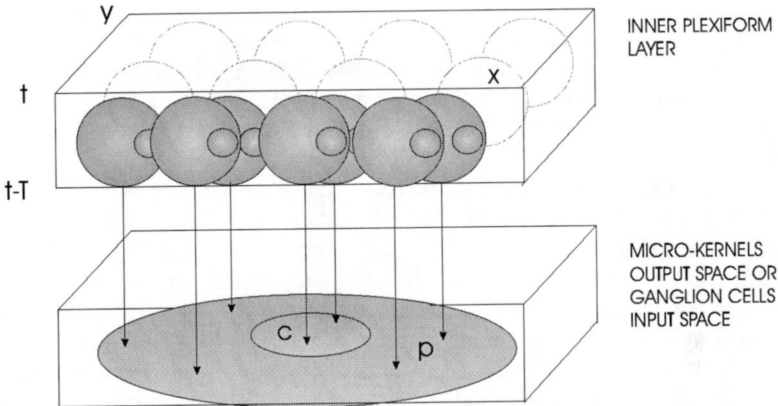

FIGURE 9. Functions at the inner plexiform layer

5. FEEDBACK AND RECURSION

Finally we consider feedback from output spaces to input spaces [11]. The role of feedback in the behavior of the retina is not very clear, except for adaptation mechanisms. From the computational point of view, it has been suggested, for example, that centrifugal fibers from the frog's tectum could help to maintain the firing of group-2 retinal ganglion cells when a small moving object stops in the retinal receptive field. Also, feedback changes the excitatory and inhibitory receptive field of ganglion cells [12].

In any case, the study of feedback in layered computation is of interest and provides clues that can be applied to the modeling of retinal computation. We shall consider here some important theoretical aspects.

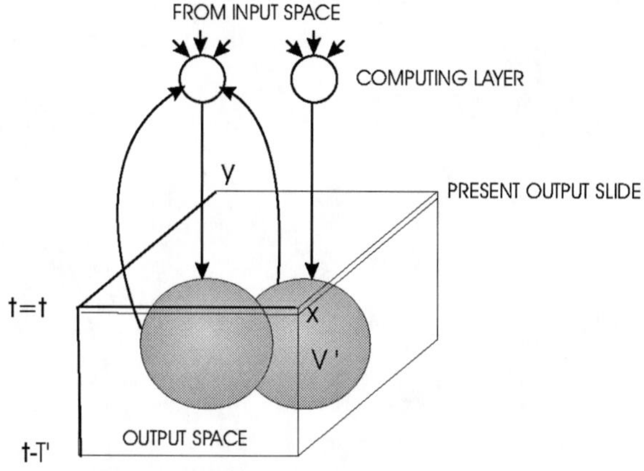

FIGURE 10. Layered feedback

The general case of layered feedback is shown in Figure 10. Past and present outputs of the layer are stacked to generate an output space. Signals from space-time volumes in the output space are fed back to the computing units of the layer to generate the output slide at time t. Obviously, feedback signals from output volumes may include slides up to $t - \Delta t$.

A homogeneous system is produced when the inputs to the computing layer are cancelled. The system will move on to behave autonomously once it is started in a generally non-zero, initial space-time output volume and switch to recursive computation.

The simplest case is a discrete time system having initially just one output slide, or image. The recursive process starts by generating new images as time goes by. This type of process is appropriate for modeling aspects of reticular formation decision making mechanisms or for wave generation and maximum activity selection. An interesting case is when the units compute an algorithm similar to lateral inhibition. In this case, each computing unit operates in overlapping spatial center-periphery regions.

One possible competitive algorithm is as follows

`Select maximum value in Center =` C_m
`Select maximum value in Periphery =` P_m

Decision rule

`If` $C_m \geq P_m \quad Y(x,y,k+1) = Y(x,y,k)$
`Otherwise` $Y(x,y,k+1) = P_m$

Such algorithms are capable of wave generation and are also useful in image processing.

6. CONCLUSION

We have proposed a framework to illustrate and explain cooperative and competitive computational processes in visual analysis in the retina. This framework is built in a Systems Theory basis and it sheds significant light on the complex processes that take place before visual information is carried to higher visual centers. The micro and macrokernel theory proposed is consistent with the latest neurophysiological findings in the retina anatomy and functioning and it contributes to explain the vast and rich type of processes that take place within it.

REFERENCES

1. F. Werblin, *J. Gen. Physiol.* **63**, 62–87 (1974).
2. P. B. Cook, and J. S. McReynolds, *Nature Neuroscience* **1, 8**, 714–719 (1998).
3. R. H. Masland, *Nature Neuroscience* **4, 9**, 877–886 (2001).
4. B. Roska, A. Molnar, and F. Werblin, *J. Neurophysiol* **95**, 3810–3822 (2006).
5. F. Werblin, *Science* **175**, 1008–1010 (1972).
6. H. Barlow, *J. Physiol. (London)* **119**, 69–88 (1953).
7. R. Moreno-Díaz, G. de Blasio, and A. Moreno-Díaz, *Lecture Notes in Computer Science* **3643**, 483–491 (2005).
8. R. Moreno-Díaz, and A. Moreno-Díaz, *Biosystems* **88, 3**, 185–190 (2007).
9. A. Moreno-Díaz, G. de Blasio, and R. Moreno-Díaz, *Lecture Notes in Computer Science* **3561**, 86–94 (2005).
10. R. Moreno-Díaz, and G. de Blasio, *Systems Analysis Modelling Simulation* **43, 9**, 1159–1171 (2003).
11. J. Reperant, D. Miceli, N. P. Vesselkin, and S. Molotchnikoff, *International Review of Citology* **118**, 115–171 (1989).
12. A. L. Byzov, and I. A. Utina, *Neurophysiology* **3, 3**, 219–224 (1971).

A Model for the Propagation of Action Potentials in Nonuniform Axons

Andreas Schierwagen* and Michael Ohme†

*Institute for Computer Science, University of Leipzig, 04109 Leipzig, Germany, E-mail: schierwa@informatik.uni-leipzig.de
†Academy of Visual Arts, 04107 Leipzig, Germany, E-mail: mohme@hgb-leipzig.de

Abstract. This paper presents a method to mathematically analyze the nerve impulse propagation in nonuniform axons. Starting from the general, nonlinear one-dimensional cable equations with spatially varying cable diameter, the problem is shown to be equivalent (under some variable transformations) to the case of uniform axons. Characterized by the same normal form, six functions for analytically treatable axon diameter variations are determined. For this class of nonuniform axons, exact solutions describing the propagation of the front of the action potential are derived. The results are used to evaluate the impact of geometric non-uniformity on the properties of propagating action potentials.

Keywords: Action potential, nonuniform axon geometry, nonlinear cable equation, analytical solutions.
PACS: 87.10.Ca, 87.19.lb, 87.19.ll

1. INTRODUCTION

A central goal of Mathematical Neuroscience is to develop explicit mathematical models of neuronal systems that enable the explanation and prediction of systems behavior. Because of the complexity of the nervous system, mathematical modeling has been used since the early years of neuroscience to facilitate the understanding of neural functions and mechanisms.

In modeling single neurons, two types of complexity must be dealt with: the intricate interplay of active conductances underlying the complex neuronal excitation dynamics, and the elaborate dendritic morphology that allows neurons to receive and process inputs from many other neurons (e.g. [1]).

In this paper, we will discuss a method that has been used to mathematically analyze the electrical signaling function of spatially complex neurons. We will begin with presenting the model assumptions underlying the general cable model for neuronal processes. We then discuss the method used to derive analytical solutions for the electrical behavior of nonuniform neuronal segments. We then show how the results can be applied to impulse propagation in nonuniform, unmyelinated axons. Elsewhere the theory has been applied to branching dendritic trees with active membrane [2].

CP1028, *Collective Dynamics: Topics on Competition and Cooperation in the Biosciences*
edited by L. M. Ricciardi, A. Buonocore, and E. Pirozzi

2. BIOLOGICAL NEURONS: MORPHOLOGY AND SIGNALING FUNCTION

The main anatomical features of neurons are as follows (Figure 1): from the cell body two kinds of processes – the dendrites and the axon – emanate. Generally, dendrites are regarded to provide receptive surfaces for input signals to the neuron. The input contacts are made by synapses which are distributed primarily over the widely branched dendritic trees. The input signals are conducted with decrement to the soma and the axon hillock. There the signals usually are converted into sequences of nerve impulses (action potentials, or spikes) which are propagated without attenuation along the axon to target cells, i. e. other neurons, muscle cells etc.

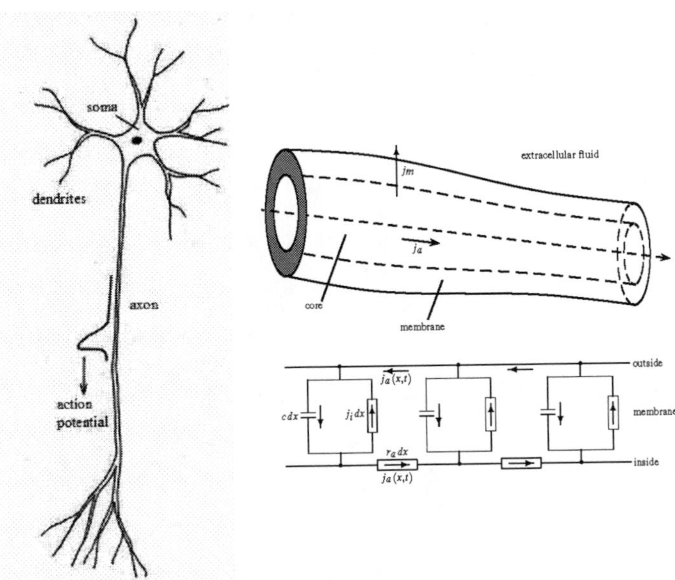

FIGURE 1. A typical textbook neuron with soma and neurites (dendrites and axon). Displayed is the current flow in a neurite segment (core conductor model) and the equivalent electrical circuit. Inside the membrane is the conducting core consisting of cell plasm, outside the extracellular fluid. In the equivalent circuit, the extracellular resistivity is neglected.

Of course, many neurons are known where this classical identification of the processing steps within a neuron must be supplemented with additional processes, such as dendritic spikes, intermittent conduction or spikeless transmission [3, 4]. Thus neurons may deviate in various ways from the above concepts which collectively comprise an idealized "standard neuron", but many of the principles seem to be common to almost all cells and probably provide the basis of neuronal operation.

The idealized neuron exhibits regionally different electrical characteristics. The soma and the dendrites have fixed ionic permeabilities, thus a change of polarization produced somewhere on the dendritic tree will spread and decay as it is conducted "electrotonically", as in a passive, leaky cable. In the axon hillock, on the other hand, the ionic permeabilities depend on the membrane potential, and the integration of the electrotonic potentials will result in the initiation of spike trains. These two kinds of membranes are referred to as passive and active. In both cases, the spatial distribution of membrane voltage can be studied with the aid of (linear or nonlinear) cable theory.

3. GENERAL MODEL FOR NEURONAL CABLES

A model of a neuronal cable (axon or dendrite) can be set up by combining the nonlinear ordinary differential equations for an excitable membrane with the parabolic partial differential equations for a core conductor. Figure 1 schematically displays a standard neuron, the core conductor model used to represent the neurite segments and its equivalent electrical circuit. The cable equation for the transmembrane potential $V(x,t)$ and the axial current $j_a(x,t)$ is as follows (x represents distance in axial direction):

$$j_a = -\frac{1}{r_a(x)}\frac{\partial V}{\partial x}, \qquad -\frac{\partial j_a}{\partial x} = j_m = j_c + j_i \tag{1}$$

where $j_m(x,t)$ denotes the membrane current consisting of a capacitive component, j_c, and a resistive one, j_i .
We assume that the current, j_i, created by the ionic channels in the membrane can be written as a product of a resting conductance $g(x)$ which depends on the membrane surface at x , and a nonlinear voltage function $f_0(V, u_1, ..., u_N)$ reflecting the threshold behavior of the voltage-dependent channels as a specific membrane property (i.e. per unit membrane surface). The latter may be time-dependent, so additional (auxiliary) variables $u_k(x,t)$ defined by first order differential equations have to be included:

$$j_c = c(x)\tfrac{\partial V}{\partial t}, \quad j_i = g(x) f_0(V, u_1, ..., u_N) \tag{2}$$

$$j_m = c(x)\frac{\partial V}{\partial t} + g(x) f_0(V, u_1, ..., u_N) \tag{3}$$

$$\frac{\partial u_k}{\partial t} = f_k(V, u_1, ..., u_N) \text{ for } 1 \leq k \leq N. \tag{4}$$

Combining equations (1) to (3) we obtain

$$\frac{\partial}{\partial x}\left(\frac{1}{r_a(x)}\frac{\partial V}{\partial x}\right) = c(x)\frac{\partial V}{\partial t} + g(x) f_0(V, u_1, ..., u_N). \tag{5}$$

3.1. Specific assumptions

In most cases the axial resistance $r_a(x)$ is assumed inversely proportional to the cross-section of the segment whereas the membrane conductance $g(x)$ and the membrane capacitance $c(x)$ (all quantities per unit length) are proportional to the membrane surface:

$$\begin{aligned} r_a(x) &= R_i \frac{4}{\pi d(x)^2} \\ c(x) &= C_m \pi d(x) \sqrt{1+\frac{1}{4}\left(\frac{\partial d}{\partial x}\right)^2} \\ g(x) &= G_m \pi d(x) \sqrt{1+\frac{1}{4}\left(\frac{\partial d}{\partial x}\right)^2} \end{aligned} \tag{6}$$

where $d(x)$ denotes the variable cable diameter, R_i the specific intracellular resistance, C_m and G_m the specific membrane capacitance and the resting conductance, respectively. The latter is the conductance in the nearly linear subthreshold range around the resting potential.
In the following, we will use the above equations for cables with circular cross-section. Further, we assume that $\sqrt{1+\frac{1}{4}\left(\frac{\partial d}{\partial x}\right)^2} \approx 1$, i.e. there is only weak taper, which should be satisfied in most cases. The results derived below are valid, however, also for more general r-, g-, and c-functions.

3.1.1. *Linear cable*

The simplest cable model (no auxiliary variables u_k, i.e., $N = 0$) is the linear one for passive cables. The voltage function reduces to $f_0(V) = V$ (resting potential set to zero) so the linear cable equations reduce to

$$\frac{\partial}{\partial x}\left(\frac{1}{r_a(x)}\frac{\partial V}{\partial x}\right) = c(x)\frac{\partial V}{\partial t} + g(x)V. \tag{7}$$

3.1.2. *Nonlinear cable*

A model of a cable with active membrane is set up by choosing suitable ionic current j_i and voltage function $\mathbf{f} = (f_1, \ldots f_N)^{\mathbf{T}}$. With such functions, system (4), (5) has been shown to simulate essential features of membrane excitation conduction. Models of increasing complexity (i.e., number of auxiliary variables) are

1. the bistable (wavefront) equation equation [5, 6] ($N = 0$)
2. the FitzHugh-Nagumo / Bonhoeffer-Van der Pol equations [7, 8] ($N = 1$)
3. the Goldstein-Rall equations [9] ($N = 2$)
4. the Hodgkin-Huxley equations [10] ($N = 3$).

Hodgkin-Huxley model. The Hodgkin and Huxley (HH) model has been used as basis of most of the conductance-based models. In the original version, the model consists of a four-dimensional system of nonlinear differential equations [10]. The ionic current, j_i, is determined by three conductances where two of them are voltage-dependent:

$$\begin{aligned} j_i &= j_{Na} + j_K + j_L \\ &= g_{Na}(V - E_{Na}) + g_K(V - E_K) + g_L(V - E_L) \\ &= m^3 h \bar{g}_{Na}(V - E_{Na}) + n^4 \bar{g}_K(V - E_K) + g_L(V - E_L), \end{aligned} \tag{8}$$

and $\bar{g}_{Na}$, $\bar{g}_K$ are the maximal conductance values of the sodium and potassium channels, and g_l is the (constant) value of the passive leak conductance. $u_1 = m$ and $u_2 = h$ are the activation and inactivation variables for sodium, $u_3 = n$ is the activation variable for potassium,$0 < m,n,h < 1$, and the following differential equations hold

$$\begin{aligned} \frac{\partial m}{\partial t} &= f_1 &= \alpha_m(1-m) - \beta_m m \\ \frac{\partial n}{\partial t} &= f_2 &= \alpha_n(1-n) - \beta_n n \\ \frac{\partial h}{\partial t} &= f_3 &= \alpha_h(1-h) - \beta_h h \end{aligned} \tag{9}$$

where $\alpha_m, \beta_m, \alpha_n, \beta_n, \alpha_h, \beta_h$, are empirical functions of the voltage V. The leak conductance $g_L(x) = G_L \pi d(x)$, according to (6). We rewrite (8) as

$$j_i = g \cdot f_0(V,m,n,h) = g(x)\left(\frac{\bar{g}_{Na}}{\bar{g}_L} m^3 h(V - E_{Na}) + \frac{\bar{g}_K}{\bar{g}_L} n^4 (V - E_K) + (V - E_L)\right). \tag{10}$$

Thus the part of the current j_i depending on cable diameter is separated from the voltage-depending part, in accordance with the general cable equation (5).

FitzHugh-Nagumo model. Below we will use the FitzHugh-Nagumo (FHN) model, a simplified version ($N = 1$) of the Hodgkin-Huxley model, to study nerve conduction. The FHN equations [7] can be written as:

$$\begin{aligned} j_m &= c(x)\frac{\partial V}{\partial t} + g f_0 = c(x)\frac{\partial V}{\partial t} + g(h(V) + u) \\ \frac{\partial u}{\partial t} &= \alpha V - \beta u \end{aligned} \tag{11}$$

with resting potential equal to zero and cubic $h(V) = V(1 - V/V_1)(1 - V/V_2)$ where $0 < V_1 < V_2$ are the roots of h . The constants α and β are positive so that u acts as a variable which takes the system from the excited state (near V_2) back to the resting state $V = 0$.

4. TRANSFORMATION INTO NORMAL FORM

We transform Eq. (5) by the variable transformation (applied already by Kelly and Ghausi [11] and later by Schierwagen [12])

$$T = \frac{t}{\tau} \quad \text{with} \quad \tau(x) = \frac{c(x)}{g(x)} \quad \text{and}$$
$$X = \int_0^x \frac{d\hat{x}}{\lambda(\hat{x})} \quad \text{with} \quad \lambda(x) = \frac{1}{\sqrt{r_a(x) g(x)}}. \tag{12}$$

For an uniform cable (i.e. c, g and r_a independent of x), λ and τ are constant and equal to the length and time constants of passive cable theory.

Now Eqs. (4), (5) can be rewritten:

$$\begin{aligned} 0 &= \frac{\partial^2 V}{\partial X^2} + Q(X)\frac{\partial V}{\partial X} - \frac{\partial V}{\partial T} - f_0(V, u_1, \ldots, u_N) \\ \frac{\partial u_k}{\partial T} &= \tau(X) f_k(V, u_1, \ldots, u_N) \quad \text{for } 1 \le k \le N \end{aligned} \tag{13}$$

with

$$Q(X) = \frac{1}{2}\frac{\mathrm{d}}{\mathrm{d}x}\ln\left(\frac{g(X)}{r_a(X)}\right). \tag{14}$$

$Q(X)$ contains all geometry-dependent parts of the transformed cable equation and can therefore be used for classifying cable geometries. For example, in the standard case of a cable with circular cross-section, we obtain

$$Q(X) = \frac{\mathrm{d}}{\mathrm{d}x}\ln D(X)^{\frac{3}{2}}. \tag{15}$$

In particular, $Q = 0$ for cylindrical cable geometries (i.e., r_a and g are constant), whereas Q remains nearly constant for a (slowly) exponentially tapering cable diameter in Eqs. (6), positive for increasing and negative for decreasing diameter.

Using the transformations

$$\begin{aligned} V(X,T) &= F(X) \cdot W(X,T) \\ u_k(X,T) &= F(X) \cdot w_k(X,T) \end{aligned} \tag{16}$$

with

$$F(X) = \exp(-\frac{1}{2}\int Q(X)\mathrm{d}X) = D(X)^{-\frac{3}{4}}, \tag{17}$$

the equations (5) or (13), respectively, in normal form read

TABLE 1. Axon geometries as defined by the solutions of the special Riccati equation $2Q' = 4P - Q^2$. The stationary solution is denoted by $Q_1 = \pm 2\sqrt{P}$ (cf. [1]). C_1 and $C_2 > 0$ are free but constant parameters.

Geometry type	P	$Q(X)$	$\frac{g(X)}{r_a(X)} = F(X)^{-4}$
uniform	$P = 0;\ Q = 0$	0	C_2
power	$P = 0;\ Q \neq 0$	$\frac{2}{X - C_1}$	$C_2(1 - \frac{X}{C_1})$
exponential	$P > 0;\ Q^2 = Q_1^2 = 4P$	Q_1	$C_2 \exp(2Q_1 X)$
hyperbolic sine	$P > 0,\ \lvert Q\rvert > Q_1$	$Q_1 \coth(\frac{X - C_1}{2})$	$C_2(\sinh(Q_1 \frac{X - C_1}{2}))^4$
hyperbolic cosine	$P > 0;\ \lvert Q\rvert < Q_1$	$Q_1 \tanh(\frac{X - C_1}{2})$	$C_2(\cosh(Q_1 \frac{X - C_1}{2}))^4$
trigonometric cosine	$P < 0$	$-\lvert Q_1\rvert \tan(\lvert Q_1\rvert \frac{X - C_1}{2})$	$C_2(\cosh(\lvert Q_1\rvert \frac{X - C_1}{2}))^4$

$$\frac{\partial^2 W}{\partial X^2} - P(X)W - \frac{f_0(FW, Fw_1, \ldots, Fw_N)}{F} - \frac{\partial W}{\partial T} = 0$$
$$\frac{\partial u_k}{\partial T} = \tau \frac{f_k(FW, Fw_1, \ldots, Fw_N)}{F} \text{ for } 1 \leq k \leq N. \tag{18}$$

For linear functions f_k $(k = 0, ..N)$, the transformation function $F(X)$ cancels in the equations. This is the case for, e.g., the bistable wavefront equation or the piecewise linear FitzHugh-Nagumo equations (see below). In the following, this case is assumed to hold.

The coefficient $P(X)$ is defined by the simple Riccati equation [1, 11, 12]:

$$P(X) = \frac{1}{2}Q'(X) + \frac{1}{4}Q(X)^2. \tag{19}$$

Various classes of nonuniform cable geometries may be obtained by choosing the function $P = P(X)$ in condition (19). The simplest class to consider are those cables for which P is a constant. This class can be determined by solving the Riccati differential equation (19). The complete solution set and the corresponding diameter functions determined from (15) are given in Table 1.

5. AXON GEOMETRIES REDUCIBLE TO THE UNIFORM CASE

If we take a closer look at the function $Q(X)$ defined by Eq. (15), we can derive explicit conditions relating the diameter of the cable in the range, $D(X)$, and that of the domain, $d(x)$. For this we calculate the back transform of the solutions from Table 1 from the range into the domain. This is not trivial because of the space-dependency of λ. Table 2 shows the transforms for the space variables $X(x)$ and $x(X)$ which follow from (12) with

$$\lambda(x) = \frac{1}{\sqrt{r_a(x)g(x)}} = \frac{\lambda_0}{d_0}\sqrt{d(x)} \tag{20}$$

where $d_0 = d(0)$, $\lambda_0 = \lambda(0)$. Appropriate transformations can be given also in the case of diameter changes which are not negligible [13].

By inverting the defining equation (17) for $F(X)$ and using the Riccati equation (19), we find after some lengthy calculations

$$P(D) = \frac{\left(D(X)^{3/4}\right)''}{\left(D(X)^{3/4}\right)} \tag{21}$$

and with $\mathrm{d}x/\mathrm{d}X = 1/\lambda(x)$ (cf. Eq. (12))

$$p(d) = \frac{3}{20\,G_m R_i}\frac{\left(d(x)^{5/4}\right)''}{d(x)^{1/4}}. \tag{22}$$

For $p(d) = 0$ we can now analytically calculate the domain solutions of the image diameter functions for the uniform and power case (see Table (3)). In the other cases, an implicit equation in $d = d(x)$ can be given:

$$\pm\frac{8\sqrt{G_m R_i p}}{3}x + c_2 = \int \frac{d^{1/4}}{d^{3/4}+c_1}\mathrm{d}d \tag{23}$$

which only for $c_1 = 0$ can be solved in closed form. In this case, the domain solution of the exponential diameter function yields the quadratic geometry type (Table 3).

6. TRAVELING WAVES IN NON-UNIFORM AXONS

6.1. General model with nonuniform geometry

In most cases, theoretical investigations of spike propagation assume uniform electrical and geometric properties along the axon. From these analyses, much insight into propagation mechanisms has been gained, suggesting constant shape and velocity of the propagating spike in axons with uniform geometry. A linear or square root relationship between velocity and axonal diameter for myelinated and unmyelinated nerve fibres, respectively was deduced [14].

TABLE 2. Transforms for the space variables $X(x)$ and $x(X)$, respectively.

Geometry type	P	$X(x)$	$x(X)$
uniform	$P=0;\ Q=0$	$\frac{x}{\lambda_0}$	$x\lambda_0$
power	$P=0;\ Q\neq 0$	$C(1-(1-\frac{5x}{3\lambda_0 C})^{\frac{3}{5}})$	$\frac{3}{5}C\lambda_0(1-(1-\frac{X}{C})^{\frac{5}{3}})$
exponential	$P>0;\ Q=Q_1$	$\frac{3}{Q_1}\ln(1+\frac{Q_1}{3\lambda_0}x)$	$\frac{3}{Q_1}\lambda_0(\exp(\frac{Q_1}{3}X)-1)$

TABLE 3. Corresponding diameter functions $d(x)$ and $D(X)$ of nonuniform cable in the domain and in the range, respectively.

Geometry type	P	$D(X)$	$d(x)$
uniform	$P=0;\ Q=0$	d_0	d_0
power	$P=0;\ Q\neq 0$	$d_0(1-\frac{X}{C})^{\frac{4}{3}}$	$d_0(1-\frac{5x}{3\lambda_0 C})^{\frac{4}{5}}$
exponential	$P>0;\ Q=Q_1$	$d_0\exp(\frac{2}{3}Q_1X)$	$d_0(\frac{Q_1x}{3\lambda_0}+1)^2$

However, experimenters have noted several effects which could not be explained with this theory. Examples are blocking of impulse conduction, frequency modulation and changes of AP shape in regions of nonuniform axon geometries (for review, see [15]). Motivated by these observations, several investigators studied the effects of changing axonal geometry upon AP propagation, both by theoretical and computational methods (e.g. [16]).

A drawback of a pure computer simulation approach has been, however, the impossibility of exploring analytically how the various physical parameters describing the inhomogeneous axon affect the solution. Goldstein and Rall [9] instead used results from dimensional analysis [17] to compare theoretical axons having different values of physical parameter but identical nonlinear membrane properties. Our analysis below is inspired by this approach.

We consider traveling wave solutions of the normalized cable equations (13). Thus, we assume that in these equations only the geometry-defining parameter Q explicitly depends on the space variable X whereas all f_k are independent of X, i.e. the voltage thresholds of all channels do not vary in space. The same should hold for the auxiliary variables $u_k(X,T)$. Then we make the ansatz $W(X,T)=W(Y)$ with $Y=X-\Theta T$, which means that a fixed voltage shape travels with constant velocity Θ along the cable (for positive Θ from left to right). Now the system of ordinary differential equations to be solved reads:

$$0=\frac{\mathrm{d}^2W}{\mathrm{d}Y^2}+(\Theta+Q)\frac{\mathrm{d}W}{\mathrm{d}Y}-f_0(W,u_1,\ldots,u_N) \tag{24}$$

$$\frac{\mathrm{d}u_k}{\mathrm{d}Y}=-\frac{\tau}{\Theta}f_k(W,u_1,\ldots,u_N)\quad\text{for } 1\leq k\leq N. \tag{25}$$

For an uniform cylindrical cable, $Q=0$, after definition of Q in (15). Only in this case Eq. (24) remains invariant with respect to the substitution $\Theta\to-\Theta$ and $Y\to-Y$ which means that for any leftwards traveling wave with velocity Θ^-, there is also a wave traveling rightwards with velocity $\Theta^+=-\Theta^-$, and vice versa.

For $Q>0$ (the analog is true for $Q<0$) in some part of the axon cable this symmetry is broken – the range of possible wave velocities will be ‘shifted’. For the general case of Eq. (24) no exact quantitative value of this shift can be given. We can explore it qualitatively, however, by looking at the velocity of the leading wave-front. Assuming that the u_k-kinetics are slower than the W-kinetics, we set $u_k=0$ during the build-up of the leading impulse front [18]. Eq. (24) then reads:

$$0 = \frac{d^2W}{dY^2} + (\Theta + Q)\frac{dW}{dY} - f_0(W). \tag{26}$$

Let the uniform cable equation with $Q = 0$ admit two traveling wave fronts (an excitation from the resting potential to some excited state) at a speed of Θ_{uni}, from right to left with $\Theta^- = -\Theta_{uni}$, and from left to right with $\Theta^+ = \Theta_{uni}$. Then the nonuniform cable admits two wave solutions with the shifted propagation velocities $\Theta^- = -(\Theta_{uni} + Q)$ and $\Theta^+ = \Theta_{uni} - Q$. For cable diameters increasing sufficiently strong from left to right (high values of the geometry parameter Q) we find two wave solutions traveling to the left ($\Theta^- < \Theta^+ < 0$) but none traveling to the right. Here the leftwards traveling front has also a much higher speed than the fronts in the uniform cable. These results demonstrate the direction-dependence of the spike propagation in non-uniform cables (see below and [2, 19]).

6.2. FHN model

As stated in Section 4, for linear functions $f_k(k = 0,..N)$, the transformation function $F(X)$ cancels in the equations (18). Using the FHN model (11) with

$$f_0(W, W_1) = H(W) + W_1 \tag{27}$$

where $H(W) = \widehat{m}_i W - n_i$ represents a piecewise linear membrane characteristic (Figure 2), we get from (18)

$$\begin{aligned} \frac{\partial^2 W}{\partial X^2} - \frac{\partial W}{\partial T} - (P(X) + \widehat{m}_i)W + n_i - w = 0 \\ \frac{\partial w}{\partial T} = \tau(\alpha W - \beta w). \end{aligned} \tag{28}$$

For $P(X) = const$ the system (28) corresponds to the basic equations of uniform cable geometry, thus the results derived by others (see [20] for review) and the present authors [2, 19] can be used. The equations then read:

$$\begin{aligned} \frac{\partial^2 W}{\partial X^2} - \frac{\partial W}{\partial T} - m_i W + n_i - w = 0 \\ \frac{\partial w}{\partial T} = \tau(\alpha W - \beta w) \end{aligned} \tag{29}$$

while we set $m_i := P + \widehat{m}_i$.

For the linear regions of $H(W)$ there are traveling wave solutions. Employing the results of Section 5, the system of ordinary differential equations (24), (25) can be written as:

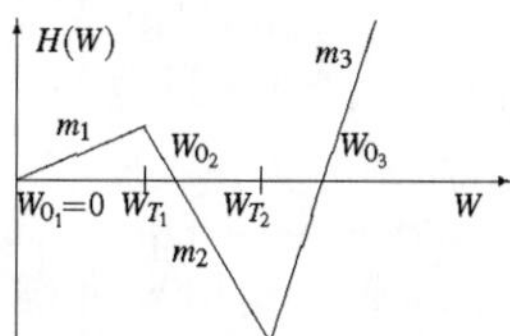

FIGURE 2. Piecewise linear approximation $H(W) = m_i W - n_i$ of the membrane current-voltage characteristic. In the general case, there are three linear regions of $H(W)$, $J_i = [W_{T_i}, W_{T_{i+1}}]$, each containg a zero.

$$\frac{d^2W}{dY^2} + \Theta\frac{dW}{dY} - m_iW + n_i - w = 0$$
$$\Theta\frac{dw}{dY} = -\tau(\alpha W - \beta w). \tag{30}$$

We will consider the bistable case; it is represented by $m_2 \to \infty$ where the middle zero, W_{0_2}, merges with the boundaries W_{T_1}, W_{T_2} to give a threshold, $W_T := W_{T_1} = W_{T_2} = W_{01}$ (see Figure 2).

Eq. (27) becomes

$$f_0(W) = \begin{cases} m_1 W & \text{for} \quad W < W_{0_2} \\ m_3(W - W_{0_3}) & \text{else} \end{cases} \,. \tag{31}$$

The simplest case to consider is $\alpha = \beta = 0$, i.e. the propagation of a wave front (Figure 3) with velocity Θ , without recovery from the excited state.

Eqs. (30) simplify to:

$$\frac{d^2W}{dY^2} + \Theta\frac{dW}{dY} - m_iW = 0. \tag{32}$$

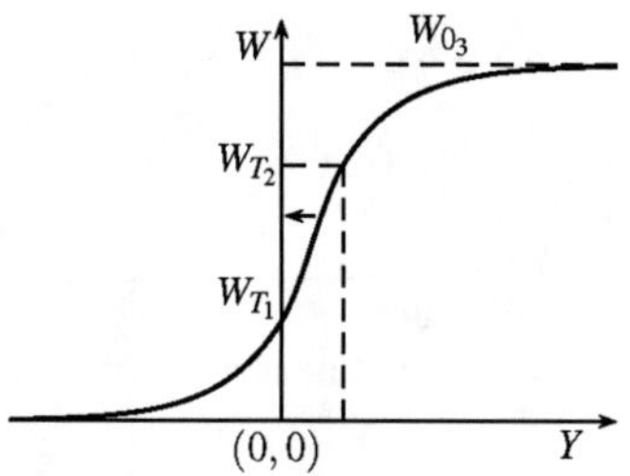

FIGURE 3. Wave front moving to the left while changing the voltage level from 0 to W_{0_3}.

The functions $W(Y)$ in the two regions separated by W_{0_2} can be specified as follows:

$$W(Y) = \begin{cases} W_{0_2} \exp(\mu_1 Y) & \text{for} \quad W < W_{0_2} \\ W_{0_3} - (W_{0_3} - W_{0_2}) \exp(\mu_2 Y) & \text{else} \end{cases} \tag{33}$$

where

$$\mu_1 = -\frac{\Theta}{2} + \sqrt{\left(\frac{\Theta}{2}\right)^2 + m_1}$$
$$\mu_2 = -\frac{\Theta}{2} - \sqrt{\left(\frac{\Theta}{2}\right)^2 + m_3}\,. \tag{34}$$

This result is obtained by connecting two solutions for linear equations where W is greater or less than W_{0_2} at the threshold boundary, $W = W_{0_2}$.

Scott ([20, p. 104] refers to work on neuristor research in the 1960s where for the model (31) , (32) the traveling-wave speed is given by the expression (translated in our notation)

$$\Theta = \frac{m_1 W_{0_2}^2 - m_3 (W_{0_3} - W_{0_2})^2}{\sqrt{(m_1 W_{0_2} + m_3 (W_{0_3} - W_{0_2}))\,(W_{0_3} - W_{0_2}) W_{0_2} W_{0_3}}}\,. \tag{35}$$

If $m_1 = m_3 = m$, we get

$$\Theta = (2W_{0_2} - W_{0_3}) \sqrt{\frac{m}{(W_{0_3} - W_{0_2}) W_{0_2}}}\,, \tag{36}$$

an expression already given in [21].

7. IMPACT OF GEOMETRIC NONUNIFORMITY ON THE PROPERTIES OF THE ACTION POTENTIAL

The results obtained so far can be used to evaluate the impact of geometric nonuniformity on the properties of the action potential. We remember that the time constant as defined in equations (12) yields – via (6) – a constant both in the domain and the range,

$$t(T) = \tau \cdot t = \frac{C_m}{G_m} \cdot t. \tag{37}$$

Thus a given duration and thus frequency of impulses in the range will not change in the domain.

In contrast, space variables transform back from range into domain by

$$x(X) = \int \lambda(X) \mathrm{d}X = \frac{1}{2\sqrt{R_i G_m}} \int \sqrt{D(X)} \mathrm{d}X, \tag{38}$$

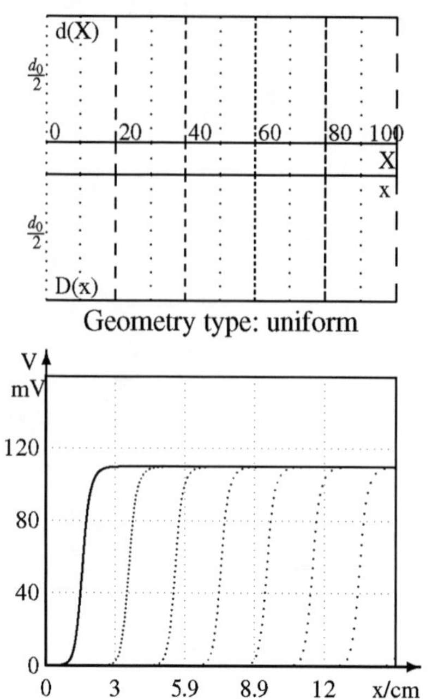

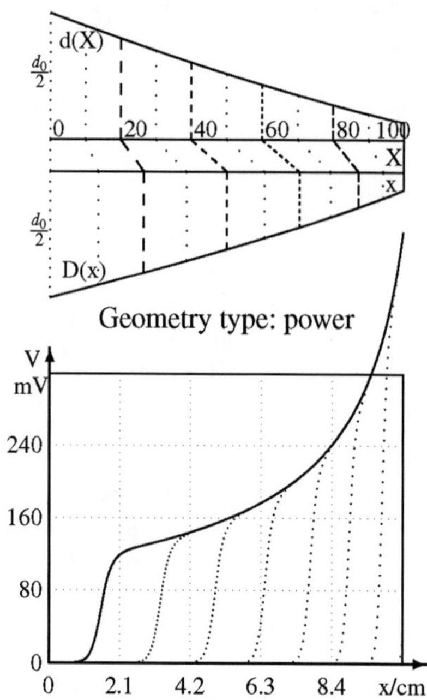

FIGURE 4. Propagation of traveling fronts along axons of the geometry types presented in Tables 1 and 3. In the subfigures, the axon diameter functions $d(x)$ and $D(X)$ in the domain and range are displayed (top), and below snapshots of traveling fronts moving leftwards are presented.

i.e., distances of equal length in the image space have in general different lengths in the domain. This is also true for potential values which transform space-dependent. A potential of amplitude W in the range yields in the domain a potential V depending on the cable diameter:

$$V = \sqrt[4]{\frac{4R_i}{\pi^2 G_m}} D(X)^{-3/4} W. \tag{39}$$

Traveling front solutions of (constant) speed Θ in the range yield excitation fronts in the domain which propagate with space-dependent speed Θ_x (Figure 4). Using Eqs. (37) and (38), we obtain

$$\Theta_x = \frac{\mathrm{d}x}{\mathrm{d}t} = \frac{\lambda}{\tau}\Theta = \frac{\Theta}{2C_m}\sqrt{\frac{G_m}{R_i}d(x)}. \tag{40}$$

In Figure 4, the propagation of traveling fronts along axon cables obeying the diameter functions given in Tables 1 or 3, respectively, is presented.

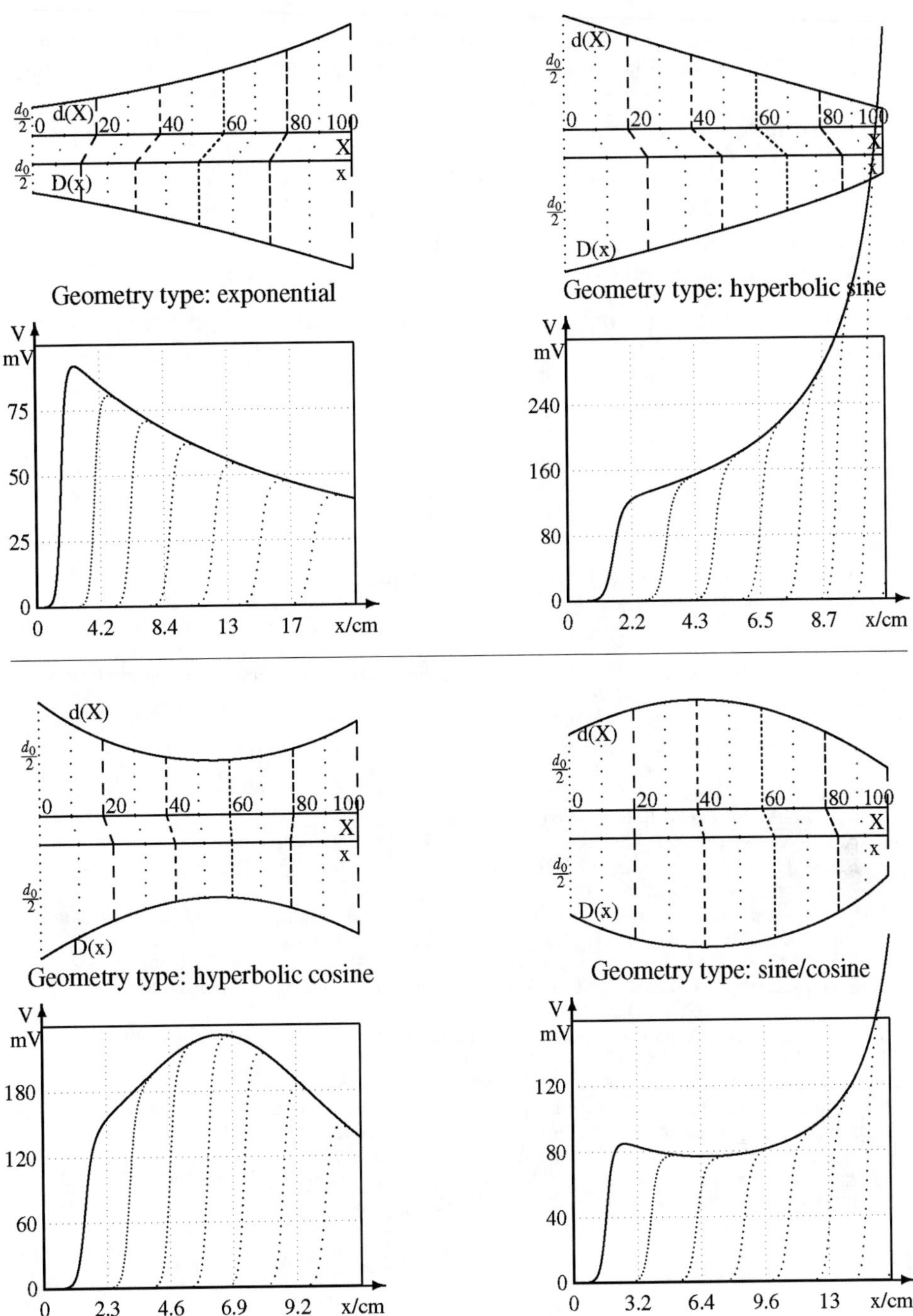

FIGURE 4. (continued).

To summarize, we can state the following conclusions on action potentials propagating along axons of the nonuniform geometry types presented in Tables 1 or 3, respectively:

If the diameter of the axon in the domain widens,

- the speed of the action potential front increases, see (40) and Figure 4,
- the length of the excited axon region increases, see (38) and Figure 4,
- the spike height decreases, see (39) and Figure 4, and
- the spike duration and frequency remain unchanged, see (37).

The method presented in this paper has been also applied to the problem of reducing a dendritic tree with active membrane to an equivalent cable [2]. In that paper, we have stated in a short note that for the piecewise linear FHN model (30),(31) traveling wave solutions can be analytically derived. We will present this case in detail in a forthcoming paper.

REFERENCES

1. A. Schierwagen, *Prog. Brain Res.* **102**, 151–167 (1994).
2. M. Ohme, and A. Schierwagen, *Biol. Cybern.* **78**, 227–243 (1998).
3. G. M. Shepherd, ed., *The Synaptic Organization of the Brain*, Oxford University Press, New York, 1998.
4. G. Stuart, N. Spruston, and M. Häusser, eds., *Dendrites*, Oxford University Press, Oxford, 1999.
5. J. P. Pawelussen, *J. Math. Biology* **15**,151–171 (1982).
6. A. C. Scott, *Rev. Mod. Phys.* **47**, 487–533 (1975).
7. R. FitzHugh: "Mathematical models of excitation and propagation in nerve", in *Biological Engineering*, edited by H. P. Schwan, McGraw-Hill, New York, 1969, pp. 1–85.
8. J. Nagumo, S. Arimoto, and S. Yoshizawa, *Proc. IRE* **50**, 2061–2070 (1962).
9. S. S. Goldstein, and W. Rall, *Biophys. J.* **14**, 731–757 (1974).
10. A. L. Hodgkin, and A. F. Huxley, *J. Physiol. (Lond.)* **117**, 500–544 (1952).
11. J. J. Kelly, and M. S. Ghausi, *IEEE Trans. Circuit Theory* **CT-12**, 554–558 (1965).
12. A. Schierwagen, *J. theor. Biol.* **141**, 159–179 (1989).
13. M. Ohme, *Modellierung der neuronalen Signalverarbeitung mittels kontinuierlicher Kabelmodelle*, Dissertation, Universität Leipzig, Fakultät für Mathematik und Informatik, 1995.
14. P. J. Hunter, P. A. McNaughton, and D. Noble: emphProg. Biophys. Molec. Biol. **30**, 99–144 (1975).
15. S. G. Waxman, J. D. Kocsis, and P. K. Stys, eds., *The Axon. Structure, Function and Pathophysiology.* Oxford University Press, Oxford, 1995.
16. A. Rabinovitch, I. Aviram, N. Gulko, E. Ovsyscher, *J. theor. Biol.* **196**, 141–154, 1999.
17. R. FitzHugh, *J. theor. Biol.* **40**, 517–541 (1973).
18. J. Rinzel, and D. Terman, *SIAM J. Appl. Math.* **42**, 1111–1136 (1982).
19. A. Schierwagen, "Traveling wave solutions of a simple nerve conduction equation for inhomogeneous axons" , in *Nonlinear Waves in Excitable Media*, edited by A. V. Holden, M. Markus, and H. Othmer, Manchester University Press, Manchester, 1991, pp. 107–114.
20. A. Scott, *Neuroscience: A Mathematical Primer*, Springer-Verlag, New York, 2002.
21. J. Rinzel and J. B. Keller, *Biophys. J.* **13**, 1313–1337 (1973).

A Design Method for Analog and Digital Silicon Neurons –Mathematical-Model-Based Method–

Takashi Kohno and Kazuyuki Aihara

Institute of Industrial Science, The University of Tokyo, Meguro-ku, Tokyo 153-8505, Japan

Abstract. Silicon neuron is electrical circuit that is analogous to biological neurons. Conventionally, it was designed mainly in the following two attitudes. One is to realize circuitry that is as close to biological neuron as possible, which enlarges circuit size and complexity terribly. Another is to realize simple and compact circuitry that can be utilized to construct large-scale silicon neural network. Because designers ignore the mechanisms underlying the neuronal phenomena, silicon neurons can be quite different from biological ones. We proposed a new design method that utilizes mathematical knowledge on neuronal phenomena, which allows us to design simple circuitry whose operating mechanism is same as biological neuron. Several types of circuits have been and are being implemented to validate its efficiency.

Keywords: Silicon neuron, nonlinear dynamics, bifurcation analysis, chaos.
PACS: 01.30.Cc, 82.40.Bj, 84.30.-r, 87.19.11

1. INTRODUCTION

Neuromorphic hardware is one inspired by any schemes and functions in biological nerve system. Silicon neuron, a kind of neuromorphic hardware, is electrical circuit that is designed to reproduce various phenomena in neuronal cells. It is studied for some different purposes. One of them is to construct hybrid system of silicon and biological neurons [1]–[6], which can be utilized to validate a neuron model by replacing a target neuron in biological neural network with silicon one and examining if the network can operate correctly. It also can be applied to implantable biomedical devices and neural network simulators. Silicon neuron is required to simulate a neuronal cell as accurately as possible in real-time. It is designed as a solver circuit for differential equations of a conductance-based neuron model utilizing analog very large-scale integrated circuit (aVLSI) technology in most cases. Several successful silicon neuron circuits whose behavior seems very close to the targeted neuronal cells have been designed according to this policy (conductance-based silicon neurons). The circuitry, however, tends to be very complex because the differential equations for ionic conductance comprise functions that are not easy to implement in electrical devices. Therefore, designers have to abandon complete implementation of the equations in order to realize fabricatable and operable circuit. Furthermore, complex circuitry generally leads to large numbers of circuit parameters to be adjusted, which makes the circuit hard to operate and unstable.

Another purpose of studying silicon neuron is to realize a circuit that can be an element to construct more complex and intelligent neuromorphic systems that mimics a part or whole of brain. In this case, silicon neuron have to be as simple, tunable, and stable as possible, because a huge number of silicon neurons have to be connected to each

CP1028, *Collective Dynamics: Topics on Competition and Cooperation in the Biosciences*
edited by L. M. Ricciardi, A. Buonocore, and E. Pirozzi

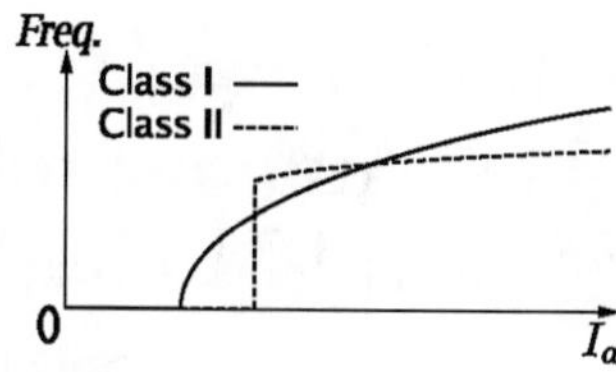

FIGURE 1. Frequency of repetitive firing in Class I and II neurons.

other. Conductance-based silicon neuron is not obviously suitable for this prerequisite. Instead, phenomenological neuron model is implemented in electrical circuit. Because such model is highly abstracted linguistically, silicon neuron can be designed utilizing characteristic curves of electrical devices, which allows us to realize simple circuitry (phenomenological silicon neurons). However, because phenomenological neuron models ignore mechanisms behind neuronal phenomena, designers have to regard a large numbers of phenomena if the silicon neuron is to be close to a neuronal cell. Circuitry gets complex eventually when designers try to implement many phenomena. Moreover, circuitry may have to be completely re-designed if "phenomenon set" to be implemented is changed. For example, leaky integrate-and-fire silicon neuron [7]–[10] cannot operate as Hodgkin-Huxley [11][12] and FitzHugh-Nagmo [13][14] type neuron. We have to re-design circuit to realize this type of silicon neuron by phenomenological approach.

Theoretical studies on conductance-based neuron models from the standpoint of nonlinear dynamics and bifurcation analysis elucidated the mechanisms behind neuronal phenomena mathematically [15]–[17]. Fundamental properties of excitable cell such as generation of overshoot and threshold phenomenon were illustrated by topological structures in phase portrait for the differential equations of the neuron model. Neuron classes determined by Hodgkin [18] are thought to be an important characteristics of neuronal cell. When neuronal cells are stimulated with sustained current, they fire repetitively when the current is sufficiently strong. Those which start firing with arbitrarily low frequency were classified as Class I, where those which start firing with a nonzero frequency were classified as Class II (see Fig. 1). It was shown that bifurcation structure in differential equations of neuron models account for this classification; saddle-node on invariant circle bifurcation causes arbitrarily low frequency (Class I) and Hopf bifurcation, nonzero frequency (Class II). If a neuronal model has some bifurcations that produce bistability between resting (stable point) and repetitive firing (stable limit cycle) states, burst firing can be generated by adding a slow negative feedback current that behave similarly to calcium-dependent potassium current (I_{K-Ca}).

It was shown that even topological structure in phase portrait that does not affect the bifurcation structures can play very important roles in neuronal phenomena. Fujii and Tsuda [19] showed that simple two-variable neuron models under some conditions can produce very complex behavior in neural network where each neuron is connected via gap junction (GJ). GJ is a physical connection between neuronal cells that works as linear resister electrically and a large number of them are shown to exist between interneurons in some regions in neocortex [20][21]. The most important condition is the existence of narrow channel in the model's phase plane when appropriate stimulus

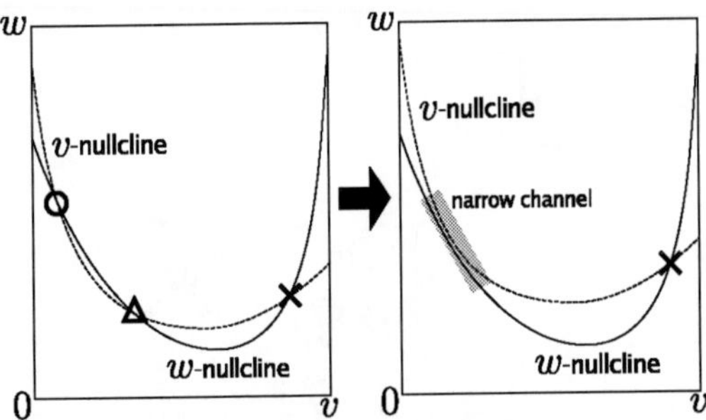

FIGURE 2. Illustration for narrow channel. The system is comprised of two variables, v and w, where the former is membrane potential and the latter recovery variable. The left figure shows the phase plane when no stimulus is applied. When an appropriate stimulus current is applied, v-nullcline shifts up and narrow channel appears (right figure).

current is applied (see Fig. 2). Narrow channel is region where the nullclines for two variables are very close to each other. Other conditions require existence of a limit cycle around a unique unstable spiral equilibrium point. Because neuron models in such situation begin to fire repetitively by arbitrarily low frequency (Class I), they are called Class I* neuron in the sense that it is a subclass of Class I. Neural network where Class I* neurons are connected each other via GJs with uniform conductance can exhibit chaotic and intermittently chaotic behaviors dependent on the conductance of GJs [19] .

Based on such mathematical knowledge on neuronal phenomena, we proposed a new design policy for silicon neuron (mathematical-model-based method) [22]–[26]. Bifurcation and phase portrait structures give us mathematical abstraction of neuronal phenomena instead of linguistic one in phenomenological models. Utilizing them, we can design simple circuitry that reproduces mechanisms behind neuronal phenomena. In this article, we introduce some of our silicon neurons.

2. SILICON NEURON CIRCUITS

Concept of mathematical-model-based design method is to construct bifurcation and phase portrait structures in an electrical circuit model utilizing functions that can be implemented easily by electrical devices. First, we design the phase portrait of the model focusing on the topological relation between nullclines and the stability of equilibrium points regarding what bifurcation should be produced in it. We can design the model whose neuron class can be changed dependent on parameters by choosing nullclines appropriately. Practically, it is quite difficult to design phase portrait whose dimension is higher than two. Thus we have to design our model so that it can be reduced into two-dimensional. The key point is that we combine silicon-native curves to construct the nullclines in this step, which allows us to implement the model with compact circuitry. Secondly, we tune the system parameters utilizing bifurcation analysis. Time constants are determined in this step.

One of the advantages in our method is that it does not depend on the implementation

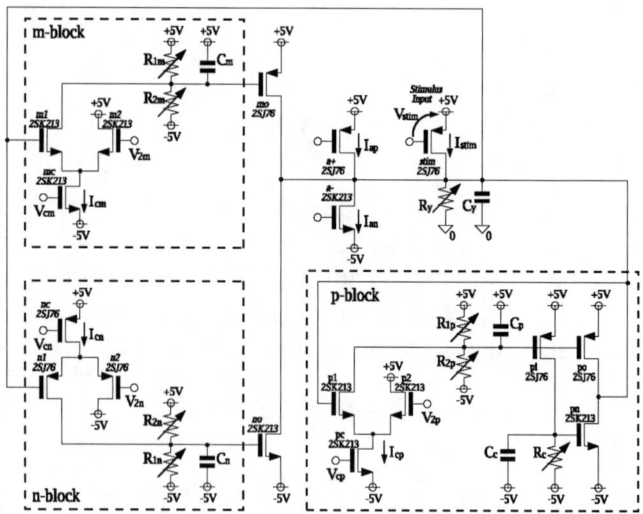

FIGURE 3. Circuit of burst-firing-capable silicon neuron for implementation with discrete MOSFET devices.

technology (e.g. analog or digital). We have studied and studying several types of silicon neurons utilizing mathematical-model-based design method. One of them is an analog circuit designed for metal oxide semiconductor field effect transistors (MOSFETs) that operate in the super-threshold region. We designed two types of analog circuits for MOSFETs that operate in the subthreshold region. One is intended to be able to realize Class I* mode. Another is designed to consume quite low-power for analog very-large-scale integrated circuit (aVLSI). We are also studying a digital silicon neuron model that can be implemented in regular digital circuit including field programmable gate arrays (FPGAs). We will see three of our silicon neuron circuits in this section.

2.1. Circuit for super-threshold operating region

This model is capable of burst firing and for implementation with discrete MOSFET devices and linear resisters [25]. The system equations were designed utilizing quadratic and sigmoidal curves that are easily implemented by grounded source and differential pair circuitries. The circuit schematic is shown in Fig. 3. It can be divided into two functional modules; the p-block and the others. The latter functions as a basic excitable system whose system equations are written as follows:

$$C_y\frac{dy}{dt} = -\frac{y}{R_y}+\frac{\beta_m}{2}m^2-\frac{\beta_n}{2}n^2+I_a+I_{stim}, \tag{1}$$

$$\frac{dm}{dt} = \frac{f_m(y)-m}{T_m}, \tag{2}$$

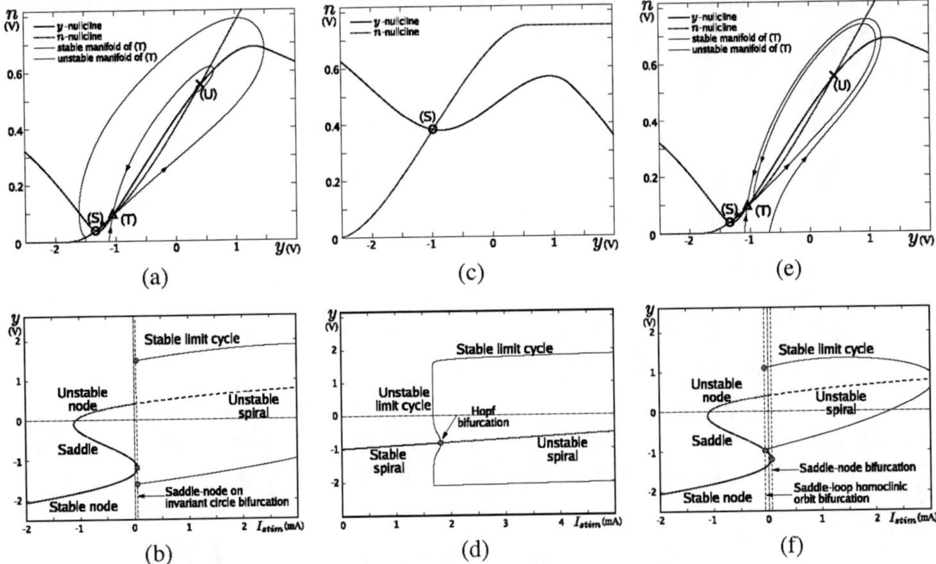

FIGURE 4. Phase planes and bifurcation diagrams for our basic system. (a), (c), and (e) are phase planes when no stimulus is applied and (b), (d), and (f) are bifurcation diagrams for Class I, II and bistability modes, respectively.

$$\frac{dn}{dt} = \frac{f_n(y) - n}{T_n}, \tag{3}$$

where y, m, and n are the membrane potential and fast and slow ionic channel variables, respectively. I_{stim} is a stimulus current given from outside. R_y is a constant resistance; C_y, membrane capacitance; I_a, a constant bias current; T_m and T_n, the time constants for m and n, respectively; β_m and β_n, the transconductance coefficients of the MOSFETs mo and no in Fig. 3 that correspond to the ionic channel currents for m and n, respectively. Functions $f_m(y)$ and $f_n(y)$ are the sigmoidal characteristic curves of differential-pair circuitry.

By assuming that the time constant for fast ionic channel variable m is sufficiently smaller than that for slow one n, we can reduce these system equations into two dimensions as follows:

$$C_y \frac{dy}{dt} = -\frac{y}{R_y} + \frac{\beta_m}{2} f_m(y)^2 - \frac{\beta_n}{2} n^2 + I_a + I_{stim}, \tag{4}$$

$$\frac{dn}{dt} = \frac{f_n(y) - n}{T_n}. \tag{5}$$

By tuning the system parameters in these equations appropriately, we can obtain typical phase plane and bifurcation structures for Class I and II (Fig. 4(a), (b), (c), and

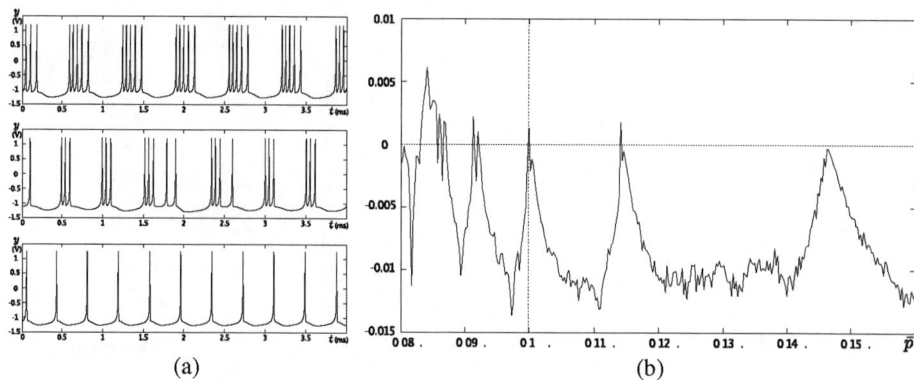

FIGURE 5. (a) Burst firing waveforms of y observed when $\bar{p} = 0.880, 0.999$, and 1.500. Regular and chaotic patterns are observed. (b) Largest Lyapunov exponent for burst firing waveforms of y. Note that it is positive for some $\bar{p}$ values. Vertical line corresponds to $\bar{p} = 0.999$.

(d)) [23][24]. In Fig. 4(a) and (c), stable equilibriums (S) produce resting potential. When a pulse stimulus current is applied, membrane capacitance is charged rapidly and the system state jumps to rightward. If it is so strong that y exceeds the stable manifold for saddle point (T) in Fig. 4(a), the system state moves to (S) along the right branch of the unstable manifold for (T) generating an action potential. If not, the system state goes directly back to (S) along the left branch of the unstable manifold for (T). The stable manifold for (T) forms separatrix that determines threshold. In Fig. 4(c), the ascending part of the y-nullcline forms separatrix for threshold. Bifurcation diagrams Fig. 4(b) and (d) illustrate how the repetitive firing emerges when we apply sustained stimulus current I_{stim}. Stable limit cycle corresponds to repetitive firing in these figures. When we increase I_{stim} in Fig. 4(b), the system state is shifted to the stable limit cycle that emerges via the saddle-node on invariant circle bifurcation. The frequency is zero at the bifurcation point and increases continuously as the current strength. Therefore the system is Class I. In Fig. 4(d), the system state jumps to the stable limit cycle when I_{stim} exceeds the Hopf bifurcation point. In this case the firing frequency cannot be low arbitrarily. This is Class II.

We performed circuit experiment for Class II mode [22] and obtained successful results that proved our circuit operated as a silicon neuron. In this work, our circuit showed chaotic responses to periodic stimulus similar to that observed in squid axon that is known to be Class II [27].

If we increase C_y or reduce T_n for Class I mode (Fig. 4(a)), an unstable manifold for (T) twists around an unstable limit cycle around equilibrium point (U) as shown in Fig. 4(e). Bifurcation analysis proves that a stable limit cycle that corresponds to repetitive firing emerges via saddle-loop homoclinic orbit bifurcation near I_{stim}=-0.05 mA (see Fig. 4(f)). There exists bistability between stable equilibrium (S) and this stable limit cycle till the former vanishes via saddle-node bifurcation near I_{stim}=0.05 mA. The p-block in Fig. 3 produces slow negative feedback current similar to I_{K-Ca}, which appends burst firing capability to the basic system with bistability described above. The equations for

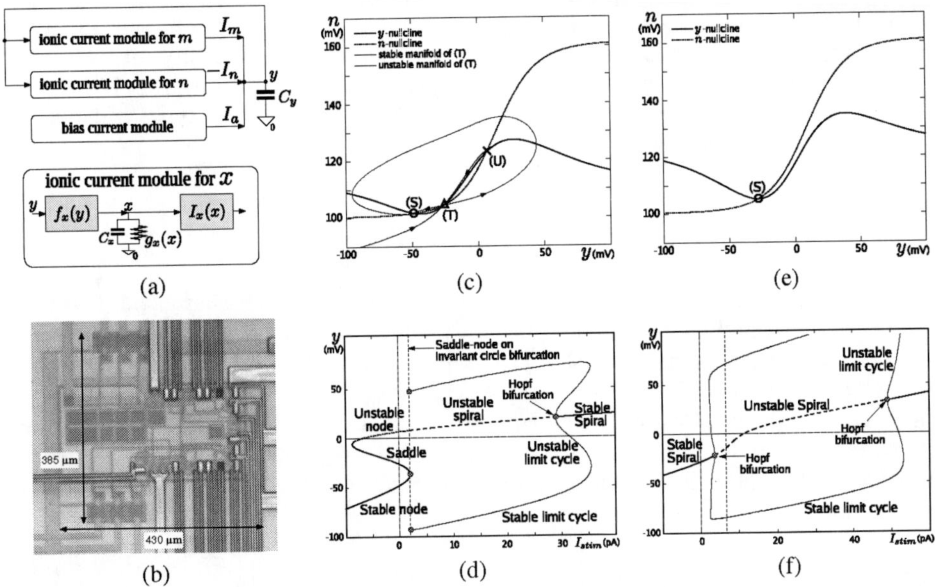

FIGURE 6. Silicon neuron for aVLSI implementation described in equations (10)–(16). (a) Block diagram for our basic silicon neuron for aVLSI. (b) Chip fabricated by TSMC CMOS .35μ. (c) and (e) are phase planes and (d) and (f) are bifurcation diagrams for Class I and II modes, respectively.

this block are as follows:

$$\frac{dV_{ca}}{dt} = \frac{1}{C_{ca}}(I_{Ca} - \frac{V_{ca}}{R_{ca}}), \tag{6}$$

$$\frac{dp}{dt} = \frac{1}{T_p}(f_P(y) - p), \tag{7}$$

$$I_{Ca} = \frac{\beta_p}{2}p^2, \tag{8}$$

$$I_{K-Ca} = \frac{\beta_k}{2}V_{ca}^2, \tag{9}$$

where p is the conductance variable for the calcium channel; T_p, the time constant for p; $f_P(y)$, a sigmoidal curve generated by MOSFET differential pair circuitry; and β_p and β_k, the transconductance coefficients of the MOSFETs for I_{Ca} and I_{K-Ca}, respectively.

Figure 5(a) shows the waveforms observed when the parameter $\bar{p}$ is changed while other parameters are fixed at values appropriate for burst firing. Both regular bursting and chaotic patterns were observed that alternated for several times as $\bar{p}$ increased [25]. In Fig. 5(b), we show the largest Lyapunov exponent for value of $\bar{p}$. Positive largest Lyapunov exponent indicates that the system dynamics is chaotic. It is quite similar phenomenon to ones reported in biologically realistic models for the same type of bursting [28][29]. It indicates that our silicon neuron circuit not only operates as a bursting neuron but also reproduces deeper properties of biological neuron. By tuning circuit parameters, this circuit was shown to be able to realize other types of bursting patterns such as elliptic bursting [25].

2.2. Circuit for aVLSI

For effective implementation of large scale network of silicon neurons, circuit size and power consumption have to be sufficiently small. We designed a circuit suitable for implementation in analog very large scale integrated circuit (aVLSI) that operates in subthreshold region of MOSFET where the current consumption is extremely low. It is designed by mathematical-model-based design method in the same way as the previous circuit, but the basic nonlinear function is sigmoidal (hyperbolic tangent) curve that is easy to implement with MOSFETs under subthreshold condition. The system equations are as follows:

$$C_y\frac{dy}{dt} = -g_y(y)+I_m(m)-I_n(n)+I_a+I_{stim}, \tag{10}$$
$$C_m\frac{dm}{dt} = f_m(y)-g_m(m), \tag{11}$$
$$C_n\frac{dn}{dt} = f_n(y)-g_n(n), \tag{12}$$

where y, m, and n are the membrane potential and fast and slow ionic channel variables, respectively. I_{stim} is a stimulus current given from outside. I_a is a constant bias current; C_y, membrane capacitance; C_m and C_n, the capacitance for m and n that determine their time constants, respectively. Functions f_x, g_x, I_x for $x = m$ and n are sigmoidal curves whose equations are as follows:

$$f_x(y) \equiv \bar{x}\frac{1}{1+\exp(-\frac{\kappa}{U_T}h_x(y))}, \tag{13}$$
$$h_x(y) \equiv \varepsilon_x\frac{1-\exp(-\frac{\kappa}{U_T}(y-\delta_x))}{1+\exp(-\frac{\kappa}{U_T}(y-\delta_x))}, \tag{14}$$
$$g_x(y) \equiv s_x\frac{1-\exp(-\frac{\kappa}{U_T}(y-\theta_x))}{1+\exp(-\frac{\kappa}{U_T}(y-\theta_x))}, \tag{15}$$
$$I_x(y) \equiv I_{0x}\frac{1}{1+\exp(-\frac{\kappa}{U_T}(y-I_{ofx}))}, \tag{16}$$

where κ and U_T are capacitive coupling ratio and thermal voltage for MOSFETs. $\delta_x, \theta_x, I_{ofx}$ are system parameters given from outside.

These equations can be implemented by relatively simple circuit whose block diagram is illustrated in Fig. 6(a). They can be reduced into two dimensions in the same way as the previous ones. We can tune the system parameters so that the reduced system's phase

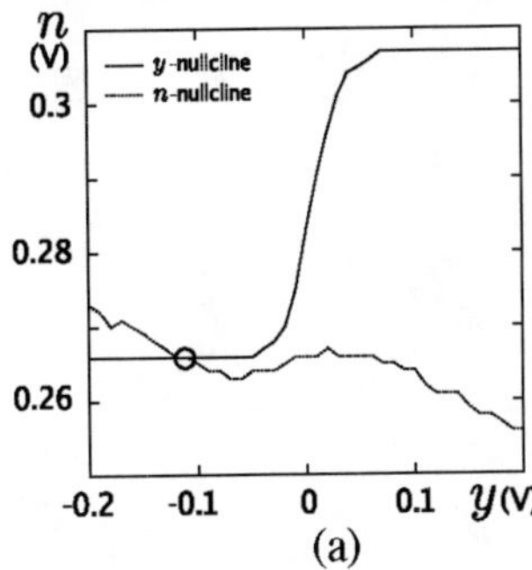

(a)

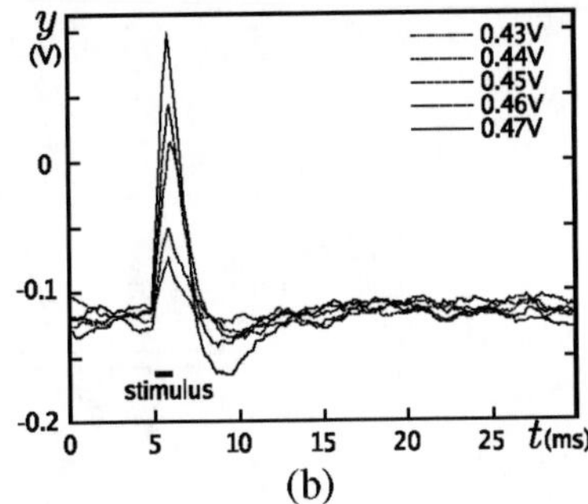

(b)

FIGURE 7. Circuit experiment results for our aVLSI silicon neuron. (a) Measured phase plane. It is similar to Class II mode shown in Fig. 6(e). (b) Waveforms of y in response to pulse stimulus. We give stimulus by voltage V_{stim}, which is converted to I_{stim} in the VLSI chip. Overshoot and threshold (between V_{stim}=0.44 V and 0.45 V) were observed.

plane and bifurcation structures are for Class I and II (see Fig. 6(c), (d), (e), and (f)). Exact shape of the curves in these figures are different from in the figures for the silicon neuron in the previous subsection (Fig. 4), but they share topology.

Fabrication was completed for this circuit (see Fig. 6(b)). In this chip, we implemented an additional feedback circuit that clamps the membrane potential y to any specified value. When enabled, it allows us to view phase plane of our silicon neuron in the same way as voltage clamp technique in biophysical experiments. We performed preliminary experiment of applying pulse stimulus (see Fig. 7). This figure shows measured phase plane in (a) and waveforms of y in response to pulse stimulus in (b). Phase plane structure seems to be similar to Class II mode shown in Fig. 6(e). In the circuit experiment, we apply stimulus voltage V_{stim} that is converted to I_{stim} in the VLSI chip. We observed overshoot when the stimulus is sufficiently strong. Threshold seems to be in between V_{stim}=0.44 V and 0.45 V. The phase plane is very noisy. We will construct more stable experiment system and perform regular experiment. Its results will appear in the future publications.

Currently, we are also working with other silicon neurons for aVLSI that is capable of burst firing. One of them utilizes current-mode integrator to solve the system equations. It is comprised of a basic excitable system and a slow feedback current generator analogous to p-block in the previous subsection. The former system is similar to one described above in this subsection but designed to reduce complexity. It was designed to have fewer parameters to tune, which will also improve stability of the circuit operation. We are getting successful simulation results for square-wave burst mode on HSPICE circuit simulator (Fig. 8).

2.3. Digital spiking silicon neuron

As mentioned above, mathematical-model-based design method is applicable to any device technologies. We are also studying a silicon neuron for regular digital circuit

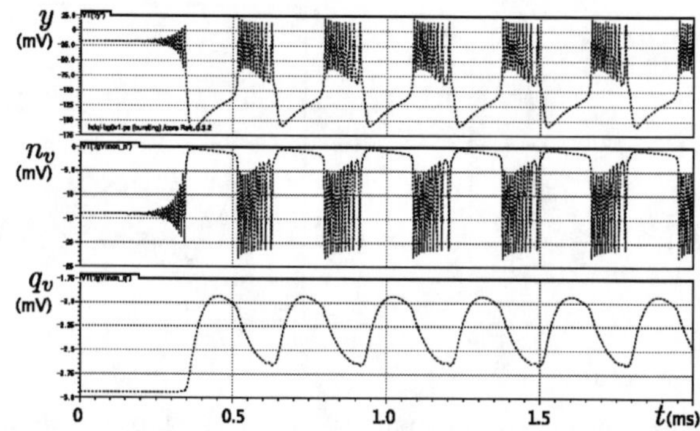

FIGURE 8. Bursting waveforms of a new generation silicon neuron for aVLSI. (HSPICE simulation result). y is membrane potential. n_v and q_v are monitoring output for slow ionic channel variable and slow negative feedback current respectively.

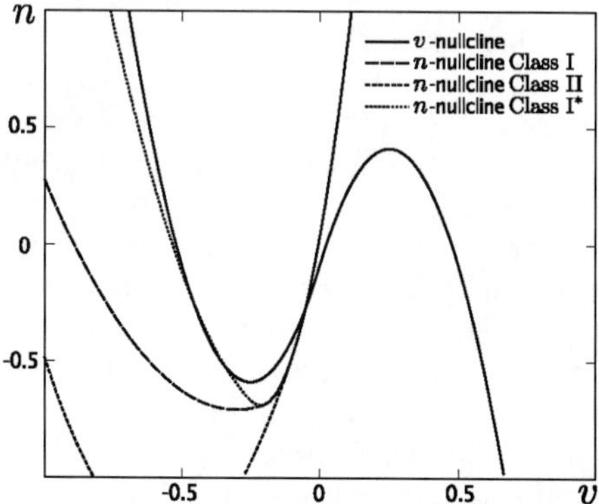

FIGURE 9. Nullclines for our DSSN model. We can construct phase plane structures for Class I, II, and I* modes by tuning n-nullclines.

including field programmable gate arrays (FPGAs) [26]. The advantages of digital circuit are its stability against external perturbations including noises on power lines, maturation of fabrication technology and easiness of design in comparison to analog circuit, and low cost in budget and time aspects (for example, FPGAs allows us to implement our circuit in a few tens of seconds). Particularly, digital circuit operates completely in the same way as simulation, which will advance the application of silicon neural networks.

Our digital spiking silicon neuron (DSSN) is essentially digital arithmetic circuit that solves differential equations. It is different from other digital silicon neurons [30][31] in that action potential is produced in the form of time series of digital binary data. In other words, it is a dedicated simulation hardware for a neuron model. The key point lies in the equations of the neuron model. We design them so that silicon neuron can operate appropriately with as simple and compact circuit as possible.

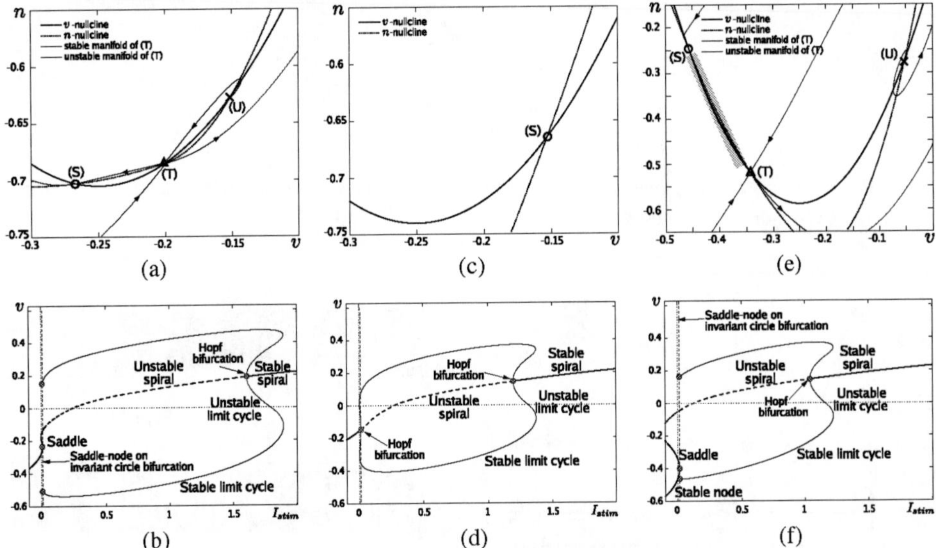

FIGURE 10. Phase planes and bifurcation diagrams for our DSSN model. (a), (c), and (e) are phase planes when no stimulus is applied (b), (d), and (f) are bifurcation diagrams for Class I, II and Class I* modes, respectively. Critical structures in phase planes are zoomed up in (a), (c), and (e).

The most simple nonlinear operation in digital circuit is switching, which provides us piecewise-linear function. However, because piecewise-linear nullclines make bifurcation point singular, we cannot invoke some bifurcations in their strict senses. Multiplication is one of the most basic nonlinear operation in digital arithmetic circuit, which provides us quadratic function. It is considerably costly in comparison to switching, but is very common and some FPGAs offer advanced support for its implementation. By combining these two nonlinear operations, we designed a two-dimensional model. The equations are:

$$\frac{dv}{dt} = \frac{\phi}{\tau}(f(v) - n + I_0 + I_{stim}), \tag{17}$$

$$\frac{dn}{dt} = \frac{1}{\tau}(g(v) - n), \tag{18}$$

where

$$f(v) \equiv \begin{cases} a_n(v+b_n)^2 - c_n & \text{when} v < 0, \\ -a_p(v-b_p)^2 + c_p & \text{when} v \geq 0, \end{cases} \tag{19}$$

$$g(v) \equiv \begin{cases} k_n(v-p_n)^2 + q_n & \text{when} v < r, \\ k_p(v-p_p)^2 + q_p & \text{when} v \geq r. \end{cases} \tag{20}$$

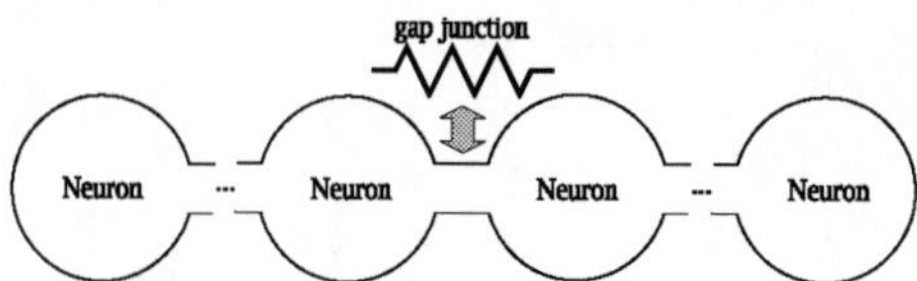

FIGURE 11. Topology of one-dimensional GJ-coupled network. Neuron models are interconnected with two nearest neighbors via GJs

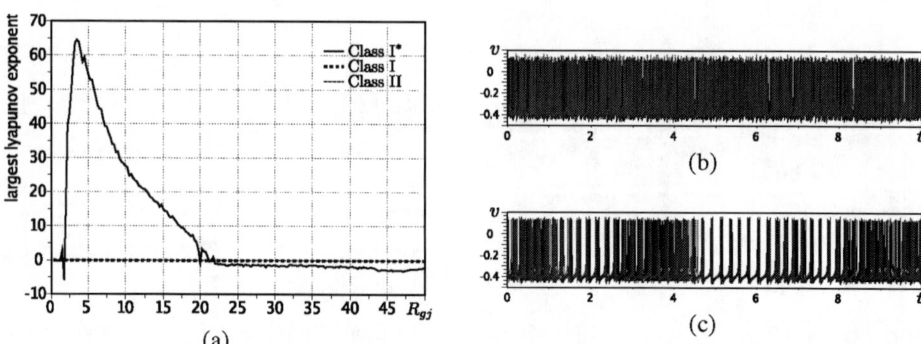

FIGURE 12. Simulation results for one-dimensional GJ-coupled network of our DSSN model. It is comprised of 20 of DSSN and each of them has uniform parameter value. R_{gj} is also uniform. (a) Largest Lyapunov exponent for time series observed in the network. (b) and (c) Superimposed waveforms for v when DSSN is Class I* mode. R_{gj} is 10 and 19.5, respectively. Intermittently chaotic behavior is observed in (c), when the largest Lyapunov exponent is relatively small.

Variables v and n are membrane potential and a slow ionic channel variable, respectively. Parameters a_x, b_x, c_x, k_x, p_x, q_x for $x = n$ and p, and r determine the form of the nullclines. ϕ and τ are time constant parameters. Multiplication between a parameter and a variable can be implemented by shift operation if we select the parameter from $\{2^n | n \in \mathbf{Z}\}$. In this case, these equations require multiplier only for calculation of v^2. It allows us to implement these equations with minimum hardware resources.

The nullclines for these equations are piecewise-quadratic curves whose shape can be cubic or parabolic. By selecting parameters appropriately, we can obtain phase plane structure for Class I, II, and I* as shown in Fig. 9 and Fig. 10(a), (c), and (e). Bifurcation analysis proves that they are Class I, II, and I, respectively. System parameters are selected so that the system variables are between -1 and 1.

To validate Class I* mode, we have to examine behavior of its GJ-coupled network. In [19], they report that simple one- and two-dimensional GJ-coupled network (see Fig. 11) of Class I* neurons exhibited chaotic and intermittently chaotic behaviors. We performed simulation for a one-dimensional network where 20 of our DSSN model with the same parameter values are connected via GJs with uniform strength R_{gj}. The current applied to the i-th DSSN model through GJ (I^i_{gj}) is given as follows:

$$I^i_{gj} = (v_{i+1} + v_{i-1} - 2v_i)/R_{gj}, \tag{21}$$

where i is the index number for DSSN model (from 1 to 20); v_i, v for the i-th DSSN model ($v_0 \equiv v_1$ and $v_{21} \equiv v_{20}$); and R_{gj}, the resistance of the GJ. These currents are added to I_{stim} for neuron i.

Simulation was performed for various R_{gj} values and synchronous, chaotic, and intermittently chaotic behaviors were observed. In Fig. 12 (a), the largest Lyapunov exponents for the Class I, II, and I* modes are shown. It is large when R_{gj} is approximately between 2.5 and 20 when the DSSN model is in Class I* mode, whereas it is approximately zero, independent of R_{gj} when the model is Class I and II modes. In the Class I* case, the network is synchronous when R_{gj} is sufficiently small, becomes chaotic (Fig. 12 (b)) as R_{gj} increases, and then returns to synchronous when R_{gj} is larger. Intermittently chaotic behaviors (Fig. 12 (c)) were observed before R_{gj} got sufficiently large to produce synchronous behavior.

Simulation results described above were calculated by LSODE solver with double-precision floating point operation. Therefore, if we design dedicated hardware for our DSSN model with the same algorithm and precision, we obtain silicon neuron circuit that operates completely identical to these simulation results. Because it requires considerably complex circuitry, simplification have to be pursued by utilizing Euler's method and fixed point operation with fewer bits. We performed simulations of the same neural network as above utilizing Euler's method with fixed point operation of various precision bits. We can utilize fixed point operation with no difficulty because the differential equations are designed so that the system variables are between -1 and 1. Very similar complex behaviors with those shown in Fig. 12(b) and (c) were observed when the precision was 28bit and $\Delta t = 10^{-5}$. Waveforms that correspond to them are shown in Fig. 13. With this precision bit, we also observed chaotic responses to some periodic pulse stimuli when the DSSN model was in Class II mode. It is quite a difficult problem to give objective criteria for determining the required minimum precision. This is because interconnected neurons and even a single neuron are complex systems and the assessment of their dynamical behavior is a complicated subject.

We are working on digital silicon neural network system whose block diagram is shown in Fig. 14. Euler's method allows architecture of DSSN to be simple (see the neuron module in the figure). It leads to high operating frequency and implementation density, which facilitate realization of larger scale of neural network. Single neuron module is operational. Currently, digital silicon synapse model is studied.

3. CONCLUDING REMARK

We introduced briefly the concept of mathematical-model-based design method for silicon neuron, and three silicon neuron circuits designed by it. Mathematical abstraction of neuronal phenomena allows us to implement simple circuitry that operates by the essentially equivalent mechanisms as biological neuron models. Because mathematical structures that produce neuronal phenomena are constructed utilizing device-native

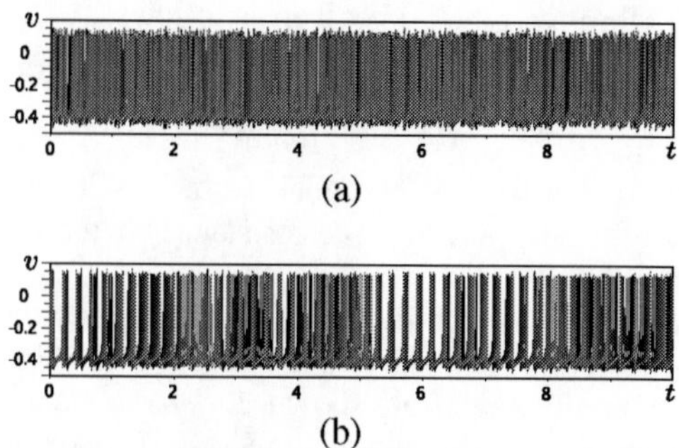

FIGURE 13. Behaviors of a GJ-coupled network of 20 DSSNs. (a) R_{gj} = 10. (b) R_{gj} = 19.5. Numerical integration was performed by Euler's method and precision was 28 bit fixed-point. They correspond to Fig. 12 (b) and (c).

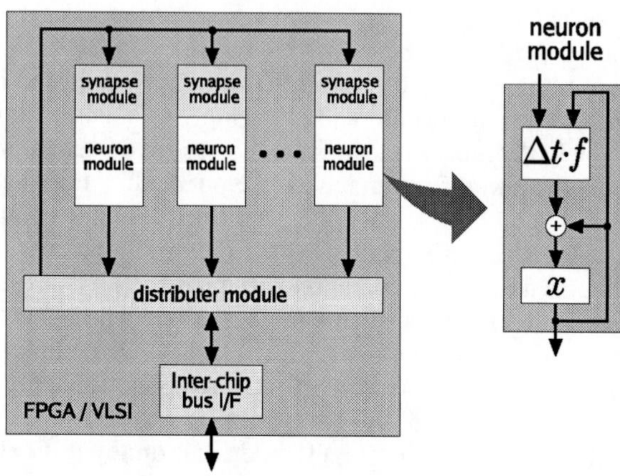

FIGURE 14. Block diagram of digital silicon neural network. Neuron module implements DSSN.

curves, the circuitry can be far simpler than conductance-based ones. The reproduction of mathematical structure instead of individual neuronal phenomena allows the silicon neuron to be far closer to biological ones than phenomenological ones. The designed silicon neuron model produces any neuronal phenomena supported by mathematical knowledge on it. Another advantage of our design method is that it can be applied to any device technologies that can produce nonlinearity. Actually, we introduced silicon neurons based on analog and digital circuits.

One goal of our study is to construct silicon neural networks that can operate with stability and execute some tasks. Our analog silicon neuron circuit for VLSI is very compact and consumes very low power. Analog circuit, however, is generally sensitive to noises, and it applies to our silicon neuron. We will have to make more try-and-errors to establish stable silicon neuron circuit that can operate appropriately in large-scale neural network. Currently, we are trying to construct small silicon neural network that can generate some rhythmic firing patterns similar to that of central pattern generator (CPG). CPG is a relatively small neural network that produces motion patterns in biological system. We still do not have and are working on silicon synapse circuit that interconnects silicon neurons. In contrast, digital circuit consumes relatively large power in comparison to our analog circuit, but it is quite resistant to noises. A large-scale recursive silicon neural network that can learn some firing patterns will be constructed in the near future. We are studying digital silicon synapse circuit that is capable of spike timing dependent plasticity (STDP), which will be incorporated into our silicon neural network.

ACKNOWLEDGMENTS

The circuit experiments in subsection "Circuit for aVLSI" in section "Silicon Neuron Circuits" was performed by Dr. Munehisa Sekikawa. This study is partially supported by Aihara Complexity Modelling Project, ERATO, JST, Japan.

REFERENCES

1. A. Laflaquiere, S. Le Masson, D. Dupeyron, and G. Le Messon, "Analog circuits emulating biological neurons in real-time experiments," *Proc. 19th Int. Conf. IEEE Engineering in Medicine and Biology Soc.*, pp. 2035–2038, 1997.
2. S. Wolpert, W. Friesen, and A. Laffely, "A silicon model of the hirudo swim oscillator," *IEEE Eng. Med. Biol. Mag.*, Vol. 19, pp. 64–75, 2000.
3. R. Jung, E. Brauer, and J. Abbas, "Real-time interaction between a neuromorphic electronic circuit and the spinal cord," *IEEE Trans. Neural Syst. Rehab. Eng.*, Vol. 9, pp. 319–326, 2001.
4. M. Simoni, G. Cymbalyuk, M. Sorensen, R. Calabrese, and S. DeWeerth "A Multiconductance Silicon Neuron With Biologically Matched Dynamics," *IEEE Trans. Biomed. Eng.*, Vol. 51, No. 2, pp. 342–354, Feb., 2004.
5. M. Sorensen, S. DeWeerth, G. Cymbalyuk, and R. Calabrese, "Using a Hybrid Neural System to Reveal Regulation of Neuronal Network Activity by an Intrinsic Current," *The Journal of Neuroscience*, Vol. 24, No. 23, pp. 5427-5438, 2004.
6. S. Renaud-Le Masson, G. Le Masson, L. Alvado, S. Saighi, J. Tomas, "A neural simulation system based on biologically realistic electronic neurons," *Information Sciences*, Vol 161, pp. 57–69, 2004.
7. C. Mead, Analog VLSI and Neural Systems. Reading, MA: Addison-Wesley, 1989.
8. K. Hynna and K. Boahen, "Space-rate coding in an adaptive silicon neuron," *Neural Networks*, Vol. 14, pp. 645-656, 2001.
9. G. Indiveri, "A low-power adaptive integrate-and-fire neuron circuit," *Proc. IEEE International Symposium on Circuits and Systems*, Vol. 4, IV-820–VI-823, May., IEEE, 2003.
10. D. Rubin, E. Chicca, and G. Indiveri, "Characterizing the firing properties of an adaptive analog VLSI neuron," *Proc. Bio-ADIT 2004, Lausanne*, pp. 314–327, 2004.
11. A. Hodgkin and A. Huxley, "Currents Carried by Sodium and Potassium Ions through the Membrane of the Giant Axon of Loligo," *J. Physiol.*, Vol. 116, pp. 449–472, 1952.

12. A. Hodgkin and A. Huxley, "A quantitative description of membrane current and its application to conduction and excitation in nerve," *J. Physiol.*, Vol. 117, pp. 500–544, 1952.
13. R. FitzHugh, "Thresholds and plateaus in the Hodgkin-Huxley nerve equations," *J. Gen. Physiology*, Vol. 43, pp. 867–896, 1960.
14. R. FitzHugh, *"Mathematical Models of Excitation and Propagation in Nerve," in "Biological Engineering",ed. H. P. Schwan*, pp. 1–85, McGraw-Hill, 1969.
15. E. M. Izhikevich, "Neural Excitability, Spiking, and Bursting," *International Journal of Bifurcation and Chaos*, Vol. 10, pp. 1171–1266, 2000.
16. J. Rinzel, "Excitation dynamics: insights from simplified membrane models," *Fed. Proc.*, Vol. 44, No. 15, pp. 2944–2946, Dec., 1985.
17. J. Rinzel and B. Ermentrout, *"Analysis of Neural Excitability and Oscillations," in "Methods in Neural Modeling", ed. C. Koch and I. Segev*, pp. 251–291, MIT Press, 1998.
18. A. Hodgkin, "The local electric changes associated with repetitive action in a non-medullated axon," *J. Physiol.*, Vol. 107, pp. 165–181, 1948.
19. H. Fujii and I. Tsuda, "Itinerant Dynamics of Class I* Neurons Coupled by Gap Junctions," *Computational Neuroscience: Cortical Dynamics Lecture Notes In Computer Science*, Vol. 3146, pp. 140-160, 2004.
20. M. Galarreta and S. Hestrin, "A network of fastspiking cells in the neocortex connected by electrical synapses," *Nature*, Vol. 402, pp. 72–75, 1999.
21. J. Gibson, M. Beierlein, and S. Hestrin, "Two networks of electrically coupled inhibitory neurons in neocortex," *Nature*, Vol. 402, pp. 75–79, 1999.
22. Takashi Kohno and Kazuyuki Aihara, "A MOSFET-based model of a Class 2 Nerve membrane," *IEEE Trans. Neural Networks*, Vol. 16, No. 3, pp. 754–773, May 2005.
23. Takashi Kohno and Kazuyuki Aihara, "Parameter tuning of a MOSFET-based nerve membrane," *Proc. Int. Symp. Artifical Life and Robotics 2005*, pp. 91–94, 2005.
24. Takashi Kohno and Kazuyuki Aihara, "Bottom-up design of Class 2 silicon nerve membrane," *J. Intelligent & Fuzzy Systems*, Vol. 18, No. 5, pp. 465–475, 2007.
25. Takashi Kohno and Kazuyuki Aihara, "Mathematical-model-based design method of silicon burst neurons," *in press.*
26. Takashi Kohno and Kazuyuki Aihara, "Digital Spiking Silicon Neuron: Concept and Behaviors in GJ-coupled Network," *Proc. Int. Symp. Artificial Life and Robotics 2007*, OS3-6, 2007.
27. G. Matsumoto, K. Aihara, Y. Hanyu, N. Takahashi, S. Yoshizawa, and J. Nagumo, "Chaos and phase locking in normal squid axons," Phys. Lett. A, vol. 123, pp. 162–166, 1987.
28. T. Chay and Y. Fan, "Bursting, Spiking, Chaos, Fractals, and Universality in Biological Rhythms," *International Journal of Bifurcation and Chaos*, Vol. 5, No. 3, pp. 595–635, 1995.
29. T. Chay, "Modeling Slowly Bursting Neurons via Calcium Store and Voltage-Independent Calcium Current," *Neural Computation*, Vol. 8, pp. 951–978, 1996.
30. Dan Hammerstrom, "Digital VLSI for Neural Networks," in *"The Handbook of Brain Theory and Neural Networks second ed.", ed. M. A. Arbib*, pp. 349–353, MIT Press, 2003.
31. *"FPGA Implementations of Neural Networks", ed. A. R. Omondi and J. C. Rajapakse*, Springer, 2006.

A Model for Axon Guidance: Sensing, Transduction and Movement

Giacomo Aletti*, Paola Causin† and Giovanni Naldi**

*Dipartimento di Matematica "F. Enriques", Università degli Studi di Milano, via Saldini 50, 20133 Milano, Italy, E-mail: giacomo.aletti@unimi.it
†Dipartimento di Matematica "F. Enriques", Università degli Studi di Milano, via Saldini 50, 20133 Milano, Italy, E-mail: causin@mat.unimi.it
**Dipartimento di Matematica "F. Enriques", Università degli Studi di Milano, via Saldini 50, 20133 Milano, Italy, E-mail: giovanni.naldi@mat.unimi.it

Abstract. Axon guidance by graded diffusible ligands plays a crucial role in the developing nervous system. In this paper, we extend the mathematical description of the growth cone transduction cascade of [1] by adding a model of the gradient sensing process related to the theory of [2]. The resulting model is composed by a series of subsystems characterized by suitable input/output relations. The study of the transmission of the noise-to-signal ratio allows to predict the variability of the gradient assay as a function of experimental parameters as the ligand concentration, both in the single and in the multiple ligand tests. For this latter condition, we address the biologically relevant case of silencing in commissural axons. We also consider a phenomenological model which reproduces the results of the experiments of [3]. This simple model allows to test hypotheses on receptor functions and regulation in time.

Keywords: Growth cone pathfinding, commissural axons, chemotaxis, gradient sensing, mathematical model, computational model.
PACS: 87.17

1. INTRODUCTION

In the developing nervous system, axons find the targets they will innervate navigating through the extracellular environment. Pathfinding crucially relies on chemical cues and, among the others, guidance by gradients of diffusible ligands plays a key role (see, *e.g.*, [4, 5, 6]). Detection and transduction of navigational cues is mediated by the growth cone (GC), a highly dynamic structure located at the axon tip [7, 8]. The cascade that leads to motility decisions is initiated by binding of the ligand with receptors located on the GC surface and filopodia, thin filaments that protrude from the distal part of the GC.

The standard benchmark chemotaxis assay studies *in vitro* the response of GCs exposed to steady graded concentrations of a single attractive/repulsive ligand [9, 10, 11]. Axon turning angles are measured after a certain time interval from the onset of the gradient. Different mathematical and computational models have been developed to model this phenomena. In [12] and in the successive paper [13], the differential receptor binding across the GC diameter is connected to the likelihood of generating new filopodia. Filopodia production is enhanced/inhibited in the angular sector facing the attractant/repellent source. This effect represents a positive feedback mechanism. The new orientation of the GC is a combination of the previous orientation plus a function of the actual angle of maximum receptor binding. The two contributions are weighted the 97%

CP1028, *Collective Dynamics: Topics on Competition and Cooperation in the Biosciences* edited by L. M. Ricciardi, A. Buonocore, and E. Pirozzi and 3%, respectively, thus introducing an inertial (memory) effect. In [14], a step toward the introduction of intracellular mechanisms is carried out by relating the angular distribution of filopodia to the angular variation of ionic calcium diffused in the periphery of the GC. In [15, 16], guidance is driven by steady–state diffusible chemoattractants and chemorepellants, as well as by homophilic axon–to–axon attraction, for which a diffusive mechanism is supposed to exist as well. In [17], a system of ordinary differential equations describes the deterministic macroscopic motion of the GC and the dynamical evolution of its internal state, respectively.

In this article, we consider a model which extends the one proposed in [1]. The chemotactic GC system is described as a series of functional subsystems, ranging from gradient sensing to signal transduction, down to motion actuation. A characteristic time is singled out for each subsystem, representing the fact that independent concentration measures by receptors, internal reorganization preceding motion and discernible axon turning act on separated temporal scales, from the smaller to the larger one. The mathematical model describes input/output relations of signals of each functional subsystem, without reproducing intracellular chemical processes. The biological situation we address is the *in vitro* exposure to multiple diffusible cues, which interaction substantially modifies the GC response. This setting is representative of the case of commissural axons (see Fig. 1). In a first phase, these axons are attracted to the nervous system midline by a gradient of the protein netrin−1, but, after crossing the midline, receptors for the repellent Slit protein are upregulated and loss of response to netrin occurs, despite the fact that expression of the receptor for this latter ligand is maintained. This sequence of events leads axons to definitely depart from the midline to which they were attracted before (see [18, 19, 20, 21]). At our knowledge, guidance for commissural axons has been mathematically dealt with only in the paper [22], but a different aspect than the present work has been considered there. Namely, in [22] a theoretical model is proposed to explain sorting of commissural axons after crossing the midline due to the expression of different subfamilies of Slit receptors. Here, we focus rather on the hypothesis of [3] that, in *Drosophila* but likely in vertebrates as well, the abrupt change of behavior of commisural axons is due to a gating effect of Slit receptors belonging to the Rondabout family [23] (Robo) on netrin receptors belonging to the Deleted in Colorectal Cancer family [24] (DCC). In [3], turning angles of axons after 1h of *in vitro* exposure to a gradient of netrin–1, Slit, or netrin–1 combined with Slit were measured. The results show that at *Drosophila* developmental stage 22, netrin–1 causes a net attraction of axons towards the source, while Slit as well as netrin–1 combined with Slit do not produce significant turnings. Moreover, at developmental stage 28, axons do not seem to be responsive any more to netrin–1, whilst they are strongly repelled by Slit and by netrin–1 combined with Slit. The conceptual idea to explain this behavior is that the silencing effect of Robo receptors on DCC receptors is partly of entirely responsible for the loss of responsiveness of commissural axons to netrin–1 (see, in particular, [3, Discussion]). This amounts to say that the different responsiveness is to be related to events occurring in the very early stages of the transduction chain.

As a matter of fact, mechanisms like the one illustrated above are still far to be completely unveiled. For example, how precisely ligand–receptor binding is converted into an intracellular signal is still a research issue. At present, receptor activation can be only monitored by observing biological responses, such as changes in neurite outgrowth [23].

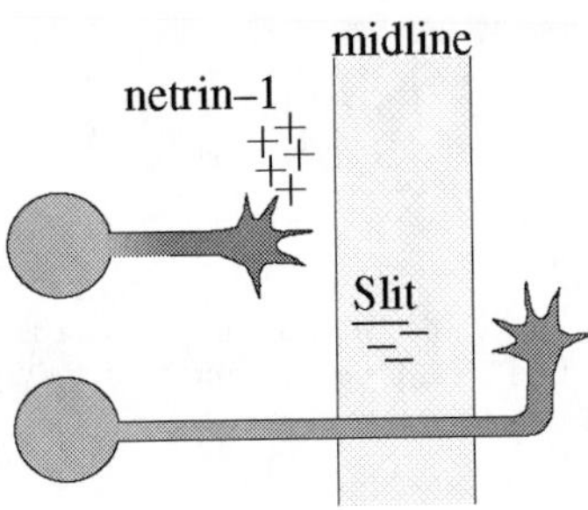

FIGURE 1. Commissural axons are first attracted by netrin proteins secreted by the nervous midline cells. After crossing the midline, Slit receptors are upregulated and loss of response to netrin occurs, despite maintaining expression of DCC receptors. Axons are eventually repelled by the midline [18, 19, 20, 21].

This motivates the use of theoretical and computational models. In particular, the emerging area of the analysis of the cellular transduction system as a device that has to make decisions based on imperfect information about the environment can provide hints about the characteristics of the hidden processes of the transduction process. Imperfect information arises due to fluctuations in the field signalling molecules as well as throughout the entire GC intracellular network (for a discussion on this topic in eukaryotic cells or bacteria chemotaxis, see, *e.g.*, [25, 26, 27, 28]). In [1], a study has been carried out on the propagation throughout the GC transduction cascade of the main statistical indexes relating signal and noise in guidance. Here, we extend this technique introducing a more detailed modeling of ligand–receptor binding and relating this process to the transduction mechanism. The study of the transmission of the noise-to-signal ratio allows, on the one hand, to predict the variability of the gradient assay as a function of experimental parameters as the ligand concentration, both in the single and in the multiple ligand tests. Experimental settings are indicated that produce most significant results (for example in the differentiation of responses). On the other hand, this approach provides mathematical tools to address the issue of whether receptor silencing in commissural axons can explain loss of organized turning response.

In this work, we also consider a phenomenological model which reproduces the macroscopic outcome (turning angles) of the experiments of [3] on commissural axons. This simple model allows to test hypotheses on receptor functions and regulation in time.

The rest of the paper is organized as follows. In Sect. 2, we illustrate the model adopted for describing the axon chemotaxis, discussing the mathematical representation of the Sensing Device, Intracellular Transduction and Motor Actuator subsystems. In Sect. 3, we introduce the statistical indexes that will be used to characterize the system performance. In Sect. 4, we perform numerical simulations of the single and multiple ligand chemotactic assays, and we discuss the results. In Sect. 5, we present a simplified phenomenological model of axon response in presence of multiple cues that macroscopically reproduces the behavior of commissural axons.

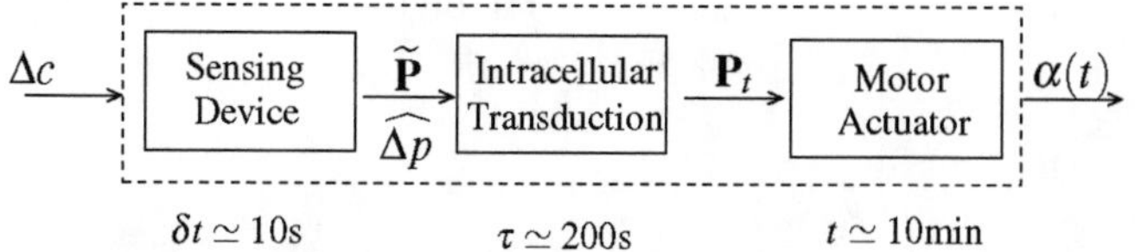

FIGURE 2. Functional subsystems of the GC transduction cascade: Sensing Device, Intracellular Transduction and Motor Actuator functions. Input and output quantities are defined later in the paper. Characteristics times are indicated under the respective box.

2. MODEL OF AXON CHEMOTAXIS

The model of axon chemotaxis we consider, introduced in [1], provides a synthetic mathematical representation of the transduction cascade of the GC. Different subsystems are identified, which lead from sensing of ligand concentration gradients to motion (see Fig. 2). The model is especially tailored for studying *2D in vitro* gradient assays, but it can be easily extended to deal with the *3D in vivo* conditions. Measures of concentration differences in the environment are produced by the Sensing Device Subsystem (SDSys). The Intracellular Transduction Subsystem (ITSys) processes the input from the SDSys producing a signal which, through the Motor Actuator Subsystem (MASys), causes the deviation of the GC trajectory. The gradient sensing process takes place in a time of the order of tenths of seconds, signal transduction and internal reorganization in a time of the order of a few minutes, trajectory deviations in a time of the order of tenth of minutes.

2.1. Model of the Gradient Sensing function

Gradient sensing in axon guidance seems to fit a spatial mechanism (on this issue, see the discussion of [29]): GCs compare spatial differences in ligand concentration (*e.g.*, front versus rear part) to determine the direction and the intensity of the gradient. The model we propose stems from the work of Berg and Purcell [2] on small sensing devices. Receptors, that play the role of sensing devices, are distributed all along the GC surface and the filopodia. Here we suppose that N_1 receptors are concentrated on the side of the GC facing the ligand source and N_2 receptors lie on the other side (see Fig. 3, left). If each receptor has a binding site capable of binding one molecule of ligand at a time, the expected time average occupation $\overline{p}$ of the receptor itself is linked to the local ligand concentration c by

$$\overline{p} = \frac{c}{c + k_D}, \tag{1}$$

where k_D is the ligand dissociation constant. The history of the i-th site located on side $j = 1$ or $j = 2$ is described by a function $p_j^{(i)}(t)$ that assumes value 1 when the site is occupied and 0 when it is empty. The information about the surrounding concentration is then represented by the processes $p_j^{(i)}(t)$ recorded for a sampling time δt. An

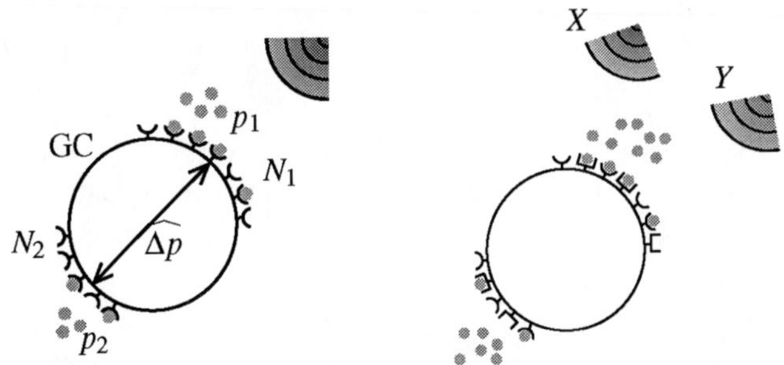

FIGURE 3. Left: a gradient of chemoattractant is established across the GC sides 1 and 2. The binding state (time average occupation) of the N_1 and N_2 receptors provides an estimate of the concentration difference. Right: ligands X and Y are present at the same time and they bind to their respective receptors (represented with a different symbol). Hierarchical interaction of receptors may take place and alter the GC turning response.

approximation of $\overline{p}$ on each side of the GC may then be obtained as

$$p_j = \frac{1}{N_i \delta t} \int_{\bar{t}}^{\bar{t}+\delta t} \sum_{i=1}^{N_j} p_j^{(i)}(t)\, dt = \frac{1}{N_i \delta t} \sum_{i=1}^{N_j} \int_{\bar{t}}^{\bar{t}+\delta t} p_j^{(i)}(t)\, dt, \qquad j = 1,2. \tag{2}$$

The difference in time occupancy $\widehat{\Delta p} = p_1 - p_2$ provides an estimate of the difference of concentration across sides 1 and 2. In Sect. 2.1.1 and 2.1.2, we provide a model for the processes $p_j^{(i)}$ in the case of single and multiple ligands, respectively.

2.1.1. The receptor binding process with one ligand

The binding of each receptor to a molecule of ligand is assumed as in [2] to be a continuous Markov chain on the state space $S = \{f, b\}$ (f =unbound, b =bound) with transition rate matrix

$$Q = \begin{pmatrix} -\frac{1}{\tau_f} & \frac{1}{\tau_f} \\ \frac{1}{\tau_b} & -\frac{1}{\tau_b} \end{pmatrix}, \tag{3}$$

where τ_b (resp.τ_f) is the average time the receptor remains bound to (resp. unbound from) a molecule of ligand. The time τ_b is estimated in [2, Eq.44] as

$$\tau_b = (4Dsk_D)^{-1}, \tag{4}$$

D being the ligand diffusion constant and s the effective radius of the receptor. The state of a receptor is supposed to be statistically independent on the processes taking place at the other binding sites. A receptor switching from state f to b (resp. from state b to f) will remain in that state for a random time distributed as an exponential law of parameter τ_b (resp. τ_f), independent on its previous history. The characteristic times τ_f and τ_b are related to the average occupancy via the Ergodic Theorem $\overline{p} = \tau_b/(\tau_b + \tau_f)$.

2.1.2. *The receptor binding process with multiple ligands: silencing effect*

When more ligands are present but non interacting, the binding processes are independent. Each process can be studied as in indicated in the previous section. We consider in the following the case of two ligands X and Y. The transition matrix (3) of the joint process becomes

$$Q_{XY} = \begin{pmatrix} -(\frac{1}{\tau_f^Y}+\frac{1}{\tau_f^X}) & \frac{1}{\tau_f^X} & \frac{1}{\tau_f^Y} & 0 \\ \frac{1}{\tau_b^X} & -(\frac{1}{\tau_f^Y}+\frac{1}{\tau_b^X}) & 0 & \frac{1}{\tau_f^Y} \\ \frac{1}{\tau_b^Y} & 0 & -(\frac{1}{\tau_b^Y}+\frac{1}{\tau_f^X}) & \frac{1}{\tau_f^X} \\ 0 & \frac{1}{\tau_b^Y} & \frac{1}{\tau_b^X} & -(\frac{1}{\tau_b^Y}+\frac{1}{\tau_b^X}) \end{pmatrix}, \tag{5}$$

with state space $S = \{f_X f_Y, b_X f_Y, f_X b_Y, b_X b_Y\}$. If one receptor of type Y in bound state can silence one receptor of type X in bound state, in the resulting process, the state $b_X b_Y$ is then split into the two states $b_X^s b_Y$ and $b_X^u b_Y$, where the apex s (resp. u) stands for silenced (resp. unsilenced). The time that would have been spent in the state $b_X b_Y$ in the non interacting case is now spent switching between the two states $b_X^s b_Y$ and $b_X^u b_Y$ with exponential laws of time parameter τ_{on} and τ_{off}, respectively. The corresponding transition matrix on the state space $S = \{f_X f_Y, b_X f_Y, f_X b_Y, b_X^s b_Y, b_X^u b_Y\}$ reads

$$Q_{XY}^{\text{int}} = \begin{pmatrix} -(\frac{1}{\tau_f^Y}+\frac{1}{\tau_f^X}) & \frac{1}{\tau_f^X} & \frac{1}{\tau_f^Y} & 0 & 0 \\ \frac{1}{\tau_b^X} & -(\frac{1}{\tau_f^Y}+\frac{1}{\tau_b^X}) & 0 & 0 & \frac{1}{\tau_f^Y} \\ \frac{1}{\tau_b^Y} & 0 & -(\frac{1}{\tau_b^Y}+\frac{1}{\tau_f^X}) & 0 & \frac{1}{\tau_f^X} \\ 0 & \frac{1}{\tau_b^Y} & \frac{1}{\tau_b^X} & -(\frac{1}{\tau_b^Y}+\frac{1}{\tau_b^X}+\frac{1}{\tau_{\text{on}}}) & \frac{1}{\tau_{\text{on}}} \\ 0 & \frac{1}{\tau_b^Y} & \frac{1}{\tau_b^X} & \frac{1}{\tau_{\text{off}}} & -(\frac{1}{\tau_b^Y}+\frac{1}{\tau_b^X}+\frac{1}{\tau_{\text{off}}}) \end{pmatrix}. \tag{6}$$

2.2. Model of the Intracellular Transduction

Intracellular transduction is a highly complex network. Here, we do not consider single physical processes, but we directly model the input/output relation of the subsystem. A gradient of chemoattractant (resp. chemorepellant) concentration orients the GC motion toward (resp. away from) the direction of the concentration source. According to a mechanical description, we ascribe the trajectory deviation to an equivalent action vector $\mathbf{P}_t$. This latter quantity is continuously compared against a vector $\widetilde{\mathbf{P}}$ produced by the SDSys, directed along the maximum gradient and related to the difference in receptor occupancy $\widehat{\Delta p}$ (this latter issue will be addressed more in detail later). The output of the ITSys results from the new information $\widetilde{\mathbf{P}}$ and from a memory effect which damps the response

$$\mathbf{P}_{t+\delta t} = (1-\lambda)\mathbf{P}_t + \lambda(\widetilde{\mathbf{P}} + \widetilde{\eta}) \tag{7}$$

where λ is a weighting factor (memory effect, see also [13, Sect. Mathematical models]), $\widetilde{\mathbf{P}} = \widehat{\mathbf{P}} + \sigma_s\sqrt{\frac{2}{\lambda}}\mathbf{Z}_s$, $\widetilde{\eta} = \sigma_\eta\sqrt{\frac{2}{\lambda}}\mathbf{Z}_\eta$, where $\widehat{\mathbf{P}} = \mathbb{E}(\widetilde{\mathbf{P}})$, $\mathbf{Z}_s$ and $\mathbf{Z}_\eta$ are two–dimensional standardized random vectors independent of $\mathbf{P}_t$, σ_s and σ_η are volatility parameters and the factor $\sqrt{2/\lambda}$ appears due to a normalization choice. The contribution $\widetilde{\eta}$ represents a noise term purely coming from the intracellular transduction mechanism and it is not dependent on the external concentration field. The time interval δt is chosen to be long enough so that the new contribution $\widetilde{\mathbf{P}} + \widetilde{\eta}$ can be assumed statistically independent from $\mathbf{P}_t$.

Let $\lambda = \dfrac{\delta t}{\tau}$, τ being a persistence time characteristic of internal signal transduction. Then, Eq. 7 can be reformulated as

$$\delta\mathbf{P} = -\frac{\mathbf{P}_t - \widehat{\mathbf{P}}}{\tau}\,\delta t + \sigma\sqrt{\delta t}\sqrt{\frac{2}{\tau}}\mathbf{Z}, \tag{8}$$

where $\sigma = \sqrt{\sigma_s^2 + \sigma_\eta^2}$ is the process volatility. The term $\delta\mathbf{P} = \mathbf{P}_{t+\delta t} - \mathbf{P}_t$ represents an incremental "kick" on the trajectory (see also [30]). Since the characteristic time scale of ITSys is larger than δt, we can adopt as a model of the input/output relation of this subsystem the following continuous generalized Ornstein–Uhlenbeck (OU) process, which is assumed to obey to Ito calculus (see, *e.g.*, [31])

$$d\mathbf{P}_t = -\frac{\mathbf{P}_t - \widehat{\mathbf{P}}}{\tau}dt + \sigma\sqrt{\frac{2}{\tau}}\,d\boldsymbol{W}_t, \tag{9}$$

where $\boldsymbol{W}_t$ denotes a two–dimensional Wiener process. When $\widehat{\mathbf{P}}$ does not depend on time, σ is constant and the solution of (9) reads

$$\mathbf{P}_t = \widehat{\mathbf{P}} + (\mathbf{P}_0 - \widehat{\mathbf{P}})e^{-t/\tau} + \int_0^t \sigma\sqrt{\frac{2}{\tau}}e^{(s-t)/\tau}d\boldsymbol{W}_s. \tag{10}$$

Relation (10) shows that the mean value of $\mathbf{P}_t$ tends exponentially fast in time to $\widehat{\mathbf{P}}$. Moreover, Ito's lemma [31] implies that, at steady–state, $\mathbf{P}_t$ is a bivariate Gaussian distribution subjected to an isotropic random perturbation of bounded variance σ^2.

2.3. Model of the Motor Actuation

We suppose that the action $\mathbf{P}_t$ coming from Eq. (9) induces an acceleration only along the direction transversal to the trajectory (see also [17] for a similar hypothesis). The law of the GC motion can be written as:

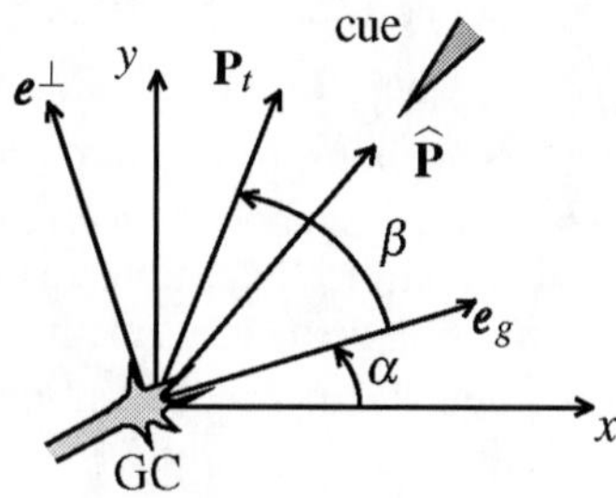

FIGURE 4. Notation for the Motor Actuation subsystem.

find for $0 \leq t \leq T$ the GC position $\boldsymbol{x}_g = \boldsymbol{x}_g(t)$, such that

$$\begin{aligned} \dot{\boldsymbol{x}}_g &= \boldsymbol{v}_g, \\ \dot{\boldsymbol{v}}_g &= |\mathbf{P}_t| \sin\beta \boldsymbol{e}^{\perp}, \\ d\mathbf{P}_t &= -\frac{\mathbf{P}_t - \widehat{\mathbf{P}}}{\tau} dt + \sigma\sqrt{\frac{2}{\tau}} d\boldsymbol{W}_t, \\ \boldsymbol{x}_g(0) &= \boldsymbol{x}_g^0, \\ \boldsymbol{v}_g(0) &= v_g \mathbf{e}_g^0, \\ \mathbf{P}_0 &= \mathbf{P}^0, \end{aligned} \tag{11}$$

where $\boldsymbol{v}_g = v_g \boldsymbol{e}_g$ is the axon velocity vector ($v_g = 20 - 30$ μm/h, [32]), $\boldsymbol{e}^{\perp}$ is the unit vector perpendicular to $\boldsymbol{e}_g$, $\boldsymbol{x}_g^0$ and v_g^0 are the given initial position and direction of the axon GC, $\widehat{\mathbf{P}}^0$ is the given initial equivalent force, m is the GC equivalent mass, α is the angle that $\boldsymbol{e}_g$ forms with the horizontal direction and β is the angle between $\boldsymbol{e}_g$ and $\mathbf{P}_t$ (see Fig. 4).

Notice that in system (11), the effect of external gradient only concerns the direction of the unit vector $\boldsymbol{e}_g$, leaving the velocity modulus constant. Moreover, notice that when $\widehat{\mathbf{P}} = \mathbf{0}$, the trajectory will not be deterministically deviated. This reproduces the physical fact that, in absence of ligand gradients and at least on *in vitro* experiments, axons tend to follow noised trajectories with no significant bias from their initial growth direction [33].

Eq. $(11)_3$ models the ITSys input/output relation, while Eq. $(11)_{1,2}$ model the MASys input/output relation. Eqs. (11) constitute a stochastic differential system and can be numerically solved by using a stochastic Runge–Kutta integration scheme (see for example [34]). Parameters of the model and their meaning are given in Table 1.

TABLE 1. Model parameters and their meaning. For parameter values used to derive numerical results, see text.

Parameter	Meaning
$p_j^{(i)}$	time occupancy process of i-th receptor on side j of the GC
$\widehat{\Delta p}$	estimate of difference in receptor time occupancy across the GC
D	Diffusion constant
s	Effective receptor radius
k_D	Dissociation constant
σ	Volatility parameter
τ_b, τ_f	average time of receptor in bound or free state
$\widetilde{\mathbf{P}}$	New contribution due to the external gradient
$\widehat{\mathbf{P}}$	Expected value of $\widetilde{\mathbf{P}}$
$\mathbf{P}_t$	Actual value of the contribution due to the external gradient with memory effect
$\widetilde{\boldsymbol{\eta}}$	Noise due to the intracellular transduction
λ	Memory effect parameter
δt	Characteristic time of gradient sensing
τ	Characteristic time of intracellular transduction
$\boldsymbol{v}_g$	GC velocity vector (v_g velocity modulus, $\boldsymbol{e}_g$ velocity direction)
α	Angle between the GC direction and the x-axis
β	Angle between the GC direction and $\mathbf{P}_t$
γ	Measure of turning angle in benchmark experiments

3. STUDY OF THE COEFFICIENTS OF VARIATION IN THE TRANSDUCTION CHAIN

In this section, we introduce statistical indexes that allow to characterize the degree of organization of the signal throughout the steps of the chemotactic system. The macroscopically observable quantities (as the turning angles in the gradient assay) allow to compute the indexes pertaining to the output of the last part of the chain. Starting from these values and using the mathematical model, we compute the statistical indexes of the earlier compartments of the system. In particular, we use the coefficient of variation, defined as the ratio between the standard deviation and the expected value of a stochastic distribution, to assess the weight of the fluctuating over the deterministic part of a signal arising from a subsystem input/output relation.

Experimentally recorded distributions of turning angles γ provide data to estimate the coefficient of variation

$$\mathrm{CV}_\gamma = \frac{\mathrm{std}(\gamma)}{\mathbb{E}(\gamma)}, \tag{12}$$

which represents an information about the degree of organization of the GC macroscopic behavior. When $\mathrm{CV}_\gamma \gg 1$, noise prevails on the signal and the motion is just a random walk. When $\mathrm{CV}_\gamma \ll 1$, noise plays a very minor role and the motion is a deterministic path. When CV_γ approaches values of the order of the unity, which is generally the case in GC chemotactic assays, noise and signal have almost the same weight. Local fluctuations do exist, but trajectories show a significant and detectable bias.

We now proceed back in the chain, relating CV_γ to the coefficient of variation of $\mathbf{P}_t$, (which is an hidden process) as

$$\mathrm{CV}_{\mathbf{P}_t} \approx \sqrt{\frac{t}{2\tau}} \mathrm{CV}_\gamma, \tag{13}$$

τ being the time parameter of the ITSys (see [1] for a detailed derivation of Eq. (13)). Observing that

$$\mathbb{E}(\widetilde{\mathbf{P}}+\widetilde{\eta}) = \mathbb{E}(\widetilde{\mathbf{P}}) = \mathbb{E}(\mathbf{P}_t), \qquad \mathrm{Var}(\widetilde{\mathbf{P}}+\widetilde{\eta}) = \frac{2\tau}{t}\sigma^2 = \frac{2\tau}{t}\mathrm{Var}(\mathbf{P}_t), \tag{14}$$

we further have

$$\mathrm{CV}_{\widetilde{\mathbf{P}}+\widetilde{\eta}} = \frac{\mathrm{std}(\widetilde{\mathbf{P}}+\widetilde{\eta})}{|\mathbb{E}(\widetilde{\mathbf{P}})|} = \sqrt{\frac{2\tau}{\delta t}} \mathrm{CV}_{\mathbf{P}_t} \approx \sqrt{\frac{t}{\delta t}} \mathrm{CV}_\gamma, \tag{15}$$

where the time scale factor $\sqrt{t/\delta t} = \sqrt{1/\lambda}$ keeps into account the memory effect which weights the signal $\widetilde{\mathbf{P}}+\widetilde{\eta}$ in $\mathbf{P}_t$ (see Eq. (7)). We can now investigate the properties of the sensing function. With this aim, we introduce the quantity

$$\ell^2 := \frac{\mathrm{Var}_{\widetilde{\mathbf{P}}+\widetilde{\eta}}}{\mathrm{Var}_{\widetilde{\mathbf{P}}}} = 1 + \frac{\mathrm{Var}_{\widetilde{\eta}}}{\mathrm{Var}_{\widetilde{\mathbf{P}}}}, \tag{16}$$

where the second relation at the right hand stems from the statistical independence of $\widetilde{\mathbf{P}}$ and $\widetilde{\eta}$. Observe that $\ell^2 - 1 = \frac{\mathrm{Var}_{\widetilde{\eta}}}{\mathrm{Var}_{\widetilde{\mathbf{P}}}}$ represents the ratio between the variability of the intracellular transduction subsystem and the variability of the gradient sensing subsystem. Using relations (15), (16) and the properties (14), we get

$$\mathrm{CV}_{\widetilde{\mathbf{P}}} = \frac{\mathrm{CV}_{\widetilde{\mathbf{P}}+\widetilde{\eta}}}{\ell} \approx \frac{1}{\ell}\sqrt{\frac{t}{\delta t}} \mathrm{CV}_\gamma, \tag{17}$$

which connects the experimentally observable coefficient of variation CV_γ with the coefficient of variation of the sensing process $\mathrm{CV}_{\widetilde{\mathbf{P}}}$. A link can be further drawn between the statistical indexes of $\widetilde{\mathbf{P}}$ and the processes of single receptors introduced in Sect.2.1. Let σ_1 and σ_2 be the variances of a typical $\frac{1}{\delta t}\int p_1^{(i)}dt$ or $\frac{1}{\delta t}\int p_2^{(i)}dt$ process on side 1 or 2 of the GC. We have that $\mathrm{Var}_{\widehat{\Delta p}} = \mathrm{Var}_{p_1} + \mathrm{Var}_{p_2} = \sigma_1^2/N_1 + \sigma_2^2/N_2$, and hence

$$\mathrm{CV}_{\widehat{\Delta p}} = \frac{\sqrt{\sigma_1^2/N_1 + \sigma_2^2/N_2}}{\mathbb{E}(\widehat{\Delta p})}.$$

When $N_1 = N_2 = N$, the above relation reads

$$\mathrm{CV}_{\widehat{\Delta p}} = \frac{1}{\sqrt{N}} \frac{\sqrt{\sigma_1^2 + \sigma_2^2}}{\mathbb{E}(\widehat{\Delta p})} \propto \frac{1}{\sqrt{N}}. \tag{18}$$

We assume that $\widetilde{\mathbf{P}} \propto \widehat{\Delta p}$. Then, by Eq. (17), we can further obtain

$$\ell = \sqrt{\frac{Nt}{\delta t}} \frac{\mathbb{E}(\widehat{\Delta p})}{\sqrt{\sigma_1^2 + \sigma_2^2}} \mathrm{CV}_\gamma. \tag{19}$$

Numerical simulations based on the gradient sensing model can be used to evaluate quantities referred to $\widehat{\Delta p}$ (see Sect. 4).

3.1. Statistical indexes for different experimental settings

The quantity CV_γ is macroscopically observable, but, generally, data are reported in literature in correspondence of specific experimental conditions. At our knowledge, only in [35] a systematic study is carried out testing the axon response for different concentrations and gradient steepness.

A sole fixed ligand concentration close to k_D is instead used in [3] in the gradient assays. For example, with ligand netrin–1 at stage 22, the value $\mathrm{CV}_\gamma \simeq 0.25$ is obtained [3, Fig.2B]. By Eq. (19), we can compute

$$\ell_{|k_D} = \sqrt{\frac{tN}{\delta t}} \frac{\mathrm{CV}_{\gamma_{|k_D}}}{\mathrm{CV}(\widehat{\Delta p}_{|k_D})} \simeq 5\sqrt{N} \frac{|\mathbb{E}(\widehat{\Delta p}_{|k_D})|}{\sqrt{\sigma^2_{1\,|k_D} + \sigma^2_{2\,|k_D}}}. \tag{20}$$

As a matter of fact, using Eq. (15) and (20), the statistical indexes of different experimental settings can be predicted. For example, we can consider the case where a concentration of ligand different from k_D is considered,

$$\left(\frac{\mathrm{CV}_{\gamma_{|x}}}{\mathrm{CV}_{\gamma_{|k_D}}}\right)^2 = \frac{\mathrm{Var}_{\widetilde{\mathbf{P}}+\widetilde{\eta}\,|x}}{\mathrm{Var}_{\widetilde{\mathbf{P}}+\widetilde{\eta}\,|k_D}} \left(\frac{\mathbb{E}(\widetilde{\mathbf{P}}_{|x})}{\mathbb{E}(\widetilde{\mathbf{P}}_{|k_D})}\right)^2$$

$$= \frac{\frac{\mathrm{Var}_{\widetilde{\mathbf{P}}|x}}{\mathrm{Var}_{\widetilde{\mathbf{P}}|k_D}} + (\ell^2_{|k_D} - 1)}{\ell^2_{|k_D}} \left(\frac{\mathbb{E}(\widetilde{\mathbf{P}}_{|x})}{\mathbb{E}(\widetilde{\mathbf{P}}_{|k_D})}\right)^2. \tag{21}$$

We can also consider the case where arbitrary concentrations x and y of ligand of type X and Y are present and compute the ratio $\mathrm{CV}_{\gamma_{|(x,y)}}/\mathrm{CV}_{\gamma_{|(x,0)}}$, which allows to quantify the effect of silencing with respect to the unsilenced case for arbitrary concentrations.

4. NUMERICAL SIMULATIONS

In this section, we carry out numerical simulations using the mathematical model introduced in the Sect 2. Then, we compute the statistical indexes of the input/output relations of the subsystem as discussed in Sect. 3.

4.1. The single ligand case

We study the properties of the sensing process $\widehat{\Delta p}$. We set in the model of Sect. 2.1.1 $\tau_b = 0.83$s, concentration at the center of the GC ranging from 0.1nM to 100nM and gradient steepness of 2% across the GC diameter. The value of τ_b is computed from Eq. (4) considering the netrin–1 parameters $D = 10^{-7}$cm^2/s [36], $k_D = 10$nM [24, 37] and $s = 10$Å [29].

In Fig. 5, left, we plot the coefficient of variation $\mathrm{CV}_{\widehat{\Delta p}}$ as a function of the ligand concentration. The variance is alternatively computed from:

1. the status $(0,1)$ of $N_1 = N_2 = 2000$ receptors for a number of binding events $\delta t/(\tau_b + \tau_f)$ (red bars)
2. the approximation of Berg and Purcell [2, Eq.(50)] ($N_1 = N_2 = 2000$ receptors, each with variance proportional to $\overline{p}(1-\overline{p})^2$, orange bars)
3. a MonteCarlo simulation (500,000 simulations) of the model of Sect. 2.1.1 (yellow bars)

We superpose in the same graph the quantity $\mathbb{E}(\widehat{\Delta p}) = \mathbb{E}(p_1) - \mathbb{E}(p_2)$, which is analytically computed from the non–negative eigenvector of the transition matrix of the process, related to the probability at steadiness of each receptor to be in a certain state [31]. In Fig. 5, right, we plot the ratio $\mathrm{CV}_{\gamma|_x}/\mathrm{CV}_{\gamma|_{k_D}}$ obtained from Eq. (21). The plot is parametrized by the number of receptors (see the legend).

4.2. The multiple ligand case

We simulate the processes $p^{(i)}, i = 1,\ldots,N_1+N_2$ for the receptors for ligand X under the silencing effect of receptors for ligand Y. We use the model in Sect. 2.1.2 with $\tau_b^X = \tau_b^Y = 0.83$s, concentration at the center of the GC ranging from 0.1nM to 100nM for both ligands and gradient steepness of 2% across the GC diameter. The value of τ_b^X and τ_b^Y are computed from Eq. (4) considering for both netrin and Slit $D = 10^{-7}$cm^2/s, $k_D = 10$nM and $s = 10$Å.

In Fig. 6 (resp. Fig. 7), we plot the ratio $\mathrm{CV}_{\gamma|_{(x,y)}}/\mathrm{CV}_{\gamma|_{(k_D,0)}}$ (resp. $\mathrm{CV}_{\gamma|_{(x,y)}}/\mathrm{CV}_{\gamma|_{(x,0)}}$) obtained from Eq. (21). The top graduation of each panel represents Slit concentration, while the bottom graduation represents netrin–1 concentration. Dependence on the number of receptors $N_1^X = N_2^X = N_1^Y = N_2^Y$ is shown in each group of bars (same legend as in Fig. 5, right panel).

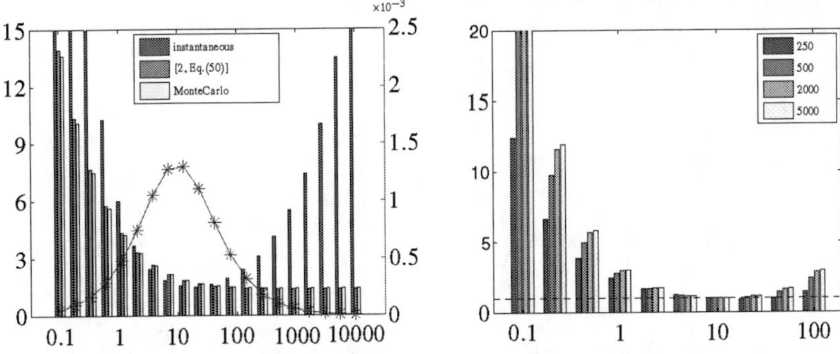

FIGURE 5. Left: coefficient of variation $CV_{\widehat{\Delta p}}$ with variance computed from: *i)* the instantaneous binding status $(0,1)$ of $N_1 = N_2 = 2000$ receptors for $n_m = \delta t/(\tau_b + \tau_f)$ times (red bars); *ii)* the approximation of Berg and Purcell [2, Eq.(50)] ($N_1 = N_2 = 2000$ receptors, each with variance proportional to $\overline{p}(1-\overline{p})^2$, orange bars); *iii)* a MonteCarlo simulation (500,000 simulations) of the model of Sect. 2.1.1 (yellow bars). Right: ratio $CV_{\gamma_{|x}}/CV_{\gamma_{|k_D}}$ obtained from Eq. (21). The plot is parametrized by the number of receptors (see the legend). Both panels have in abscissa ligand concentration c_m (in nM, logarithmic scale).

4.3. Discussion of the results

We have analyzed the performance of the gradient sensing subsystem by introducing a model based on the Berg and Purcell theory. Following this model, the maximum expected value of the output signal is attained at a concentration value corresponding to the dissociation constant (see Eq. (1)). However, the noise-to-signal ratio represented by $CV_{\widehat{\Delta p}}$ is monotonically decreasing (yellow bars of Fig. 5, left panel). An explanation of this trend may be found in the fact that, while τ_b is a chemical property independent on the ligand concentration, τ_f is instead a function of the concentration. The time $\tau_b + \tau_f$ represents the average time between two consecutive binding events. The quantity $\delta t/(\tau_b + \tau_f)$ relates to the number of binding events during time δt. Then, if one records the binding status of each receptor $\delta t/(\tau_b + \tau_f)$ times independently or performs a time average of the binding processes, very different behaviors are obtained (red bars vs. yellow bars of Fig. 5, left panel). In fact, the variance of a single binding measure is proportional to $\overline{p}(1-\overline{p})$ (Bernoulli random variable), while the variance of the time average measure is proportional to $\overline{p}(1-\overline{p})^2$ (orange bars of Fig. 5, left panel, see also [2]).

As for the outcome of the entire system, *in vitro* experiments suggest that the highest guidance is observed for a ligand concentration equal to k_D (see, *e.g.*, [35, Fig.3c]), for which the most organized motion (most efficient response) is displayed. The ratio $CV_{\gamma_{|x}}/CV_{\gamma_{|k_D}}$ obtained from MonteCarlo simulations reproduces this property (see Fig. 5, right). The lack of monotonicity of this curve is due to the effect of the intracellular noise $\widetilde{\eta}$, that we have supposed to be independent on the concentration field, produced in the transduction step. In the neighborhood of the dissociation constant, the sensing process signal is high, as well its variability ($CV_{\widehat{\Delta p}} \approx 1$). For ligand concentrations greater than the dissociation constant, the strength of the signal decreases, but the

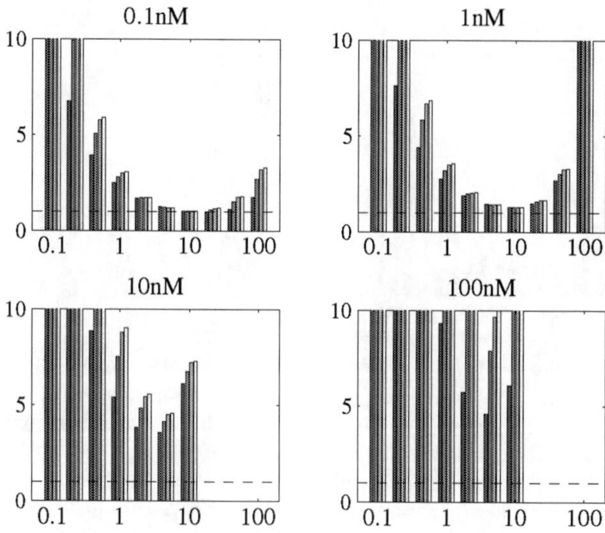

FIGURE 6. Ratio $\mathrm{CV}_{\gamma|_{(x,y)}}/\mathrm{CV}_{\gamma|_{(k_D,0)}}$. The top graduation of each panel represents Slit concentration, while the bottom graduation represents netrin–1 concentration (values in nM, logarithmic scale). Each group of bars contains results for different number of receptors (the same convention as in Fig. 5, right panel, is used). Groups of bars that do not appear, as in the rightmost part of the two panels of the bottom row, indicate that every form of organized directionality is lost.

coefficient of variation remains of the same order. This means that the variability has decreased. The contribution of the constant factor $\widetilde{\eta}$ impacts more significantly for $c > k_D$ than for $c \simeq k_D$, producing the uprise of CV_γ in the tail. Moreover, the parametrization with respect to the number of receptors suggests that a lower number of receptors produces a more reliable mechanism, in the sense that the quality of the signal does not change too much for different concentrations (cf. [35, Fig.3c]). A lower bound on the number of receptors, $\simeq 200$ with the data [3, Fig.2B], is provided by the relation $\ell^2 > 1$, which amounts to say that the intracellular contribution is adding noise (and not filtering out the signal).

The results obtained in the case of interacting ligands predict the behavior of the coefficient of variation of the turning angle in different conditions. The mathematical model makes the assumption that the silencing process is entirely due to the gradient sensing phase. At the same time, the model suggests the most favorable experimental settings in order to validate this hypothesis. More precisely, too large coefficients of variation should be avoided, since they do not carry precise information about the origin of the noise. Figs. 6 and 7 allow to select a couple of netrin–1 and Slit concentrations, which lead to a coefficient of variation ≈ 1. In such a condition the hypothesis underlying the mathematical model is verifiable by *in vitro* experiments.

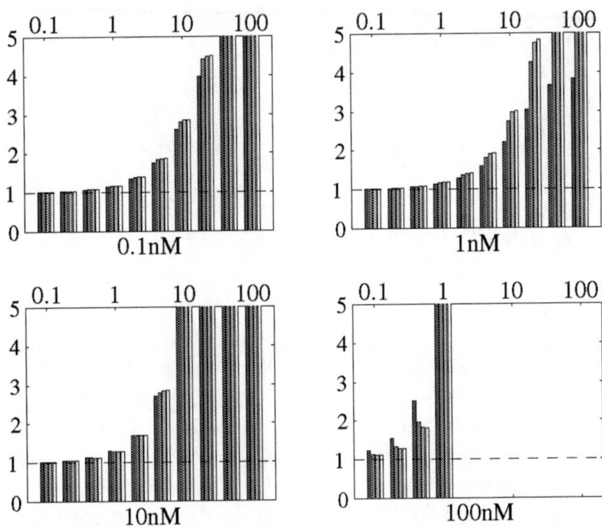

FIGURE 7. Ratio $CV_{\gamma|_{(x,y)}}/CV_{\gamma|_{(x,0)}}$. The top graduation of each panel represents Slit concentration, while the bottom graduation represents netrin–1 concentration (values in nM, logarithmic scale). Each group of bars contains results for different number of receptors (the same convention as in Fig. 5, right panel, is used). Groups of bars that do not appear, as in the rightmost part of the last panel in the bottom row, indicate that every form of organized directionality is lost.

5. PHENOMENOLOGICAL STUDY OF AXON RESPONSE TO MULTIPLE CUES

In this section, we propose a simplified phenomenological model to reproduce axon response in presence of multiple cues. This model does not consider the detail of the gradient sensing process. We deal with the case where two cues X and Y are interacting. In order to model the resulting contribution $\widetilde{\mathbf{P}}$ in Eq. (7), we introduce the time dependent weights w_X and w_Y, related to the activity of receptors binding to cue X and Y, respectively, and we set

$$\widetilde{\mathbf{P}} = w_X \widetilde{\mathbf{P}}_X + w_Y \widetilde{\mathbf{P}}_Y, \tag{22}$$

where $\widetilde{\mathbf{P}}_X$, $\widetilde{\mathbf{P}}_Y$ arise from the single X and Y cues.

The time functional dependence of the weights w_X and w_Y is supposed to be governed by the following differential system which represents the evolution at different developmental stages of the receptor activation

$$
\begin{cases}
\theta_X \dfrac{\partial w_X}{\partial t} = \begin{cases} 0 & \text{for } 0 < t < t_{a,X}, \\ \operatorname{sign}(t_{d,X} - t) w_X^\beta (1 - w_X)^\beta - a w_Y^\beta w_X^\beta, & \text{for } t_{a,X} \le t, \end{cases} \\
\theta_Y \dfrac{\partial w_Y}{\partial t} = \begin{cases} 0 & \text{for } 0 < t < t_{a,Y}, \\ \operatorname{sign}(t_{d,Y} - t) w_Y^\beta (1 - w_Y)^\beta - b w_X^\beta w_Y^\beta, & \text{for } t_{a,Y} \le t, \end{cases} \\
w_X(0) = w_{X0}, \\
w_Y(0) = w_{Y0},
\end{cases}
\tag{23}
$$

where θ_X, θ_Y are characteristic times for receptors of ligand X and Y, $t_{a,X}$, $t_{a,Y}$ and $t_{d,X}$, $t_{d,Y}$, are receptor activation and de–activation times, respectively, and where a, b are influence coefficients that represent the competition effects (if $a > 0$, signals from type–Y receptors down-regulate signals from type–X receptors, if $a < 0$, signals from type–Y receptors up-regulate signals from type–X receptors, and the same for b). The parameter β is an exponent in the range $(0,1)$. Notice that in this range the solution of the above system is not unique, the sets $(t,0)$ and $(t,1)$ being made of branching points. In any point of lack of uniqueness of the solution, the branch of the bifurcation in the image range $(0,1)$ is chosen. The choice of $\beta \in (0,1)$ allows for obtaining a finite growth time for any initial positive condition. Analogously, for $t > t_d$, a finite decay time is observed if $y(t_d) \le 1$; a corner point appears when $y(t_d) < 1$.

We use system (23) in the model (22) to study the behavior of commissural axons (substance X being netrin–1 and substance Y Slit). As illustrated in [3], in the case of sole netrin–1 gradient, the decay time must occur between stage 22 and stage 28; in the case of sole Slit gradient, full activation of the receptors occurs after stage 22. These conditions may be reproduced by choosing $\beta = 0.5$, $\theta_X = 0.5, \theta_Y = 2$, $t_{a,X} = 4, t_{a,Y} = 20.5$ and $t_{d,X} = 25, t_{d,Y} = 30$. The behavior of sole netrin–1 and Slit receptors is represented in Fig. 8, continuous lines. Both w_X and w_Y follow a power-law growth with saturation till the respective decay time In case of interaction, Slit silences netrin–1 while no effect is exerted by netrin–1 on Slit. We choose here for the simulation $a = 5$, $b = 0$. The silenced behavior of netrin–1 receptors is represented in Fig. 8, dashed line. In this case, the decay time of w_X occurs much before due to the silencing effect of Y receptors.

The relation between an assigned concentration gradient and the vector $\widehat{\mathbf{P}}$ can be found by experimental results using the relation [1]

$$
R_{c,min} = \frac{v_g^2}{|\widehat{\mathbf{P}}|},
\tag{24}
$$

where $R_{c,min}$ is the minimum radius of curvature of the trajectory (see also [17, Eq. (28)]).

The calibration of the model is obtained by solving problem (11) coupled with system (23) using the test setting of [3]. Final turning angles after 1h of exposure to the gradient, computed with parameters corresponding to stage 22 and stage 28, are reported in Fig. 9.

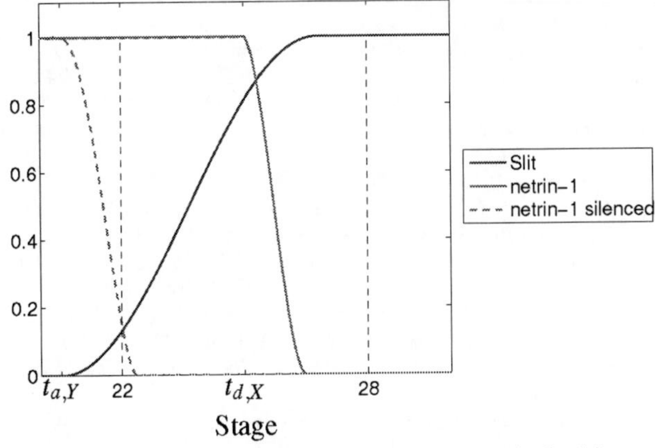

FIGURE 8. Time evolution of the weight factors w_X and w_Y obtained from system (23).

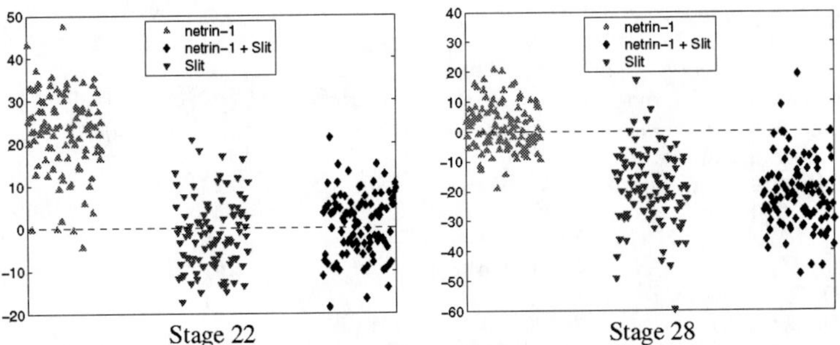

FIGURE 9. Simulated final axon turning angles (°) in response to netrin–1, Slit and combined netrin–1 and Slit at different development stages (cf. [3, Fig.2B,2E]).

This simple phenomenological model indicates that at stage 22 the repellent effect of Slit is weak (corresponding to a low weighting factor), whilst its silencing effect is relevant. This condition suggests that Slit activates slowly, but significantly before stage 22, because it needs a suitable time to cause decrease of the netrin–1 weight by hierarchical interaction.

REFERENCES

1. G. Aletti, and P. Causin, *IET Sys. Biol.* **to appear** (2008).
2. H. Berg, and E. Purcell, *Biophys. J.* **20**, 193–219 (1977).
3. E. Stein, and M. Tessier-Lavigne, *Science* **291(5510)**, 1928–1938 (2001).
4. M. Tessier-Lavigne, and C. Goodman, *Science* **274**, 1123–1133 (1996).
5. B. Mueller, *Annu. Rev. Neurosci.* **22**, 351–601 (1999).
6. H. Song, and M.-M. Poo, *Nat. Cell Biol.* **3**, E81–E88 (2001).

7. K. Guan, and Y. Rao, *Nature Rev. Neurosci.* **4**, 941–956 (2003).
8. P. Gordon-Weeks, *Neuronal growth cones*, Cambridge University Press, 2000.
9. J. Q. Zheng, M. Felder, J. A. Connor, and M. Poo, *Nature* **368**, 140–144 (1994).
10. J. Q. Zheng, J. Wan, and M. Poo, *J. Neurosci.* **16(3)**, 1140–1149 (1996).
11. G.-l. Ming, H.-j. Song, B. Berninger, C. Holt, M. Tessier-Lavigne, and M. Poo, *Neuron* **19**, 1225–1235 (1997).
12. G. J. Goodhill, M. Gu, and J. S. Urbach, *Neural Comp.* **16**, 2221–2243 (2004).
13. J. Xu, W. Rosoff, J. Urbach, and G. Goodhill, *Development* **132**, 4545–4562 (2005).
14. M. Aeschlimann, and L. Tettoni, *Neurocomputing* **38–40**, 87–92 (2001).
15. H. G. E. Hentschel, and A. van Ooyen, *Proc. R.Soc. Lond. B* **266**, 2231–2238 (1999).
16. H. G. E. Hentschel, and A. van Ooyen, *Physica A* **288**, 369–379 (2000).
17. J. K. Krottje, and A. van Ooyen, *Bull. Math. Biol.* **69**, 3–31 (2007).
18. Z. Kaprielian, R. Imondi, and E. Runko, *Anat. Rec.* **261**, 176–197 (2000).
19. K. Brose, K. Bland, K. Wang, D. Arnott, W. Henzel, C. Goodman, M. Tessier-Lavigne, and T. Kidd, *Cell* **96(11)**, 795–806 (1999).
20. T. E. Kennedy, T. Serafini, J. de la Torre, and M. Tessier-Lavigne, *Cell* **78**, 425–435 (1994).
21. T. Kidd, K. Bland, and C. Goodman, *Cell* **96**, 785–594 (1999).
22. G. J. Goodhill, *Neural Comp.* **15**, 549–564 (2003).
23. E. Hohenester, S. Hussain, and J. Howitt, *Biochem. Soc. Trans.* **34**, 418–421 (2006).
24. K. Keino-Masu, M. Masu, L. Hinck, E. Leonardo, S. Chan, J. Culotti, and M. Tessier-Lavigne, *Cell* **87(2)**, 175–185 (1996).
25. E. Korobkova, T. Emonet, J. M. G. Vilar, T. S. Shimizu, and P. Cluzel, *Nature* **428**, 574–578 (2004).
26. C. Rao, D. Wolf, and A. Arkin, *Nature* **420**, 231–237 (2002).
27. B. Andrews, T.-M. Yi, and P. Iglesias, *PLoS Computational Biology* **2(11)**, 1407–1418 (2006).
28. B. Andrews, and P. Iglesias, *PLoS Computational Biology* **3(8)**, 1489–1497 (2007).
29. G. J. Goodhill, and J. S. Urbach, *J. Neurobiol.* **41**, 230–241 (1999).
30. H. M. Buettner, R. N. Pittman, and J. Ivins, *Dev. Biol.* **163**, 407–422 (1994).
31. F. Beichelt, *Stochastic processes in science, engineering and finance*, Chapman & Hall/CRC, Boca Raton, FL, 2006, ISBN 978-1-58488-493-4; 1-58488-493-2.
32. A. Tullio, P. Bridgman, N. Tresser, C. Chan, M. Conti, R. Adelstein, and Y. Hara, *Jour.Comp. Neur.* **433(1)**, 62–74 (2001).
33. J. L. Goldberg, *Genes and development* **17**, 941–958 (2003).
34. P. E. Kloeden, and E. Platen, *Numerical solution of stochastic differential equations*, Springer Berlin, 1992.
35. W. J. Rosoff, J. S. Urbach, M. A. Esrick, R. McAllister, L. Richards, and G. Goodhill, *Nat. Neurosci.* **7(6)**, 678–82 (2004).
36. G. J. Goodhill, *Eur. J. Neur.* **9**, 1414–1421 (1997).
37. T. Serafini, T. Kennedy, M. Galko, C. Mirzayan, T. Jessell, and M. Tessier-Lavigne, *Cell* **78(3)**, 409–24 (1994).

The Adaptation of the Moth Pheromone Receptor Neuron to its Natural Stimulus

Lubomir Kostal*, Petr Lansky† and Jean-Pierre Rospars**

*Institute of Physiology, Academy of Sciences, Videnska 1083, 142 20 Prague 4, Czech Republic, E-mail: kostal@biomed.cas.cz
†Institute of Physiology, Academy of Sciences, Videnska 1083, 142 20 Prague 4, Czech Republic
**INRA, UMR 1272 Physiologie de l'insecte, F-78000 Versailles, France

Abstract. We analyze the first phase of information transduction in the model of the olfactory receptor neuron of the male moth *Antheraea polyphemus*. We predict such stimulus characteristics that enable the system to perform optimally, i.e., to transfer as much information as possible. Few *a priori* constraints on the nature of stimulus and stimulus-to-signal transduction are assumed. The results are given in terms of stimulus distributions and intermittency factors which makes direct comparison with experimental data possible. Optimal stimulus is approximatelly described by exponential or log-normal probability density function which is in agreement with experiment and the predicted intermittency factors fall within the lowest range of observed values. The results are discussed with respect to electroantennogram measurements and behavioral observations.

Keywords: Pheromone receptor, information processing, *antheraea polyphemus*.
PACS: 87.19.lt, 87.19.ls

1. INTRODUCTION

The main task of neuronal sensory systems is to "encode" information about the animal's environment into its internal representation. Physiological reasons limit the range of neuronal responses and consequently not all stimulus states can be encoded with equal reliability. The stimulus-response relation describes the reliability of encoding and thus implicitly provides such stimulus characteristics that maximize the information capacity of the neuron. One of the first studies of stimulus-response function with respect to maximizing the information gain was done on large monopolar cells (LMC) in the compound eye of the fly [1]. The LMC is a graded potential cell which codes the contrast fluctuations. The contrast levels in natural fly's habitat were measured by objective methods (photodiode) and the resulting characteristics were compared with those predicted from the stimulus-response curve. It was shown that LMCs are adapted to the animal's ecology as the natural stimulus maximizes the cells' performance. The following studies, e.g., Atick [2], Bialek and Owen [3], Hateren [4], Hornstein et al. [5], Laughlin [6], confirmed that the natural signals are processed optimally by sensory systems. Nevertheless, the majority of available studies consider the visual system only. In this paper we parallel the pioneering work by Laughlin [1], adapting the method to suit the specificity of invertebrate olfactory system.

Orientation towards food and mate, especially in insects, is an olfactory-controlled behavior which relies on the detection of odorant molecules delivered from the source. The atmospheric turbulence causes strong mixing of the air and creates a wide spectrum

CP1028, *Collective Dynamics: Topics on Competition and Cooperation in the Biosciences*
edited by L. M. Ricciardi, A. Buonocore, and E. Pirozzi

of spatio-temporal variations in the signal. The largest eddies are hundreds of meters in extent and may take minutes to pass a fixed point, while the smallest spatial variations are less than a millimeter in size and lasts for miliseconds only [7, 8]. The mean concentration of the odorant decreases monotonically with the distance from the source, however, the relation for concentration fluctuations and thus for instantaneous magnitude of the signal is more complicated. Due to the inhomogenous mixing very high concentration values can by found in a wide range of distances from the source, though their frequency decreases with distance [7]. An important characteristics of the detected signal is its intermittency, i.e., the fraction of time during which non-zero concentrations are detected. It has been shown [9, 10] that the natural signal is highly intermittent in a wide range of experimental conditions. The signal is present less than 50 % of the total time, usually even smaller intermittency is detected, e.g., Murlis et al. [10] report 20 % in measurements of pheromone dispersion in natural conditions close to the source. Various types of ion detectors are usually employed for measurements, though Baker and Haynes [11], Murlis et al. [10] have also used electroantennogram responses. The description of the complicated and inhomogeneous structure of the detected odorant concentrations requires an approximative approach and statistical methods are usually employed. The probability density function over the whole stimulus range is the most convenient descriptor of the signal [7, 8, 10, 12, 9].

The variations in the concentration of the stimulus are essential for the insect to locate the source of the stimulus. The animal loses direction to the source and its upwind flight gets "arrested" if it gets into a cloud of homogeneously distributed pheromone [13, 14]. Experiments in tunnels have shown that characteristics like frequency and intensity of the intermittent stimulus play a key role in maintaining the proper direction of flight [15]. The insect's sensory system differs from the ion detector and thus the level of temporal and spatial detail the receptor neuron perceives is limited by both physical and biochemical reasons [11, 16, 17]. In other words, not all the information pheromone signal potentially carries can be processed. We analyze the first phase of information transduction in the olfactory receptor of the male moth *Antheraea polyphemus*. The external stimulus (the odorant) is given by the temporal concentration of the major component of the sex pheromone, the (E,Z)-6,11-hexadecadienyl acetate. The response of the system (the internal signal) is the graded concentration of activated receptor molecules. This process of transduction represents the first stage in the cascade of events finally leading to generation of action potential. The detailed analysis of the first phase provides insight into the information processing at the single-receptor level. We may paraphrase the fundamental data processing inequality [18]: if some information does not pass the first stage, it cannot reappear in any sequential stage of the processing. The first stage of transduction cascade therefore sets constraints on the final performance of the receptor.

The goal of this paper is to characterize the performance of the stimulus-to-response transformation, namely to find and describe the optimal stimulus (or the class of optimal stimuli) that maximizes the performance. Mathematical basis for this task is provided by the statistical theory of information and the proposed method can be used in similar or more general situations. Similarly to LMCs studied in [1] the response of the first-stage information transduction in the olfactory neuron is a graded signal. Likewise, the experimental measurements of odorant plume concentration characteristics in the animal's habitat were performed by objective devices (ion detectors). The comparison of predicted and natural stimulus reveals how well the receptor is adapted or "tuned" to the signals it encounters most often.

2. METHODS

2.1. The model of the odorant receptor

The first stage of information processing in the olfactory sensory neuron is described by the transformation of the external signal (the odorant concentration in the air) to the internal signal (the concentration of activated receptors). The model of odorant receptor we consider here was developed by Kaissling and Rospars [19] and represents a modified version of the original model developed by Kaissling [16]. The modification has no impact on the obtained results (verified numerically) though it simplifies the original model in terms of required parameters and variables. The chemical reactions form the following chain:

$$\mathrm{L_{air}} \xrightarrow{k_i} \mathrm{L} \tag{1}$$

$$\mathrm{L+R} \underset{k_{-3}}{\overset{k_3}{\rightleftharpoons}} \mathrm{R_L} \underset{k_{-4}}{\overset{k_4}{\rightleftharpoons}} \mathrm{R^*} \tag{2}$$

$$\mathrm{L+N} \underset{k_{-5}}{\overset{k_5}{\rightleftharpoons}} \mathrm{N_L} \xrightarrow{k_6} \mathrm{P+N}. \tag{3}$$

The network (1)–(3) includes the external ligand (the odorant) $\mathrm{L_{air}}$, its uptake L and reversible binding to a receptor R, the reversible change of the complex $\mathrm{R_L}$ to an activated state $\mathrm{R^*}$ (the internal signal), a reversible binding of L to a deactivating enzyme N (see Kaissling and Rospars [19] for details) and an irreversible odorant deactivation by changing of the complex $\mathrm{N_L}$ to P+N. The concentrations of the eight species involved are denoted by square brackets and the values are functions of time. For simplicity we omit to denote the explicit dependence on the time variable t and adopt the following notation for the individual concentrations: $L_{air} = [\mathrm{L_{air}}](\mathrm{t})$, $L = [\mathrm{L}](\mathrm{t})$, $R = [\mathrm{R}](\mathrm{t})$, $R_L = [\mathrm{R_L}](\mathrm{t})$, $R^* = [\mathrm{R^*}](\mathrm{t})$, $N = [\mathrm{N}](\mathrm{t})$, $P = [\mathrm{P}](\mathrm{t})$ and $N_L = [\mathrm{N_L}](\mathrm{t})$.

The total concentration of the receptor molecules, $R_{tot} = R + R_L + R^*$, does not change over time as well as the total concentration of the deactivating enzyme, $N_{tot} = N + L_N$, remains constant. The evolution of the reactions (1)–(3) in time given the external signal L_{air} is fully described by five first order ordinary differential equations (4)–(8) and two algebraic equations (9) and (10):

$$\frac{dL}{dt} = k_i L_{\text{air}} - k_3 LR + k_{-3} R_L - k_5 LN + k_{-5} L_N \quad (4)$$

$$\frac{dR_L}{dt} = k_3 LR - k_{-3} L_R - k_4 L_R + k_{-4} R^* \quad (5)$$

$$\frac{dR^*}{dt} = k_4 R_L - k_{-4} R^* \quad (6)$$

$$\frac{dL_N}{dt} = k_5 LN - k_{-5} L_N - k_6 L_N \quad (7)$$

$$\frac{dP}{dt} = k_6 L_N \quad (8)$$

$$R = R_{tot} - R_L - R^* \quad (9)$$

$$N = N_{tot} - L_N. \quad (10)$$

The state of the system at any given time, $S(t) = \{L(t), R_L(t), R^*(t), L_N(t), P(t)\}$, is given by the actual values of the involved variables and we assume that at $t = 0$ the concentrations L, R_L, R^*, L_N and P are zero. The values of parameters were determined by Kaissling [16] and Kaissling and Rospars [19], we summarize them in Tab. 1.

TABLE 1. Summary of the odorant receptor model parameters [16, 19].

k_3	=	$0.209\ \text{s}^{-1}\mu\text{M}^{-1}$	k_{-3}	=	$7.9\ \text{s}^{-1}$
k_4	=	$16.8\ \text{s}^{-1}$	k_{-4}	=	$98\ \text{s}^{-1}$
k_5	=	$4\ \text{s}^{-1}\mu\text{M}^{-1}$	k_{-5}	=	$98.9\ \text{s}^{-1}$
k_6	=	$29.7\ \text{s}^{-1}$	k_i	=	$29000\ \text{s}^{-1}$
R_{tot}	=	$1.64\ \mu\text{M}$	N_{tot}	=	$1\ \mu\text{M}$

The differential equations (4)–(8) follow the law of mass action for chemical reactions. In reality, the response of the system is not deterministic. The value fluctuates due to the stochastic effects like spatial inhomogeneities in the distribution of reactants. If the concentrations of reactants are high enough above single-molecular levels then the fluctuations are relatively small and can be neglected. However, for small doses the situation is more complicated and the stochastic effects have to be described properly [20]. In this paper we thus do not investigate the effect of extremely small odorant doses. The value of R^* corresponding to one activated receptor molecule per neuron is approximately $10^{-6.2}\ \mu$M [19] which is far below the values considered in this paper.

2.2. Optimal stimulus reconstruction

The main task of the first-stage of signal processing in the olfactory receptor neuron is to transform the input signal (the odorant concentration) into its internal representation (the concentration of activated receptors). The neuron performs optimally if it preserves as much information about the input as possible. According to the information theory, information is transmitted only if the input varies randomly [18]. From this point of view a homogeneous cloud of odorant would carry zero information. The exact amount

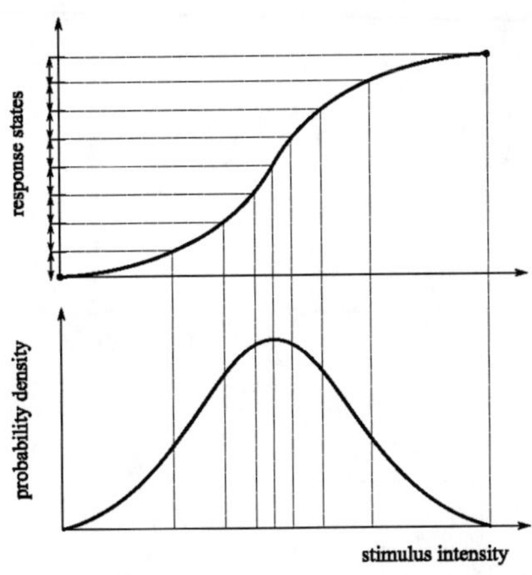

FIGURE 1. The amount of information the neuron can transfer is limited by the finite range of possible response states. Furthermore, the minimal stimulus increment that changes the response state is not constant over whole the stimulus range. The amount of transferred information therefore depends both on the range of stimulus and on the frequency with which particular concentration values occur. If the neuron performs optimally then all response states have to be used equally likely, which in turn uniquely determines the optimal stimulus probability density function (adapted from Laughlin [1]).

of transferred information is determined from the stimulus-response relationship. The problem therefore lies in relating two signals, L_{air} (the stimulus) and R^* (the response). The information, however, is not transmitted at a time instant, rather we assume that it is gained within a time interval. Therefore we divide the time axis into "windows" or "bins" of length Δt, i.e., we do not to consider any temporal details below Δt. The stimulus is represented by a constant value of concentration L_{air} in the time window and the response, ρ, to such stimulation is the average value of activated receptors taken over the corresponding time window $\rho = \langle R^* \rangle_{\Delta t}$. This simple set-up allows us to test the performance of the receptor model at different levels of temporal resolution.

The most important factor limiting the information transfer is the bounded range of responses ρ due to finite number of receptor molecules per neuron. Once the maximum number of receptor molecules is activated no higher stimulus concentration can be encoded. Furthermore, the neuron can perceive a change in stimulus value differently depending on the basal stimulus concentration. The amount of transferred information therefore depends both on the range of stimulus L_{air} and on the frequency with which particular concentrations values occur, see Fig. 1. In other words, the description of L_{air} in each selected time window is given in terms of probability density function $f(L_{\mathrm{air}})$.

The information theory [18, 21, 1] describes the optimal stimulus characteristics implicitly: the system performs optimally if all possible response values are used with equal frequency. In the following we describe the "step-wise" method to obtain such stimulus characteristics that equalize the output usage. First we compute the reaction of the system to all possible stimuli in the first time window, $(0, \Delta t)$, given the zero initial condition at $t = 0$, see Fig. 2a. For each stimulus the response ρ is the average number of activated receptors in $(0, \Delta t)$, Fig. 2b. The relation between L_{air} and ρ describes the

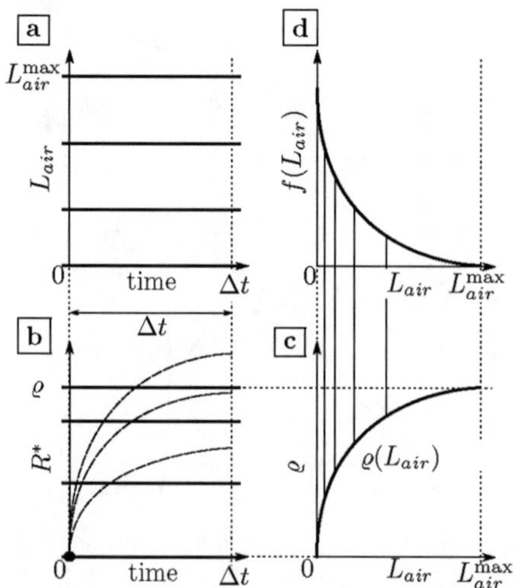

FIGURE 2. Stimulus optimization in the first time window $(0, \Delta t)$. (a) The stimulus $L_{\rm air}$ is a constant function bounded inside $[0, L_{\rm air}^{\rm max}]$ (several examples shown). (b) Starting from zero at $t = 0$ the time development of R^* (dashed line) is averaged (ρ) over the first time window (solid line). (c) ρ "encodes" the stimulus value so the stimulus-response curve $\rho(L_{\rm air})$ can be constructed. (d) The stimulus-response curve uniquely determines such stimulus probability density function $f(L_{\rm air})$ that all responses ρ are used equally likely.

stimulus-response curve, $\rho(L_{\rm air})$, Fig. 2c. Conditioned that all responses have to be used with equal frequency, the stimulus cumulative distribution function can be related to $\rho(L_{\rm air})$. The optimal stimulus probability density function $f(L_{\rm air})$ can be then written as

$$f(L_{\rm air}) = \frac{d}{dL_{\rm air}} \left[\frac{\rho(L_{\rm air}) - \rho_{\rm min}}{\rho_{\rm max} - \rho_{\rm min}} \right], \tag{11}$$

where $\rho_{\rm min}$ resp. $\rho_{\rm max}$ are the minimal resp. maximal response values encountered in the time window, Fig. 2d. Once the optimal stimulus probability density for the time $(0, \Delta t)$ is known we use it to select one stimulus value. The reaction of the system to this particular stimulus is computed, the state of the system S at $t = \Delta t$ is determined and we can proceed to the next time window $(\Delta t, 2\Delta t)$. Note that the information transfer in this system has a memory, i.e., the current state is affected not only by the current stimulus but also by the history of stimulation. Therefore the state of the system $S(\Delta t)$ must be taken into account for evaluation of the response in the time window $(\Delta t, 2\Delta t)$. The optimization proceeds similarly in $(\Delta t, 2\Delta t)$: we again compute the time course of the activated receptor concentration (R^*) and determine their averages (ρ) under all possible stimulus conditions, Fig. 3. The optimal stimulus probability density function in $(\Delta t, 2\Delta t)$ is determined again by employing formula (11). After selecting one random stimulus value the process continues into the following time window. The range of ρ and the shape of $f(L_{\rm air})$ may change from one window to another.

The actual amount of transferred information in each step can be estimated from the available response range. If we divide the range $(\min\rho, \max\rho)$ into n bins (that cannot

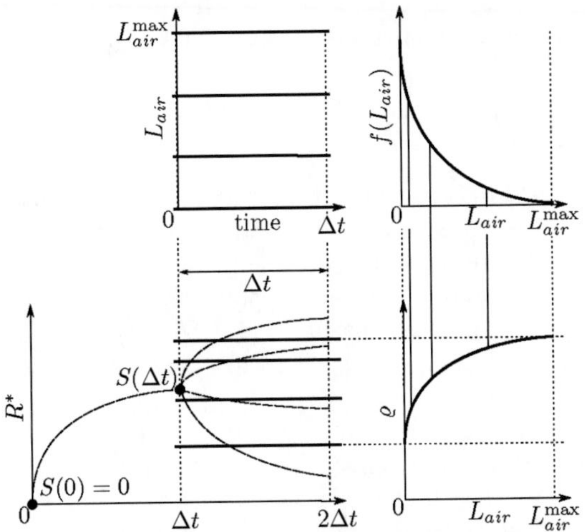

FIGURE 3. Stimulus optimization in the second time window $(\Delta t, 2\Delta t)$. One particular stimulus value is drawn randomly from $f(L_{\text{air}})$ reconstructed in the first time window $(0, \Delta t)$. The corresponding response and the state of the system $S(\Delta t)$ are computed. In the second time window $(\Delta t, 2\Delta t)$ the responses and their averages are determined again for all stimulus values (taking into account the state system at $t = \Delta t$). The stimulus-response curve and the corresponding stimulus probability density function are determined and the process is carried into the next time window.

be divided any further) the amount of information which can be transferred is $\log_2 n$ bits [18]. For each time window we thus compute the estimate of transferred information, here denoted as obtainable information, I_{obt}, in bits as

$$I_{obt} = \log_2 \left(\frac{\max \rho - \min \rho}{\Delta \rho} \right), \tag{12}$$

where the division factor $\Delta\rho$ is set prior to the the optimization process. We adopt the convention that $I_{obt} = 0$ if the coding range is smaller than $\Delta\rho$. The factor $\Delta\rho$ corresponds to the minimal number of activated or deactivated receptor molecules the system perceives as a change. We set the minimal value to 100 molecules which consequently gives $\Delta\rho = 10^{-4.2}\,\mu$M. Substituting for $\Delta\rho$ into formula (12) and taking into account that the maximal concentration of activated receptors is $R^* \approx 0.24\,\mu$M [19] yields the maximum information gain $I_{obt} \approx 12$ bits. We furthermore assume that $\Delta\rho$ does not depend on the length of the time window.

3. RESULTS

3.1. Single-pulse stimulation

First we examine the behavior of the model under the stimulation with a single pulse of unlimited duration. Setting the left-hand sides of equations (4)–(8) equal to zero gives the asymptotic value of R^* as a function of the constant L_{air},

$$R^* = (1 - Q_4)R_{tot}\left[\frac{K_{d3}Q_4}{K_{m5,6}}\left(\frac{k_6N_{tot}}{k_iL_{\text{air}}} - 1\right) + 1\right]^{-1}, \tag{13}$$

[19], where $K_{d3} = k_{-3}/k_3$ corresponds to the dissociation constant of ligand and receptor, $Q_4 = k_{-4}/(k_4 + k_{-4})$ and $K_{m5,6} = (k_{-5} + k_6)/k_5$ are the Michaelis constants of the ligand and the deactivating enzyme. Using the values from Tab. 1 we find that the dose-response relationship (13) is almost perfectly linear with the maximum concentration of the activated receptors $\max R^* = (1 - Q_4)R_{tot} \approx 0.24\,\mu\text{M}$ [19]. The minimal concentration of infinite duration that activates the maximal number of receptors is $L_{\text{air}} \approx 0.001\,\mu\text{M}$.

Next we examine the response of the system to a constant stimulation of limited duration. In Fig. 4 we see the time course of R^* given several different values of L_{air} from $0.0001\,\mu\text{M}$ to $0.005\,\mu\text{M}$. The stimulation starts at $t = 1$ s and lasts for 1 s. We see, that the system responds differently even for stimulus concentrations higher than the minimum concentration which evokes asymptotically maximum number of activated receptors (the asymptotic maximum $L_{\text{air}} = 0.001\,\mu\text{M}$). In other words, values $L_{\text{air}} > 0.001\,\mu\text{M}$ can be distinguished only if the duration of the stimulus pulse decreases. However, the duration of the falling phase gets progressively longer which has important consequences on distinguishing details in sequences of large stimulus values. During the simulations we avoid extremely small doses of odorant due to the validity of the mass action law. For the same reason we do not set the length of the time window Δt close to zero. The smallest value we allow is $\Delta t = 0.2$ s which is near the upper value of the experimentally observed range [17]

3.2. Multi-pulse stimulation

First we employ the optimization process directly, i.e., under the condition of stimulus being permanent but varying. In the first example we set the level of temporal detail to $\Delta t = 0.2$ s. The upper bound on stimulus value is $L_{\text{air}}^{\max} = 0.1\,\mu\text{M}$, which is $1000\times$ the concentration sufficient to reach the asymptotic maximum and therefore the stimulus range can be considered unrestricted.

The results are presented in Fig. 5. The plots show the state of the system in each time window (time is on the horizontal axis). The first row shows the optimized stimulus value which is randomly drawn from the optimal probability density function in each time window. The chosen stimulus value in turn determines the behavior of the system in the next time window due to the memory effect. The second row is the reaction of the system to the optimized stimulus and the third row shows the transferred (or obtainable) information. We see, that the performance of the system is not stable in time, i.e., the obtainable information I_{obt} monotonically decreases. The reason lies in the prolongation of the falling phase of R^*, see Fig. 4. The response range is initially bounded from below by $\min \rho = 0\,\mu\text{M}$ (we start from zero initial condition) but due to the memory effect the actual value of $\min \rho$ increases in subsequent time windows. The upper limit of responses, $\max \rho$, does not change because it is given by the physical properties of

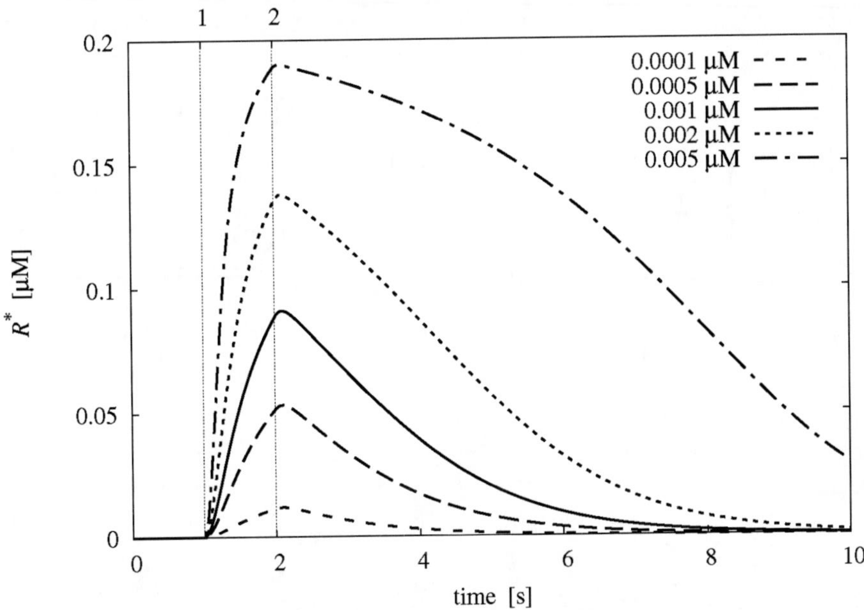

FIGURE 4. Response R^* of the system to pulsed stimulation (from $t = 1$ s to $t = 2$ s) of varying intensity. The limited duration of the pulse allows the system to detect L_{air} higher than 0.001 μM (solid line), for which the system saturates asymptotically. Note the significant prolongation of the falling phase with increasing stimulus value.

the system. Consequently, the response range decreases and I_{obt} finally (and inevitably) reaches zero no matter how large is the stimulus value. The instability of the system is reflected also in the time development of optimal stimulus probability density function, $f(L_{\text{air}})$, see Fig. 6. Stimulus probability density function $f(L_{\text{air}})$ in the first time window, which corresponds to maximal I_{obt}, can be approximated by the exponential probability density function

$$f(L_{\text{air}}) = \frac{1}{\lambda}\exp(-L_{\text{air}}/\lambda), \tag{14}$$

with mean value $\lambda = 0.03\,\mu$M. A better fit, in this case, is provided by the log-normal distribution

$$f(L_{\text{air}}) = \frac{1}{x\sigma\sqrt{2\pi}}\exp\left[-\frac{(\ln x - \mu)^2}{2\sigma^2}\right], \tag{15}$$

with $\sigma = 1.5$ $\mu = -3.6$ and mean value $0.08\,\mu$M. In subsequent time windows $f(L_{\text{air}})$ transforms into uniform distribution over the whole stimulus range meaning that there is no stimulus value preference once no information can be encoded.

The system can be stabilized by limiting the upper stimulus range to its asymptotic maximum $L_{\text{air}}^{\max} = 0.001$, i.e., the response range is zero if stimulated constantly by $L_{\text{air}}^{\max}$. The result is presented in Fig. 7. The temporal detail of the stimulus (the time window

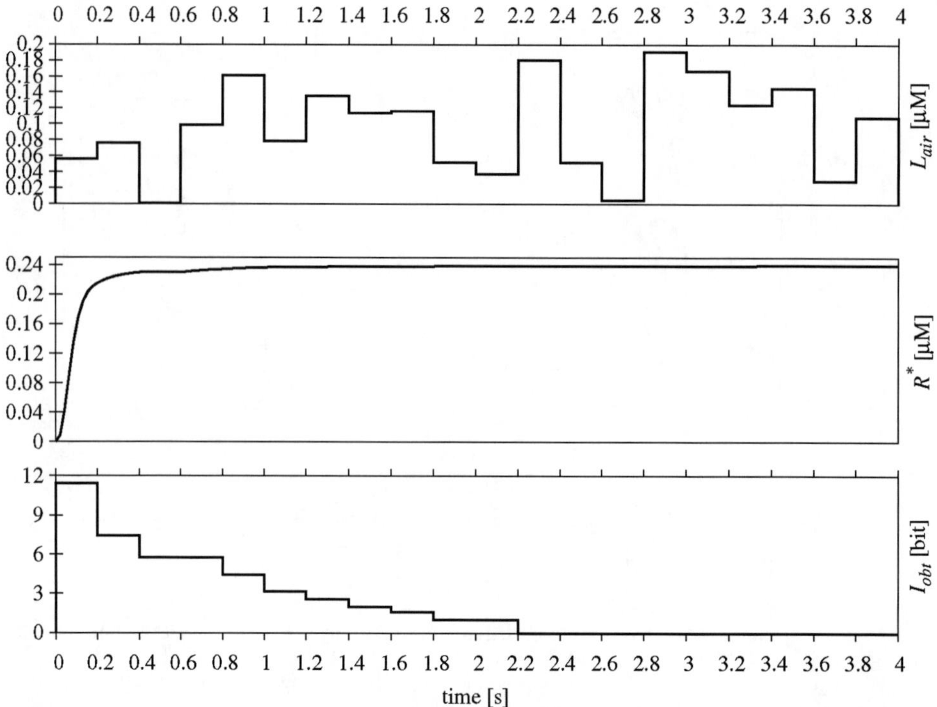

FIGURE 5. Optimal stimulus reconstruction: $\Delta t = 0.2\,\text{s}$ and $\max L_{\text{air}} = 0.1\,\mu\text{M}$. The three plots show (from top): the sample stimulus course, the response of the system R^*, and the corresponding obtainable information I_{obt}. The maximum stimulus value is chosen high enough to show the range accepted by the system at the selected temporal detail level. Initially, the optimal stimulus probability density function $f(L_{\text{air}})$ coincides with the exponential probability density but changes towards uniform distribution. The obtainable information I_{obt} decreases quickly due to the effect of memory. After $t = 2.2\,\text{s}$ no information is encoded. The effect of memory therefore disables high-precision coding for a prolonged period of time.

Δt) is set to 0.4 s. We see that though the performance is stable now the obtainable information is always below 12 bits and the full encoding capacity is never used. The optimal stimulus probability density is also stable in time. Its shape resembles the uniform probability density function, nevertheless it is slightly skewed towards higher values.

3.3. Intermittent stimulation

The sample optimization process illustrated in the previous examples was carried out under the condition of signal presented in every time window. Another possibility to obtain stable performance and to avoid saturation effects is to leave the stimulus range virtually unrestricted and let some time windows to contain no signal The fraction of

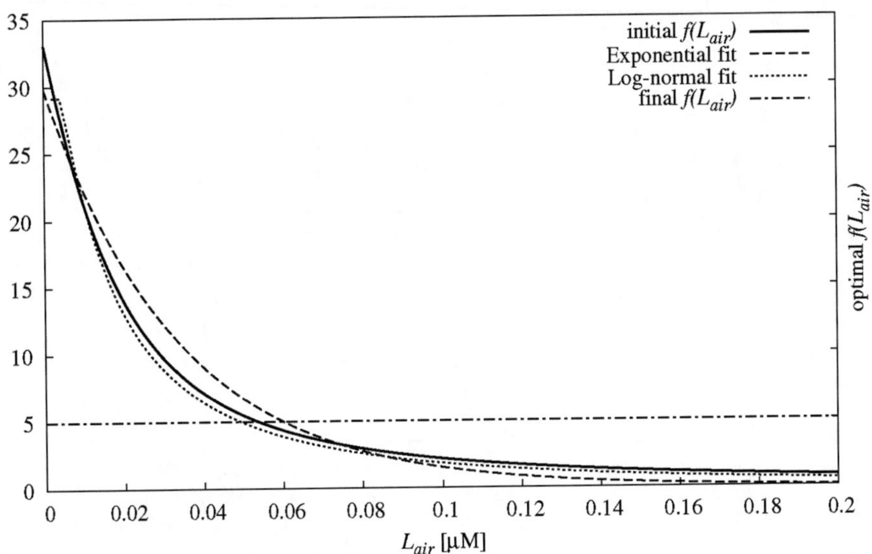

FIGURE 6. Optimal stimulus probability density functions $f(L_{\text{air}})$ for the case of persistent stimulation with unrestricted concentration values in Fig. 5. The initial $f(L_{\text{air}})$ which corresponds to maximal information transfer can be roughly approximated by the exponential or better by the log-normal probability density function. The final $f(L_{\text{air}})$ when no information is transferred is uniform over the whole stimulus range.

the total recording time where the signal is present is called intermittency and it is well known that for natural signals its value is very low, almost always less than 50 % [7, 8]. Murlis et al. [10] report intermittency of naturally dispersed odorant plume as low as 10 % or 20 % in the range of meters from the source.

The final example shows a possibility to predict the optimal intermittency value. We are interested in encoding the signal with maximum sensitivity whenever possible. This condition sets the limits on the recovery time needed after one particular stimulus is presented. Whenever the obtainable information decreases below 11 bits (the threshold for optimal performance) we let the following time windows contain no signal until $I_{obt} > 11.5$ bits again. The intermittency allows the system to "reset" and perform in optimal state again. The result for $\Delta t = 0.4$ s and $L_{\text{air}}^{\text{max}} = 0.03\,\mu$M is shown in Fig. 8. The optimal stimulus probability density function is stable in time (whenever the stimulus is present) and can be described by the exponential probability density function (14), this time with $\lambda \approx 0.005\,\mu$M. The intermittency predicted in this case is 7 %. However, the intermittency value is directly dependent on the threshold value of I_{obt} for optimal performance and the threshold value for zero signal.

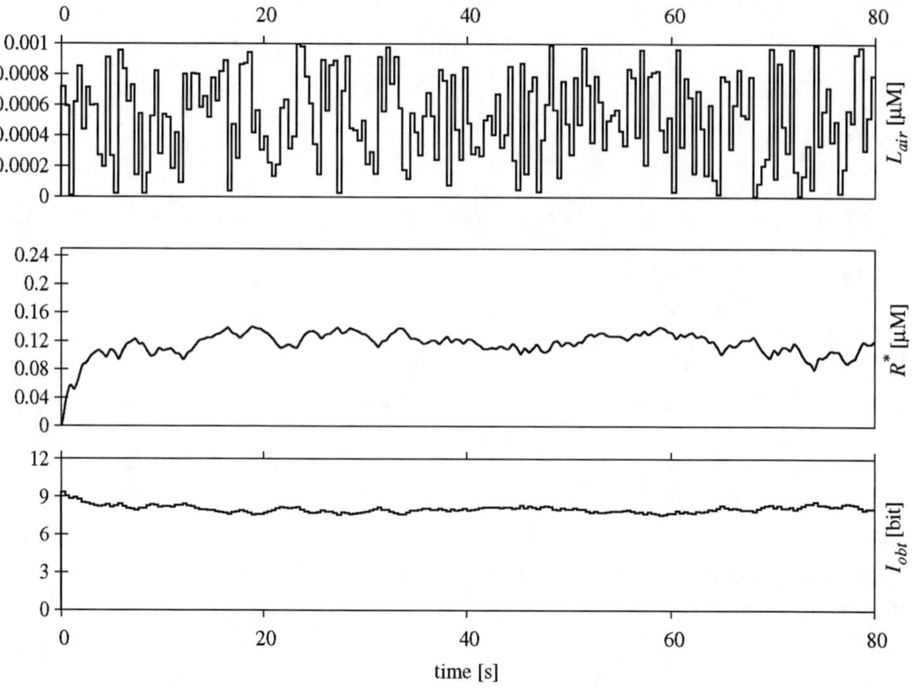

FIGURE 7. Reconstruction of optimal stimulus : $\Delta t = 0.4$ s and $L_{air}^{max} = 0.1\,\mu$M. The upper range of stimulus is limited by the asymptotic maximum in order to stabilize the performance of the system under persistent stimulation. However, the obtainable information is always below 12 bits and the full encoding capacity is never used.

4. DISCUSSION

It is impossible to characterize the optimal stimulus distribution without restricting either the maximum pheromone concentration or by considering the intermittent nature of stimulation, since the performance of the system may be unstable in time, see Fig. 5. We discuss the two possible considerations separately:

1. The signal is present all the time but its upper bound is limited, which prevents the responses from saturation (Fig. 7). The optimal probability density function is skewed towards higher stimulus values depending on the level of temporal detail Δt. However, continuous stimulation with higher stimulus may lead to the saturation of activated states, decrease in transferred information and consequent loss of correct direction of flight for the insect. Such situations have been observed close to the source [22] leading to flight "arrestment" of the male moths. This behavioral change can be explained using the obtained results. Once the transferred information decreases to zero, there is no way for the animal to tell whether it

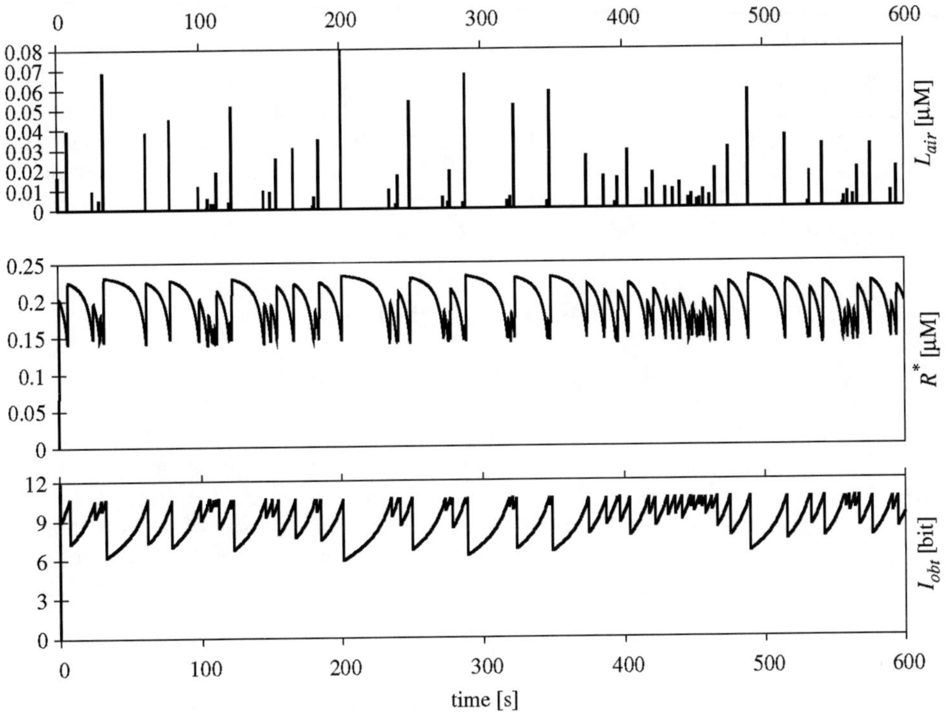

FIGURE 8. Reconstruction of optimal stimulus including intermittency: $\Delta t = 0.4\,\mathrm{s}$ and $\max L_{\mathrm{air}} = 0.3\,\mu\mathrm{M}$. The periods of zero signal are predicted by the model under the condition of optimal signal coding, so the system performs optimally and $f(L_{\mathrm{air}})$ is stable in time (whenever the signal is present). The optimal stimulus probability density function is described by the exponential probability density. The resulting intermittency factor is 7 %.

is flying in the odorant plume or in a clean air. The observed zigzag pattern and counterturns then lead to decrease in the stimulus value and to the reset of the coding process. It has been observed [23] that after extremely strong pheromone stimulation the recovery time may take up to minutes.

2. The upper bound of the signal is unrestricted but the signal is highly intermittent (Fig. 8). We found that the shape of optimal stimulus probability density can be well approximated either by the exponential or log-normal models. This is in full agreement with experimental data [7, 12, 9]. The intermittency is characteristic for natural signals [10]. The values of intermittency predicted by our model are rather in the lower range usually observed in experimental data (10–20 % reported by Jones [7], 10–40 % by Murlis et al. [10] depending on the experimental conditions). The value obtained by our method (≈ 7 %) is mainly due to the condition of optimal performance whenever non-zero stimulus is present. The threshold value for zero signal also affects the intermittency factor greatly (this affects the experimental data as well [7, 9]). Another (hypothetic) possibility lies in more rapid deactivation of

the receptors.

The coding range is widest if the initial concentration of activated receptors R^* is zero. In other words the first stimulus encounter is coded with finest precision and the situation progressively worsens (Fig. 5). This observation is also confirmed experimentally, though not directly, from the electroantennographic measurements of pheromone plume structure. Baker and Haynes [11] found, that after the burst of stimulus is encountered only several first peaks may be followed by the response.

5. CONCLUSIONS

We find that the optimization of information transfer in the model of the odorant receptor is complicated by the "memory effect" and the response saturation. If we leave the range of possible stimulus intensities unrestricted then intermittency must be taken into account. The stimulus probability density function that maximizes the information transfer can be well approximated by the exponential model which is in agreement with experimental data. The predicted intermittency is in the lower range of experimentally observed values. The obtained results are put into correspondence with behavioral observations, namely the upwind flight arrestment reported in homogeneous plume clouds or very close to the source.

5.1. Acknowledgements

This work was supported by Marie-Curie fellowship HPMT-CT-2001-00244 to L.K., by ECO-NET 12644PF from French Ministère des Affaires Étrangères, by Research project AV0Z50110509, Centre for Neuroscience LC554 and by Academy of Sciences of the Czech Republic Grants (1ET400110401 and KJB100110701).

REFERENCES

1. S. Laughlin, *Z Naturforsch* **36**, 910–912 (1981).
2. J. Atick, *Network: Comp Neur Sys* **3**, 213–251 (1992).
3. W. Bialek, and W. G. Owen, *Biophys J* **58**, 1227–1233 (1990).
4. J. Hateren, *J Comp Physiol A* **171**, 157–170 (1992).
5. E. P. Hornstein, D. C. O'Carroll, J. C. Anderson, and S. B. Laughlin, *Proc Biol Sci* **267**, 2111–2117 (2000).
6. S. B. Laughlin, *Vision Res* **36**, 1529–1541 (1996).
7. C. Jones, *J Hazard Mat* **7**, 87–112 (1983).
8. J. Murlis, "Odor plumes and the signal they provide," in *Insect Pheromone Research: New Directions*, edited by R. Carde, and A. Minks, Chapman and Hall, New York, 1996, pp. 221–231.
9. K. Mylne, and P. Mason, *Q J Roy Meteo Soc* **117**, 177–206 (1991).
10. J. Murlis, M. Willis, and R. Cardé, *Physiol Entomol* **25**, 211–222 (2000).
11. T. Baker, and K. Haynes, *Physiol Entomol* **14**, 1–12 (1989).
12. K. Mylne, "Experimental Measurements of Concentration Fluctuations," in *Air Pollution Modelling and Its Application VII*, edited by H. van Dopp, Plenum Press, New York, 1988, pp. 555–565.
13. J. Kennedy, A. Ludlow, and C. Sanders, *Nature* **288**, 475–477 (1980).
14. M. Willis, and T. Baker, *Physiol Entomol* **9**, 341–358 (1984).

15. N. Vickers, and T. Baker, *J Insect Behavior* **5**, 669–687 (1992).
16. K. E. Kaissling, *Chem Senses* **26**, 125–150 (2001).
17. B. Kodadová, *J Comp Physiol A* **179**, 301–310 (1996).
18. T. Cover, and J. Thomas, *Elements of information theory*, Wiley, New York, 1991.
19. K. E. Kaissling, and J.-P. Rospars, *Chem Senses* **29**, 529–531 (2004).
20. K. G. Gurevich, P. S. Agutter, and D. N. Wheatley, *Cell Signal* **15**, 447–453 (2003).
21. P. Dayan, and L. Abbott, *Theoretical neuroscience: computational and mathematical modeling of neural systems*, MIT Press, 2001.
22. T. Baker, M. Willis, K. Haynes, and P. Phelan, *Physiological entomology* **10**, 257–265 (1985).
23. C. Zack, *Sensory Adaptation in the Sex Pheromone Receptor Cells of Saturniid Moths*, Dissertation, Ludwig-Maximilians-Universität, 1979.

Interaction and Integration of Synaptic Activity on Dendritic Tree

Vito Di Maio

Istituto di Cibernetica "E. Caianiello" del CNR
Via Campi Flegrei, 34, 80078 Pozzuoli (Na), Italy
E-mails: vdm@biocib.cib.na.cnr.it vito.dimaio@cnr.it
web: http://www-biocib.cib.na.cnr.it/DiMaio/dimaio.html

Abstract. The interaction among synapses on dendritic tree of pyramidal neurons, combined with the variability of dendritic passive membrane properties, produces a complex system which regulates both the transfer of information among neurons and between different dendritic areas of the same neuron. A non linear mechanism, based on the excitatory reversal potential which behaves like a threshold, can act as a computational system improving the computational ability of the single neuron. Some examples of inter-synaptic interaction and of synaptic interaction with the dendritic tree are given and the new concept of *"Competition for Plasticity"* among synapses is proposed.

Keywords: Synaptic Interaction, synaptic Integration, neuronal computation, neuronal modeling.
PACS: 87.18.Nq, 87.18.Sn, 87.18.Mp, 87.19.Im

1. INTRODUCTION

The complex structure of the dendritic tree of pyramidal neurons, with tens of thousands of synaptic connections, might play an important role in the integration of the incoming information to the neuron. Some non liner mechanisms, generated by the interaction between the synaptic activity and some dendritic properties [1, 2, 3, 4, 5, 6, 7, 8, 9], can also participate in computational tasks which can be the base for the computational ability of the single neuron. Non linear processes can occur at two main levels: i) interaction among excitatory synapses; ii) interaction between synapses and dendrites. The interaction among synapses is essentially based on the ability of each active excitatory synapse to reduce the driving force which produce the Excitatory Post Synaptic Current (EPSC) of other excitatory synapses. This driving force is the difference between the membrane potential and the equilibrium potential (also called reversal potential) of the ions involved in the EPSC generation. The effect of interaction among synapses depends on the mutual distance [1, 2, 3, 4] and on the activation time [5, 6, 7, 10, among others]. A paradox effect can be noted in the interaction and integration of synaptic activity. Active excitatory synapses, reducing the driving force which produces the EPSC, act as *"inhibitory"* of the excitation reducing (or preventing) the current generated by other excitatory synapses (autoregulation of the excitation [6]). The inhibitory synapses have the opposite effect because the Inhibitory Post Synaptic Current (IPSP), flowing in the opposite direction, restores the driving force of excitatory synapses acting like *"excitatory"*. These "paradox effects" should suggest a rethinking of the meaning of words like *"excitatory"* and *"inhibitory"* when considering synaptic activity in dendrites.

CP1028, *Collective Dynamics: Topics on Competition and Cooperation in the Biosciences*
edited by L. M. Ricciardi, A. Buonocore, and E. Pirozzi

The interactions between synapses and dendrites, depend both on passive and active mechanisms.
The input resistance of the dendritic branch, where the synapses are located, is one of the most important passive dendritic properties which can influence synaptic interaction. Different values of the input resistance, in fact, can produce different Excitatory Post Synaptic Potentials (EPSPs) for the same change of synaptic conductance [10]. It seems likely that the far dendritic branches, being thiner than branches located closer to the soma, can have an higher input resistance [1, 2, 3, 4] and, consequently, a greater EPSP for the same variation of synaptic conductance [10]. A consequence of this is that the degree of synaptic interaction depends on the location of synapses in the dendritic tree.
Additionally, dendrites can have active mechanisms based on Ca^{++} and Na^{+} voltage dependent ionic channels [see for example, 11, 12]. These channels, triggered by local membrane depolarization due to the excitatory synaptic activity, produce spikes which, diffusing in the dendritic tree, induce non localized effects on the EPSCs generation.
A complete description of the synaptic dynamics in the dendritic tree is not the goal of the present paper. Here, the effects of the interaction among synapses, and their influence on the synaptic integration, are analyzed by considering only the modification of the driving force for the EPSC depending on the passive dendritic properties. In addition, by discussing some results, will be proposed the new concept of *"Synaptic Competition For Plasticity"* as a phenomenon which could depend on the activation sequence of synapses located close the one to each other.

2. COMPUTATIONAL PROCEDURE

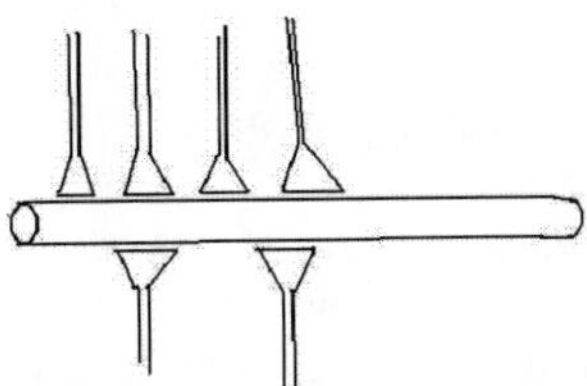

FIGURE 1. A schematic representation of a small dendritic branch with several synapses

To show the synaptic interaction on dendritic tree, the computational procedure will be restricted to a terminal dendritic branch as shown in fig.1. However, this configuration will be further simplified and synapses will be modelled as located to the same point so that all of them experience the same events simultaneously. This is the same to reduce the dendritic branch of fig.1 to a single point.
Although glutamatergic synapses have co-localized AMPA and NMDA receptors, for simplicity only the fast EPSC produced by the opening of the AMPA receptors will be considered. In addition, the same input resistance will be used for excitatory and inhibitory synapses irrespectively of the fact that the former are connected to the dendritic spines and the latter to the dendritic shaft [13, 14]. The membrane potential (V_m) is as-

sumed to be at the resting level (V_r) at the starting time t_0, (i.e., $V_m(t_0) = V_r = -70mV$), the dendritic branch to be isopotentail $\left(\frac{dV_m(x)}{dx} = 0\right)$, and no current flows in any direction.

Any synapse connected to this dendritic branch at any time $t \geq t_0$ produces a PSC ($I_s(t)$) according to

$$I_s(t) = \begin{cases} g_e(t)\left(V_m(t) - E_e\right) & \text{if synapse is excitatory} \\ g_i(t)\left(V_m(t) - E_i\right) & \text{if synapse is inhibitory} \\ 0 & \text{if synapse is inactive} \end{cases} \tag{1}$$

where $g_e(t)$ is time course of the excitatory synaptic conductance due to the opening of the ionic receptors (in our case AMPA receptors); $g_i(t)$ is the time course of the inhibitory conductance produced by the opening of GABA receptors; E_e and E_i are the reversal potential for the excitatory and inhibitory synapses, respectively. The difference $(V_m - E_{syn})$ is the driving force which produces the EPSC or the IPSC if $E_{syn} = E_e$ or $E_{syn}E_i$, respectively.

The EPSC at the resting level depolarizes the membrane (increases V_m) because E_e is close to $0mV$ and this reduces the difference $(V_m - E_e)$ which is the force generating the EPSC [see Results and 10, for numerical examples]. The IPSC repolarizes the membrane because E_i is close (if not more negative) to the resting potential and this will increase the difference $(V_m - E_e)$ increasing the force which produce the EPSC.

Although the amplitude of the EPSCs produced by AMPA receptors is in the order of 10^{-12} Amperes (on the average $\simeq 25pA$, [15]), the high input resistance of the membrane (R_i) of the far, thin, small dendrite is in the order of $10^9 Ohms$ ($G\Omega$) [3, 4, 8, 9] and consequently the EPSPs will be in the order of $10^{-3}V$ (which is the order of magnitude for all the electrical activities of the neurons). This means that: i) the EPSP produced by an active excitatory synapse can influence (reduce) the activity of any other excitatory synapse although the degree of influence depends on the distance [1, 2, 3, 4]; ii) few excitatory synapses, active at the same time, can easily produce a depolarization such to drive the membrane potential close to E_e. If this occurs (i.e., if $V_m = E_e$), the driving force becomes null and no current will flow (*"veto"* effect). Even more, if $V_m > E_e$ the current would invert the direction becoming a repolarizing current.

Synaptic conductance time course ($g(t)$), both for excitatory and inhibitory synapses, can be fitted by the difference of two exponentials.

$$g(t) = k\left(e^{-\frac{t}{\tau_2}} - e^{-\frac{t}{\tau_1}}\right), \tag{2}$$

where k is a constant related to the peak value of the current while τ_1 and τ_2 are respectively the rising and decay time constants. For the present computational experiments the values of τ_1 and τ_2 for the AMPA conductance was respectively 0.2 and $2.0ms$ while k was chosen and adjusted in dependence of the input resistance so to have the EPSCs with peak amplitudes comparable with data obtained experimentally [15, 16] and by computational models of synaptic transmission in pyramidal neurons [18, 19, 20].

By following the classical notation of the cable theory, the current produced on a dendritic branch like the one proposed in fig. 1 should be of the form

$$I_m(t) = C_m\left(\frac{dV_m(t)}{dt}\right) + g_e(t)(V_m(t) - E_e) + g_i(t)(V_m(t) - E_i) - \left(\frac{V_m(t) - V_r}{R_m}\right), \quad (3)$$

where C_m is the membrane capacitance and R_m the membrane resistance. The term $C_m\left(\frac{dV_m(t)}{dt}\right)$ represent the capacitative component of the membrane and the term $\left(\frac{V_m(t)-V_r}{R_m}\right)$ is a leaky current depending on the surface of the dendrite. For simplicity, herein, will be assumed that synapses are so close the one to each other that all of them furnish its current (EPSC or IPSC) to the same point. This assumption reduces the capacitance to zero and the membrane resistance can be considered as an input resistance R_i. The leaky current can be neglected because the high value of R_i. With this simplifications, the system under consideration becomes a purely resistive system the current of which can be computted as

$$I_m(t) = (V_m(t) - E_e)\sum_{i=0}^{n} g_{e_i}(t) + (V_m(t) - E_i)\sum_{j=0}^{m} g_{i_j}(t), \quad (4)$$

where $g_{e_i}(t)$ and $g_{i_j}(t)$ are the conductance at time t of the i_{th} excitatory and j_{th} inhibitory synapse respectively.
This current applied across an input resistance, R_i, will give a variation of potential by the Ohm's law ($V(t) = R_i I_m(t)$) and this variation of potential applied to the membrane resting level (V_r) will give the membrane potential $V_m(t)$.

$$V_m(t) = V_r - R_i\left\{(V_m(t) - E_e)\sum_{i=0}^{n} g_{e_i}(t) + (V_m(t) - E_i)\sum_{j=0}^{m} g_{i_j}(t)\right\} = V_r - R_i I_m(t), \quad (5)$$

The membrane potential computed in equation (5) was then used for each synapse in equation (4). This loop continued for the total simulation time which for all the computational experiments was $30ms$ with a time step of $1\mu s$.
This equation represents a great simplification of the reality. In a more realistic case, the leaky of current along the axial resistance, the leaky and capacitative components of equation 3, the distance between synapses and the possible current arriving from other dendritic branches should be considered according to the dendritic cable equation [1, 2, 3, 4].
For the analysis of the cooperative activity of synapses as described herein, the time integral of the membrane voltage can be a good estimator [7] and then

$$X = \int_0^t V_m(x)dx \quad (6)$$

where V_m is the EPSP as given by equation 5 was computed.
The time integral of the membrane potential can be considered as the marker of the energy produced by the synaptic activation.
In the present computational experiments, the time integral normalized to the time

integral of the single EPSP (or, for the case of simultaneous activation, normalized to the time integral of two synchronous EPSPs) was also calculated as

$$X' = \frac{\int_0^t V'_m(x)dx}{X}, \tag{7}$$

where V'_m is the EPSP or the composite EPSP (cEPSP) and X is the value of the integral of the single EPSP (eq. 6) or alternatively the integral of two synchronous EPSPs. The normalized time integral gives an immediate estimation of the changes occurring as a function of the considered parameters.

3. RESULTS

A first factor which can influence the mutual interaction between synapses located close the one to each other is the input resistance of the dendritic branch where they are located. The input resistance can vary between different dendritic areas so that for the

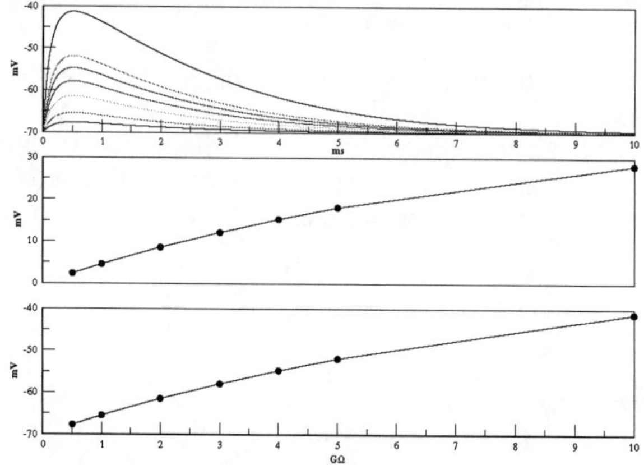

FIGURE 2. Dependence of the EPSP on the dendritic input resistance. The results have been obtained by changing R_i (ranging $0.5-10G\Omega$) for the same $g(t)$. Top panel shows the EPSPs time course. The middle and lower panels show respectively the peak amplitudes and the peak values of the membrane potential

same change of the AMPA conductance ($g_e(t)$) a different EPSC and a different EPSP can be obtained. Figure 2 shows the effect of the input resistance on the EPSP for the same variation of conductance.

Another important factor is the activation sequence of the synapses. The EPSP produced by one or more (cEPSP) synapses activated at a time t can influence the EPSC of synapses activated at a time $t_s = t + \delta : \{t \leq t_s < t + \Delta\}$, being Δ the total duration of the EPSP produced by the synapses activated at time t and $\delta < \Delta$. This is because the depolarization produced by any excitatory synapse contributes to the reduction of the driving force producing the EPSC of any other active synapse.

The fig. 3 shows the results of the interaction effect of several synapses activated synchronously (i.e., $t = t_s$) which is the limiting condition with $\delta = 0$. This limit

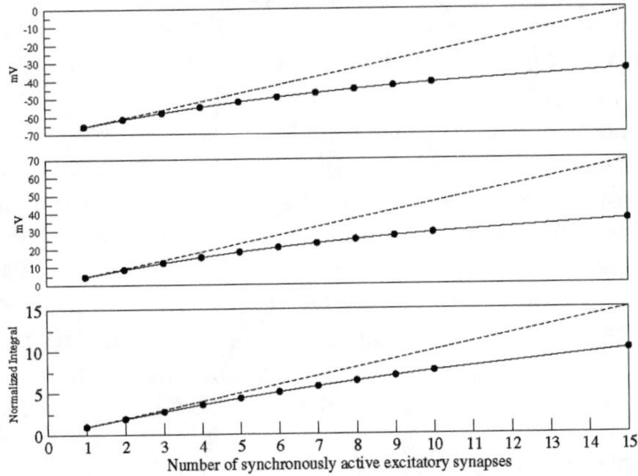

FIGURE 3. Cooperative effect of synchronous synapses (solid lines and symbols). The dashed line represent the values obtained if linear summation would occur. The top and middle panels show respectively the peak amplitudes and the peaks of the membrane potential for the composite EPSP. The bottom panel shows the values of the time integral normalized to the time integral of the single EPSP.

case shows the maximal interaction possible and from fig. 3 appears clear that non linear summation occurs for multiple synchronous synaptic activations. The same input resistance ($R_i = 5.0G\Omega$) and the same conductance ($g_e(t)$) were used across all the experiments for any active synapse. The difference between solid and dashed lines of fir.3 gives the size of the non linearity in the summation process when several synapses are simultaneously active and represents the digree of interaction.

For values of $\delta > 0$ (synapses simultaneously but not synchronously active), different degrees of interaction can be obtained. Figure 4 shows an example of interaction of two synapses for different values of δ. The value of (R_i) and the variation of conductance

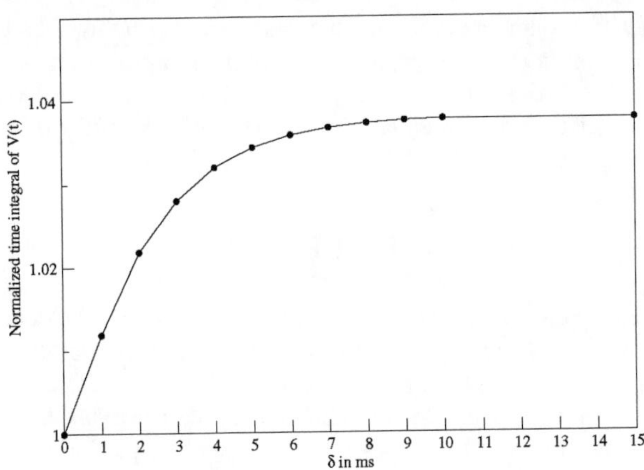

FIGURE 4. Time integral of the membrane voltage of two excitatory synapses as a function of δ, normalized to the time integral of the two synapses activated synchronously ($\delta = 0$).

($g_e(t)$) used for the computational experiments of fig. 4 are the same of those used for the experiments of fig. 3. An important comment on fig. 4 is that the maximal value of the time integral is obtained when the effect of the two synapses becomes completely independent which is for $\delta \geqslant \Delta$. In terms of energy we can say that when two synapses are activated synchronously ($\delta = 0$), the combined signal produces a minimal energy which becomes maximal when the two synapses act independently ($\delta \geqslant \Delta$). The normalized time integral appears very useful both to define the degree of non linearity in the summation and in defining the correct time window (the value of Δ) in which the activity of the two synapses can be considered cooperative or better "not independent". The value of Δ will be, in fact, the value of δ which gives the maximal value of the normalized integral. Finally, just as example, fig 5 shows the interaction of an excitatory and an inhibitory synapse. The action of the inhibitory synapses depends clearly on

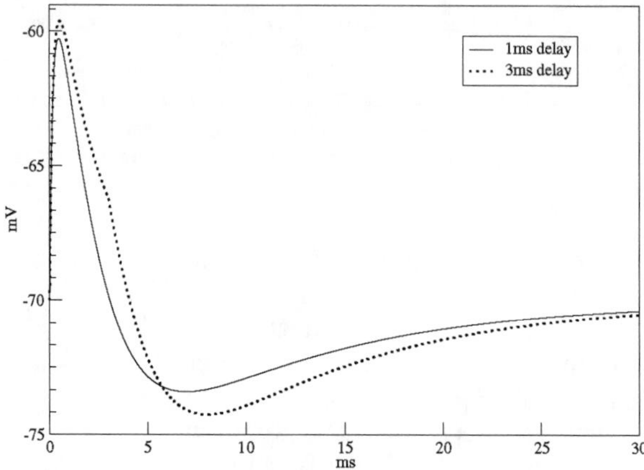

FIGURE 5. Example of interaction between an excitatory and an inhibitory synapse. Straight line $\delta = 1ms$ and dashed line $\delta = 3ms$

the time of activation. However, also in the case of the synchronous activation of an excitatory and an inhibitory synapses, the action of the inhibitory synapses is not simply to abolish the EPSP. The IPSP has a time course different from the one of the EPSP because the values of τ_1 and τ_2 of equation (2) are usually one order of magnitude larger for the inhibitory conductance than for the AMPA conductance.

4. DISCUSSION

In the present paper, some aspects of the synaptic interaction on the dendritic tree of the pyramidal neurons have been investigated by using simple computational experiments. A main goal has been to show that the synaptic activities on the dendritic tree is not only a simple integration process but a more complex regulatory system which tunes the information passing from a neuron to another, and may be a part of a complex computational system.

The degree of interaction among synapses depends on several factors. One of this is the input resistance of the dendritic branch where synapses are located. As shown in fig. 2,

the same variation of conductance ($g_e(t)$) can produce different effects depending on the input resistance.
Another important factor to consider is that synaptic events do not give independent responses (see fig.s 3 and 4). If one or more excitatory synapses are active in a dendritic area, then, the depolarization they produce reduces the driving force generating the EPSC and this produces two main effects: i) the activity of each synapse is limited because the driving force is reduced by the action of the others (auto limitation of the excitation, [6]); ii) a group of excitatory synapse, simultaneously active in the same dendritic area, can depolarize the membrane up to the reversal potential value (E_e). When this occur, the current produced by any active excitatory synapse become zero (*"veto"* effect).
The action of inhibitory synapses located on the same dendritic area can reduce or abolish the *"veto"* effect allowing the EPSC formation by some synapses that otherwise will not be able to transfer their information to the postsynaptic neuron.
By combining the interaction between active synapses located on the same dendritic area, the different input resistances in different dendritic branches [1, 2, 3, 4], the complex mechanisms of presynaptic regulation of the neurotransmitter release [for a review see 17, among many others], the complex mechanism of pre- and postsynaptic regulation of the EPSC [18, 19, 20, among others], the activity of the Ca^{++} and $Na+$ voltage dependent channels [11, 12], and the interplay between excitatory and inhibitory synapses, a complex scenario of interactions can be seen in the passage of information in the synaptic tree.
In summary, from this scenario, it appear clear that: i) synapses communicate with other synapses announcing that they are transmitting information and trying to prevent the simultaneous passage of the information by the other synapses; ii) the strength of the information passed between synapses depends on the distance and on the activation time; iii) the passage of information among synapses and among different branches of the dendritic tree, depends on the different passive (input resistance and diameter, for example) and active (voltage dependent ionic channels) dendritic properties.
The dependence of synaptic interaction on the distance suggests that the dendritic organization of pyramidal neurons, probably, follows a geometry defined by the need of communication between synapses such that synapses which carry information completely independent are segregates far the one to the others while those carrying competitive or cooperative information are situated close each other and maybe on the same dendritic branch.
The dependence of the outputs on the activation time, suggests that probably there is a sort of competition in the passage of information which can consequently result in a *"Competition For Plasticity"* among synapses located close the one to each other. Synapses activated sooner have a greater probability to pass their information than those activated later (because the first reduce the force for the latter and can also *"veto"* their activity) and this increase their probability to gain plasticity which is activity dependent.
The different input resistance of dendrites and its dependence on the distance from the soma, creates a hierarchy along the dendritic tree. This hierarchy combined with other passive properties, depending on the dendritic size (diameter), and in cooperation with the Ca^{++} and Na^{+} voltage dependent channels, defines the rules for the passage of information among different dendritic districts and finally to the hillock contributing to

the neuronal behavior (spike generation).
The interplay between excitatory and inhibitory synapses constitute a fine tuning system for information passing between neurons and among different dendritic branches.
As shown above, the reversal potential of excitatory synapses behaves like a threshold point and then represent a point of non linearity in the neuronal dynamics. This important threshold mechanism, combined with some intrasynaptic interactions and with synaptic-dendritic interactions, produces some operations similar to those used in the Boolean algebra computation [6].
In conclusion, the single pyramidal neuron, very likely, has to be considered as a device which performs operations of information processing and complex computational tasks. Probably, its computational ability make it more like a *"computer"* than like a simple linear integrator of analog inputs as earlier proposed.[21].

REFERENCES

1. W. Rall, *Exp. Neurol.*, **1**, 491–527 (1959).
2. W. Rall, "Theoretical significance of dendritic trees for neuronal input-output relationship".In *Neural theory and Modeling*, edited by R.F. Reis , Stanford University Press, Palo Alto, 1964.
3. W. Rall and J. Rinzel, *Biophys. J.*, **13**, 648–688 (1973)
4. J. Rinzel, J.and W. Rall, *Biophys. J.*, **14**, 759–790 (1974).
5. C. Koch, T. Poggio and V. Torre, *Proc. Natl. Acad. Sci. USA*, **80**, 2799–2802 (1983).
6. M. London and M. Häusser, *Annu. Rev. Neurosi*, **28**, 503–532 (2005).
7. I. Segev, and I. Parnas, *Biophys. J.*, **41**, 41–50, (1983).
8. I. Segev, J. Rinzel and G.M. Shepherd, *The theoretical foundation of dendritic function*, The MIT Press, Cambridge, London (1995).
9. I. Segev and M. London, *Science*, **290**, 744–750 (2000).
10. V. Di Maio. "Excitatory synaptic interaction on the dendritic tree". in *Lectur notes in computer sciences : Advances in Brain Vision and Artificial Intelligence*, edited by F. Mele et al., Springer-Verlag, Berlin, Heidelberg, 2007, pp 388–397.
11. N.L. Golding and N. Spruston, *Neuron*, **21**, 1189–1200 (1998).
12. N.L. Goldin, N.P. Staff and N. Spruston, *Nature*, **418**, 326–331 (2002).
13. N. Ishizuka, W.M. Cowan and D.G. Amaral, *J. Comp. Neurol.*, **362**, 17–45 (1995).
14. M. Megías, Z.S. Emri, T.F. Freund and A.I. Gulyás *Neurosci*, **102**, 527–540 (2001).
15. L. Forti, M. Bossi, A. Bergamaschi, A. Villa and A. Malgaroli, *Nature*, **338**, 874–878 (1997).
16. P. Jonas, G. Major and B. Sakman, *J. Physiol.*, **472**, 615–663 (1993).
17. V. Di Maio, *Brain Res., (2008, in press).*
18. F. Ventriglia and V. Di Maio, *Biosystem*, **67**, 287–294 (2002).
19. F. Ventriglia and V. Di Maio, *Biol. Cybernet*, **88**, 201–209 (2003).
20. F. Ventriglia and V. Di Maio, *Biosystem*, **71**, 195–204 (2003).
21. W.S. McCulloch and W.H. Pitts, *Bull. Math. Biophys*, **5**,115–133 (1943).

Comparison of Statistical Methods for Estimation of the Input Parameters in the Ornstein-Uhlenbeck Neuronal Model from First-Passage Times Data

Susanne Ditlevsen* and Petr Lansky†

*Department of Mathematical Sciences, University of Copenhagen, Denmark
†Institute of Physiology, Academy of Sciences of the Czech Republic, Prague, Czech Republic

Abstract. The Ornstein-Uhlenbeck neuronal model is reviewed, and estimation of the input parameters from first-passage times are being discussed. Three methods previously suggested in the literature are compared through simulations; namely what we have denoted the Moment Method, the Laplace Transform method, and the Integral Equation method. Finally the methods are also applied to experimental data, where the membrane potential is also recorded intracellularly. This permits to compare the methods based on the limited information contained in first-passage times only to evaluations from more complete observations of the otherwise hidden membrane potential.

Keywords: Parameter estimation, Laplace Transform, moment method, Fortet Integral equation, interspike interval, spontaneous firing.
PACS: 87.19.La, 02.50.–r,05.40.–a

1. INTRODUCTION

Mathematical models of a single neuron is a class of models which result from more or less dramatic simplifications of so-called biophysical neuronal models. These attempts to simplify are sometimes criticized, but there are studies showing that even substantial reductions of the Hodgkin-Huxley model mimics the reality quite well ([1, 2]). From a biophysical point of view, the models reflect electrical properties of neuronal membranes. These circuit models are written as a system of differential equations. Usually by reduction/simplification of these models an integrate-and-fire type model is obtained, which also can be seen as a very simple electrical circuit. This type of simplification implies that the shape of the action potential (spike) is neglected and all neuronal output is represented by a point event. Thus, the models describe the dynamics of interspike intervals only and they are based on a one-dimensional representation of the time evolution of the neuronal membrane depolarization.

The most common integrate-and-fire model is the deterministic leaky integrate-and fire (LIF) model, also known as the Lapicque model or RC-circuit (for a review see [3]),

$$\frac{dx(t)}{dt} = -\frac{x(t)-x_0}{\tau} + \mu(t), x(0) = x_0, \tag{1}$$

where $x(t)$ represents the cell membrane voltage, x_0 is the initial voltage after spike generation, $\mu(t)$ is an input signal, and $\tau > 0$ is a time constant governing the spontaneous

CP1028, *Collective Dynamics: Topics on Competition and Cooperation in the Biosciences*
edited by L. M. Ricciardi, A. Buonocore, and E. Pirozzi

decay of the voltage back to a resting level, which is set to the initial voltage. The solution of Eq. (1) includes an integral of the input signal $\mu(t)$ with exponentially decaying effect, which is why the model is named "leaky-integrator". Note that $\mu(t)$ appearing in (1) is a representation of an external signal (light, sound, odorant, or a sequence of incoming action potentials) transformed into an internal generator potential, a quantity having dimension of voltage per time. Following the electrical circuit representation implies that $\tau = RC$, where R is the membrane resistance and C is the capacitance, and thus $\mu(t) = I(t)/C$, where $I(t)$ is the incoming current to the neuron.

Due to the simplicity of model (1), the action potential generation is not an inherent part of the model like in more complex ones and the firing threshold S has to be imposed, where $S > x_0$. The model neuron fires whenever the threshold is reached and then the voltage $x(t)$ is reset to its initial value. The reset following the threshold crossing introduces a strong nonlinearity into the model. For a constant input $\mu(t) = \mu > S/\tau$, the neuron fires regularly, whereas for $\mu \leq S/\tau$, the model never reaches the threshold S and the neuron remains silent. This defines sub– and suprathreshold signal in model (1).

An input to a neuron is either a sequence of pulses coming from other neurons or a stimulation arriving from an external environment. Specification of the input as well as of characteristics of the neuron are important tasks. It creates two interconnected problems, generally denoted parameter identification, and it can be solved if either the trajectory of $x(t)$ or only the interspike intervals, i.e. the time intervals between the crossings of the threshold S, are available. The first problem is identification of the model parameters (S, x_0, τ), which have their biophysical meaning despite the simplicity of the model. In this case the input $\mu(t)$ is assumed known. Once the model parameters (S, x_0, τ) are found, it can be checked how well the model predicts spiking activity if the same input is applied to a model neuron as to a real neuron. The second problem is to identify the signal $\mu(t)$ impinging upon the neuron, assuming the parameters (S, x_0, τ) known. In this case the signal to the neuron is attempted reconstructed from the neuronal behavior.

The experimental data recorded from different neuronal structures and under different experimental conditions suggest that stochasticity is present in neuronal activity. A phenomenological way of how to introduce stochasticity into the deterministic LIF model is simply by assuming an additional noise term in model (1),

$$\frac{dX(t)}{dt} = -\frac{X(t) - x_0}{\tau} + \mu(t) + \sigma(t)F(t),\, X(0) = x_0, \tag{2}$$

where $F(t)$ represents Gaussian and δ–correlated noise with zero mean. Arguments can be given that $\sigma(t)$ is a part of the signal, not a biophysical parameter of the model ([4]). Model (2) with constant μ and σ and $F(t)$ a Gaussian white noise, is well known as an Ornstein-Uhlenbeck model ([3]). Similarly to the deterministic model (1), also in stochastic LIF models, the firing is not an intrinsic part of the model and a firing threshold has to be imposed. In these models a stochastic process $X(t)$ (the capital X is used to distinguish formally between deterministic and stochastic models) describing the membrane depolarization makes random excursions to the firing threshold S. When the threshold is reached, a firing event occurs and the membrane depolarization is reset deterministically to its starting point $X(0)$. The interspike intervals are identified with the first-passage times of X across S.

Analogously to the deterministic model, one can try to estimate the signal assuming the neuronal parameters known, or reversely estimate the model parameters assuming the signal known, in both cases from data consisting of either the trajectory of $X(t)$ or from the interspike interval times only. This contribution is devoted to the methods for signal estimation in model (2) when the signal is constant, i.e. $\mu(t) = \mu$ and $\sigma(t) = \sigma$, and when only the interspike interval times are available. The intrinsic parameters (S, x_0, τ) of the model are assumed to be known. In Section 2 the model is presented, and probabilistic properties of the first-passage time needed in the development of the estimation methods are reviewed. In Section 3 three estimation methods; the Moment, the Laplace Transform and the Integral Equation method, are reviewed. In Section 4 the three methods are applied to simulated data in sub-, supra- and threshold regimes, and to experimental data from auditory neurons, and the performances of the methods are compared. Finally, in Section 5 the results, and advantages and drawbacks of each method are discussed.

2. MODEL AND PARAMETERS

The changes in the membrane potential between two consecutive neuronal firings are represented by a stochastic process X_t indexed by the time t. The reference level for the membrane potential is taken to be the resting potential. The initial voltage (the reset value following a spike) is assumed to be equal to the resting potential and set to zero, $X_0 = x_0 = 0$. An action potential is produced when the membrane voltage X_t exceeds a voltage threshold for the first time, for simplicity assumed to be equal to a constant $S > 0$. Formally, the interspike interval T is identified with the first-passage time of the threshold,

$$T = \inf\{t > 0 : X_t \geq S\}. \tag{3}$$

It follows from the model assumptions that for time-homogeneous input containing either a Poissonian or white noise only, the interspike intervals form a renewal process and the initial time following a spike can always be identified with zero. Here we consider a Gaussian white noise input and thus X_t is a diffusion process. A scalar diffusion process $X = \{X_t; t \geq 0\}$ can be described by the stochastic differential equation

$$\mathrm{d}X_t = \mu(X_t,t)\,\mathrm{d}t + \sigma(X_t,t)\,\mathrm{d}W_t, \qquad X_0 = x_0 \tag{4}$$

where $W = \{W_t; t \geq 0\}$ is a standard Wiener process and $\mu(\cdot)$ and $\sigma^2(\cdot)$ are real-valued functions of their arguments called the infinitesimal mean and variance. For LIF the diffusion process given in Eq. (4) specified by the infinitesimal mean

$$\mu(X_t,t) = -\frac{X_t}{\tau} + \mu \tag{5}$$

is considered, where constant μ characterizes the neuronal input and $\tau > 0$ reflects spontaneous voltage decay (the membrane time constant) in absence of input. Moreover, the constant square root of the infinitesimal variance,

$$\sigma(X_t,t) = \sigma > 0 \tag{6}$$

is the second input parameter, see also [5]. The diffusion process (4) with the infinitesimal moments given by (5) and (6) defines the Ornstein-Uhlenbeck (OU) diffusion process:

$$dX_t = \left(-\frac{X_t}{\tau} + \mu\right) dt + \sigma \, dW_t \; ; \; X_0 = x_0 = 0. \quad (7)$$

The parameters appearing in (3) and (7) can be divided into two groups: parameters characterizing the input, μ and σ, and intrinsic parameters, τ, x_0 and S, which describe the neuron irrespectively of the incoming signal ([6]). Note that compared with the deterministic LIF given by Eq. (1), an additional parameter σ appears in Eq. (7).

The transition probability density function of model (7) in absence of a threshold is normal with mean M_t and variance V_t ([7]), and given by

$$f(x,t) = (2\pi V_t)^{-\frac{1}{2}} \exp\left\{-\frac{(x-M_t)^2}{2V_t}\right\} \quad (8)$$

where

$$M_t = \mu\tau(1-e^{-t/\tau}), \quad (9)$$

$$V_t = \frac{\sigma^2\tau}{2}(1-e^{-2t/\tau}). \quad (10)$$

Analogously to the deterministic LIF described in the Introduction, two distinct firing regimes, usually called sub- and suprathreshold, can be established for the OU model. In the suprathreshold regime, the asymptotic mean depolarization $\mu\tau$ given by (9) is far above the firing threshold S and the ISIs are relatively regular (deterministic firing - which means that the neuron is active also in the absence of noise). In the subthreshold regime, $\mu\tau \ll S$ and firing is caused only by random fluctuations of the depolarization (stochastic or Poissonian firing). The term "Poissonian firing" indicates that when the threshold is far above the steady-state depolarization $\mu\tau$ (relatively to σ), the firing achieves characteristics of a Poisson point process ([8, 9]). For our purposes, let us denote the third regime, when $\mu\tau \approx S$, as the threshold regime. Division of the firing regimes in three parts was already proposed in [9].

The properties of the random variable T including its probability density function $g(t \mid x_0, S) = g(t)$ have been extensively studied ([10, 8, 11, 12, 13, 14, 15]). The distribution $g(t)$ is only known for the specific situation $\mu\tau = S$, where the first-passage time density of the OU process across the boundary S is ([16, 7]):

$$g(t) = \frac{2S\exp(2t/\tau)}{\sqrt{\pi\tau^3\sigma^2}(\exp(2t/\tau)-1)^{\frac{3}{2}}} \exp\left\{-\frac{S^2}{\sigma^2\tau(\exp(2t/\tau)-1)}\right\}. \quad (11)$$

When $\mu\tau \neq S$ approximation techniques have been devised ([12]), of which many are based on the renewal equation, the so-called Fortet's equation ([17, 18]) relating the first passage-time density and the transition density $f(\cdot)$ for $x \geq S$,

$$f(x,t \mid x_0) = \int_0^t f(x,t-u \mid S) g(u \mid x_0, S) \, du. \quad (12)$$

We write $F(x,t-s\,|\,x_s)=\int^x f(v,t-s\,|\,x_s)\,\mathrm{d}v$ for the corresponding transition distribution function. Also the Laplace transform of T has been used to find characteristics of the first-passage time distribution $g(t)$. A representation is given by

$$E\left[e^{kT/\tau}\right]=\frac{\exp\{\frac{(\mu\tau)^2}{2\tau\sigma^2}\}D_k\left(\frac{\sqrt{2}\mu\tau}{\sqrt{\tau}\sigma}\right)}{\exp\{\frac{(\mu\tau-S)^2}{2\tau\sigma^2}\}D_k\left(\frac{\sqrt{2}(\mu\tau-S)}{\sqrt{\tau}\sigma}\right)}=\frac{H_k\left(\frac{\mu\tau}{\sqrt{\tau}\sigma}\right)}{H_k\left(\frac{(\mu\tau-S)}{\sqrt{\tau}\sigma}\right)} \tag{13}$$

for $k<0$, where $D_k(\cdot)$ and $H_k(\cdot)$ are parabolic cylinder and Hermite functions, respectively, see [19, 20]. Eq. (13) can be extended to $k>0$ when $(\tau,\mu,\sigma)=\theta\in\Theta^{(k)}=\{\theta\,|\,\mu\tau>S,\sqrt{\tau\sigma^2}<(\mu\tau-S)/\lambda^{(k)}\}$ to ensure that $E\left[e^{kT/\tau}\right]<\infty$ ([15]). Here $\lambda^{(k)}$ is the largest root of the k'th Hermite polynomial.

In [11] moments of T were given, the first two being

$$E[T] = \tau(\phi_1(\eta)-\phi_1(\xi)) \tag{14}$$

$$E[T^2] = \tau^2\left(2\phi_1^2(\eta)-\phi_2(\eta)-2\phi_1(\eta)\phi_1(\xi)+\phi_2(\xi)\right) \tag{15}$$

where

$$\xi = -\mu\tau\sqrt{2/\sigma^2\tau} \tag{16}$$

$$\eta = (S-\mu\tau)\sqrt{2/\sigma^2\tau} \tag{17}$$

$$\phi_1(z) = \frac{1}{2}\sum_{n=1}^{\infty}\frac{(\sqrt{2}z)^n}{n!}\Gamma\left(\frac{n}{2}\right) \tag{18}$$

$$\phi_2(z) = \phi_1(z)(\psi(n/2)-\psi(1)) \tag{19}$$

and $\Gamma(\cdot)$ and $\psi(\cdot)$ denote the gamma and the digamma function, respectively.

It is sometimes convenient to reformulate model (3) and (7) to the equivalent dimensionless form

$$\mathrm{d}Y_s=(-Y_s+\alpha)\,\mathrm{d}s+\beta\,\mathrm{d}W_s,\qquad Y_0=0 \tag{20}$$

where

$$s=\frac{t}{\tau},\,Y_s=\frac{X_t}{S},\,W_s=\frac{W_t}{\sqrt{\tau}},\,\alpha=\frac{\mu\tau}{S},\,\beta=\frac{\sigma\sqrt{\tau}}{S} \tag{21}$$

and $T/\tau=\inf\{s>0:Y_s\geq 1\}$. Note however that the model now operates on the timescale of $s=t/\tau$, not on the original measured timescale. All observed ISIs thus have to be transformed by dividing by τ.

3. PARAMETER ESTIMATION

Estimation of μ and σ when discrete observations of the trajectory of $X(t)$ are available can be done by the maximum likelihood method. In this case it is straightforward to estimate τ also, and is for completeness included in the likelihood equations below. Assume $X(t)$ observed at equidistant time points $i\Delta$, $i=0,1,\ldots,n$, for some $\Delta>0$. Let

X_i be the observation of $X(t)$ at time $i\Delta$. The maximum likelihood estimators are given by the equations

$$\hat{\mu}\hat{\tau} = \frac{\sum_{i=1}^{n}(X_i - X_{i-1}e^{-\Delta/\hat{\tau}})}{n(1-e^{-\Delta/\hat{\tau}})} \tag{22}$$

$$e^{-\Delta/\hat{\tau}} = \frac{\sum_{i=1}^{n}(X_i - \hat{\mu}\hat{\tau})(X_{i-1} - \hat{\mu}\hat{\tau})}{\sum_{i=1}^{n}(X_{i-1} - \hat{\mu}\hat{\tau})^2} \tag{23}$$

$$\hat{\sigma}^2 = \frac{2\sum_{i=1}^{n}(X_i - \hat{\mu}\hat{\tau} - (X_{i-1} - \hat{\mu}\hat{\tau})e^{-\Delta/\hat{\tau}})^2}{n(1-e^{-2\Delta/\hat{\tau}})\hat{\tau}} \tag{24}$$

(see e.g. [21, p. 152] for $\mu = 0$). The maximum likelihood estimator exists only if $\sum_{i=1}^{n}(X_i - \hat{\mu}\hat{\tau})(X_{i-1} - \hat{\mu}\hat{\tau}) > 0$. In the case when τ is known, the likelihood equations become particularly simple, and the estimators are explicit and exist always. See also [22, 23, 24, 25] for estimation from intracellular data in the neuronal context.

If first-passage times only are available, the attempts to solve the estimation problem are rare, and the membrane time constant has to be assumed, or otherwise estimated from other types of data. Some references are [26, 27, 5, 28, 29, 30].

In this paper three approaches to the estimation problem from first-passage time data are compared through simulations. Assume n observations of the random variable $T : t_i, i = 1, \ldots, n$. The first approach we call the Moment Method, which was proposed by [26], where they applied Eqs. (14) and (15). The next approach we call the Laplace Transform method, proposed by [5], where the extensions to the positive half plane of Eq. (13) was applied. Finally, the Integral Equation method, proposed in [29] and applied in [28], based on Eq. (12), is treated. The three methods are also illustrated on experimental data.

3.1. The Moment method

In [26] it is proposed to equate the sample moments

$$m_1 = \frac{1}{n}\sum_{i=1}^{n} t_i \tag{25}$$

$$m_2 = \frac{1}{n}\sum_{i=1}^{n} t_i^2 \tag{26}$$

with expressions (14) and (15), thus yielding estimates of η and ξ, which provides estimates of μ and σ through (16) and (17). However, (18) and (19) are only useful for numerical calculations whenever $z \geq 0$ and $|z|$ is small. Otherwise approximations of (18) and (19) must be used. These expressions have to be evaluated at $z = \xi$ and $z = \eta$. Note that ξ indicates the distance between reset value and asymptotic mean, and η indicates the distance between the threshold and the asymptotic mean measured in units of the asymptotic standard deviation of the membrane potential. Here $\eta < 0$ corresponds to suprathreshold regime, and when $|\eta|$ is large the model is far away from threshold regime. Thus this methods works best in moderate subthreshold regime. In [26] tables are provided of the estimated parameters as functions of the sample first moment m_1/τ and the sample coefficient of variation (CV). The tables cover sample values of $0.5 \leq m_1/\tau \leq 50$ and $0.1 \leq \text{CV} \leq 2$, and are calculated for $\tau = 5$ ms and $S - x_0 = 15$ mV. It is straightforward to transform the estimates from the tables to relevant values of the intrinsic parameters.

3.2. The Laplace Transform method

By defining suitable martingales and applying Doob's Optional-Stopping Theorem, it is possible to find closed expressions for the Laplace transform $E[e^{kT/\tau}]$, $k = 1, 2$, in the suprathreshold regime and with certain restrictions on the size of σ. These expressions can then be applied to define estimators for this parameter region. In [5] it is shown that

$$E[e^{T/\tau}] = \frac{\mu\tau}{\mu\tau - S}, \tag{27}$$

$$E[e^{2T/\tau}] = \frac{2(\mu\tau)^2 - \tau\sigma^2}{2(\mu\tau - S)^2 - \tau\sigma^2}, \tag{28}$$

if

$$\sigma^2 < \frac{2(\mu\tau - S)^2}{\tau}. \tag{29}$$

Condition (29) means that the asymptotic standard deviation of X_t is smaller than the distance between the threshold and the asymptotic mean of X_t. Straightforward estimators of $E[e^{T/\tau}]$ and $E[e^{2T/\tau}]$ are obtained from the empirical moments:

$$Z_1 = \hat{E}[e^{T/\tau}] = \frac{1}{n}\sum_{i=1}^{n} e^{t_i/\tau} \tag{30}$$

$$Z_2 = \hat{E}[e^{2T/\tau}] = \frac{1}{n}\sum_{i=1}^{n} e^{2t_i/\tau}. \tag{31}$$

Moment estimators of the parameters, assuming that the data are in the allowed parameter region, are then obtained from equations (27) and (28)

$$\hat{\mu} = \frac{SZ_1}{\tau(Z_1 - 1)} \tag{32}$$

and

$$\hat{\sigma}^2 = \frac{2S^2(Z_2 - Z_1^2)}{\tau(Z_2 - 1)(Z_1 - 1)^2}. \tag{33}$$

Note that the asymptotic depolarization will always be estimated to be suprathreshold ($\hat{\mu}\tau > S$), and that $0 < \hat{\sigma}^2 < 2(\hat{\mu}\tau - S)^2/\tau$.

3.3. The integral equation method

In [29] it is proposed to apply the integral equation (12) in the following way. Define $\theta = (\alpha, \beta)$, where α and β are given by Eq. (21). The probability

$$P[X_t > S \mid X_0 = x_0] \quad = \quad 1 - F_\theta(S, t|x_0) \tag{34}$$

can alternatively be calculated by the transition integral

$$P[X_t > S|X_0 = x_0] = \int_0^t g_\theta(u)(1 - F_\theta(S, t-u|S))\,\mathrm{d}u. \tag{35}$$

For fixed θ, the probability expressed by the right hand side of (34) is a function of t and can be calculated directly using Eq. (8). For the same value of θ, the probability expressed by the right hand side of (35) can be estimated at t from the sample by the average

$$\frac{1}{n}\sum_{i=1}^{n}(1 - F_\theta(S, t - t_i|S))\,1_{\{t_i \le t\}} \tag{36}$$

where 1_A is the indicator function of the set A, since it is the expected value of

$$1_{T\in[0,t]}\,(1 - F_\theta(S, t - T\,|S)) \tag{37}$$

with respect to the distribution of T. A statistical error measure is then defined as the maximum over t of the distance between (34) and (36), suitably normalized by dividing by $\omega(\theta) = \sup_{t>0}(1 - F_\theta(S, t|x_0))$ so that (34) will vary between 0 and 1 for all θ. To find the maximum over t a grid on the positive real line has to be chosen. A good choice for fixed θ is the set $\{t \in \mathbf{R}_+ : (1 - F_\theta(S, t|x_0))/\omega(\theta) = i/N, i = 1, \ldots, N-1\}$ for some reasonably large number N. In the applications below $N = 100$. The estimator of θ is finally obtained by minimizing this error function over the parameter space. The estimate is denoted $\hat{\theta}$.

Let $\Phi(\cdot)$ be the normal cumulative distribution function. Combining (34) and (8) and applying the transformations (20) and (21) we obtain

$$P[Y_s > 1\,|\,Y_0 = 0] \quad = \quad \Phi\left(\frac{\alpha(1 - e^{-s}) - 1}{\sqrt{1 - e^{-2s}}\,\beta/\sqrt{2}}\right), \tag{38}$$

which we estimate from the sample using (36) by

$$\frac{1}{n}\sum_{i=1}^{n}\Phi\left(\frac{\alpha - 1}{\beta/\sqrt{2}}\sqrt{\frac{1 - e^{-(s-s_i)}}{1 + e^{-(s-s_i)}}}\right)1_{\{s_i \le s\}}, \tag{39}$$

where $s_i = t_i/\tau$. The normalizing constant is given by $\Phi[(\alpha - 1)/(\beta/\sqrt{2})]$ for $\alpha \ge 0$ and $\Phi[-\sqrt{1 - 2\alpha}/(\beta/\sqrt{2})]$ for $\alpha < 0$. Then $\hat{\alpha}$ and $\hat{\beta}$ can be transformed to estimates of μ and σ through (21).

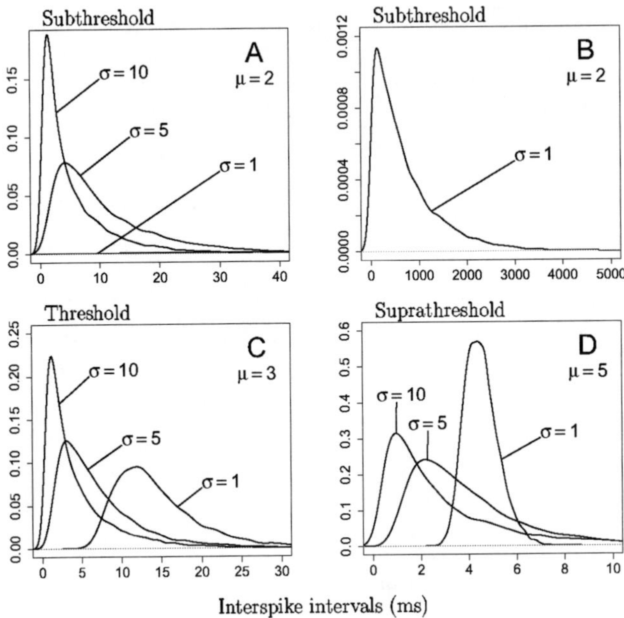

FIGURE 1. Empirical densities from simulated data, each density corresponds to 10.000 ISIs measured in ms. A: Subthreshold regime, $\mu = 2$ V/s and $\sigma = 1, 5$ or 10 mV/$\sqrt{\text{ms}}$. B: Subthreshold regime, $\mu = 2$ V/s and $\sigma = 1$ mV/$\sqrt{\text{ms}}$, enlarged from panel A. C: Threshold regime, $\mu = 3$ V/s and $\sigma = 1, 5$ or 10 mV/$\sqrt{\text{ms}}$. D: Suprathreshold regime, $\mu = 5$ V/s and $\sigma = 1, 5$ or 10 mV/$\sqrt{\text{ms}}$. In all cases $\tau = 5$ ms and $S = 15$ mV.

4. APPLICATIONS

4.1. Simulated data

Trajectories from the OU process were simulated according to the Euler scheme with a stepsize of 0.001 ms for different input parameter values. In all simulations the intrinsic parameters were set to $S - x_0 = 15$ mV and $\tau = 5$ ms. The process was run until reaching the threshold S where the time was recorded. This was repeated 10000 times, and the simulated observations were split into 100 data sets of 100 ISIs each. Nine different sets of input parameter values were obtained by combining 3 different values of μ, namely 2 (subthreshold), 3 (threshold) and 5 V/s (suprathreshold), and of $\sigma = 1, 5$ and 10 mV/$\sqrt{\text{ms}}$. Empirical densities of the simulated data are given in Fig. 1 for the nine combinations of parameter values. The densities are highly skewed, except in suprathreshold regime with small variance, where the distribution is approximately Normal.

On the data sets obtained, μ and σ were estimated by the three estimation methods. Note that the assumptions for the Laplace Transform method is not fulfilled in subthreshold and in threshold regime. In suprathreshold regime the assumptions are always met

TABLE 1. Values of μ and σ used in the simulations of the ISIs, and the corresponding averages of the samples of 100 parameter estimates $\pm$ the sample standard deviation (SSD), using the different estimation methods. All first-passage time samples contain 100 simulated observations.

	True values		statistics of 100 estimates: average $\pm$ SSD	
regime	μ	σ	$\hat{\mu}$	$\hat{\sigma}$
Moment method				
	2	1	0.55 ± 2.53	2.99 ± 3.06
subthreshold	2	5	2.04 ± 0.40	4.84 ± 0.90
	2	10	2.11 ± 0.77	10.06 ± 1.68
	3	1	2.97 ± 0.06	1.02 ± 0.17
threshold	3	5	3.02 ± 0.32	5.05 ± 0.68
	3	10	3.47 ± 0.84	9.75 ± 1.67
	5	1	5.25 ± 0.08	1.03 ± 0.08
suprathreshold	5	5	5.39 ± 0.32	5.26 ± 0.61
	5	10	4.88 ± 1.89	10.02 ± 1.61
Laplace transform method				
	2	1	2.94 ± 0.42	0.00 ± 0.00
subthreshold	2	5	3.03 ± 0.03	0.09 ± 0.10
	2	10	3.31 ± 0.23	0.92 ± 0.67
	3	1	3.06 ± 0.03	0.18 ± 0.08
threshold	3	5	3.32 ± 0.17	0.93 ± 0.46
	3	10	3.83 ± 0.46	2.30 ± 1.09
	5	1	5.00 ± 0.05	0.99 ± 0.09
suprathreshold	5	5	4.95 ± 0.28	3.97 ± 0.49
	5	10	5.22 ± 0.66	5.32 ± 1.11
Integral equation method				
	2	1	1.80 ± 0.78	1.42 ± 0.80
subthreshold	2	5	2.14 ± 0.32	4.70 ± 0.60
	2	10	2.51 ± 0.57	9.43 ± 1.02
	3	1	3.00 ± 0.06	0.99 ± 0.13
threshold	3	5	3.04 ± 0.28	4.88 ± 0.56
	3	10	3.34 ± 0.65	9.42 ± 1.01
	5	1	5.00 ± 0.05	1.00 ± 0.09
suprathreshold	5	5	4.95 ± 0.29	5.01 ± 0.54
	5	10	5.14 ± 0.62	9.75 ± 0.80

for estimation of μ, but only met for estimation of σ when the true $\sigma < 6.3$. Results are reported in Table 1, where averages and sample standard deviations are given for the 100 estimates within each parameter combination and estimation method. To evaluate the shape of the distribution and correlation of estimates, in Fig. 2 (subthreshold), 3 (threshold), and 4 (suprathreshold) the estimates of σ are plotted against the estimates of

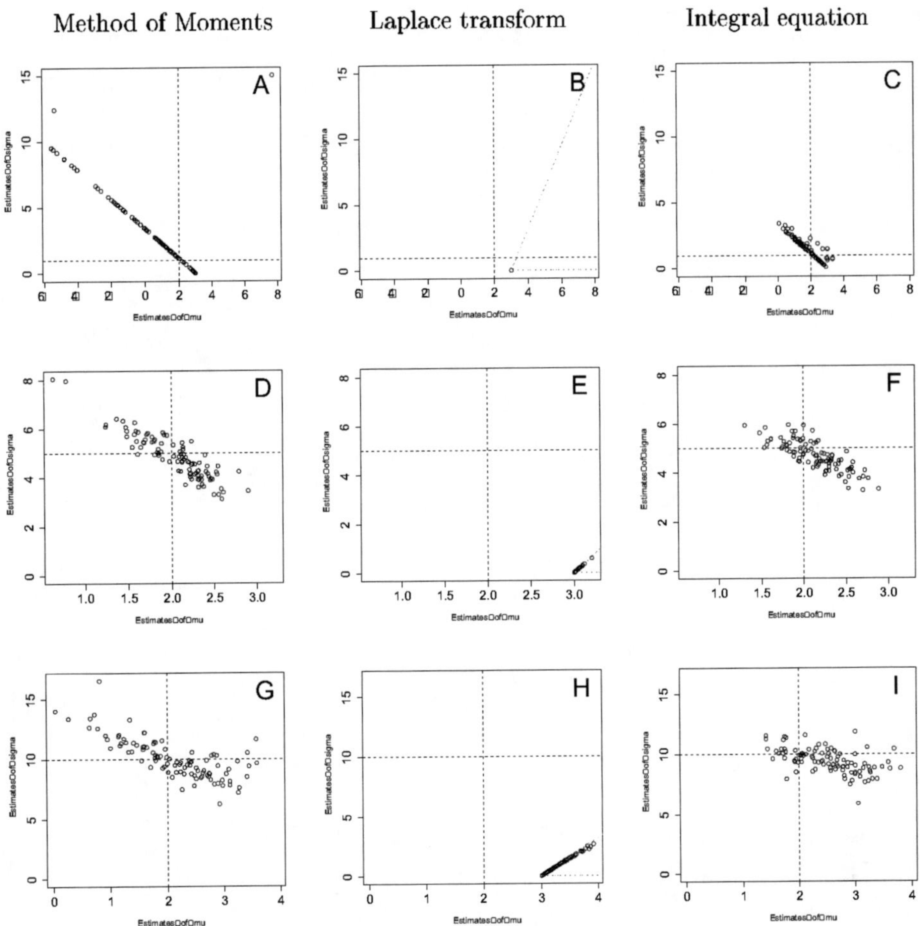

FIGURE 2. Estimates of σ plotted against estimates of μ in subthreshold regime for the method of moments (A, D and G), method of Laplace transform (B, E and H), and method of integral equation (C, F and I). True values in simulated data sets are $\mu = 2$ V/s (all panels), $\sigma = 1$ mV/$\sqrt{\text{ms}}$ (A, B and C), $\sigma = 5$ mV/$\sqrt{\text{ms}}$ (D, E and F), and $\sigma = 10$ mV/$\sqrt{\text{ms}}$ (G, H and I). In all cases $\tau = 5$ ms and $S = 15$ mV. Note that the Laplace transform method is not valid in subthreshold regime, and can only estimate to the right of the gray dashed lines (B, E and H).

μ. Estimates from the Moment and from the Integral Equation methods appear strongly negatively correlated in sub- and threshold regime. The Moment method is biased in suprathreshold regime, more pronounced for smaller σ. In nearly all cases the Integral Equation method has smaller variance than the Moment method, ranging from slightly to considerably smaller. If the assumptions for the Laplace Transform method are not met, the estimates are seen to gather on the border of their definition area. When the process

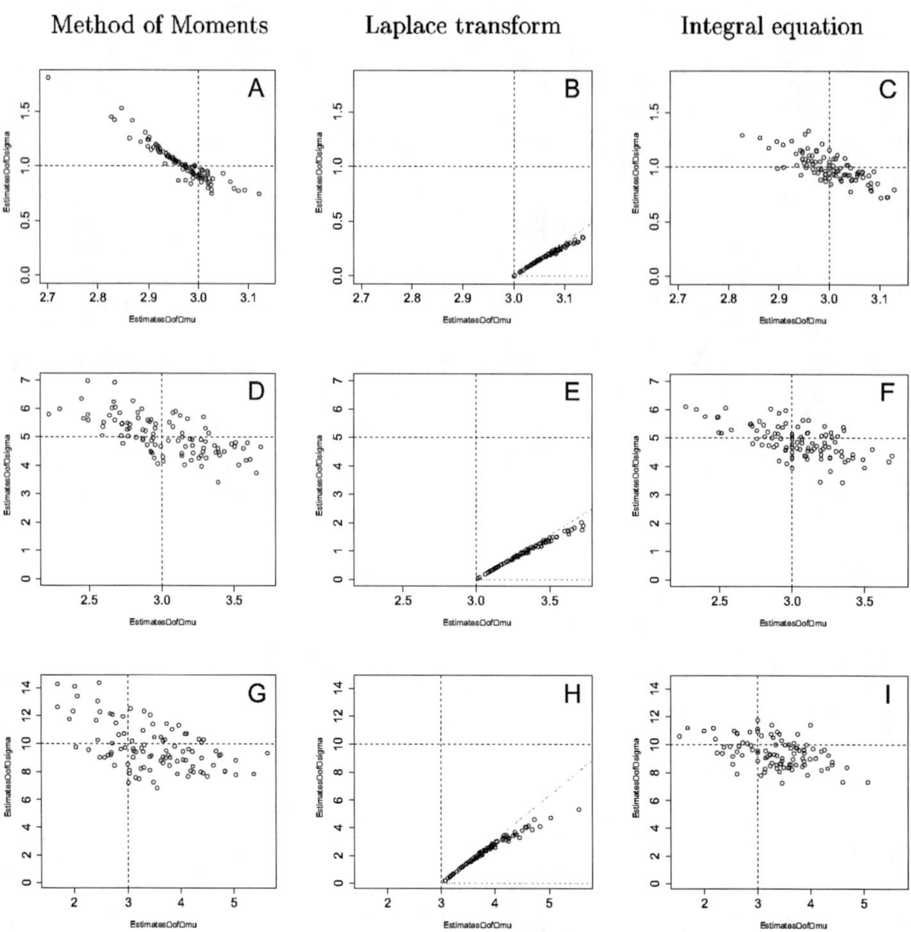

FIGURE 3. Estimates of σ plotted against estimates of μ in threshold regime for the method of moments (A, D and G), method of Laplace transform (B, E and H), and method of integral equation (C, F and I). True values in simulated data sets are $\mu = 3$ V/s (all panels), $\sigma = 1$ mV/$\sqrt{\text{ms}}$ (A, B and C), $\sigma = 5$ mV/$\sqrt{\text{ms}}$ (D, E and F), and $\sigma = 10$ mV/$\sqrt{\text{ms}}$ (G, H and I). In all cases $\tau = 5$ ms and $S = 15$ mV. Note that the Laplace transform method for σ is not valid in threshold regime, and can only estimate to the right of the gray dashed lines (B, E and H).

is suprathreshold and the variance is small, the Laplace Transform method performs as well as any of the others.

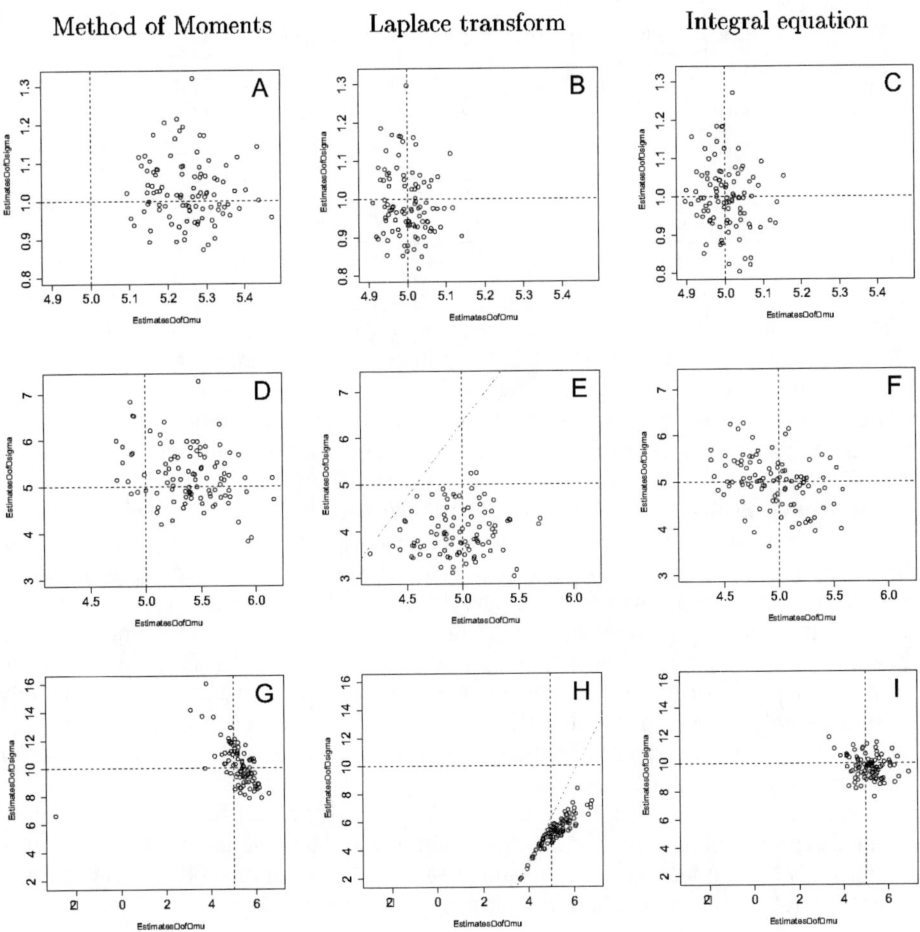

FIGURE 4. Estimates of σ plotted against estimates of μ in suprathreshold regime for the method of moments (A, D and G), method of Laplace transform (B, E and H), and method of integral equation (C, F and I). True values in simulated data sets are $\mu = 5$ V/s (all panels), $\sigma = 1$ mV/$\sqrt{\text{ms}}$ (A, B and C), $\sigma = 5$ mV/$\sqrt{\text{ms}}$ (D, E and F), and $\sigma = 10$ mV/$\sqrt{\text{ms}}$ (G, H and I). In all cases $\tau = 5$ ms and $S = 15$ mV. Note that the Laplace transform method for $\sigma = 10$ is not valid, and can only estimate to the right of the gray dashed lines (B, E and H).

4.2. Experimental data

The two sets of experimental ISI data were recorded intracellularly from the auditory system of a guinea pig (for details on stimulation protocol, data acquisition and processing see [31]). The first set consists of 312 ISIs during spontaneous activity and being based on the intracellular data (not only the ISIs), the parameters were also estimated

TABLE 2. Estimates from experimental data by different methods.

	Spontaneous record		Stimulated record	
Method	μ [V/s]	σ [V/$\sqrt{s}$]	μ [V/s]	σ [V/$\sqrt{s}$]
From intracellular data	0.278	0.014	1.344	0.028
Moment method	0.182	0.012	0.700	0.027
Laplace transform method	0.282	≈ 0	1.347	0.041
Integral equation method	0.240	0.005	1.351	0.035

using all information contained in the trajectory ([23, 25]). This permits us to evaluate the estimation procedures. Moreover, intracellular data gives "exact" information about the threshold value, the reset value and the membrane time constant. The second set, obtained from the same neuron, consists of 83 ISIs within the stimulation period. The intrinsic parameters were considered to be the same for both spontaneous and stimulated activity, and set to $S - x_0 = 11$ mV and $\tau = 39$ ms ([23]).

Estimation results are shown in Table 2. Note that the value of μ, which reflects intensity of stimulation, is 5–6 times larger than for the spontaneous activity. The estimates using the extra information contained in the samples of the membrane potential should be trusted more than the estimates from only the ISI times. Thus, comparing to the estimation from intracellular recordings, the Integral Equation method behaves better for μ than the Moment method, and conversely for σ. The Laplace Transform method cannot be trusted during spontaneous activity, since the neuron is most probably in subthreshold regime. In fact, note that $S/\tau = 11$ mV /39 ms = 0.282 V/s is exactly the estimate for μ that the Laplace Transform method gives, thus it estimates on the border of the definition area (the threshold).

When the neuron is stimulated, the Laplace Transform and the Integral Equation methods agree, whereas the Moment method gives smaller estimates. From the simulation study the Laplace Transform and the Integral Equation methods should be more trusted than the Moment method in suprathreshold regime.

5. CONCLUSIONS

As anticipated from the theoretical results in Section 3, the Moment method works best in moderate subthreshold regime, the Laplace Transform method only works in suprathreshold regime, and the Integral Equation method works in the entire parameter space.

The first problem one faces when estimating the input parameters in the Ornstein-Uhlenbeck neuronal model when only observations of first-passage times are available is determination of the regime. This is not a solvable task, and the Integral Equation method should be applied. Both the Moment method and the Integral Equation method can be computationally cumbersome, at least for researchers with limited mathematical and computational skills. The Laplace transform method, on the other hand, is easily implemented, but has the drawback of being restricted to a subset of the parameter space.

The simulation study reveals that in nearly all cases the Integral Equation estimator has small variance compared to the other approaches and appear to have limited bias. It also seems reasonable to assume an asymptotic normal distribution of the estimator. Due to these good statistical properties, this approach is recommended in most problems. Only in the case when the regime is known to be suprathreshold, and μ is the key parameter of interest, the Laplace transform method can compete, and should be preferred because it is easier to implement.

ACKNOWLEDGMENTS

The authors thank J.F. He for making the experimental data available. Supported by grants from the Danish Medical Research Council and the Lundbeck Foundation to S. Ditlevsen, and the Center for Neurosciences LC554, AV0Z50110509 and Academy of Sciences of the Czech Republic (Information Society, 1ET400110401) to P. Lansky.

REFERENCES

1. W. Kistler, W. Gerstner, and J. van Hemmen, *Neural Comput.* **9**, 1015–1045 (1997).
2. R. Jolivet, T. Lewis, and W. Gerstner, *J. Neurophysiology* **92**, 959–976 (2004).
3. H. Tuckwell, *Introduction to theoretical neurobiology, Vol.2: Nonlinear and stochastic theories*, Cambridge Univ. Press, Cambridge, 1988.
4. P. Lansky, and L. Sacerdote, *Physics Letters A* **285**, 132–140 (2001).
5. S. Ditlevsen, and P. Lansky, *Phys. Rev. E* **71**, Art. No. 011907 (2005).
6. H. Tuckwell, and W. Richter, *J. Theor. Biol.* **71**, 167–180 (1978).
7. L. Ricciardi, *Diffusion processes and related topics in biology*, Springer, Berlin, 1977.
8. A. Nobile, L. Ricciardi, and L. Sacerdote, *J. Appl. Prob.* **22**, 360–369 (1985).
9. F. Wan, and H. Tuckwell, *J. Theoret. Neurobiol.* **1**, 197–218 (1982).
10. L. Ricciardi, and L. Sacerdote, *Biol. Cybern.* **35**, 1–9 (1979).
11. L. Ricciardi, and S. Sato, *J. Appl. Prob.* **25**, 43–57 (1988).
12. L. Ricciardi, A. Di Crescenzo, V. Giorno, and A. Nobile, *Math. Japonica* **50**, 247–322 (1999).
13. L. Alili, P. Patie, and J. Pedersen, *Stochastic Models* **21**, 967–980 (2005).
14. O. Aalen, and H. Gjessing, *Lifetime data analysis* **10**, 407–423 (2004).
15. S. Ditlevsen, *Statistics and Probability Letters* **77**, 1744–1749 (2007).
16. A. Bulsara, T. Elston, C. Doering, S. Lowen, and K. Lindberg, *Phys. Rev. E* **53**, 3958–3969 (1996).
17. J. Durbin, *J. Appl. Prob.* **8**, 431–453 (1971).
18. R. Fortet, *Journal de mathématiques pures et appliquées* **22**, 177–243 (1943).
19. A. Borodin, and P. Salminen, *Handbook of Brownian motion - Facts and Formulae*, Probability and its applications, Birkhauser Verlag, Basel, 2002.
20. N. Lebedev, *Special functions and their applications*, Dover Publications, New York, 1972.
21. B. Prakasa Rao, *Statistical inference for diffusion type processes*, Arnold, 1999.
22. P. Lansky, *Math. Biosci.* **67**, 247–260 (1983).
23. P. Lansky, P. Sanda, and J. He, *J. Comput. Neurosci.* **21**, 211–223 (2006).
24. R. Höpfner, *Math. Biosci.* **207**, 275–301 (2007).
25. U. Picchini, P. Lansky, A. De Gaetano, and S. Ditlevsen, *Neural Computation* (2008), to appear.
26. J. Inoue, S. Sato, and L. Ricciardi, *Biol. Cybern.* **73**, 209–221 (1995).
27. S. Shinomoto, Y. Sakai, and S. Funahashi, *Neural Comput.* **11**, 935–951 (1999).
28. S. Ditlevsen, and P. Lansky, *Phys. Rev. E* **76**, Art. No. 041906 (2007).
29. S. Ditlevsen, and O. Ditlevsen, *Prob. Eng. Mech.* (2008), in press.
30. P. Lansky, and S. Ditlevsen, *Biological Cybernetics* (2008), to appear.
31. Y. Yu, Y. Xiong, Y. Chan, and J. He, *J. Neurosci.* **24**, 3060–3069 (2004).

Overdispersion in the Place Cell Discharge - Stochastic Modelling and Inference

Viktor Beneš*, Blažena Frcalová*, Daniel Klement† and Petr Lánský†

*Department of Probability and Mathematical Statistics, Faculty of Mathematics and Physics, Charles University, Sokolovská 83, 18675 Prague, Czech Republic
†Institute of Physiology, Academy of Sciences of the Czech Republic, Vídeňská 1083, 14000 Prague, Czech Republic

Abstract. The paper is a case study monitoring the spiking activity of a place cell of hippocampus of a rat moving in an arena. Real data are evaluated using a new statistical methodology. Experimentally observed overdispersion suggests a doubly stochastic spatio-temporal point process model of the time of spikes and the location of the rat. The inference of the driving intensity leads to a nonlinear filtering problem. Jump processes are used as parametric models of the driving intensity which enables the solution of the filtering problem by means of Bayesian Markov chain Monte Carlo methods. Simultaneously the parameters of the model are estimated. Model selection, numerical results and receptive field plasticity are discussed.

Keywords: Filtering, overdispersion, place cell discharge, spatio-temporal point process.
PACS: 60G55, 92C20

1. INTRODUCTION

Importance of the hippocampus for solving difficult spatial problems is well known (O'Keefe and Nadel, 1978; Morris et al. 1982). In addition pyramidal cells in the hippocampus display spatial specific activity. It is high in a small portion of the rat's environment called a firing field and low elsewhere. It was shown by Fenton and Muller (1998) that the spiking activity during passes through the firing field is characterized not only by a high firing rate, but also by its very high variability (overdispersion). (Olypher et al, 2002) propose that overdispersion reflects switches of the rats attention between different spatial reference frames. Theta cycle and direction of rat's movement were considered as covariates in a study of (Brown et al., 1998). A conjecture that with increasing speed of the rat the spiking intensity increases was studied by (Csurko et al, 1999). Despite intensive research, our understanding of underlying mechanisms leading to spatially organized activity of place cells is incomplete.

In this paper experimental data (Klement, 2006) of times of occurences of action potentials of a hippocampal neuron together with the track of a rat are investigated. The question is whether a global stochastic model is able to describe the measurements of the experiment in space and time. While a Poisson process is a model for complete randomness, in our situation a doubly stochastic (Cox) spatio-temporal Poisson process is a suitable model, see (Lánský and Vaillant, 2000). These authors studied theoretical properties of the model. In the present paper parameters of a spatio-temporal Cox point process model are estimated and the degree of fit is quantified. Conditionally on data the driving intensity of the Cox process is evaluated using nonlinear filtering and its

CP1028, *Collective Dynamics: Topics on Competition and Cooperation in the Biosciences*
edited by L. M. Ricciardi, A. Buonocore, and E. Pirozzi

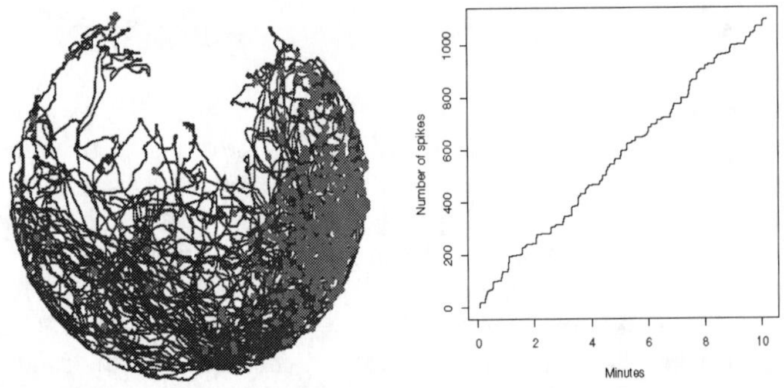

FIGURE 1. Positional firing of the hippocampal neuron. Left (a) - rat's track in the arena is displayed by the blue line. Places at which the neuron fired are indicated by red dots. Right (b) - the temporal evolution of spikes with graphical presentation as a point process realization. At each spike time the graph is increased by 1 so that at each time it corresponds to the total number of spikes.

characteristics estimated. An adaptive filtering was applied to hippocampal neurons by (Brown et al., 2001) who use a parametric modelling of the conditional intensity.

Spatio-temporal point processes (Daley and Vere-Jones, 1988), (Schoenberg et al., 2002) are the main tool. A general background of Lévy based Cox process on a curve (Frcalová and Beneš, 2007) was developed. Here we use a driving intensity which is derived from a compound Poisson process. Filtering and estimation of model parameters is available in a hierarchical Bayesian model using Metropolis within Gibbs algorithm, cf. (Roberts et al. 2004). We can estimate further characteristics, e.g. the Fano factor describing overdispersion. The posterior predictive distributions enable to select a model. The model sensitivity can be studied in various ways (with respect to priors, discretization step, etc.). Finally it is shown that the residual analysis for the point process is available as an alternative tool for the degree of fit testing as well as for the adaptive filtering.

2. MATERIALS AND METHODS

2.1. Data

The experimental data investigated are times of occurences of action potentials (spikes) of a hippocampal neuron. The shape of action potentials is considered to be irrelevant. Therefore the pulses may be seen as a realization of a spatio-temporal point process. The spikes were recorded with 0.1μ sec precision from a rat searching for food and at the same time avoiding a northern part of a 75 cm wide circular arena. Each $\frac{1}{60}$s the location of the rat was monitored. The recording lasted 10 minutes and 14 seconds, there were 1096 spikes observed, the average firing rate was 1.79 Hz. In Fig.1a there is a planar plot of the measurement in space. The neuron fires mostly when the rat visits the east part of the arena. The temporal behaviour of the recorded neuron is such that short periods of high activity are separated by longer periods of small activity (Fig.1b).

2.2. INTENSITY AND OVERDISPERSION

The Poisson process in the d–dimensional Euclidean space $\mathbb{R}^d$ is such that in each bounded measurable set $B \subset \mathbb{R}^d$ the number of events has Poisson distribution with mean $\lambda_m(B)$, where λ_m is the intensity measure. Moreover the numbers of events in disjoint sets are stochastically independent. If experimental data suggest that the variability of the point process is higher than that of the Poisson process (overdispersion) a Cox process X in $\mathbb{R}^d$ can be used. It means that the intensity λ_m is replaced by random driving measure Λ_m. Also we say that conditionally given $\Lambda_m = \lambda_m$ we have that X is a Poisson process with intensity measure λ_m. For the Cox process, denoting $X(B)$ the random counting measure (number of events in B), we have the Fano factor F (Fano, 1947)

$$F = \frac{varX(B)}{\mathbb{E}X(B)} = 1 + \frac{var[\Lambda_m(B)]}{\mathbb{E}[\Lambda_m(B)]}$$

which is a quantity equal to 1 for the Poisson process (deterministic Λ_m) and larger otherwise.

Consider a bounded arena $\mathscr{A} \subset \mathbb{R}^2$, a positive number T and a continuous function y_t, $t \in [0,T]$ in $\mathscr{A}$ which describes the position y_t of a rat at time t. Further consider a locally integrable nonnegative random function $\Lambda = \{\Lambda(t,v),\ v \in \mathscr{A},\ t \geq 0\}$ with the corresponding measure $\Lambda_m(D) = \int_D \Lambda(t,v)dvdt$. Then we define a Cox point process X on the track $y = \{y_t,\ 0 \leq t \leq T\}$ so that conditionally given $\Lambda = \lambda$ the number of spikes on $y \cap B$, $B \subset \mathscr{A}$ from time t_1 to t_2, $0 \leq t_1 < t_2 \leq T$, is Poisson distributed with mean $\int_{t_1}^{t_2} \mathbf{1}_B(y_t)\lambda(t,y_t)dt$.

Following (Lánský and Vaillant, 2000) the mean number of spikes during $[0,T]$ fired in $B \subset \mathscr{A}$ given intensity Λ and a track y is

$$\mathbb{E}(X(B) \mid \Lambda) = \int_0^T \mathbf{1}_B(y_t)\Lambda(t,y_t)dt$$

and

$$\begin{aligned} var(X(B)) &= \mathbb{E}(var(X(B) \mid \Lambda)) + var(\mathbb{E}(X(B) \mid \Lambda)) \\ &= \mathbb{E}(X(B)) + var\left(\int_0^T \mathbf{1}_B(y_t)\Lambda(t,y_t)dt\right). \end{aligned}$$

The Fano factor is then for $B \subset \mathscr{A}$

$$F(B) = 1 + \frac{var\left(\int_0^T \mathbf{1}_B(y_t)\Lambda(t,y_t)dt\right)}{\int_0^T \mathbf{1}_B(y_t)\mathbb{E}\Lambda(t,y_t)dt} \tag{1}$$

which is the event number variance to mean ratio, equal to 1 for homogeneous Poisson process.

For spontaneously active place cells, as a first approximation, we may assume that Λ is homogeneous in time and inhomogeneous in space, i.e. $\mathbb{E}(\Lambda(t,v)) = \mu_v$ and $var(\Lambda(t,v)) = \sigma_v^2$, $v \in \mathscr{A}$, do not depend on t. If $\mathscr{A}$ is divided in l boxes where μ_v is piecewise constant denote μ_i intensity mean in i–th box. Consider experimental data in a form (t_{ij}, n_{ij}), $i = 1, \ldots, l$; $j = 1, \ldots, k_i$, where t_{ij} is the duration of j–th stay of the rat in i–th box and n_{ij} is the number of spikes during this stay. A natural estimator of the expected intensity μ_i is then

$$\hat{\mu}_i = \frac{\sum_j n_{ij}}{\sum_j t_{ij}}. \tag{2}$$

Estimation of further characteristics of Λ, e.g. the variance, is not at all so easy. We proceed therefore in another way.

2.3. FILTERING AND ESTIMATION

A rigorous approach to the problem of the inference of the driving intensity Λ and its characteristics, cf. (2), is the filtering, see (Fishman and Snyder, 1976), (Brix and Diggle, 2001). Given a realization of a spatio-temporal Cox point process X of spikes and given the track, the solution of the nonlinear filtering problem is the conditional expectation $\mathbb{E}[\Lambda|X]$ which is not explicitly available. Nevertheless it is possible to express the corresponding conditional distribution using the Bayes formula for probability densities

$$f(\lambda|x) \propto f(x|\lambda) f(\lambda),$$

where from the definition of the Cox process $f(x|\lambda)$ is a density of an inhomogeneous Poisson process with intensity λ.

We consider a stochastic model of the driving intensity

$$\Lambda(\xi) = \sum_j w_j g(\xi, \eta_j),\ \xi \in \mathbb{R}^3, \tag{3}$$

where η_j are events of an auxiliary spatio-temporal Poisson process Φ (possibly inhomogeneous with intensity function ρ) and $w_j > 0$ are independent identically distributed jump sizes, independent of Φ. That means we have a compound Poisson process $\psi = \{\eta_j, w_j\}$. The function g is deterministic nonnegative such that realizations λ of Λ are integrable on any bounded set. Because of the jump character of the model (3) an approach to filtering based on finite point process theory (Møller and Waagepetersen, 2003) is available. Simultaneously the parameters in a parametric model are estimated, e.g. the parameters of the model of intensity ρ of Φ, the parameters of the jump size distribution, etc. Let $W = \mathscr{A} \times [0, T]$ be the window where the experiment is observed, denote the data $x = \{\tau_j\}$ where each τ_j reflects time and location of a spike on y. We have now

$$f(\psi, b|x) \propto f(x|\psi, b) f(\psi|b) f(b), \tag{4}$$

where $\psi = \{(t_j, z_j), w_j\}$ represents the point process Φ with jumps at times t_j and locations z_j and b is a vector of unknown parameters. Here the likelihood

$$f(x|\psi, b) = \exp\left(-\int_0^T \lambda(t, y_t)\,dt\right) \prod_{\tau_i \in x} \lambda(\tau_i) \tag{5}$$

corresponds to a density (with respect to a unit Poisson process in time) of an inhomogeneous spatio-temporal Poisson process on y given $\Lambda = \lambda$. In fact two point process densities compete in formula (4), the second one is

$$f(\psi|b) \propto \exp\left(-\int \rho(v)dv\right) \prod_{(t_j,z_j,w_j)\in\psi} \rho(t_j,z_j)h(w_j),$$

where h is the probability density of the jump size. Finally $f(b)$ is a prior distribution of parameters.

2.4. COMPLEMENTARY VARIABLES AND MODEL SELECTION

The dependence of the driving intensity on the covariates can be modelled in various ways, cf. (Beneš et al., 2005) for a spatial log-Gaussian Cox process. For the spatio-temporal Cox point process on the track there are further possibilities to modify the model. Consider p covariates $v_i(t,u)$, $i = 1,\ldots,p$ as measurable functions on W. To guarantee the nonnegativity we suggest the driving intensity on y including covariates in a form

$$\Lambda(t,y_t)\exp\left\{\sum_{i=1}^{p}\beta_i v_i(t,y_t)\right\}, \tag{6}$$

where β_i, $i = 1,\ldots p$ are real parameters.

The estimation of these parameters by means of posterior mean is obtained naturally including (6) in the Bayesian MCMC filtering algorithm. That means that formula (5) takes form

$$f(x|\psi,b) = \tag{7}$$

$$= \exp\left(-\int_0^T \lambda(y_t,t)\exp\left\{\sum_{i=1}^{p}\beta_i v_i(t,y_t)\right\}dt\right)\prod_{\tau_i\in x}\lambda(\tau_i)\exp\left\{\sum_{j=1}^{p}\beta_j v_j(\tau_i)\right\}$$

and the other formulas remain unchanged.

An important question from the statistical point of view is whether there is a good fit between the model and data. The model selection is discussed here by means of posterior predictive distributions. Let us have a summary statistics $U(x)$ computed from the data and its counterpart $U(X)$, where X is considered with estimated parameters. We take for U counts, i.e. number of points in a set $C \subset A$. Another possible choice is the the factorial second moment measure

$$\alpha^{(2)}(C,D) = \mathbb{E}\sum_{\xi,\eta\in X}^{\neq}\mathbf{1}_{[\xi\in C,\eta\in D]}$$

for sets $C,D \subset \mathscr{A}$. Formulas for the posterior predictive distributions of these two types of summary statistics are derived in (Frcalová and Beneš, 2007). They can be evaluated and compared with those obtained from the data. Monte Carlo statistical tests are available based on simulating realizations of the Cox process with estimated parameters, estimating summary statistics and constructing confidence bounds, cf. (Diggle, 2003).

3. RESULTS

3.1. A SPECIAL MODEL

We use a special form of the function g in (3) suggested by (Barndorff-Nielsen and Schmiegel, 2004) to obtain

$$\lambda(t,v) = \sum_{t_j \le t} w_j e^{t_j - t}\, \mathbf{1}_{B_{t_j-t}(v)}(z_j), \tag{8}$$

where $B_s(v)$, $s \le 0$, is a circle centered in v with radius $-us$ for a parameter $u > 0$. The infinite sum in (8) is in practice approximated, we have to consider a bounded window W_0 such that $W \subset W_0$ to get also negative t_j, see Fig.2. The model is built by means of an auxiliary point process $\psi = \{w_j, t_j, z_j\}$, cf. Fig.3, with intensity function ρ which will be estimated from data.

Let the jump size w have exponential distribution with mean $\alpha > 0$. Consider a cubic subdivision of the window W_0, denote the cubes $A_{ijk} = A_i \times A_{jk}$, A_i is a time interval, $A_{jk} \subset \mathbb{R}^2$. The model

$$\rho(\xi) = \sum_{ijk} \rho_{ijk} \mathbf{1}_{A_{ijk}}(\xi), \quad \xi \in \mathbb{R}^3$$

of a piecewise constant ρ is used. Then the vector of parameters is

$$b = (\alpha, u, \{\rho_{ijk}, i, j, k = 1, \ldots, n\}).$$

The prior distributions are chosen flat, e.g. one-dimensional exponential with fixed hyperparameters $l_\alpha, l_u, l_{ijk} >> 0$ for random parameters α, u, ρ_{ijk}, respectively.

Under these assumptions we can re-write (4) as (covariates not involved, vol is the volume of a set in $\mathbb{R}^3$, N_ψ is the number of points of ψ in W_0)

$$f(\psi, b|x) \propto \exp\left(-\int_0^T \sum_{t_j \le t} w_j e^{t_j - t} \mathbf{1}_{B_{t_j-t}(y_t)}(z_j)dt\right) \times \tag{9}$$

$$\times \prod_{\tau_m \in x} \lambda(\tau_m) \left(\prod_{ijk} \exp(-\rho_{ijk}\, vol(A_{ijk} \cap W_0))\right) \alpha^{-N_\psi} \exp\left(-\sum_i \frac{w_i}{\alpha}\right) \times$$

$$\times \left(\prod_{(t,z)\in\psi} \sum_{jlk} \rho_{jlk} \mathbf{1}_{A_{jlk}}(t,z)\right) l_\alpha^{-1} e^{-\frac{\alpha}{l_\alpha}} l_u^{-1} e^{-\frac{u}{l_u}} \prod_{ijk} l_{ijk}^{-1} e^{-\frac{\rho_{ijk}}{l_{ijk}}}.$$

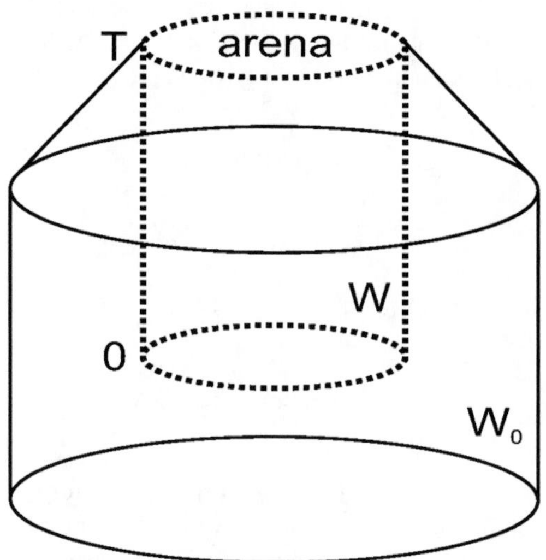

FIGURE 2. The window W where the spikes are observed within arena $\mathscr{A}$ and the time interval $[0,T]$ is enlarged to a set W_0 is which an auxiliary point process ψ lives, cf. (8).

3.2. NUMERICAL EVALUATION

The Metropolis within Gibbs algorithm can be used to simulate an MCMC chain $(\psi,b)^{(l)}$, $l=0,\ldots,J$, the distribution of which tends to the desired conditional distribution (9). For the point set variable ψ the birth-death Metropolis-Hastings algorithm is used (Moller and Waagepetersen, 2003), while the remaining parameters are iterated by the standard Gaussian random walk.

Using ergodicity properties of the MCMC chain we can estimate statistical characteristics of Λ. Denote $\Lambda^{(l)}(t,v)$ from (8) the $l-$th iteration of the intensity (conditioned on a realization of x) of the MCMC chain. J is the number of iterations, $0<K<J$ the burn-in of the chain, put $k=J-K$. The desired conditional expectation of Λ is estimated by the average value

$$\hat{\Lambda}(t,v)=\frac{1}{k}\sum_{l=K+1}^{J}\Lambda^{(l)}(t,v), \tag{10}$$

analogously we get estimators of higher moments and conditional variance of Λ. Fix M large integer, let $\triangle=\frac{T}{M}$. Denote

$$\mathscr{E}(B)=\frac{\triangle}{M}\sum_{q=0}^{M-1}\hat{\Lambda}(q\triangle,y_{q\triangle})\mathbf{1}_{[y_{q\triangle}\in B]},\ B\subset\mathscr{A}$$

and

$$\mathscr{V}(B)=\frac{1}{k}\sum_{l=K+1}^{J}\left(\frac{\triangle}{M}\sum_{q=0}^{M-1}\Lambda(q\triangle,y_{q\triangle})\mathbf{1}_{[y_{q\triangle}\in B]}\right)^2-\mathscr{E}(B)^2.$$

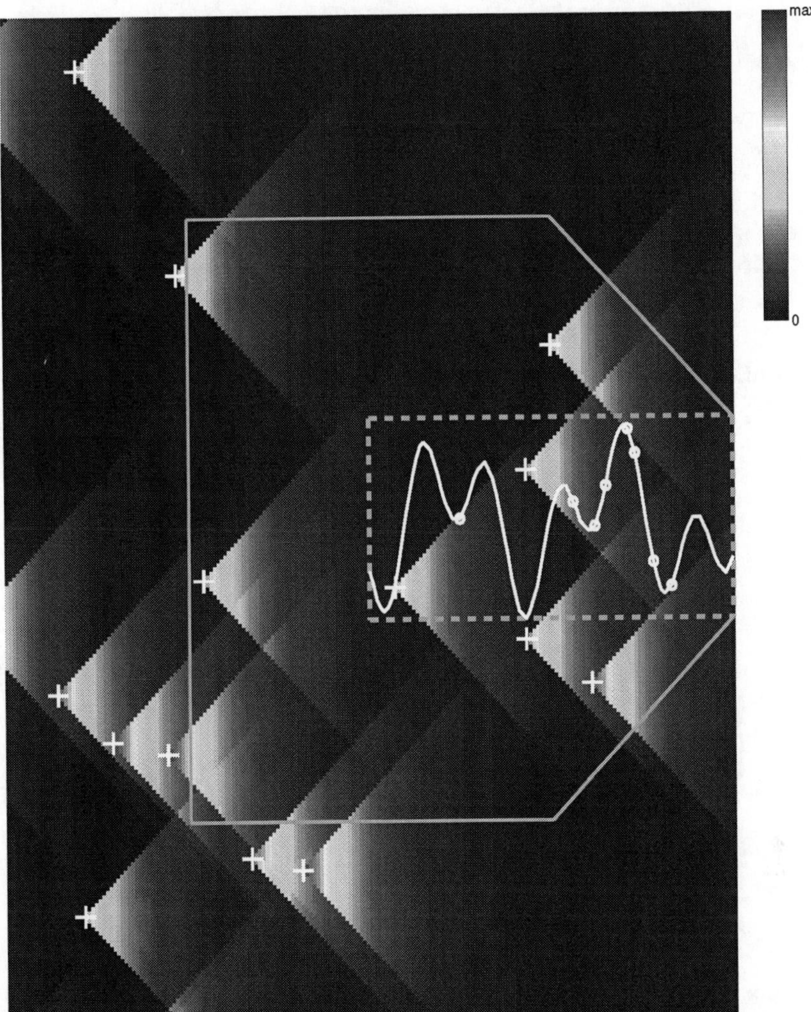

FIGURE 3. A simplified representation of the model (8). Here the horizontal axis presents time and the vertical axis space. In the window W (the rectangle delimited by the dotted white line) the track with spikes (circles) is drawn. The crosses denote events of the auxiliary point process ψ, which lie theoretically within the whole space and time. For numerical evaluation they are limited to W_0 (region delimited by the white full line), cf. Fig.2. The numerical contribution of each event of ψ to the intensity (8) is expressed by spectral colours.

Then we estimate the Fano factor as

$$\hat{F}(B) = 1 + \frac{\mathscr{V}(B)}{\mathscr{E}(B)}. \tag{11}$$

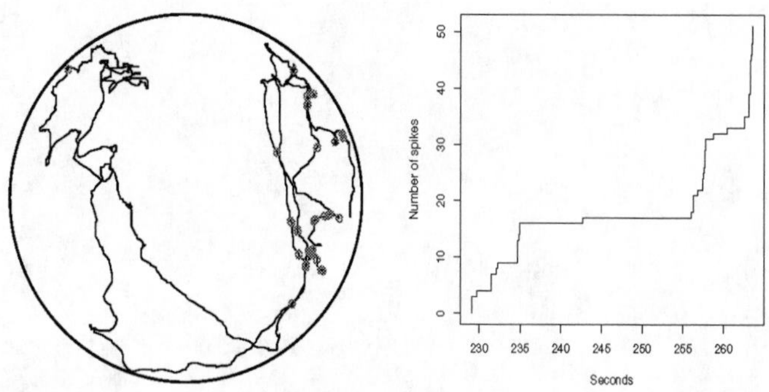

FIGURE 4. Data for the numerical evaluation. A part of the track in the arena with spikes (red crosses) - left, the temporal evolution - right

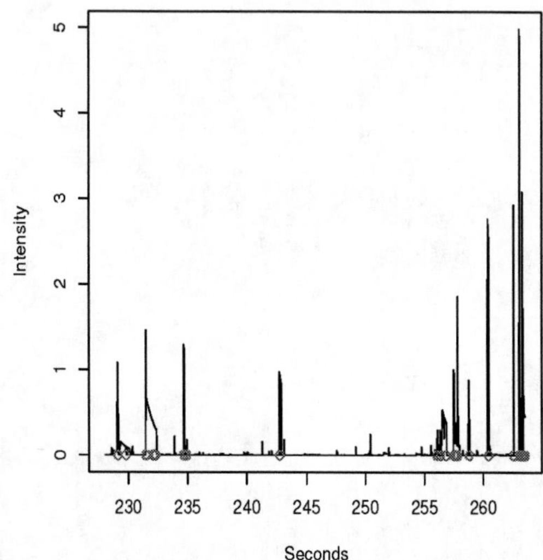

FIGURE 5. Circles on the horizontal (temporal) axis denote spikes from Fig.4, the results of filtering are presented by a graph of $\hat{\Lambda}(t, y_t)$, cf. (10).

3.3. REAL DATA ANALYSIS

A subrecord of the record from Fig.1 is analysed, see Fig.4. The covariate available in data is the rat's velocity, which can be split onto a scalar speed and the direction of movement. Comparing the histograms of these quantities for the whole track and for the times of spikes does not lead to significant differences in this experiment, therefore covariates are not considered within the spatio-temporal point process model here.

The method of filtering the driving intensity and estimation of parameters of the model was used. The estimated conditional intensity is in Fig. 5. We can observe how the temporal projection peaks at spike cluster times. The model selection results

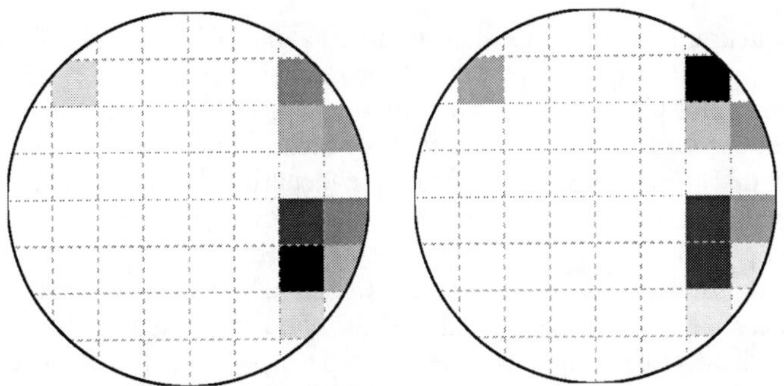

FIGURE 6. Posterior predictive distributions from MCMC and counts. Left (a) - data counts (increasing with grey level), right (b) - evaluation of $M(C_{ij})$, C_{ij} quadrats of $\mathscr{A}$.

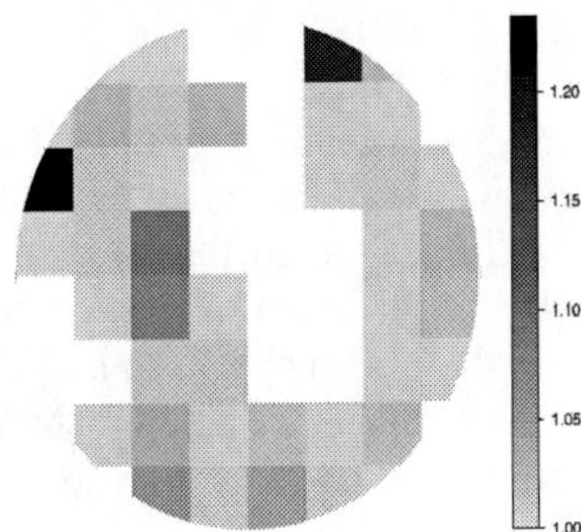

FIGURE 7. Fano factor map in quadrats, those which do not hit the track remain white - with no value.

using posterior predictive distribution are in Fig. 6. The diagram of counts in Fig. 6a corresponds to Fig. 4a. The diagram of first moment measures in quadrats obtained from MCMC, see Fig. 6b, is in a good agreement in spatial distribution with Fig. 6a. Finally the Fano factors are estimated in Fig. 7, they are close to 1 in the subsets of the firing field.

4. DISCUSSION

We presented a global approach to the stochastic modelling and inference of an experiment monitoring the spiking activity of a place cell of hippocampus of a rat moving in an arena. Global means here that the whole time interval $[0,T]$ is evaluated simultaneously. A spatio-temporal Cox point process was investigated which yields an acceptable model. We can observe how the parameters of the auxiliary point process Φ change in time.

(Brown et al, 2001), (Eden et al, 2004), (Ergun et al, 2007) are using point process techniques in this field leading to adaptive filtering. This situation where the model parameters change subsequently in time is called receptive field plasticity and it can express how neural systems adapt their representation of biological information. The stochastic model used in these papers is the conditional intensity (Daley and Vere-Jones, 1988) of a temporal point process in a parametric form. Their development tested in simulation studies leads to sequential methods which is undoubtly the right direction.

We remark that one can also realize time dependent (in all parameters) filtering within our context which is based on the fact that the conditional intensity of a Cox process on the track is $\lambda_s^* = \mathbb{E}[\Lambda(s, y_s)|N_u, u < s]$, where $N_u = X([0,u] \times \mathscr{A})$ is the number of events up to time u. This is a quantity which we obtain by means of the filtering method described in the paper, considering data and the window limited by time s (increasing s follows the history of the process). This approach presents also the background for the residual analysis of point processes (Ogata, 1998; Zhuang, 2006), a classical method describing the fit between a model and data. Here separate MCMC results are needed for each desired time of estimation of the conditional intensity, which is computationaly demanding. Some improvements in MCMC may help, e.g. the reparametrization in the sense of (Papaspiliopoulos et al., 2003).

However even from the numerical analysis presented in this paper we can study the neuron receptive field plasticity. An empirical approach is to start from Fig. 5, choose times of extremes of $\hat{\Lambda}$ and investigate whether the corresponding locations change randomly or systematically in time. For 20 largest peaks of Fig. 5 we evaluated the coefficient of multiple correlation of their times and planar locations to obtain 0.326 (little systematic change). Here the global approach is again used. When overdispersion is expected, the Cox process model may have a better chance of a good fit with data than the models based on conditional intensity.

ACKNOWLEDGMENTS

Research supported by grants of Ministery of Education (MSM0021620839) and Academy of Sciences (IAA101120604), Czech Republic.

REFERENCES

1. Barndorff-Nielsen OE, Schmiegel J (2004). Lévy based tempo-spatial modelling; with applications to turbulence. Usp Mat Nauk 159, 63–90.
2. Beneš V, Bodlák K, Møller J, Waagepetersen R (2005). Application of log-Gaussian Cox processes in disease mapping. Image Anal & Stereol 26, 115–123.
3. Brix A, Diggle P (2001). Spatio-temporal prediction for log-Gaussian Cox processes. J Royal Statist Soc B 63, 823–841.
4. Brown EN, Frank LM, Tang D, Quirk MC, Wilson MA (1998). A statistical paradigm for neural spike train decoding applied to position prediction from ensemble firing patterns of rat hippocampal place cells. J Neuroscience 18, 7411–25.
5. Brown EN, Nguyen DP, Frank LM, Wilson MA, Solo V (2001). An analysis of neural receptive field plasticity by point process adaptive filtering. Neurobiology 98, 21, 12261–66.
6. Czurko A, Hirace H, Csicsvari J, Bursaki G (1999). Sustained activation of hippocampal pyramidal cells by 'space clumping' in a running field. Europ J of Neuroscience 11, 344–352.
7. Daley DJ, Vere-Jones D (1988). An introduction to the theory of point processes. New York: Springer.
8. Eden UT, Frank LM, Barbieri R, Solo V, Brown EN (2004) Dynamic analysis of neural encoding by point process adaptive filtering. Neural Comp 16, 971–998.

9. Ergun A, Barbieri R, Eden UT, Wilson MA, Brown EN (2007) Construction of point process adaptive filter algorithms for neural system using sequential Monte Carlo methods. IEEE Transactions on Biomedical Engineering 54, 3, 419–28.
10. Fano U (1947). Ionization yields radiations. II. The fluctuations of the number of ions. Phys Rev 72, 26–29.
11. Fenton AA, Muller RU (1998). Place cell discharge is extremely variable during individual passes of the rat through the firing field. Proc Nat Acad Sci USA 95, 3182–3187.
12. Fishman PM, Snyder D (1976). The statistical analysis of space-time point processes. IEEE Trans on Inf Theory 22, 257–274.
13. Frcalová B, Beneš V (2007). Filtering and inference in spatio-temporal Cox point processes on a curve. Submitted.
14. Klement D (2006). Stochastic models in neurophysiology. Thesis. Charles University, Faculty of Math and Physics, Prague.
15. Lánský P, Vaillant J (2000). Stochastic model of the overdispersion in the place cell discharge. BioSystems 58, 27–32.
16. Møller J, Waagepetersen R (2003). Statistics and simulations of spatial point processes. Singapore: World Sci.
17. Ogata Y (1998). Space-time point process models for eartquake occurences, Annals Inst. Stat. Math. 50, 379–402.
18. O'Keefe J, Nadel L (1978). The Hippocampus as a Cognitive Map. Clarendon Press. Oxford.
19. Olypher AV, Lánský P, Fenton AA (2002). Properties of the extra-positional signal in hippocampal place cell discharge derived from the overdispersion in location specific firing. Neuroscience 111, 3, 553–566.
20. Papaspiliopoulos O, Roberts G, Skoeld M (2003). Non-centered parametrisations for hierarchical models and data augmentation, Bayesian Statistics 7, 307–326.
21. Roberts GO, Papaspiliopoulos O and Dellaportas P (2004). Bayesian inference for non-Gaussian Ornstein-Uhlenbeck stochastic volatility processes. J. Royal Statist. Soc. B 66, 2, 369-393.
22. Schoenberg FP, Brillinger DR, Guttorp PM (2002). Point processes, spatial-temporal. In: El-Shaarawi A, Piegorsch W, eds. Encyclopedia of Environmetrics, vol. 3. New York: Wiley, 1573–1577.
23. Zhuang J (2006). Second-order residual analysis of spatiotemporal point processes and applications in model evaluation, J. Royal Statist. Soc. B 68, 635–653.

A Stochastic Model for Infective Events in Operating Room Caused by Air Contamination

Paolo Abundo*, Nicola Rosato† and Mario Abundo**

*Servizio Ingegneria Medica, Policlinico Tor Vergata, viale Oxford 81, 00133 Roma, Italy
†Dipartimento di Medicina Sperimentale e Scienze Biochimiche, Università Tor Vergata, via Montpellier 1, 00133 Roma, Italy
**Dipartimento di Matematica, Università Tor Vergata, via della Ricerca Scientifica, 00133 Roma, Italy

Abstract. We propose a simple stochastic model for the movement of a potentially infective particle in operating room in which the local air contamination level is reduced by using a double laminar flow. Numerical simulation is used to obtain qualitative scenario analysis, in order to prevent infection, i.e. impact of the infective particle with the surgical wound, during the operation.

Keywords: Stochastic differential equation, infective event, operating room, air contamination.
PACS: 05.40.Jc, 02.50.Ey, 02.60.Cb

1. INTRODUCTION

In operating rooms, the air is required to have suitable characteristics to preserve the optimal conditions for well-being of the medical staff and above all of the patients which undergo surgical operations; when these conditions occur we are in the presence of "controlled contamination implant" [1, 2, 3] . The microorganisms giving rise to surgical infections are multiresistant bacteria, such as Staphylococcus aureus, Pseudomonas, Escherichia coli. These infections make the patient stay in hospital dramatically worse; in particular, in the case of haunch or knee arthroprosthesis, they can even lead to the mobilization of the prosthesis. A simple solution to the problem of infections in operating room endowed with a laminar air flow, consists of applying an additional sterile air flow, which contrasts that coming from the top of the room. Having in mind the control of infection (i.e. impact of an infective particle with the surgical wound, during the operation), we propose here a stochastic model for the movement of an air-carried particle fluctuating in the air zone between the ceiling and the surgical table. Potentially, this particle carries germs or infective agents in the operating room. In particular, we study as the height of the particle (i.e. its distance from the surgical table and therefore from the wound) varies, as a function of time, by considering two cases of scenario: the case of a single sterile air flow (generally present in operating room), and that of double flow. What we propose here is a very simple, one-dimensional model; however it appears to give a satisfactory qualitative description of the phenomenon. We consider also the possibility of a temperature gradient inside the operating room. A more complicated description of the movement of the particle could be given, by taking into account e.g. the mean size of the particle, the air resistance, etc.; this would result in a model depending on a large number of parameters. On the contrary, the peculiarity of our model

CP1028, *Collective Dynamics: Topics on Competition and Cooperation in the Biosciences*
edited by L. M. Ricciardi, A. Buonocore, and E. Pirozzi

consists in the fact that, though it depends only on few parameters, it is able to capture the essence of the phenomenon under study. Moreover, to our knowledge, this kind of description appears to be novel in medical and clinical literature. Scenario analysis, in the dependence of different values of air flow, have been obtained by numerical simulation.

The (random) times at which the particle carrying the infection, reaches the surgical table, turn out to be rather different, in the cases of single and double air flow: in the second case the particle fluctuation remains very long; indeed, to reach the surgical table, it takes on the average from 500% to 2000% of the time spent in the case of single air flow. Our analysis shows that the presence of the double air flow can result in a dejection of surgical infections, by extending the time necessary to an infective particle to reach the work zone, this leading to a smaller probability of infection.

2. THE MODEL

We suppose that the motion of a potentially infective particle in the operating room takes place between the ceiling of the room (conventionally set at height $z = 1$) and the surgical table (at height $z = 0$), and it starts from the position $0 < z_0 < 1$. Really, the particle is subject both to the gravitational force and the perturbation action of the laminar air flows, which are present in the room. We disregard the mass of the particle and we assume that it comes down with a velocity, whose vertical component at height z has mean intensity $v(z)$. Thus, we suppose that the projection of the motion of the particle on the vertical axes is described by the stochastic differential equation (SDE):

$$\begin{cases} dz(t) = -v(z(t))dt + \sigma(z(t))dW t \\ z(0) = z_0 \in (0,1] \end{cases} \tag{1}$$

where W_t is Brownian motion (BM), i.e. W_t is a random variable normally distributed with mean 0 and variance t, $v(z)$ represents the mean vertical component of the velocity at position z, while the diffusion term $\sigma(z)$ takes into account the random fluctuations of the position of the particle, and it possibly depends also on the temperature in the operating room, at the actual position z. In the presence of a single air flow, we can suppose, at a first approximation, that the particle comes down with constant mean velocity $\lambda > 0$, independent of the position; if we introduce a transversal air flow pointing toward the top of the room, whose vertical component has mean μ, then the vertical upward component of the additional flow opposes to the motion of the particle towards the surgical table, and so the mean downward velocity becomes $v = \lambda - \mu$. Notice that the equation (1) describes only the vertical component of the motion of the particle, and it does not take into account the value of its mass; indeed, in this way we give a simple qualitative description of the motion of the particle, by means of one-dimensional parameters. Nevertheless, our description is able to show the qualitative behavior of the phenomenon, and numerical simulation of the solutions of (1) permits to obtain scenario analysis for the first (random) instant at which the particle reaches the surgical wound (at height $z = 0$), in dependence of the starting position z_0 and the parameters λ and μ.

In the case when both $v(\cdot)$ and $\sigma(\cdot)$ are state independent, i.e. $v(z) = v$, $\sigma(z) = \sigma$, $Z(t)$ turns out to be Brownian motion with drift, i.e.

$$Z(t) = z_0 - vt + \sigma W_t \tag{2}$$

If the operating table is at height $z = 0$, we denote by τ_{z_0} the first (random) instant at which the particle reaches the surgical wound, with the condition that it has started from position z_0, i.e.:

$$\tau_{z_0} = \inf\{t > 0 : Z(t) \leq 0\} \tag{3}$$

it is called the first-passage-time (FPT) of the process $Z(t)$ through the boundary $\{z = 0\}$ (i.e. the operating table), starting from z_0. From (2) and (3), we obtain:

$$\tau_{z_0} = \inf\{t > 0 : z_0 - vt + \sigma W_t \leq 0\} = \inf\left\{t > 0 : W_t \leq \frac{vt - z_0}{\sigma}\right\} \tag{4}$$

Thus the FPT τ_{z_0} of $Z(t)$ through the boundary $\{z = 0\}$ becomes the FPT of the Brownian motion W_t below the linear boundary

$$z = \alpha t + \beta \tag{5}$$

where $\alpha = \frac{v}{\sigma}$ and $\beta = -\frac{z_0}{\sigma}$.
Now, set $\sigma = 1$ for simplicity, and let us consider the two diffusion processes, $Z(t)$ and $Z^1(t)$, obtained by taking $v = \lambda$ and $v = \lambda - \mu$ in (2), i.e.:

$$\begin{cases} Z(t) = z_0 - \lambda t + W_t \\ Z^1(t) = z_0 - (\lambda - \mu)t + W_t \end{cases} \tag{6}$$

The FPTs of the two processes through the boundary $\{z = 0\}$ are therefore:

$$\begin{cases} \tau_{z_0} = \inf\{t > 0 : z_0 + W_t \leq \lambda t\} \\ \tau^1_{z_0} = \inf\{t > 0 : z_0 + W_t \leq (\lambda - \mu)t\} \end{cases} \tag{7}$$

Since $-W_t$ has the same distribution of W_t, we have also:

$$\begin{cases} \tau_{z_0} = \inf\{t > 0 : W_t \geq z_0 - \lambda t\} \\ \tau^1_{z_0} = \inf\{t > 0 : W_t \geq z_0 - (\lambda - \mu)t\} \end{cases} \tag{7'}$$

So τ_{z_0} is the FPT of BM over the straight line $z = z_0 - \lambda t$, while $\tau^1_{z_0}$ is the FPT of BM over the straight line $z = z_0 - (\lambda - \mu)t$. We remark that $P(\tau_{z_0} < \infty) = 1$, as well $P(\tau^1_{z_0} < \infty) = 1$. The drift coefficients of the diffusions $Z(t)$ and $Z^1(t)$ are, respectively, $-\lambda$ and $-(\lambda - \mu) > -\lambda$, being $\mu > 0$. Note that $Z^1(t)$ has the same diffusion coefficient as that of $Z(t)$ and the drift coefficient larger than that of $Z(t)$. Thus, by a standard comparison theorem for SDE (see e.g. [4]) it follows that:

$$Z^1(t) \geq Z(t) \ , \forall t \geq 0 \tag{8}$$

and therefore

$$\tau_{z_0} \leq \tau^1_{z_0} \tag{9}$$

Now we recall that the distribution function of the FPT of BM over the straight line $z = z_0 - \lambda t$ is given by the so-called Bachelier-Levy formula (see e.g. [5]):

$$P(\tau_{z_0} \leq t) = 1 - \Phi\left(\frac{z_0}{\sqrt{t}} - \lambda\sqrt{t}\right) + \exp(2\lambda z_0)\Phi\left(-\lambda\sqrt{t} - \frac{z_0}{\sqrt{t}}\right) \tag{10}$$

where

$$\Phi(x) = \int_{-\infty}^{x} \phi(t)dt \tag{11}$$

and

$$\phi(t) = \frac{1}{\sqrt{2\pi}}\, e^{-\frac{x^2}{2}} \tag{12}$$

is the standard Gaussian density.
Taking the derivative in (10), we obtain the density of τ_{z_0} :

$$f(t, z_0) = \frac{z_0}{t\sqrt{t}}\, \phi\left(\frac{\lambda t - z_0}{\sqrt{t}}\right) = \frac{z_0}{t\sqrt{2\pi t}}\, e^{-\frac{1}{2t}(\lambda t - z_0)^2} \tag{13}$$

This means that

$$P(\tau_{z_0} \in (t, t+dt)) = f(t, z_0)dt \tag{14}$$

In analogous way, the density of $\tau^1_{z_0}$ is:

$$f^1(t, z_0) = \frac{z_0}{t\sqrt{t}}\, \phi\left(\frac{(\lambda - \mu)t - z_0}{\sqrt{t}}\right) = \frac{z_0}{t\sqrt{2\pi t}}\, e^{-\frac{1}{2t}((\lambda - \mu)t - z_0)^2} \tag{15}$$

For fixed z_0, the functions f and f^1 tend to zero, as $t \to 0^+$, and they have a peak at times $\bar{t}$ and $\overline{t^1}$, respectively. These instants represent the more probable values of τ_{z_0} and $\tau^1_{z_0}$.

The mean value and the variance of τ_{z_0} can be calculated by the formulae:

$$E(\tau_{z_0}) = \int_0^{+\infty} t f(t, z_0)dt\ , \ Var(\tau_{z_0}) = E(\tau^2_{z_0}) - E(\tau_{z_0})^2 \tag{16}$$

where

$$E(\tau_{z_0})^2 = \int_0^{+\infty} t^2 f(t, z_0)dt \tag{17}$$

and $f(t, z_0)$ is given by (13). Analogous formulae hold for $\tau^1_{z_0}$. Unfortunately, the integrals in (16), (17) cannot be calculated elementarily. However, to obtain the first two moments of τ_{z_0} we can proceed as it follows. Note that τ_{z_0} is nothing but the FPT of the BM with drift, $W_t + \lambda t$, over the level z_0; the Laplace transform of τ_{z_0} is explicitly known and it is given by (see [6]):

$$\Psi(\theta) = E\left(e^{-\theta\tau_{z_0}}\right) = \exp[-z_0(\sqrt{\lambda^2 + 2\theta} - \lambda)] \tag{18}$$

By calculating the first and second derivatives of Ψ at $\theta = 0$, we obtain:

$$E(\tau_{z0}) = -\Psi'(0) = \frac{z_0}{\lambda} \text{ and } E(\tau_{z0}^2) = \Psi''(0) = \frac{z_0}{\lambda^2}\left(z_0 + \frac{1}{\lambda}\right) \tag{19}$$

In analogous way we obtain:

$$E(\tau_{z0}^1) = \frac{z_0}{\lambda - \mu} \text{ and } E((\tau_{z0}^1)^2) = \frac{z_0}{(\lambda - \mu)^2}\left(z_0 + \frac{1}{\lambda - \mu}\right) \tag{19'}$$

If $\sigma \neq 1$, the Laplace transform of τ_{z0} is $\exp[-(z_0/\sigma^2)(\sqrt{\lambda^2 + 2\theta\sigma^2} - \lambda)]$ and we obtain

$$E(\tau_{z0}) = \frac{z_0}{\lambda} \text{ and } E(\tau_{z0}^2) = \frac{z_0}{\lambda^2}\left(z_0 + \frac{\sigma^2}{\lambda}\right) \tag{19''}$$

For instance, in the case when $\sigma = 1$, $\lambda = 0.1$, $z_0 = 0.9$, one gets:

$$\begin{cases} E(\tau_{z0}) = 9. \\ E(\tau_{z0}^2) = 981. \\ Var(\tau_{z0}) = E(\tau_{z0}^2) - E(\tau_{z0})^2 = 900. \end{cases} \tag{20}$$

These results have been confirmed by calculating the integrals in (16), (17) by a numerical procedure.

3. INTRODUCING THE TEMPERATURE IN THE MODEL

First, we consider the case in which the temperature in the operating room is a constant T. We replace σ with $K = T/24$ (the temperature is measured in Celsius degrees); then, the SDE (1) becomes:

$$\begin{cases} dZ(t) = -vdt + KdW_t \\ Z(0) = z_0 \end{cases} \tag{21}$$

A further step consists of considering a gradient of temperature in the operating room; to this end we take in the SDE (1):

$$\sigma(z) = \frac{T_0 + z(T_1 - T_0)}{T_0}, \quad T_1 > T_0 > 0 \tag{22}$$

where T_0 is the temperature at height 0, i.e. at the level of the operating table, and $T_1 > T_0$ is the temperature at height 1, i.e. at the level of the ceiling of the room. Notice that the temperature is supposed here to be an increasing function of the height (analogously, the case when $T_1 < T_0$ may be considered). Of course, the admissible variation of the temperature inside the surgical room is typically 2 or 3 Celsius degrees, so at a first approximation, one can suppose the temperature to be constant. As easily seen, from (23) it follows that for $z = 0$, it results $\sigma(z) = 1$ and so the preceding case is obtained again (fluctuation proportional to BM). On the contrary, at height $z = 1$, we have: $\sigma(Z(t)) = \frac{T_1}{T_0} > 1$ i.e. the fluctuations at high level are greater than those at low level. However, for any $0 \le z \le 1$, we have

$$1 \le \sigma(z) \le \frac{T_1}{T_0} \tag{23}$$

The study of the SDE (1) with $\sigma(\cdot)$ given by (22) is more complicated than that of the SDE with constant diffusion coefficient, because now the FPT of the process $Z(t)$ through the boundary $\{z = 0\}$ cannot be reduced to the FPT of a BM through a straight line. Thus, analytical expressions neither for $E(\tau_{z_0})$ and $Var(\tau_{z_0})$, nor for the probability density of τ_{x_0} can be found. However, lower and upper bounds to the first two moments of τ_{z0} can be obtained by using the techniques of [8]. Indeed, the process $Z(t)$ can be written as

$$Z(t) = z_0 - vt + \int_0^t \sigma(Z(s))dW_s$$

with $\sigma(z)$ given by (22). The quadratic variation of $Z(t)$ is

$$\rho(t) \doteq \int_0^t \sigma^2(Z(s))ds$$

and, thanks to (23), it satisfies for $t \ge 0$:

$$t \le \rho(t) \le \left(\frac{T_1}{T_0}\right)^2 t \tag{24}$$

and it results $\rho(+\infty) = +\infty$. So, by using a random-time change (see e.g. [7]) the stochastic integral above can be written as $\tilde{W}_t$, where $\tilde{W}_t$ is a suitable BM. Then, we have:

$$\tau_{z_0} = \inf\{> 0 : z_0 - vt + \tilde{W}_{\rho(t)} \le 0\} = \inf\{> 0 : \tilde{W}_{\rho(t)} \le -z_0 + vt\} \tag{25}$$

Moreover:

$$\begin{aligned} \rho(\tau_{z_0}) &= \inf\{t > 0 : \tilde{W}_t \le -z_0 + vA(t)\} = \\ &= \inf\{t > 0 : \tilde{W}_t \ge z_0 - vA(t)\} \doteq \tilde{\tau}_{z0} \end{aligned} \tag{26}$$

where A denotes the "inverse" of the random increasing process $\rho(t)$, i.e. $A(t) = \inf\{s > 0 : \rho(s) > t\}$. Thanks to (24), we have

$$\left(\frac{T_0}{T_1}\right)^2 t \le A(t) \le t \tag{27}$$

Now, let $\tilde{\tau}_1 = \inf\{t > 0 : \tilde{W}_t \ge z_0 - vt\}$ and $\tilde{\tau}_2 = \inf\{t > 0 : \tilde{W}_t \ge z_0 - v\left(\frac{T_0}{T_1}\right)^2 t\}$. Reasoning in analogous way as done in [8] (note that now $\tilde{\tau}_i$ are FPTs of BM over decreasing boundaries), we get:

$$\tilde{\tau}_1 \le \tilde{\tau}_{z_0} = \rho(\tau_{z_0}) \le \tilde{\tau}_2 \tag{28}$$

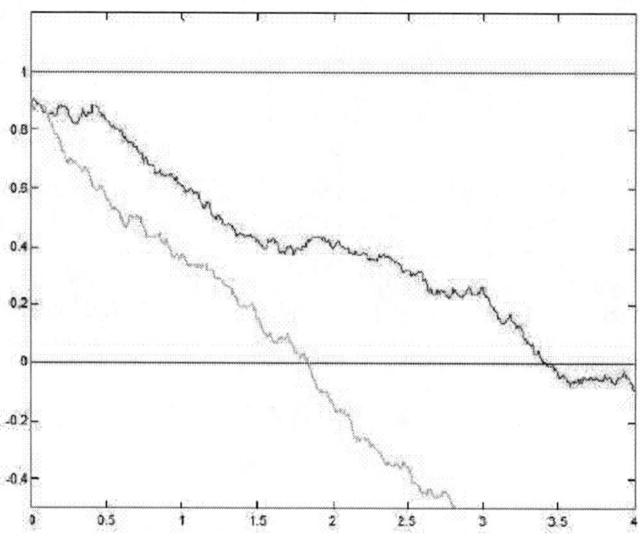

FIGURE 1. Simulation run of the trajectories of $Z(t)$ (lower line), $Z_1(t)$ (upper line) and the FPTs at constant temperature for $z_0 = 0.9$, $\lambda = 0.45$, $\mu = 0.15$.

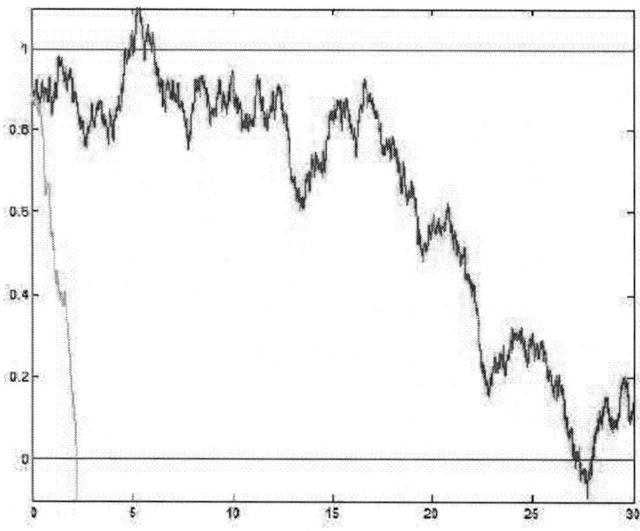

FIGURE 2. Simulation run of the trajectories of $Z(t)$ (lower line), $Z_1(t)$ (upper line) and the FPTs at constant temperature for $z_0 = 0.9$, $\lambda = 0.45$, $\mu = 0.40$.

and so

$$\left(\frac{T_0}{T_1}\right)^2 \tilde{\tau}_1 \leq A(\tilde{\tau}_1) \leq \tau_{z_0} \leq A(\tilde{\tau}_2) \leq \tilde{\tau}_2 \tag{29}$$

By taking expectation in (29), and using the definitions of $\tilde{\tau}_i$ and formulae (19) found in the case when $\sigma(\cdot)$ is independent of z, it finally follows:

$$\left(\frac{T_0}{T_1}\right)^2 \frac{z_0}{v} \leq E(\tau_{z_0}) \leq \left(\frac{T_1}{T_0}\right)^2 \frac{z_0}{v} \tag{30}$$

$$\left(\frac{T_0}{T_1}\right)^4 \frac{z_0}{v^2}\left(z_0 + \frac{1}{v}\right) \leq E(\tau_{z_0}^2) \leq \left(\frac{T_1}{T_0}\right)^2 \frac{z_0}{v^2}\left(z_0 + \left(\frac{T_1}{T_0}\right)^2 \frac{1}{v}\right) \tag{31}$$

4. SIMULATION AND NUMERICAL RESULTS

We have performed numerical simulation, to obtain a qualitative description of the behavior of the model, as a function of the parameters involved in it. The SDEs have been solved numerically by the iterative Euler method, by implementing a user friendly MATLAB program.

The Euler method to solve a SDE such as

$$\begin{cases} X(t) = f(X(t))dt + g(X(t))dW_t \\ X(0) = x_0 \end{cases}$$

consists of finding recursively the first order approximation of the solution at times $t_i = ih$, in the following way:

$$\begin{cases} X_{n+1} = f(X_n)h + g(X_n)\triangle W_n \; i = 1,2,\ldots \\ X_0 = x_0 \end{cases}$$

where $h = t_{i+1} - t_i$ is the integration step, $\triangle W_n = W_{t_{n+1}} - W_{t_n}$, and $X(t_i) = X_i + o(h)$, $i = 1,2,\ldots$.

In the Table 1 and 2 we report the results of some simulation runs concerning a set of values of the parameters λ and μ, in the two cases of constant and height-varying temperature.

TABLE 1. Values of τ_{z_0} and $\tau^1_{z_0}$ obtained in several simulation runs, for various values of the parameters λ and μ, in the case of constant temperature. The initial value is $z_0 = 0.9$ in all the simulations.

λ	μ	τ_{z_0}	$\tau^1_{z_0}$
0.10	0.05	9.2	12.9
0.5	0.1	1.8	2.35
0.45	0.15	1.8	3.4
0.45	0.25	1.75	6.3
0.45	0.30	2.10	6.5
0.45	0.40	2.48	26.8
0.45	0.41	2.49	28.9
0.45	0.42	2.5	29.1
0.45	0.43	2.6	33.2
0.45	0.44	2.1	32.6

TABLE 2. Values of τ_{z_0} and $\tau^1_{z_0}$ obtained in several simulation runs, for various values of the parameters λ and μ, in the case of height-varying temperature, for $T_1 = 24$, $T_0 = 23$ (temperature measured in Celsius degrees). The last row regards a run in which $T_1 >> T_0$. The initial value is $z_0 = 0.9$ in all the simulations.

λ	μ	τ_{z_0}	$\tau^1_{z_0}$
0.10	0.05	9.2	12.9
0.5	0.1	1.8	2.35
0.45	0.15	1.8	3.4
0.45	0.25	1.75	6.3
0.45	0.30	2.10	6.5
0.45	0.40	2.48	26.8
0.45	0.41	2.49	28.9
0.45	0.42	2.5	29.1
0.45	0.43	2.6	33.2
0.45	0.44	2.1	32.6

The results of some simulation runs are also graphically illustrated in the Figures 1–5.

In the Figures 1 – 2, which refer to constant temperature in the operating room, the shapes of $Z(t)$ (lower line) and $Z^1(t)$ (upper line) are reported as function of time t, for $z_0 = 0.9$, and the corresponding FPTs (τ_{z_0} and $\tau^1_{z_0}$) through the boundary $z = 0$ (operating table) are evidenced. The values of the basic air flow (λ) and the additional air flow (μ) are indicated in the figure captions.

Figures 3 – 4 refer to height-varying temperature in the operating room, with the

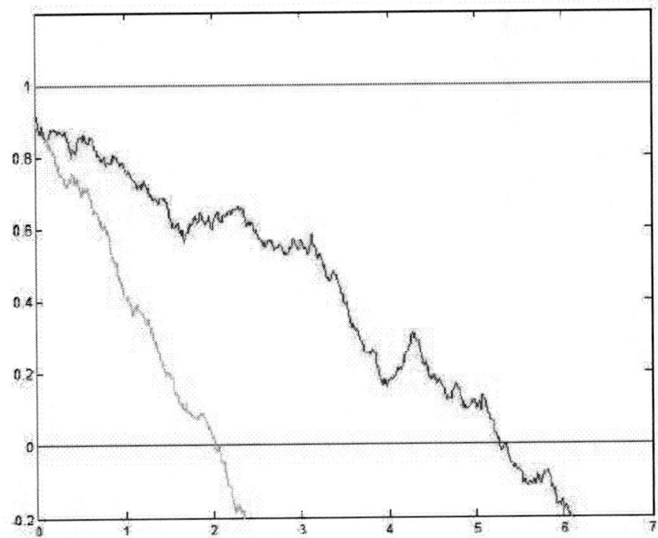

FIGURE 3. Simulation run of the trajectories of $Z(t)$ (lower line), $Z_1(t)$ (upper line) and the FPTs at height-varying temperature for $T_1 = 24, T_0 = 23$, $z_0 = 0.9$, $\lambda = 0.45$, $\mu = 0.25$.

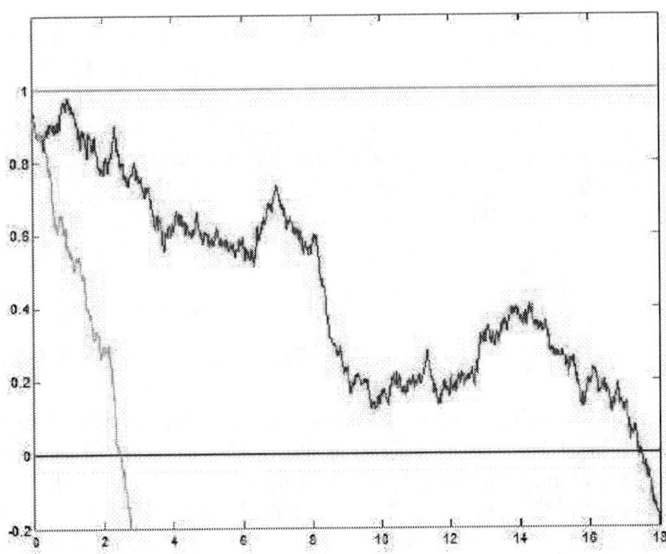

FIGURE 4. Simulation run of the trajectories of $Z(t)$ (lower line), $Z_1(t)$ (upper line) and the FPTs at height-varying temperature for $T_1 = 24, T_0 = 23$, $z_0 = 0.9$, $\lambda = 0.45$, $\mu = 0.40$.

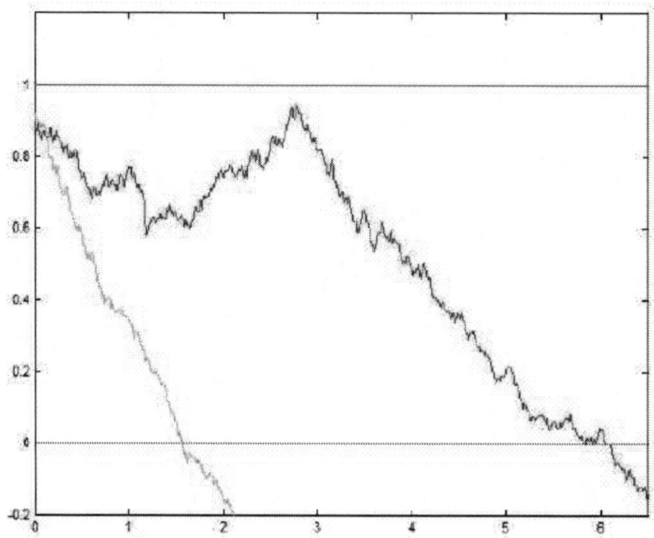

FIGURE 5. Simulation run of the trajectories of $Z(t)$ (lower line), $Z_1(t)$ (upper line) and the FPTs at height-varying temperature for $T_1 >> T_0$, $z_0 = 0.9$, $\lambda = 0.45$, $\mu = 0.30$.

temperature (T_1) at the ceiling level $(z = 1)$ greater than the temperature (T_0) at the level of the operating table $(z = 0)$; precisely $T_1 = 24$, $T_0 = 23$ (temperature measured in Celsius degrees). The shapes of $Z(t)$ (lower line) and $Z^1(t)$ (upper line) are reported as function of time t, for $z_0 = 0.9$, and the corresponding FPTs (τ_{z_0} and $\tau^1_{z_0}$) through the boundary $z = 0$ are evidenced. The values of the basic air flow (λ) and the additional air flow (μ) are indicated in the figure captions.

Finally, Figure 5 refers to a simulation run relative to height-varying temperature with $T_1 >> T_0$.

5. ANALYSIS OF THE SIMULATION AND CONCLUSIONS

The simulations relative to the two cases considered, i.e. homogeneous temperature and height-varying temperature in the operating room, show that, for large values of the downward air flow λ and small values of the additional upward air flow μ, the downward vertical flow is prevalent. Thus, the FPTs through the boundary $\{z = 0\}$ (operating table) are rather close each the other. For smaller values of $\lambda - \mu$ substantial differences are detected between the FPTs τ_{z_0} and $\tau^1_{z_0}$, the maximum difference appearing when $\lambda - \mu \simeq 0$. The difference $\tau^1_{z_0} - \tau_{z_0}$ is very large when the additional air flow is substantial; $\tau^1_{z_0}$ takes values in a range going from 15 to 20 times the value of τ_{z_0}. From the practical point of view, this means that the application of an additional (sufficiently large) air flow makes the patient less likely to contract infection during surgical operation, in the sense that it extends the time necessary to an infective particle to reach the operation zone; in other words there is a smaller probability that an infective particle can reach the surgical wound, before the operation has been completed.

Analyzing the simulation runs relative to height-varying temperature in the operating room, we have noted an increasing of the fluctuations in the random motion of the particle, when it crosses the zone with larger temperature $(z = 1)$. However, this phenomenon appears to be negligible, since the temperature at different heights in the operating room varies typically from 1 to 2 C^o. This means that, to capture the qualitative behavior of the phenomenon under study, it is not necessary, at a first instance, to suppose that the temperature can vary in the room. In order to validate furthermore our model, we have done a simulation run in a case (realistically unlikely) when $T_1 \gg T_0$ (see Fig. 5 and Table 2 -last row).

REFERENCES

1. R.H. Kruse, W. H. Puckett and J.H. Richardson, *Biological Safety Cabinetry, Clinical Microbiology Reviews*, American society for Microbiology, 1991.
2. L. Sabatini, "Conventional Clean Room project and control dimensionless approach," *11-th ICCS International Symposium*, 2000.
3. W.J. Whitfield, *A new approach to clean room design* , Sandia Corp. Technical report no. Sc-4673. Sandia Corp., Albuquerque, 1962.
4. N. Ikeda and S. Watanabe, "Stochastic differential equations and diffusion processes," North-Holland Publishing Co., 1981.
5. T. H. Scheike, *J. Appl. Prob* **29**, 448–453 (1992).
6. S. Karlin and H.M. Taylor, "A first course in stochastic processes," Academic Press, London, 1975.
7. D. Revuz and M Yor, "Continuous martingale and Brownian motion," Springer-Verlag, Berlin, 1991.
8. M. Abundo, *Stochastic Anal. Appl.* **24**, 1119–1145 (2006).

Flow Rate Driven by Peristaltic Movement in Plasmodial Tube of *Physarum Polycephalum*

Hiroyasu Yamada* and Toshiyuki Nakagaki†

*Harukidai 5-1-2, Togo, Aichi 470-0161, Japan, and Bio-Mimetic Control Research Center, The Institute of Physical and Chemical Research (RIKEN), Nagoya 463–0003, Japan
†Research Institute for Electronic Science, Hokkaido University, Sapporo, 060-0812, Japan, and Creative Research Initiative "SOUSEI", Hokkaido University, Sapporo, 001-0021, Japan

Abstract. We report a theoretical analysis of protoplasmic streaming driven by peristaltic movement in an elastic tube of an amoeba-like organism. The plasmodium of *Physarum polycephalum*, a true slime mold, is a large amoeboid organism that adopts a sheet-like form with a tubular network. The network extends throughout the plasmodium and enables the transport and circulation of chemical signals and nutrients. This tubular flow is driven by periodically propagating waves of active contraction of the tube cortex, a process known as peristaltic movement. We derive the relationship between the phase velocity of the contraction wave and the flow rate, and we discuss the physiological implications of this relationship.

Keywords: Physarum, peristaltic movement, fluid mechanics, circulation.
PACS: 47.70.Fw, 87.17.Aa, 47.63.-b

1. INTRODUCTION

The body of the plasmodium of *Physarum polycephalum* consists of a network of tubular elements by means of which nutrients, chemical signals and protoplasm circulate through the organism. The motive force of this streaming is hydrodynamic pressure, which is produced by the propagation of waves of contraction activity along a tube wall. This process is known as peristaltic movement and it allows several important functions to take place in the plasmodium: 1) the organism can migrate because body mass is transported by the peristaltic movement; 2) chemical signals can be communicated through the tubular network; 3) nutrients can be delivered throughout the organism. Here we report a hydrodynamical analysis of the flow rate that results from peristaltic movement. The theoretical results obtained are compared with the behavior observed experimentally in the plasmodium.

2. BASIC EQUATIONS FOR PERISTALTIC MOVEMENT AND APPROXIMATIONS FOR PLASMODIAL FLOW

We consider a flow of incompressible fluid in a circular, cylindrical, elastic tube of infinite length. For the sake of simplicity, we assume that the thickness of the tube wall is zero. We begin with the law of conservation of mass and momentum, as found in standard textbooks[1]. When $A(x,t)$ is the cross-sectional area of the tube and $u(x,t)$ is the mean velocity of flow over the cross-section of the tube, both of which are functions

CP1028, *Collective Dynamics: Topics on Competition and Cooperation in the Biosciences*
edited by L. M. Ricciardi, A. Buonocore, and E. Pirozzi

of the axial coordinate x and time t, then the conservation of mass is given by

$$\frac{\partial A}{\partial t}+\frac{\partial}{\partial x}(uA)=0. \tag{1}$$

When $p(x,t)$ is the pressure of the fluid inside the tube, the conservation of momentum is expressed as

$$\frac{\partial u}{\partial t}+u\frac{\partial u}{\partial x}=-\frac{1}{\rho}\frac{\partial p}{\partial x}-Ru, \tag{2}$$

where ρ is the (constant) density of the fluid and R is the friction coefficient of the fluid in contact with the tube wall. For simplicity, we assume that R is constant; however, in general it is a function of A.

Next, we consider $p(x,t)$, which is a function of A. We assume the relationship

$$p-p_{ex}=f(A), \tag{3}$$

where p_{ex} is the external pressure that gives rise to peristaltic movement. The function f represents the elasticity of the tube when it expands and contracts laterally, and is a monotone increasing function of A. Equation 3 implies that the dynamics between p_{ex} and A is replaced by an equation of statics and that the mass of the tube wall is negligible. We note that equations 1 - 3 were originally defined in a previous study of blood vessels [1].

In order to describe the peristaltic movement in *Physarum*, we express the external pressure p_{ex} as

$$p_{ex}=p_0e^{i(kx-\omega t)}. \tag{4}$$

Equations 1 - 4 are then written in the vector form

$$\frac{\partial}{\partial t}\begin{bmatrix} A \\ u \end{bmatrix}+\begin{bmatrix} u & A \\ f'/\rho & u \end{bmatrix}\frac{\partial}{\partial x}\begin{bmatrix} A \\ u \end{bmatrix}=\begin{bmatrix} 0 \\ -(1/\rho)\partial p_{ex}/\partial x-Ru \end{bmatrix}, \tag{5}$$

where the first and second terms on the right-hand side are the injected momentum and the momentum dissipated by friction, respectively. These equations are of the hyperbolic type. The coefficient matrix on the left-hand side contains the eigenvalues $\lambda_{\pm}=u\pm c$, which correspond to forward and backward waves, where c is the sonic speed of the elastic tube, $c=\sqrt{Af'/\rho}$. We rescale the equations by the dimensionless parameters,

$$\tau=kc_0t,\xi=kx,\alpha=\frac{A}{A_0},\nu=\frac{u}{c_0},\Pi=\frac{p}{\rho c_0^2}, \tag{6}$$

where c_0 is the sonic speed for the mean cross-sectional area; c is a function of the cross-sectional area and is given by $c_0=\sqrt{A_0f'(A_0)/\rho}$. As above, the sonic speed is chosen as the rescaling factor. We then obtain

$$\frac{\partial\alpha}{\partial\tau}+\frac{\partial}{\partial\xi}(\nu\alpha)=0, \tag{7}$$

$$\frac{\partial\nu}{\partial\tau}+\frac{\partial\nu}{\partial\xi}=-\frac{\partial\Pi}{\partial\xi}-\gamma\nu, \tag{8}$$

where

$$\Pi = F(\alpha) + \Pi_0 e^{i(\xi-\mu\tau)}, F(\alpha) = \frac{f(A_0)\alpha}{A_0 f'(A_0)}, \tag{9}$$

$$\Pi_0 = \frac{p_0}{\rho c_0^2}, \mu = \frac{\omega/k}{c_0}, \gamma = \frac{R}{kc_0}. \tag{10}$$

In this dimensionless system, the sonic speed is normalized because $\sqrt{F'(1)} = 1$ for the normalized cross-sectional area, $\alpha = 1$. The parameter μ is important because it expresses the ratio of the sonic speed to the wave speed of the peristaltic movement. In the actual organism, we can assume that ω is nearly constant. We solve this system of equations numerically.

3. SIMULATION RESULTS

Figure 1 shows the relationship between pressure and flux for different values of μ and γ when $\frac{dF(\alpha)}{d\alpha}$ is a non-zero constant (stress is proportional to the cross-section). When a contraction wave passes through a tube, the pressure varies from maximum to minimum in cyclical fashion. The flux also changes with the pressure. Thus, the relationship between pressure and flux takes the form of a closed curve, as shown in Fig.1. The flux reaches its maximum and minimum values close to where pressure=1 because the pressure gradient of the sinusoidal wave is maximum and minimum at these points, respectively. The flux is zero close to the maximum and minimum pressures because the gradient is zero here. The flow direction is both negative and positive during the course of a cycle, but the net flux is positive with respect to the propagation direction of the contraction wave. In general, the pressure-flux curves shrink as the wave propagation becomes slower (μ is smaller). Therefore, more protoplasmic sol moves back and forth during a flow cycle when the wave propagates faster. This implies that communication via chemical signals can occur more densely, because the rate of exchange of chemical material between neighboring elements along a tube increases.

However, it is not trivial to establish whether the net flux is also a function of the wave propagation speed. Figure 2 shows the net flux per peristaltic cycle and per unit time, both as a function of peristaltic speed. The plot of net flux per cycle is bell-shaped and the flux reaches an optimum value at a finite peristaltic speed before decreasing again. The net flux per unit time is approximately proportional to the peristaltic speed when the peristaltic speed is small, but reaches a constant value at higher speeds.

4. ESTIMATION OF PERISTALTIC SPEED IN PLASMODIUM

Here we estimate the peristaltic speed μ in the plasmodium, where $\mu = (\omega/k)/(c_0)$ and $c_0 = \sqrt{A_0 f'(A_0)/\rho}$. The parameter μ is approximately 2×10^{-4} for the following experimentally measured values: $\omega = 10^{-2}$ sec^{-1}, $k = 10^{-1}$ m^{-1}, $\Delta P = 5\times10^3$ Pa, $\Delta A = 10^{-7}$ m^2, $A_0 = 8\times10^{-7}$ m^2 and $\rho = 10^3$ kg m^{-3}. We assume that the function $f(A)$ is proportional to the cross-section A and is obtained as $(\Delta P)/(\Delta A)$. As the size of

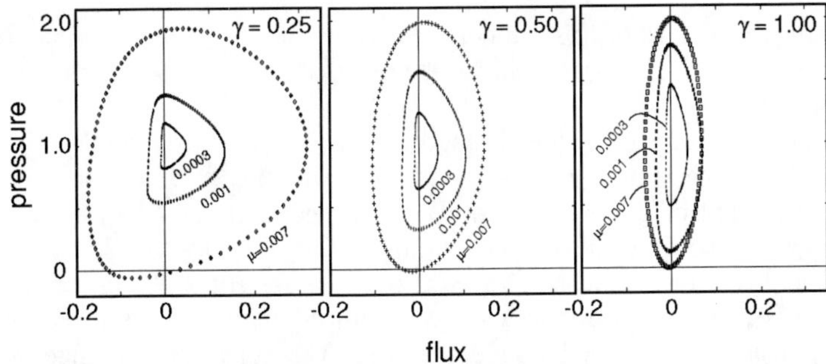

FIGURE 1. Flux-pressure relationship for different values of γ and μ. Left: $\gamma = 0.25$. Center: $\gamma = 0.50$. Right: $\gamma = 1.00$. In each panel, plots for $\mu = 0.0003$ (small loop), 0.0010 (medium loop) and 0.0070 (large loop) are shown. In all cases, $F(\alpha) = \alpha$ and $\Pi_0 = 1.0$.

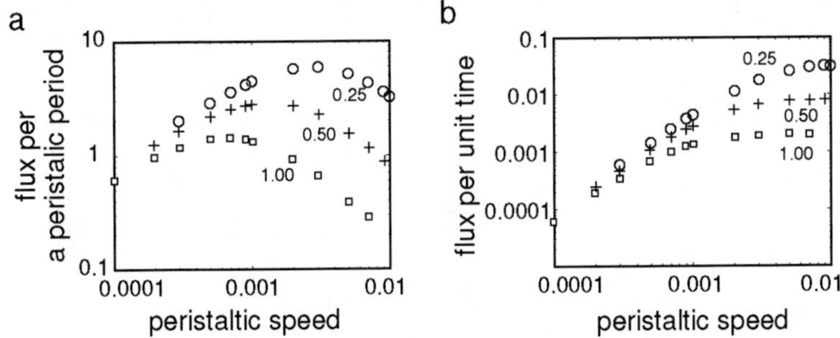

FIGURE 2. Flow rate as a function of peristaltic speed: (a) Flux per peristaltic period; (b) Flux per unit time. Plots are shown for $\gamma = 0.25$, 0.50 and 1.00. In all cases, $F(\alpha) = \alpha$ and $\Pi_0 = 1.0$.

the plasmodium increases from 1 cm to 100 cm, μ increases from 2×10^{-5} to 2×10^{-3}. These μ values are near the optimum values of flux per peristaltic cycle and are also near the saturation values of flux per unit time. This implies that the actual peristaltic speed in the organism is effective for the transportation of protoplasmic sol. According to this analysis, we can conclude that the flow rate is greater in a large plasmodium (100 cm) than in a small one (less than 1cm), which demonstrates the merit of large systems.

5. DISCUSSION

We now address the physiological implications of the flow rate. The maximum flow rate per peristaltic period, at the top of the bell-shaped curve in Fig. 2a, corresponded to an optimum flow speed of ($\mu_1^\star$). The flow rate per unit time increased with respect to flow speed and saturated at the flow speed $\mu_2^\star$ (Fig. 2b). The values of $\mu_1^\star$ and $\mu_2^\star$ were similar, and we thus denote them by $\mu^\star$. The flow rate $\mu^\star$ is convenient for physiological purposes because the power increases as the flow speed increases. A balance must be reached between the cost of achieving a high power and the benefit of obtaining a high net flow rate. Therefore, it is optimal for the organism that the benefit is maximized at minimum cost.

The measured flow speed of the plasmodium was less than $\mu^\star$, and thus the flow rate did not reach the maximum possible value. For example, the flow rate was 2×10^{-5} for the plasmodium of diameter 1 cm. The flow speed increased as the plasmodium became larger and became close to the optimum value of $\mu^\star$ for the plasmodium of 100 cm diameter. Therefore, increasing the size of the organism to the order of one meter is beneficial because the flow rate can be optimized. This may be a reason why the plasmodium can extend to such large dimensions. This effective flow enables chemical and physical signals to be transported efficiently throughout the organism.

REFERENCES

1. Y. C. Fung, *Biomechanics -Circulation-* (2nd edition), Springer-verlag, New York, 1997, pp. 140–143.

On the Influence of Quorum Sensing in the Competition Between Bacteria and Immune System of Invertebrates

Paolo Fergola*, Juan Zhang†, Marianna Cerasuolo*,* and Zhien Ma**

*University of Naples Federico II, Department of Mathematics "R.Caccioppoli", Italy
†North China Electric Power University, Beijing 102206, China
**Department of Mathematics, Xi'an Jiaotong University, Xi'an 710049, China

Abstract. The competition between bacteria and innate immune system of invertebrate animals is described by means of ODEs. Two different systems are considered corresponding to the absence or the presence of Quorum Sensing (Q.S.) mechanism. Qualitative properties of the solutions of both systems as well as the stability of their meaningful equilibria are analyzed. By constructing suitable Lyapunov functions, global asymptotic stability results have been proved when the quorum sensing is absent. In order to better illustrate the dynamics of competition, some numerical simulations, obtained by means of MATHEMATICA (Wolfram Research, 1989) are presented.

Keywords: Invertebrate species, innate immunity, bacterial infections, quorum sensing, asymptotic stability.
PACS: 87.19.xb, 87.23.Cc

1. INTRODUCTION

The immune system of invertebrate animals only consists of the innate response which reacts to the presence of pathogens (bacteria, virus, fungi) (see [1]). So, for instance, bacterial infections in these species give rise to a competition between bacteria and their immune system. By assuming in our mathematical modeling approach a macroscopic point of view, we will give a qualitative description of such competition by means of an ODEs system. The innate response is non-specific because the strategy is irrespective of the bacterial species. The immune system can suffice to clear the pathogen in most cases, but sometimes it is insufficient. Bacteria can indeed be able to overcome the innate response and successfully colonize and infect the host. The mathematical model suggested to describe such a competition is built by keeping the same lines of modeling used in [2]. It is constituted by four ODE equations involving the four time-dependent functions X_U (uninfected target cells), X_I (infected cells), B (bacteria), I_R (phenomenological variable representing innate responses).

Uninfected target cells have a natural turnover S_U and half-life μ_{X_U} and can become infected (mass-action term $\alpha_1 X_U B$). Infected cells can be cleared by the immune response (mass action term $\alpha_2 X_I I_R$) and die (half-life μ_{X_I}):

$$\begin{cases} \frac{dX_U}{dt} = S_U - \alpha_1 X_U B - \mu_{X_U} X_U \\ \frac{dX_I}{dt} = \alpha_1 X_U B - \alpha_2 X_I I_R - \mu_{X_I} X_I \end{cases} \tag{1}$$

CP1028, *Collective Dynamics: Topics on Competition and Cooperation in the Biosciences*
edited by L. M. Ricciardi, A. Buonocore, and E. Pirozzi

The innate response has a source term and a half-life term denoted by S_{I_R}, μ_{I_R}. S_{I_R} includes a wide range of cells and is enhanced and sustained by signals that we have captured by bacteria load:

$$\frac{\mathrm{d}I_R}{\mathrm{d}t} = S_{I_R} + \beta_1 B - \mu_{I_R} I_R \tag{2}$$

The bacteria population has a net growth term, represented by a logistic function $\alpha_{20}B(1-\frac{B}{\sigma})$ and it is also cleared by innate immunity through a mass action term $\alpha_3 BI_R$.

$$\frac{\mathrm{d}B}{\mathrm{d}t} = \alpha_{20}B\left(1-\frac{B}{\sigma}\right) - \alpha_3 BI_R \tag{3}$$

We observe that the model constituted by equations (1)-(3) represents the instantaneous version of the first four equations of model (7) of [2]. In particular, $\beta_1 B$ of equation (2) substitutes the delayed term of $(7)_4$ of [2].

The qualitative properties of the solutions of the system constituted by equations (1)-(3), its equilibria and their stability properties are analyzed. In particular the global asymptotic stability of two equilibria respectively infected and uninfected, has been proved and some numerical simulations have been performed. Afterwards, taking into account several recent experiments that have shown how the dynamics of many bacterial infections is influenced by the quorum sensing mechanism ([3] - [6]), we have represented the role of this mechanism by modifying equation (3) as follows:

$$\frac{\mathrm{d}B}{\mathrm{d}t} = \alpha_{20}B\left(1-\frac{B}{\sigma}\right) - \alpha_3\frac{BI_R}{A+B} + \frac{\gamma B}{A+B}. \tag{4}$$

where a new term has been added for the bacterial population and the interaction between bacterial and immune system cells has been modeled through an Holling-type II term. Also for the new system constituted by (1), (2) and (4) the qualitative properties of the solutions can be proved by standard techniques. Moreover, the steady states solutions have been found and their local stability properties have been determined.

The numerical simulations exhibit a rich dynamics including a saddle-node and a transcritical bifurcation. Finally the last Section is devoted to the discussion.

2. THE MATHEMATICAL MODEL IN ABSENCE OF QUORUM SENSING

Taking into account equations (1), (2) and (3) our model can be written as follows:

$$\begin{cases} \frac{\mathrm{d}X_U}{\mathrm{d}t} = S_U - \alpha_1 X_U B - \mu_{X_U} X_U, \\ \frac{\mathrm{d}X_I}{\mathrm{d}t} = \alpha_1 X_U B - \alpha_2 X_I I_R - \mu_{X_I} X_I, \\ \frac{\mathrm{d}B}{\mathrm{d}t} = \alpha_{20}B\left(1-\frac{B}{\sigma}\right) - \alpha_3 BI_R, \\ \frac{\mathrm{d}I_R}{\mathrm{d}t} = S_{I_R} + \beta_1 B - \mu_{I_R} I_R, \end{cases} \tag{5}$$

where all the parameters are supposed to be positive and constant.

2.1. Qualitative properties of the solutions of system 5

Denoting by $x(t) = (X_U(t), X_I(t), B(t), I_R(t))$ a solution of system (5) and by R^4_+ the positive cone

$$\Re^4_+ = \{x = (x_1, x_2, x_3, x_4) \in \Re^4 | x_i > 0,\, i = 1,2,3,4\}$$

we can prove the following lemmas:

Lemma 1 *Any solution $x(t)$ of (5) with $x(0) \in \Re^4_+$ remains positive whenever it exists, i.e. $x(t) \in \Re^4_+$.*

Proof Consider the third equation in (5)

$$\frac{dB}{dt} = B\left[\alpha_{20}\left(1 - \frac{B}{\sigma}\right) - \alpha_3 I_R\right]$$

with $B(0) > 0$. Then

$$B(t) = B(0)\exp\left\{\int_0^t \left[\alpha_{20}\left(1 - \frac{B}{\sigma}\right) - \alpha_3 I_R\right] dr\right\} > 0, t \geq 0.$$

Of course, if $B(0) = 0$, $B(t) = 0\ \forall t \geq 0$. By considering the first equation of (5)

$$\frac{dX_U}{dt} = S_U - X_U\left[\alpha_1 B + \mu_{X_U}\right]$$

we obtain

$$\frac{dX_U}{dt} \geq -X_U\left[\alpha_1 B + \mu_{X_U}\right]$$

therefore if $X_U(0) > 0$, then

$$X_U(t) \geq X_U(0)\exp\left\{-\int_0^t \left[\alpha_1 B + \mu_{X_U}\right] dr\right\} > 0, t \geq 0.$$

Of course, if $X_U(0) = 0$, $X_U(t) = 0\ \forall t \geq 0$.
By considering the second equation of (5)

$$\frac{dX_I}{dt} = \alpha_1 X_U B - \alpha_2 X_I I_R - \mu_{X_I} X_I$$

since $B(t), X_U(t) > 0,\ \forall t \geq 0$ we obtain

$$\frac{dX_I}{dt} > -\left[\alpha_2 I_R + \mu_{X_I}\right] X_I$$

therefore if $X_I(0) > 0$, then

$$X_I(t) \geq X_I(0)\exp\left\{-\int_0^t \left[\alpha_2 I_R + \mu_{X_I}\right] dr\right\} > 0, t \geq 0.$$

Of course, also in this case, if $X_I(0) = 0, X_I(t) = 0\,\forall t \geq 0$. By analyzing the forth equation we obtain

$$\frac{\mathrm{d}I_R}{\mathrm{d}t} \geq -\mu_{I_R} I_R$$

which imply that if $I_R(0) > 0$, then

$$I_R(t) \geq I_R(0)e^{-\mu_{I_R} t} > 0,\, t \geq 0.$$

This completes the proof of the positivity.

Lemma 2 *Any solution $x(t)$ of (5) is bounded.*

Proof From the third equation, it is easy to see that

$$\frac{\mathrm{d}B}{\mathrm{d}t} \leq \alpha_{20} B(1 - \frac{B}{\sigma})$$

which implies that $\limsup_{t\to+\infty} B(t) \leq \sigma\, \forall t > 0$. In other word, $\{B | B \leq \sigma\}$ is the positive invariant set with respect to the equation $\frac{\mathrm{d}B}{\mathrm{d}t} = \alpha_{20} B(1 - \frac{B}{\sigma})$.
Let $\mu = \min\{\mu_{X_U}, \mu_{X_I}, \mu_{I_R}\}$, $\Sigma = X_U + X_I + I_R$ and $S_1 = S_U + S_{I_R} + \beta_1\sigma$, then

$$\frac{\mathrm{d}\Sigma}{\mathrm{d}t} \leq S_U + S_{I_R} + \beta_1 B - \mu\Sigma \leq S_1 - \mu\Sigma.$$

By integrating from 0 to t we obtain

$$\Sigma(t) < \Sigma(0)e^{-\mu t} + \frac{S_1}{\mu}(1 - e^{-\mu t})$$

and then

$$\Sigma(t) < +\infty\, \forall t \in [0, +\infty[.$$

This, together with $B \leq \sigma$, proves the boundedness of $x(t)$.

Remark 1 *The closed set*

$$\Gamma_1 = \{(X_U, X_I, B, I_R) \in R^4_+ : \quad B \leq \sigma,\ X_U + X_I + I_R \leq S_1/\mu\}$$

is positively invariant for the system (5) since no positive path inside this cone leaves through any boundary of the region Γ_1. We denote by $\overset{\circ}{\Gamma}_1$ the interior of Γ_1 in R^4_+. In this section, we only need to study the system (5) in the closed set Γ_1.

2.2. Equilibria of system (5) and their local asymptotic stability

Setting the right side of each of the four differential equations of (5) equal to zero, equilibria can be obtained by solving the algebraic system

$$\begin{cases} S_U - \alpha_1 X_U B - \mu_{X_U} X_U = 0, \\ \alpha_1 X_U B - \alpha_2 X_I I_R - \mu_{X_I} X_I = 0, \\ B\left[\alpha_{20}\left(1 - \frac{B}{\sigma}\right) - \alpha_3 I_R\right] = 0, \\ S_{I_R} + \beta_1 B - \mu_{I_R} I_R = 0. \end{cases} \quad (6)$$

Let us denote by $\bar{E} = (\bar{X}_U, \bar{X}_I, \bar{B}, \bar{I}_R)$, where $\bar{X}_U$, $\bar{X}_I$, $\bar{B}$ and $\bar{I}_R$ are non-negative constants, the generic steady-state solution to the system (5).

From the third equation, we see that if $B = 0$ the system admits the equilibrium $E_0 = \left(\frac{S_U}{\mu_{X_U}}, 0, 0, \frac{S_{I_R}}{\mu_{I_R}}\right)$ called uninfected equilibrium. If $B \neq 0$, we have $I_R = \frac{\alpha_{20}}{\alpha_3}\left(1 - \frac{B}{\sigma}\right)$ and provided that $\alpha_{20}\mu_{I_R} > \alpha_3 S_{I_R}$ system (5) admits the equilibrium $E^* = (X_U^*, X_I^*, B^*, I_R^*)$ called infected equilibrium, with

$$\begin{aligned} B^* &= \frac{(\frac{\alpha_{20}}{\alpha_3}\mu_{I_R} - S_{I_R})\sigma}{\frac{\alpha_{20}}{\alpha_3}\mu_{I_R} + \beta_1\sigma}, \qquad X_U^* = \frac{S_U}{\alpha_1 B^* + \mu_{X_U}}, \\ X_I^* &= \frac{\alpha_1 B^* X_U^*}{\mu_{X_I} + \alpha_2 I_R^*}, \qquad I_R^* = \frac{\alpha_{20}}{\alpha_3}\left(1 - \frac{B^*}{\sigma}\right) = \frac{S_{I_R} + \beta_1 B^*}{\mu_{I_R}}. \end{aligned} \tag{7}$$

Therefore, we have proved the following:

Theorem 1 *System (5) admits two steady-state solutions:*

1. *For all positive values of the parameters the uninfected equilibrium* $E_0 = \left(\frac{S_U}{\mu_{X_U}}, 0, 0, \frac{S_{I_R}}{\mu_{I_R}}\right)$*;*
2. *If* $\alpha_{20}\mu_{I_R} > \alpha_3 S_{I_R}$*, the infected equilibrium* E^* *whose components are given in (7).*

The stability analysis of the generic equilibrium $\bar{E} = (\bar{X}_U, \bar{X}_I, \bar{B}, \bar{I}_R)$ can be performed by means of the characteristic equation, associated to the linearized system of (5) in $\bar{E}$, that can be written as follows:

$$\left\| \begin{array}{cccc} -\alpha_1\bar{B} - \mu_{X_U} - \lambda & 0 & -\alpha_1\bar{X}_U & 0 \\ \alpha_1\bar{B} & -\alpha_2\bar{I}_R - \mu_{X_I} - \lambda & \alpha_1\bar{X}_U & -\alpha_2\bar{X}_I \\ 0 & 0 & -\alpha_3\bar{I}_R + \alpha_{20}\left(1 - \frac{2\bar{B}}{\sigma}\right) - \lambda & -\alpha_3\bar{B} \\ 0 & 0 & \beta_1 & -\mu_{I_R} - \lambda \end{array} \right\| = 0. \tag{8}$$

Local stability properties of the equilibria E_0 and E^* are proved in the following:

Theorem 2 *The following statements hold true*

1. *If* $\alpha_{20}\mu_{I_R} < \alpha_3 S_{I_R}$ *then the equilibrium* E_0 *is locally asymptotically stable otherwise, if this inequality is reversed, it is unstable.*
2. *If the equilibrium* E^* *exists then it is locally asymptotically stable.*

Proof (1) By computing (8) in E_0 it is easy to find that if $\alpha_{20} - \alpha_3\frac{S_{I_R}}{\mu_{I_R}} < 0$, then all the five eigenvalues are negative and, therefore, E_0 is locally asymptotically stable (it is unstable if the foregoing inequality is reversed), in this case the equilibrium E^* does not exist.
(2) By computing (8) in E^*, it can be immediately checked that: $\lambda_1 = -\alpha_2 I_R^* - \mu_{X_I}$, $\lambda_2 = -\alpha_1 B^* - \mu_{X_U}$, are two negative eigenvalues. Therefore, to prove the statement it is sufficient to study the properties of the matrix

$$\bar{J}(E^*) = \begin{bmatrix} -\alpha_3 I_R^* + \alpha_{20}\left(1 - \frac{2B^*}{\sigma}\right) & -\alpha_3 B^* \\ \beta_1 & -\mu_{I_R} \end{bmatrix}.$$

Using (7), the trace of the matrix $\bar{J}(E^*)$ is $\mathrm{tr}(\bar{J}(E^*)) = -\mu_{I_R} - \alpha_{20}\frac{B^*}{\sigma} < 0$, and the determinant $\det(\bar{J}(E^*)) = \alpha_3(\frac{\alpha_{20}}{\alpha_3}\mu_{I_R} - S_{I_R})$ that is $\det(\bar{J}(E^*)) > 0$ if $\alpha_{20}\mu_{I_R} > \alpha_3 S_{I_R}$. Therefore, by noting that E^* does not exist if $\alpha_{20}\mu_{I_R} \leq \alpha_3 S_{I_R}$, we can say that E^* is locally asymptotically stable if and only if it exists. In this case, E_0 is unstable.

2.3. Global Asymptotical Stability of Equilibria of system (5)

In this subsection, we establish that E_0 is globally asymptotically stable in Γ_1 when $\alpha_{20}\mu_{I_R} \leq \alpha_3 S_{I_R}$ and E^* is globally asymptotically stable in $\mathring{\Gamma}_1$ when $\alpha_{20}\mu_{I_R} > \alpha_3 S_{I_R}$. First, let us consider the system

$$\begin{cases} \frac{dB}{dt} = \alpha_{20}B\left(1-\frac{B}{\sigma}\right) - \alpha_3 B I_R \\ \frac{dI_R}{dt} = S_{I_R} + \beta_1 B - \mu_{I_R} I_R \end{cases} \tag{9}$$

The same procedure used in the proof of Theorem 1 and 2 allows to prove the following:

Lemma 3 *The following statements hold true*

1. *For all positive values of the parameters system (9) admits the boundary equilibrium $\widetilde{E}_0 = \left(0, \frac{S_{I_R}}{\mu_{I_R}}\right)$. $\widetilde{E}_0$ is locally asymptotically stable if $\alpha_{20}\mu_{I_R} < \alpha_3 S_{I_R}$ and it is unstable if this inequality is reversed.*
2. *If $\alpha_{20}\mu_{I_R} > \alpha_3 S_{I_R}$ then there is a unique internal equilibrium $\widetilde{E}^* = (B^*, I_R^*)$ to (9), with B^* and I_R^* given in (7), which is locally asymptotically stable if and only if it exists.*

By means of the Lyapunov function argument, we can obtain the following:

Lemma 4 *(1) The boundary equilibrium $\widetilde{E}_0$ to (9) is globally asymptotically stable in R^2_+ if $\alpha_{20}\mu_{I_R} \leq \alpha_3 S_{I_R}$.*

(2) The interior equilibrium $\widetilde{E}^*$ to (9) is globally asymptotically stable in R^2_+ except B-axis if and only if $\widetilde{E}^*$ exists.

Proof (1) To prove the global stability of E_0 let us consider as Lyapunov function

$$V_1 = V_1(B, I_R) = B + \frac{c_1}{2}\left(I_R - \frac{S_{I_R}}{\mu_{I_R}}\right)^2,$$

where c_1 is a positive parameter to be chosen below. First, we observe that the function V_1 is positive definite in $\Re^2_+$, that is

$$V_1(B, I_R) > 0 \quad \text{except for} \quad \tilde{E}_0$$

Its time-derivative along the solutions of system (9) is

$$\begin{aligned}
\left.\frac{dV_1}{dt}\right|_{(9)} &= \alpha_{20}B\left(1-\frac{B}{\sigma}\right)-\alpha_3 BI_R+c_1\left(I_R-\frac{S_{I_R}}{\mu_{I_R}}\right)\left(S_{I_R}+\beta_1 B-\mu_{I_R}I_R\right)\\
&= -\alpha_{20}\frac{B^2}{\sigma}+\left(\alpha_{20}-c_1\beta_1\frac{S_{I_R}}{\mu_{I_R}}\right)B-(\alpha_3-c_1\beta_1)BI_R\\
&\quad +c_1\left(I_R-\frac{S_{I_R}}{\mu_{I_R}}\right)\left(S_{I_R}-\mu_{I_R}I_R\right)\\
&= -\alpha_{20}\frac{B^2}{\sigma}+D_1+D_2+D_3
\end{aligned}$$

where

$$\begin{aligned}
D_1 &= \left(\alpha_{20}-c_1\beta_1\frac{S_{I_R}}{\mu_{I_R}}\right)B\\
D_2 &= (\alpha_3-c_1\beta_1)BI_R\\
D_3 &= c_1\left(I_R-\frac{S_{I_R}}{\mu_{I_R}}\right)\left(S_{I_R}-\mu_{I_R}I_R\right).
\end{aligned}$$

We analyze each of the three parts separately and determine the fit value of c_1 such that $\left.\frac{dV_1}{dt}\right|_{(9)}<0$. We observe that if we choose $c_1=\frac{\alpha_3}{\beta_1}$, we obtain that $D_2=0$. Furthermore D_3 becomes

$$D_3 = c_1\left(I_R-\frac{S_{I_R}}{\mu_{I_R}}\right)\left(S_{I_R}-\mu_{I_R}I_R\right)=-\frac{\alpha_3}{\beta_1}\mu_{I_R}\left(I_R-\frac{S_{I_R}}{\mu_{I_R}}\right)^2\le 0.$$

Finally, D_1 can be written as follows

$$D_1 = \left(\alpha_{20}-c_1\beta_1\frac{S_{I_R}}{\mu_{I_R}}\right)B=\left(\alpha_{20}-\alpha_3\frac{S_{I_R}}{\mu_{I_R}}\right)B=\frac{\alpha_3}{\mu_{I_R}}\left(\frac{\alpha_{20}}{\alpha_3}\mu_{I_R}-S_{I_R}\right)B$$

and it is $D_1\le 0$ provided that $\alpha_{20}\mu_{I_R}\le\alpha_3 S_{I_R}$. Furthermore, we have

$$\left.\frac{dV_1}{dt}\right|_{(9)}\le-\alpha_{20}\frac{B^2}{\sigma}-\frac{\alpha_3}{\beta_1}\mu_{I_R}\left(I_R-\frac{S_{I_R}}{\mu_{I_R}}\right)^2\le 0.$$

And $\left.\frac{dV_1}{dt}\right|_{(9)}=0$ if and only if $B=0$ and $I_R=\frac{S_{I_R}}{\mu_{I_R}}$. As a consequence the function V_1 satisfies all the hypotheses of the Lyapunov Theorem on the asymptotic stability and our proof is complete.
(2) To prove the global stability of E_1, let us take as Lyapunov function

$$V_2=V_2(B,I_R)=\int_{B^*}^{B}\frac{x-B^*}{x}dx+\frac{c_2}{2}(I_R-I_R^*)^2,$$

where c_2 is to be chosen below. Then, the derivative along the system (9) is

$$\begin{aligned}
\left.\frac{dV_2}{dt}\right|_{(9)} &= (B-B^*)\left[\alpha_{20}\left(1-\frac{B}{\sigma}\right)-\alpha_3 I_R\right]+c_2(I_R-I_R^*)\left(S_{I_R}+\beta_1 B-\mu_{I_R} I_R\right) \\
&= -\frac{\alpha_{20}}{\sigma}(B-B^*)B+\alpha_{20}(B-B^*)-c_2\mu_{I_R} I_R(I_R-I_R^*) \\
&\quad -\alpha_3 I_R(B-B^*)+c_2\beta_1 B(I_R-I_R^*)+c_2 S_{I_R}(I_R-I_R^*) \\
&= -\frac{\alpha_{20}}{\sigma}(B-B^*)^2-c_2\mu_{I_R}(I_R-I_R^*)^2 \\
&\quad -\frac{\alpha_{20}}{\sigma}(B-B^*)B^*-c_2\mu_{I_R} I_R^*(I_R-I_R^*)+\alpha_{20}(B-B^*) \\
&\quad -(\alpha_3-c_2\beta_1)BI_R+(\alpha_3 B^*+c_2 S_{I_R})I_R-c_2\beta_1 I_R^* B-c_2 S_{I_R} I_R^* \\
&= -\frac{\alpha_{20}}{\sigma}(B-B^*)^2-c_2\mu_{I_R}(I_R-I_R^*)^2+D_4.
\end{aligned}$$

We analyze D_4 separately making use, when necessary, of

$$B^*=\frac{(\frac{\alpha_{20}}{\alpha_3}\mu_{I_R}-S_{I_R})\sigma}{\frac{\alpha_{20}}{\alpha_3}\mu_{I_R}+\sigma\beta_1} \quad \text{and} \quad I_R^*=\frac{\alpha_{20}}{\alpha_3}\left(1-\frac{B^*}{\sigma}\right)=\frac{S_{I_R}+\beta_1 B^*}{\mu_{I_R}}.$$

So,

$$\begin{aligned}
D_4 &= -\frac{\alpha_{20}}{\sigma}(B-B^*)B^*-c_2\mu_{I_R} I_R^*(I_R-I_R^*)+\alpha_{20}(B-B^*) \\
&\quad +(\alpha_3 B^*+c_2 S_{I_R})I_R-c_2\beta_1 I_R^* B-c_2 S_{I_R} I_R^*-(\alpha_3-c_2\beta_1)BI_R \\
&= [\alpha_3 B^*+c_2(S_{I_R}-\mu_{I_R} I_R^*)]I_R+(-\frac{\alpha_{20}}{\sigma}B^*+\alpha_{20}-c_2\beta_1 I_R^*)B \\
&\quad +\frac{\alpha_{20}}{\sigma}(B^*)^2-\alpha_{20}B^*+c_2\mu_{I_R}(I_R^*)^2-c_2 S_{I_R} I_R^*-(\alpha_3-c_2\beta_1)BI_R \\
&= (\alpha_3-c_2\beta_1)B^* I_R+[\alpha_{20}(1-\frac{B^*}{\sigma})-c_2\beta_1 I_R^*]B \\
&\quad -\alpha_{20}(1-\frac{B^*}{\sigma})B^*+c_2 I_R^*(\mu_{I_R} I_R^*-S_{I_R})-(\alpha_3-c_2\beta_1)BI_R \\
&= (\alpha_3-c_2\beta_1)B^* I_R+(\alpha_3-c_2\beta_1)BI_R^*-\alpha_3 I_R^* B^*+c_2 I_R^*(\beta_1 B^*)-(\alpha_3-c_2\beta_1)BI_R \\
&= (\alpha_3-c_2\beta_1)(B^* I_R+BI_R^*-I_R^* B^*-BI_R).
\end{aligned}$$

We let $c_2=\frac{\alpha_3}{\beta_1}$, then c_2 is positive. Furthermore,

$$\left.\frac{dV_2}{dt}\right|_{(9)}=-\frac{\alpha_{20}}{\sigma}(B-B^*)^2-\frac{\alpha_3}{\beta_1}\mu_{I_R}(I_R-I_R^*)^2\leq 0.$$

And $\left.\frac{dV_2}{dt}\right|_{(9)}=0$ if and only if $B=B^*$ and $I_R=I_R^*$.

The function V_2 satisfies all the hypotheses of the Lyapunov Theorem on the asymptotic stability, thus the proof is complete.

Now, by means of the theory of the limit system, we can prove the following Theorem.

Theorem 3 *The following statements hold true:*

(i) The uninfected equilibrium E_0 to (5) is globally asymptotically stable in Γ_1 if $\alpha_{20}\mu_{I_R} \leq \alpha_3 S_{I_R}$.

(ii) The infected equilibrium E^* to (5) is globally asymptotically stable in $\mathring{\Gamma}_1$ if it exists.

Proof We first prove the first claim.
Suppose that $(B, I_R) = (B(t), I_R(t))$ is any solution of the system (9).
Substituting $(B, I_R) = (B(t), I_R(t))$ into the first equation of system (5), we can obtain the following non-autonomous equation

$$\frac{\mathrm{d}X_U}{\mathrm{d}t} = S_U - [\alpha_1 B(t) + \mu_{X_U}]X_U, \tag{10}$$

Since from Lemma 4, when $\alpha_{20}\mu_{I_R} \leq \alpha_3 S_{I_R}$, we have

$$\lim_{t\to+\infty} B(t) = 0 \quad \text{and} \quad \lim_{t\to+\infty} I_R(t) = \frac{S_{I_R}}{\mu_{I_R}}. \tag{11}$$

Therefore, (10) admits the following limit equation

$$\frac{\mathrm{d}X_U}{\mathrm{d}t} = S_U - \mu_{X_U} X_U \tag{12}$$

The unique equilibrium of the linear equation (12) $\frac{S_U}{\mu_{X_U}}$ is trivially globally asymptotically stable. Due to the work of Thieme [7], Corollary 8 (see Appendix), we have that when $\alpha_{20}\mu_{I_R} \leq \alpha_3 S_{I_R}$

$$\lim_{t\to+\infty} X_U(t) = \frac{S_U}{\mu_{X_U}}, \tag{13}$$

Substituting $B = B(t)$, $I_R = I_R(t)$ and $X_U = X_U(t)$ into the second equation of the system (5), we can obtain the following non-autonomous equation

$$\frac{\mathrm{d}X_I}{\mathrm{d}t} = \alpha_1 X_U(t)B(t) - (\alpha_2 I_R(t) + \mu_{X_I})X_I, \tag{14}$$

From (11) and (13), we have that (14) admits the following limit equation

$$\frac{\mathrm{d}X_I}{\mathrm{d}t} = -\left(\alpha_2 \frac{S_{I_R}}{\mu_{I_R}} + \mu_{X_I}\right) X_I, \tag{15}$$

It is obvious that the zero solution of the linear equation (15) is globally asymptotically stable. Again using the work of Thieme [7], we have that if $\alpha_{20}\mu_{I_R} \leq \alpha_3 S_{I_R}$ then

$$\lim_{t\to\infty} X_I(t) = 0. \tag{16}$$

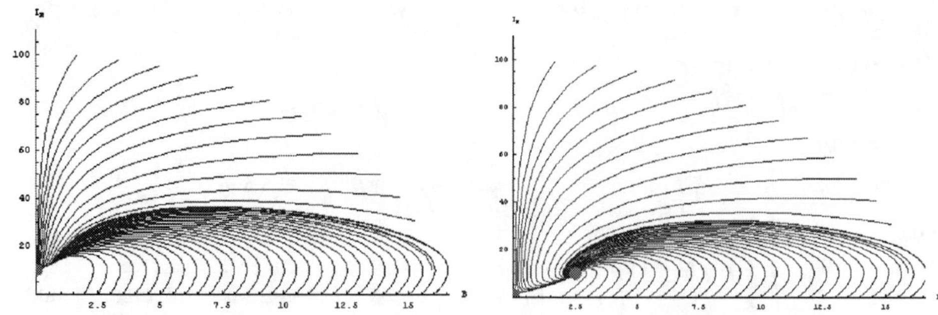

FIGURE 1. (a) $\alpha_{20}\mu_{I_R} < \alpha_3 S_{I_R}$. E_0 is g.a.s. (b) $\alpha_{20}\mu_{I_R} > \alpha_3 S_{I_R}$. E_0 is unstable, E_1 is g.a.s.

Furthermore, together with (11), (13) and (16), we can obtain that any solution of the system (5) converges to its uninfected equilibrium $E_0 = \left(\frac{S_U}{\mu_{X_U}}, 0, 0, \frac{S_{I_R}}{\mu_{I_R}}\right)$ as $t \to +\infty$ if $\alpha_{20}\mu_{I_R} \leq \alpha_3 S_{I_R}$. Together with the local asymptotical stability of E_0 (theorem 2), we can obtain the first claim.

The same procedure can be used to prove the second claim, concluding the proof of this Theorem.

2.4. Simulations of the System (5)

In order to confirm the results of Theorem 3 we performed some simulations with system (5) by using MATHEMATICA.
We can observe that the third and the fourth equations of system (5) are independent of the other ones. Hence, we had two choices: 1. analyze the subsystem (9), in the variables $B(t)$ and $I_R(t)$; 2. consider the simulations of system (5) focused on $B(t)$ and $I_R(t)$, that is consider the projection of the flow in $\mathfrak{R}^4$ to the BI_R-plane. We chose this second way.
Case I $\alpha_{20}\mu_{I_R} < \alpha_3 S_{I_R}$. In this case we are in the hypothesis of Theorem 3-(i) therefore there exists only the equilibrium E_0 and it is globally asymptotically stable.
To compute simulations we consider the following parameter values:

$$S_{I_R} = 1.1; S_U = \mu_{X_U} 10^4; \beta_1 = 0.4; \sigma = 6 \times 10^3; \alpha = 0.05;$$
$$\alpha_3 = 5 \times 10^{-3}; \alpha_1 = 10^{-3}; \alpha_2 = 10^{-3}; \mu_{I_R} = 0.11; \mu_{X_U} = 0.013; \mu_{X_I} = 0.011.$$

Thus, $\frac{\alpha_{20}}{\alpha_3}\mu_{I_R} < S_{I_R}$. The simulation results are shown in Fig.$1_{(a)}$, where, the marked point is the equilibrium E_0. Fig.$1_{(a)}$ shows that all considered trajectories, independently of their initial conditions, tend to the uninfected equilibrium.

Case II $\frac{\alpha_{20}}{\alpha_3}\mu_{I_R} > S_{I_R}$. In this case we are in the hypothesis of Theorem 3-(ii) therefore there exist two equilibria E_0 and E^*, and E^* is globally asymptotically stable in $\overset{\circ}{\Gamma}_1$.
To compute simulations we consider for the parameters the same values we have considered before except for S_{I_R} which, in this case, is $S_{I_R} = 0.1$, that implies $\frac{\alpha_{20}}{\alpha_3}\mu_{I_R} > S_{I_R}$.

The simulation results are shown in Fig.$1_{(b)}$, where the marked points are the equilibria E_0 and E^*. Fig.$1_{(b)}$ shows that all considered trajectories, independently of their initial conditions, tend to to the infectious equilibrium.

3. THE INFLUENCE OF QUORUM SENSING ON BACTERIAL DYNAMICS

Recent experimental studies have shown that some bacterial populations have the ability to sense the current state of their habitat and the presence of other bacteria and change their growth rate and the expression of their genes accordingly. This phenomenon is better known as Quorum Sensing (Q.S. see for instance [3]).

In order to model the effect of Q.S. on the bacteria growing process we consider, in this Section, a model differing by (5) in the presence of a new proliferation term of the bacterial population

$$\frac{\gamma B}{A+B} \qquad \gamma, A > 0 \tag{17}$$

which tends to saturation beyond the concentration A.
Moreover, we model the interaction between bacteria and immune system cells with an Holling-type II term instead of a Lotka-Volterra model. Here the constant A represents a given concentration of bacteria beyond which the attack capacity of the innate system tends to saturation. In this way, the bacteria dynamics is governed by the equation

$$\frac{\mathrm{d}B}{\mathrm{d}t} = \alpha_{20} B\left(1 - \frac{B}{\sigma}\right) - \alpha_3 \frac{BI_R}{A+B} + \frac{\gamma B}{A+B} \tag{18}$$

instead of $(5)_3$. Therefore our new system is:

$$\begin{cases} \frac{\mathrm{d}X_U}{\mathrm{d}t} = S_U - \alpha_1 X_U B - \mu_{X_U} X_U \\ \frac{\mathrm{d}X_I}{\mathrm{d}t} = \alpha_1 X_U B - \alpha_2 X_I I_R - \mu_{X_I} X_I \\ \frac{\mathrm{d}B}{\mathrm{d}t} = \alpha_{20} B\left(1 - \frac{B}{\sigma}\right) - \alpha_3 \frac{BI_R}{A+B} + \frac{\gamma B}{A+B} \\ \frac{\mathrm{d}I_R}{\mathrm{d}t} = S_{I_R} + \beta_1 B - \mu_{I_R} I_R \end{cases} \tag{19}$$

As before, from biological considerations, we study (19) only in the positive cone R^4_+. Also in this case, by standard techniques, it is easy to prove positivity, boundedness and existence in the future of the solutions of system (19).

3.1. Equilibria of system (19) and their local stability properties

Setting the right hand side of each of the four differential equations of (19) equal to zero, the equilibria can be obtained by solving the algebraic system

$$\begin{cases} S_U - \alpha_1 X_U B - \mu_{X_U} X_U = 0, \\ \alpha_1 X_U B - \alpha_2 X_I I_R - \mu_{X_I} X_I = 0, \\ B\left[\alpha_{20}\left(1-\frac{B}{\sigma}\right) - \alpha_3 \frac{I_R}{A+B} + \frac{\gamma}{A+B}\right] = 0, \\ S_{I_R} + \beta_1 B - \mu_{I_R} I_R = 0, \end{cases} \tag{20}$$

Let us denote by $\bar{E} = (\bar{X}_U, \bar{X}_I, \bar{B}, \bar{I}_R)$, where $\bar{X}_U$, $\bar{X}_I$, $\bar{B}$ and $\bar{I}_R$ are non-negative constants, the generic steady-state solution to the system (19). From the third equation, we see that if $B = 0$ the system admits the uninfected equilibrium $E_0 = \left(\frac{S_U}{\mu_{X_U}}, 0, 0, \frac{S_{I_R}}{\mu_{I_R}}\right)$. Therefore, we can assert that

Theorem 4 *There exists the uninfected equilibrium* $E_0 = \left(\frac{S_U}{\mu_{X_U}}, 0, 0, \frac{S_{I_R}}{\mu_{I_R}}\right)$ *to the system (19) for all positive values of the parameters of (19).*

If $B \neq 0$ from $(20)_4$, we have $I_R = \frac{S_{I_R} + \beta_1 B}{\mu_{I_R}}$, which substituted in $(20)_3$ produces the equation

$$f(B) \doteq B^2 + \left(A - \sigma + \frac{\alpha_3 \beta_1 \sigma}{\alpha_{20} \mu_{I_R}}\right) B - \sigma \left(A - \frac{\alpha_3 S_{I_R} - \gamma \mu_{I_R}}{\alpha_{20} \mu_{I_R}}\right) = 0. \tag{21}$$

The discriminant of equation (21) is

$$\begin{aligned} \triangle_f(A) &= \left(A - \sigma + \frac{\alpha_3 \beta_1 \sigma}{\alpha_{20} \mu_{I_R}}\right)^2 + 4\sigma \left(A - \frac{\alpha_3 S_{I_R} - \gamma \mu_{I_R}}{\alpha_{20} \mu_{I_R}}\right) \\ &= (A - A_3)^2 + 4\sigma (A - A_2) \\ &= A^2 + 2A(2\sigma - A_3) + (A_3^2 - 4\sigma A_2). \end{aligned} \tag{22}$$

Where

$$A_2 = \frac{\alpha_3 S_{I_R} - \gamma \mu_{I_R}}{\alpha_{20} \mu_{I_R}}, \qquad A_3 = \sigma - \frac{\alpha_3 \beta_1 \sigma}{\alpha_{20} \mu_{I_R}} = \frac{\sigma(\alpha_{20} \mu_{I_R} - \alpha_3 \beta_1)}{\alpha_{20} \mu_{I_R}}. \tag{23}$$

It is easy to see that when $\alpha_3 S_{I_R} > \gamma \mu_{I_R}$ (i.e. $A_2 > 0$),

$$(2\sigma - A_3)^2 + (4\sigma A_2 - A_3^2) = 4\sigma \left(\frac{\alpha_3 \beta_1 \sigma}{\alpha_{20} \mu_{I_R}} + A_2\right) > 0.$$

Noting that $2\sigma - A_3 = \sigma + \frac{\alpha_3\beta_1\sigma}{\alpha_{20}\mu_{I_R}} > 0$, thus, there are two zero points to the function $\triangle_f(A)$ about A, one is negative, the other is

$$\begin{aligned} A_1 &= -(2\sigma - A_3) + \sqrt{(2\sigma - A_3)^2 + (4\sigma A_2 - A_3^2)} \\ &= \frac{-\sigma(\alpha_{20}\mu_{I_R} + \alpha_3\beta_1) + 2\sqrt{\alpha_{20}\mu_{I_R}(\alpha_3 S_{I_R} - \gamma\mu_{I_R} + \alpha_3\sigma\beta_1)\sigma}}{\alpha_{20}\mu_{I_R}}. \end{aligned} \tag{24}$$

Let

$$\sigma_1 = \frac{4\alpha_{20}\mu_{I_R}(\alpha_3 S_{I_R} - \gamma\mu_{I_R})}{(\alpha_{20}\mu_{I_R} - \alpha_3\beta_1)^2}, \tag{25}$$

then, the following lemma can be easily proved.

Lemma 5 *When $\alpha_3 S_{I_R} > \gamma\mu_{I_R}$, the following statements hold true*

1. *$A_1 \geq 0$ if and only if $\sigma \leq \sigma_1$, and the equal-sign holds at the same time;*
2. *if $A_3 \neq A_2$ then $A_1 < A_2$;*
3. *$A_3 > A_1$ if and only if $A_3 > A_2$.*

Theorem 5 *System (19) admits*

(i) a unique positive equilibrium E_1^* if one of the following conditions holds:
  1. *$A_2 < 0$;*
  2. *$A_2 > 0$ and $A > A_2$;*
  3. *$A_2 > 0$, $A_3 > 0$ and $A_2 = A < A_3$.*

(ii) When $A_2 > 0$ and $A_2 < A_3$ there are two equilibria E_1^* and E_2^* if one of the following conditions holds:
  1. *$A_1 < 0$ and $0 < A < A_2$;*
  2. *$A_1 > 0$ and $A_1 < A < A_2$.*

(iii) There not exists any positive equilibrium if one of the following conditions holds:
  1. *$A_2 > 0$, $A_2 \geq A_3$ and $0 < A \leq A_2$;*
  2. *$A_2 > 0$, $A_1 > 0$, $A_2 < A_3$ and $0 < A < A_1$.*

Where $E_i^* = (X_{Ui}^*, X_{Ii}^*, B_i^*, I_{Ri}^*)$, $i = 1, 2$, with:

$$X_{Ui}^* = \frac{S_U}{\alpha_1 B_i^* + \mu_{X_U}}, \qquad X_{Ii}^* = \frac{\alpha_1 B_i^* X_{Ui}^*}{\mu_{X_I} + \alpha_2 I_{Ri}^*}, \qquad I_{Ri}^* = \frac{S_{I_R} + \beta_1 B_i^*}{\mu_{I_R}}. \tag{26}$$

and whose B_i^* components are given by

$$\begin{aligned} B_{1,2}^* &= \frac{\alpha_{20}\mu_{I_R}(\sigma - A) - \sigma\alpha_3\beta_1 \pm \sqrt{4\alpha_{20}\mu_{I_R}\sigma\left[(A\alpha_{20} + \gamma)\mu_{I_R} - \alpha_3 S_{I_R}\right] + \left[\sigma\alpha_3\beta_1 - \alpha_{20}\mu_{I_R}(\sigma - A)\right]^2}}{2\alpha_{20}\mu_{I_R}} \\ &= \frac{1}{2}\left[-(A - A_3) \pm \sqrt{(A - A_3)^2 + 4\sigma(A - A_2)}\right] \end{aligned} \tag{27}$$

The stability analysis of the generic equilibrium $\bar{E} = (\bar{X}_U, \bar{X}_I, \bar{B}, \bar{I}_R)$ can be performed by means of the Jacobian matrix of (19) in $\bar{E}$, that can be written as follows:

$$J(\bar{E}) = \begin{pmatrix} -\alpha_1\bar{B} - \mu_{X_U} & 0 & -\alpha_1\bar{X}_U & 0 \\ \alpha_1\bar{B} & -\alpha_2\bar{I}_R - \mu_{X_I} & \alpha_1\bar{X}_U & -\alpha_2\bar{X}_I \\ 0 & 0 & h(\bar{B}) & -\alpha_3\frac{\bar{B}}{A+\bar{B}} \\ 0 & 0 & \beta_1 & -\mu_{I_R} \end{pmatrix} \tag{28}$$

with

$$h(B) = \alpha_{20}\left(1 - \frac{2B}{\sigma}\right) + \frac{A}{(A+B)^2}\left(\gamma - \alpha_3\frac{S_{I_R} + \beta_1 B}{\mu_{I_R}}\right) \tag{29}$$

Noting that $h(0) = \frac{\alpha_{20}}{A}(A - A_2)$, it is easy to prove the following:

Theorem 6 *The following statements hold:*

(i) If $A_2 < 0$ then the equilibrium E_0 is unstable.
(ii) If $A < A_2$ then the equilibrium E_0 is locally asymptotically stable.

Theorem 7 *The stability properties of the internal equilibrium E_1^* and E_2^*.*

1. *When $\frac{\mu_{I_R}}{\alpha_{20}} \geq 1$ or $\frac{\mu_{I_R}}{\alpha_{20}} < 1$ and $A > A_5$, the equilibrium E_1^* is locally asymptotically stable if it exists;*
2. *If the equilibrium E_2^* exists, then it is always unstable.*

Where $A_5 = \sigma\left[1 + 3d - 2\sqrt{2d(1+d)}\right]$, $d = \frac{\mu_{I_R}}{\alpha_{20}}$.

Proof We can immediately check that $\lambda_1 = -\alpha_1 B_i^* - \mu_{X_U}$ and $\lambda_2 = -\alpha_2 A_{Ri}^* - \mu_{X_I}$ are two negative eigenvalues of the matrix $J(B_i^*)$ $(i = 1,2)$. From (29), considering with the third and fourth equations of (20), we have

$$\begin{aligned} h(B_i^*) &= \alpha_{20}\left(1 - \tfrac{2B_i^*}{\sigma}\right) + \tfrac{A}{A+B_i^*}\tfrac{\gamma - \alpha_3 I_R^*}{A+B_i^*} \\ &= \alpha_{20}\left(1 - \tfrac{2B_i^*}{\sigma}\right) + \tfrac{\alpha_{20}A}{A+B_i^*}\left(1 - \tfrac{B_i^*}{\sigma}\right) \\ &= -\tfrac{\alpha_{20}B_i^*}{\sigma(A+B_i^*)}\left(2B_i^* + A - \sigma\right) \end{aligned}$$

Therefore, the stability properties of the positive equilibria E_i^* $(i = 1,2)$ is equal to one of the following matrix

$$\bar{J}(E_i^*) = \begin{bmatrix} -\frac{\alpha_{20}B_i^*(2B_i^*+A-\sigma)}{\sigma(A+B_i^*)} & -\alpha_3\frac{B_i^*}{A+B_i^*} \\ \beta_1 & -\mu_{I_R} \end{bmatrix}.$$

The trace of the matrix $\bar{J}(E_i^*)$ is given by

$$\begin{aligned} \mathrm{tr}(\bar{J}(E_i^*)) &= -\mu_{I_R} - \tfrac{\alpha_{20}B_i^*(2B_i^*+A-\sigma)}{\sigma(A+B_i^*)} \\ &= -\tfrac{\alpha_{20}}{\sigma(A+B_i^*)}\left[2B_i^{*2} + (A - A_4)B_i^* + \sigma dA\right] \\ &= -\tfrac{\alpha_{20}}{\sigma(A+B_i^*)}g(B_i^*) \end{aligned}$$

and the determinant of the matrix $\bar{J}(E_i^*)$ by

$$\begin{aligned}\det(\bar{J}(E_i^*)) &= \frac{-\alpha_{20}\mu_{I_R}B_i^*(\sigma - A - 2B_i^*)}{\sigma(A+B_i^*)} + \alpha_3\beta_1\frac{B_i^*}{A+B_i^*} \\ &= \frac{\alpha_{20}\mu_{I_R}B_i^*}{\sigma(A+B_i^*)}(2B_i^* + A - A_3),\end{aligned}$$

Where $g(B) = 2B^2 + (A - A_4)B + \sigma dA,\ A_4 = \sigma(1-d)$.

Due to the expression of B_i^* $(i = 1,2)$, $\det(\bar{J}(E_1^*)) > 0$ and $\det(\bar{J}(E_2^*)) < 0$, consequently, E_2^* is always unstable if it exists and the stability properties of E_1^* depend on the sign of $\mathrm{tr}(\bar{J}(E_1^*))$. Thus, E_1^* is locally asymptotically stable if $g(B_1^*) > 0$ and E_1^* is unstable if $g(B_1^*) < 0$.

If $\frac{\mu_{I_R}}{\alpha_{20}} \geq 1$, then $A_4 \leq 0$, consequently, $g(B_1^*) > 0$, that is, E_1^* is locally asymptotically stable.

If $\frac{\mu_{I_R}}{\alpha_{20}} < 1$ and $A \geq A_4 > 0$, consequently, $g(B_1^*) > 0$, that is, E_1^* is locally asymptotically stable.

If $\frac{\mu_{I_R}}{\alpha_{20}} < 1$ and $A_5 < A < A_4$, noting that $0 < A_5 < A_4 < A_6 = \sigma\left[1 + 3d + 2\sqrt{2d(1+d)}\right]$, then for the equation $g(B) = 0$, its discriminant $\triangle_g(A) = (A - A_5)(A - A_6) < 0$, consequently, $g(B) > 0$, that is E_1^* is locally asymptotically stable if it exists.

3.2. Simulations of the System (19)

In this Section we present several examples of solutions of the model (19) discussed above. MATHEMATICA (Wolfram Research, 1989) has been used to solve the differential equations of the model for a variety of initial conditions in order to illustrate the nature of the dynamical system. These results are presented in the form of a plot of B versus I_R.

The dynamics of system (19) is reacher then that one of system (5) and several outcomes are possible. Among these, the most interesting is that one corresponding to the saddle-node bifurcation. To compute simulations we consider the following parameter values:

$$\begin{aligned}&\beta_1 = 0.4;\ \sigma = 6\,10^3;\ \alpha_{20} = 0.05;\ \alpha_3 = 5\,10^{-3};\ \mu_{I_R} = 0.11;\ S_{I_R} = 10^3;\ \gamma = 0.8;\\ &S_U = \mu_{X_U}10^4;\ \alpha_1 = 10^{-3};\ \alpha_2 = 10^{-3};\ \mu_{X_U} = 0.013;\ \mu_{X_I} = 0.011\end{aligned}$$

Moreover, by varying the value of the parameter A we are able to observe some different cases of Theorem 5.

In particular when $A = 300$ we are in the hypothesis of Theorem $5_{(iii)_2}$ then there exists only E_0 which is locally asymptotically stable (Fig. 2_a). Furthermore, when $A = A_1 = 408.7$ and $A = 710$ we are in the hypothesis of Theorem $5_{(ii)_2}$ a saddle-node bifurcation happens and the two internal equilibria appear, one stable the other one unstable, E_0 remains asymptotically stable (Fig. $2_{b,c}$). When $A = 900$ we are in the hypothesis of Theorem $5_{(i)_2}$ a transcritical bifurcation happens, E_0 becomes unstable and only one internal equilibrium remains which is asymptotically stable (Fig. 2_d).

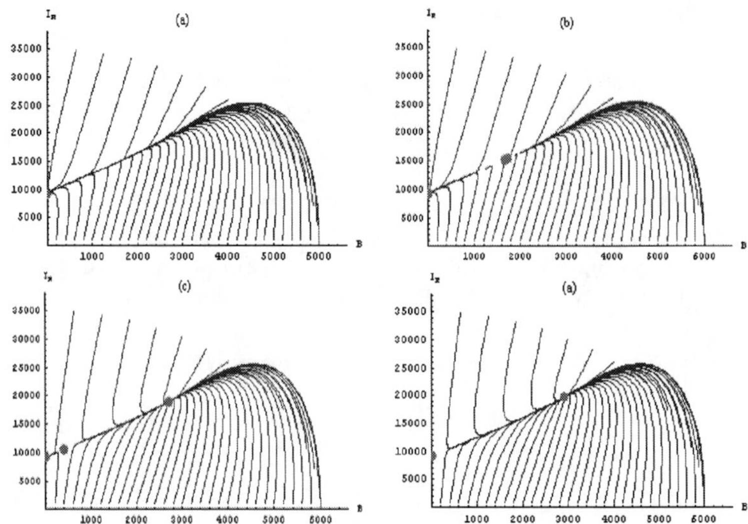

FIGURE 2. **(a)** $A = 300$ There exists only E_0 which is asymptotically stable. **(b)** $A = A_1 = 408.7$. There are three equilibria: two internal overlapped, one asymptotically stable the other one unstable, and the boundary one which remains asymptotically stable. **(c)** $A_1 < A = 710 < A_2 = 893.1$. In this case there are the same three equilibria as in (b) but the interior ones are detached. **(d)** $A(= 900) > A_2$. E_0 is unstable and only one internal equilibrium remains which is asymptotically stable.

4. CONCLUSION AND DISCUSSION

Invertebrates lack adaptive immune systems. They have indeed developed a unique system of biological host response, called innate immunity, which represents the main defense from invading bacterial, fungal and viral pathogens. This defense system is actually composed by several components with very complex kinetics at cellular, biochemical and genetic levels. In this paper we have adopted a macroscopic point of view which gives up the representation of the real processes and describes the innate system simply through the phenomenological variable I_R.

The influence of quorum sensing on the bacterial infection has been introduced in the third Section where, in order to simplify the computation, we assume that the positive constant A has a double role. It represents on one hand a "threshold concentration" beyond which the new proliferation term of the bacterial population, due to the quorum sensing, tends to saturation; on the other hand, at the same time, A represents a given value of the bacterial population beyond which the attack capability of the innate immune response begins to saturate. We observe that if we neglect the term (17), the equation $(19)_3$ becomes strongly similar to the "prey equation" of a predator-prey system discussed by May [8].

The presence of quorum sensing modifies the previous scenario because, now, we have the foregoing uninfected equilibrium but two (instead of one) infected equilibria. Of course their existence and stability properties generally depend also on the new parameters A and γ, so that no comparison is possible with the corresponding conditions obtained in absence of quorum sensing. Nevertheless, we point out that the infected equilibrium E_2^* is unstable when it exists and therefore biological harmless. Moreover, we observe that, in our model the parameters A and γ depend on the bacterial species which gives rise to the infection. Therefore, according to theorem 6 and 7_1 the same bacterial species could be cleared or become permanent for suitable initial data, depending on the host.

Finally, we observe that among the immunological interactions between bacteria and invertebrates a relevant importance have those between bacteria and Mollusca. This happens because, due to their nutritional role, the Mollusca, and in particular the bivalves, can represent the vectors of pathogen agents for human population, and this explains why they play a central role in the analysis of the alimentary risk for man [9].

APPENDIX

The following definitions and results about limit systems, which is given in [7].

$$x' = \phi(t,x) \tag{30}$$

$$y' = \psi(y), \tag{31}$$

where ϕ and ψ are continuous and locally Lipschitz in $x \in R^n$ and solutions exist for $t \geq 0$. Equation (30) is called asymptotically autonomous with limit equation (31) if $\phi(t,x) \to \psi(x)$ as $t \to \infty$ uniformly for $x \in R^n$.

Lemma 6 *Let e be a locally asymptotically stable equilibrium of (31) and ω the ω-limit set of a forward bounded solution x of (30). If ω contains a point y_0 such that the solution y of (31), with $y(0) = y_0$, converges to e for $t \to \infty$, then $\omega = \{e\}$, i.e., $x(t) \to e$, $t \to \infty$.*

Corollary 8 *If solutions of system (30) are bounded and the equilibrium e of the limit system (31) is globally asymptotically stable, then any solution $x(t)$ of (30) converges to e as $t \to \infty$.*

REFERENCES

1. S. Iwanaga, and B. L. Lee, "Recent Advances in the Innate Immunity of Invertebrate Animals", *Journal of Biochemistry and Molecular Biology*, Vol. 38, No. 2, (2005), 128-150.
2. E. Beretta, M. Carletti, D. E. Kirschner, and S. Marino, "Stability analysis of a mathematical model of the immune response with delays", *Mathematics for life science and medicine*, Springer Ed. Y. Takeuchi, Y. Isawa and K. Sato., 177-206, 2006.
3. B. L. Bassler, "How bacteria talk to each other: regulation of gene expression by quorum sensing", *Curr. Opinions Microbiol.*, 2, (1999), 582.
4. P. Fergola, F. Aurelio, M. Cerasuolo, and A. Noviello, "Influence of mathematical modelling of nutrient uptake and quorum sensing on the allelopathic competition", in *Proocedings of "WASCOM 2003" 13th Conference on Waves and Stability in Continuous Media*, (2004), 191-203.
5. P. Fergola, M. Cerasuolo, and E. Beretta, "An Allelopathic Competition Model with Quorum Sensing and Delayed Toxicant Production", *Math. Biosci. and Engin.* 3, n.1, (2006), 37-50.
6. P. Fergola, E. Beretta and M. Cerasuolo, "Some New Results on an Allelopathic competition models with quorum sensing and delayed toxicant production", *Nonlinear Anal. Real World Appl.* 7, (2006), 1081-1095.

7. H. R. Thieme, "Convergence results and a Poincaré-Bendixson trichotomy for asymptotically autonomous differential equations", *J. Math. Biol.*, 30, (1992), 755-763.
8. R. M. May, *Stability and Complexity in Model Ecosystems*, Second Edition, Princeton University Press, 1974.
9. P. G. Tiscar, and F. Mosca, "Patologie emergenti in molluschicoltura", *Webzine Sanità Pubblica Veterinaria*, n. 27, 2004.

Two Types of Coexistence in Cross-Feeding Microbial Consortia

Shinji Nakaoka* and Yasuhiro Takeuchi†

*Aihara Complexity Modelling Project, ERATO, JST, Komaba Open Laboratory, The University of Tokyo 4-6-1, Komaba, Meguro-ku, Tokyo, 153-8505, Japan
†Graduate school of Science and Engineering, Shizuoka University, Johoku 3-5-1, Hamamatsu, Shizuoka 432-8561, Japan

Abstract. Exploitative competition of two cross-feeding strains is studied. We found that two types of coexistence of two cross-feeding strains, type-I coexistence (cultivated type) and type-II coexistence (self-sufficiency type) are possible for microbial cross-feeding strains. In all cases of coexistence, trade-off in nutrient availability is required. However, trade-off is necessary but is not sufficient for the coexistence of two strains. Over-production of metabolite can induce competitive exclusion on one hand (cultivated regime) whereas do support the coexistence of two strain on the other hand (self-sufficiency regime). Coexistence of two strains is evaluated by invasibility and permanence criteria and numerical simulations.

Keywords: Microorganisms, cross-feeding, trade-off, chemostat equations, population dynamics, basic production ratio, mutually invasible, permanence.
PACS: 02.30.Hq, 02.60.Cb

1. INTRODUCTION

Microbes distribute everywhere throughout the Earth. Metabolically related bacterial populations occur within a given space. Indeed, some bacterial populations correlate strongly with each other in the expression of function [24], such an association is regarded as a microbial consortium. The importance of microbial consortia as a fundamental component of an ecosystem is demonstrated by the fact that they play a major role in a mineralization of organic compounds. It is often the case that more than two species are synergistically involved in the degradation of organic compounds by syntrophic association or cross-feeding in a consortium (in the case of nitrification, [7]). The current studies have revealed that the degradation of introduced xenobiotics is carried out by bacterial species being capable of degrading them. Two or more species are involved in the degradation of xenobiotics [2] [9], implying that the degradation is mediated effectively by multiple species, rather than a single species. Although a diverse range of studies have investigated on xenobiotics-degrading consortia in which synergetic degradations play an important role [26] [30], the current understanding of microbial consortia is limited to descriptions of the members and their inherent functions. In general, the manner of the interaction to maintain the consortium consisting of two or more species in a given metabolic sequence has yet to be understood.

CP1028, *Collective Dynamics: Topics on Competition and Cooperation in the Biosciences*
edited by L. M. Ricciardi, A. Buonocore, and E. Pirozzi

Most theoretical studies to investigate the mechanisms for the coexistence of two or more species have always faced to a problem on the limitation of the number of species to coexist. The competitive exclusion principle predicts that the number of coexistent species competing for several available resources cannot exceed the number of available resources in steady state [12]. The chemostat is an experimental apparatus which offers the ideal place to study exploitative competition of microbial species. The well operated experiments on the chemostat confirmed the establishment of competitive exclusion [14]. Various mathematical models of resource competition for the limiting resources, including the basic chemostat equations [27], have been presented and revealed that the consequences are consistent with the competitive exclusion principle [1]. The competitive exclusion principle is described, in terms of the theory on the chemostat, such that the species with the lowest break-even concentration survives [27]. Many variants of the basic chemostat equations have been proposed and identified the mechanisms underlying the coexistence of two or more species in the apparent contradictions of the competitive exclusion principle (see the standard-textbook of chemostat equations [27] or an excellent review [5]).

The current studies revealed the potential importance of cross-feeding or syntrophic associations for the maintenance of microbial consortia [10], [23], [25], [29]. Various theories have been proposed to address the question of how cross-feeding evolves in microbial communities. Pfeiffer et al. [23] showed how stable polymorphisms are maintained by cross-feeding as a consequence of optimization principles. Doebeli [10] proposed a theory that explains how cross-feeding polymorphisms can emerge from a monomorphism, based on the framework of adaptive dynamics [11]. Many experimental studies have investigated the adaptive diversification of *E. coli* populations [25]. Although all of these previous studies clearly demonstrate the feasibility of cross-feeding in microbial communities as a consequence of evolution, the mechanisms that underlies the coexistence of metabolically related microbial species is not fully investigated.

A variety of interactions could be considered among cross-feeding strains. It is often the case that no energy gain is obtained through the first and the following several steps in the degradation of organic compounds. The degradation of organic compounds can be mediated by extracellular enzymes bound to microorganisms which work outside the surface of cells. Tang and Wolkowicz [28] studied mathematical models of microbial competition between two species for nutrient substrate and its metabolite that can be utilized as two types of energy source. The degradation of nutrient substrate to metabolite is mediated by extracellular enzymes bound to microorganisms. They examined whether nutrient substrate and its metabolite support the coexistence of two species. However, coexistence of two species was never observed, although coexistent equilibrium points can exist.

An experimental study by Christensen *et al.* demonstrated that coexistence of metabolically related two species does occur within biofilm [6]. They showed that metabolite produced by benzyl alcohol-degrading strain, *Pseudomonas putida* strain

R1 (in short, strain R1) was utilized by another strain, *Acinetobacter* strain C6 (strain C6) which is the worse competitor than strain R1 in exploitative competition for benzyl alcohol. Recently, the authors [22] studied a chemostat model that is similar to one of models studied in [28]. We revealed that coexistence of two strains can occur if we assume a particular manner of interactions between two strains different from the manner of interactions assumed on the models in [28]. Thus different outcomes can be derived, depending on the manner of interactions between two strains in the metabolism of nutrient substrate and its metabolite. It is therefore expected to identify the mechanisms, or the manner of interactions, underlying the coexistence of two strains. In this paper, we consider

$$\begin{cases} R'(t) = (R^0 - R(t))D - dR(t)x_1(t) - g_1(R(t))x_1(t) - g_2(R(t))x_2(t), \\ S'(t) = (S^0 - S(t))D + dR(t)x_1(t) - f_1(S(t))x_1(t) - f_2(S(t))x_2(t), \\ x_1'(t) = x_1(t)(f_1(S(t)) + g_1(R(t)) - D), \\ x_2'(t) = x_2(t)(f_2(S(t)) + g_2(R(t)) - D) \end{cases} \tag{E$_0$}$$

with the initial condition

$$(R(0), S(0), x_1(0), x_2(0)) \in \mathbb{R}_+^4. \tag{I$_0$}$$

We hereafter call the nutrient substrate, resource. Let R, S, x_1 and x_2 denote the concentrations of resource, metabolite and microbial strain-1 and strain-2, respectively. Resource and metabolite are provided constantly by R^0 and S^0 with a constant dilution rate D, respectively. Note that $R^0 > 0$ whereas $S^0 \geq 0$. This implies that (E$_0$) includes the case of $S^0 = 0$ to allow no external supply of metabolite S. (E$_0$) accounts for the degradation of resource R into metabolite S via the metabolism of strain-1. The degradation of resource R is mediated by extracellular enzymes secreted by strain-1. For simplicity, we assume that the amount of extracellular enzyme, denoted by E, is proportional to the concentration of strain-1. Thus $E(t) = kx_1(t)$ with the proportionality constant k. We further assume that all biochemical reactions follow the well-known mass action law. Hence the amount of the produced metabolite is given by $d'RE = dRx_1$, where the parameter d measures degradation rate of resource R per unit concentrations of resource and strain-1. Note that (E$_0$) includes the case of $d = 0$. If $d = 0$, (E$_0$) is reduced to the chemostat equations with two independent available resources (c.f. studies for complementary and substitutable resources, [3], [20] and [21]). For simplicity, we assume that the yields of resource R and metabolite S are the same and moreover have been scaled out from (E$_0$). Functional responses $g_j(R)$ and $f_j(S)$ represent the growth kinetics of strain-j via the catabolism of resource R and metabolite S, respectively ($j = 1, 2$). Mathematically, these functional responses are assumed to be bounded, monotonically increasing and continuously differentiable with respect to the argument. Moreover $f_j(0) = g_j(0) = 0$. The maximum growth rate m_{fj} and m_{gj} are given by

$$m_{fj} := \lim_{S \to \infty} f_j(S) \text{ and } m_{gj} := \lim_{R \to \infty} g_j(R), \quad (j = 1, 2).$$

It is easy to show the nonnegativeness of solutions of (E_0) with initial condition (I_0). The dissipativeness of solutions of (E_0) can be proved by introducing a function V such that

$$V(R,S,x_1,x_2) := R+S+x_1+x_2. \tag{1}$$

The derivative of (1) along the solution of (E_0) is

$$\dot{V}_{(E_0)}(R,S,x_1,x_2) = D(R^0+S^0)-DV.$$

Hence system (E_0) is dissipative (for the definition of dissipative, see [13]).

The organization of this paper is as follows. In the next section, existence and stability conditions for interior equilibrium points of the subsystems of (E_0) are summarized. In section 3, we shall introduce the criteria for the permanence of (E_0) to represent the coexistence of two strains. Numerical simulations are performed to investigate the dynamics of (E_0). In section 4, we study four specific cases of (E_0) in order to figure out the manner of interactions on the utilization of two different types of energy source under which coexistence is mediated. Additional analytical results for (E_0) are summarized in Appendix.

2. PRELIMINARY RESULTS (ONE-STRAIN MODELS)

In order to derive criteria for invasibility and permanence in the next sections, we summarize (global) stability results for subsystems of (E_0), most of which have already been studied in [22] (see also Tang and Wolkowicz [28]).

2.1. Population dynamics of strain-1

One of subsystems of (E_0) is given by

$$\begin{cases} R'(t) = (R^0-R(t))D - dR(t)x_1(t), \\ S'(t) = (S^0-S(t))D + dR(t)x_1(t) - f_1(S(t))x_1(t), \\ x_1'(t) = x_1(t)(f_1(S(t))-D) \end{cases} \tag{B$_1$}$$

with the initial condition

$$(R(0),S(0),x_1(0)) \in \Omega_1, \tag{I$_1$}$$

where

$$\Omega_1 := \{(R,S,x_1) \in \mathbb{R}^3_+ | R \geq 0, S \geq 0, x_1 \geq 0\}. \tag{2}$$

We assume that strain-1 cannot utilize resource R as an energy source. If $d=0$, (B_1) is reduced to the basic chemostat equations with the independent dynamics for resource R [27, Chapter 1].

There is always *washout* equilibrium point

$$E_0 := (R^0, S^0, 0).$$

Under certain conditions, there are generically two interior equilibrium points of (B_1), denoted by

$$\bar{E}_{1+}^C := (\bar{R}_{1,C}, \bar{S}_1, \bar{x}_{1,C}) \text{ and } \bar{E}_{1+}^S := (\bar{R}_{1,S}, \bar{S}_1, \bar{x}_{1,S}).$$

Two important thresholds are introduced to derive existence and stability conditions for interior equilibrium points of (B_1). One threshold is known as the *break-even concentration,* which determines the asymptotic behavior of the basic chemostat equations [27, Chapters 1 and 2]. If $m_{f1} > D$, the break-even concentration for strain-1 is defined by the second component of the interior equilibrium point: $\bar{S}_1 = f_1^{-1}(D)$. Another threshold is the *basic production ratio,* which is defined by the average number of newly produced metabolites by one microorganism during its sojourn span in the chemostat [22]. Note that the reciprocal number of D denotes an average sojourn time of microorganisms in the chemostat. The term $d\bar{R}_{1,C}$ denotes the production rate of metabolites via the degradation of resource R by one microorganism in steady state. Hence basic production ratio $\bar{\mathscr{P}}_{1,C}$ defined for $\bar{E}_{1+}^C$ is given by

$$\bar{\mathscr{P}}_{1,C} := \frac{d\bar{R}_{1,C}}{D}.$$

The maximum basic production ratio is given by $\bar{\mathscr{P}}_{1,\max} = \frac{dR^0}{D}$. The asymptotic behavior of (B_1) is classified into two cases according to the signs of $S^0 - \bar{S}_1$ and $\bar{\mathscr{P}}_{1,\max} - 1$.

(i) Cultivated regime

Assume that

$$S^0 > \bar{S}_1.$$

There exists a unique interior equilibrium point of (B_1), simply denoted by $\bar{E}_{1+} = \bar{E}_{1+}^C$. We have the following result for the asymptotic stability of $\bar{E}_{1+}$.

Proposition 1. *[22, Theorem 1] Assume that $S^0 > \bar{S}_1$. Then $\bar{E}_{1+}$ is globally asymptotically stable.*

The asymptotic behavior of (B_1) is the same as that of the basic chemostat equations if break-even concentration $\bar{S}_1$ does not exceed an input concentration of substrate S^0.

(ii) Self-sufficiency regime

Assume that

$$S^0 < \bar{S}_1.$$

Microorganisms are inevitably washed out in terms of the consequence of the basic chemostat equations. There are two possible cases according to the sign of $\bar{\mathscr{P}}_{1,\max} - 1$.

Proposition 2. *[22, Theorem 2] Assume that $S^0 < \bar{S}_1$ and $\bar{\mathscr{P}}_{1,max} < 1$. Then washout equilibrium E_0 is globally asymptotically stable.*

Thus the asymptotic behavior of (B_1) is the same as that of the basic chemostat equations. By contrast, if $\bar{\mathscr{P}}_{1,\max} > 1$ together with the condition

$$\left\{dR^0 + d(S^0 - \bar{S}_1) + D\right\}^2 - 4dR^0D > 0, \tag{SNB$_1$}$$

microorganisms can stay in the chemostat even though $S^0 < \bar{S}_1$.

Proposition 3. *[22, Theorem 3] Assume that $S^0 < \bar{S}_1$, $\bar{\mathscr{P}}_{1,max} > 1$ and (SNB$_1$) holds. Then E_0 and $\bar{E}^C_{1+}$ are locally asymptotically stable. $\bar{E}^S_{1+}$ is saddle.*

Proposition 3 implies that bistability between washout equilibrium E_0 and persistent equilibrium $\bar{E}^C_{1+}$ occurs, depending on the initial concentration of strain-1. Hence we could observe two different outcomes. If microorganisms are enough to live on the supplied food ($S^0 > \bar{S}_1$), they do nothing other than being cultivated. If not ($S^0 < \bar{S}_1$), they have got engaged on producing an energy source (metabolite) to maintain their growth in self-sufficient manner ($\bar{\mathscr{P}}_{1,\max} > 1$).

If strain-1 can utilize resource R and metabolite S as energy sources, the model is given by

$$\begin{cases} R'(t) = (R^0 - R(t))D - dR(t)x_1 - g_1(R(t))x_1(t), \\ S'(t) = (S^0 - S(t))D + dR(t)x_1 - f_1(S(t))x_1(t), \\ x_1'(t) = x_1(t)(f_1(S(t)) + g_1(R(t)) - D) \end{cases} \tag{B$_2$}$$

with initial condition (I_1). Break-even concentration and basic production ratio again play an important role. Under certain conditions, there exist two interior equilibrium points of (B_2), denoted by (distinguished to the previous case by adding "~" for the components of an equilibrium point)

$$\tilde{E}^C_{1+} := (\tilde{R}_{1,C}, \tilde{S}_{1,C}, \tilde{x}_{1,C}) \text{ and } \tilde{E}^S_{1+} := (\tilde{R}_{1,S}, \tilde{S}_{1,S}, \tilde{x}_{1,S}).$$

Function $\phi : [0,\infty) \to [0,\infty)$ is defined by the sum of two composition functions such that

$$\phi(x_1) := f_1(S(x_1)) + g_1(R(x_1)).$$

Accordingly, we introduce the break-even concentration defined for (B_2) as follows.

$$\phi(0) = f_1(S^0) + g_1(R^0).$$

Basic production ratio $\tilde{\mathscr{P}}_{1,C}$ defined for $\tilde{E}^C_{1+}$ is given by

$$\tilde{\mathscr{P}}_{1,C} := \frac{d\tilde{R}_{1,C} + g_1(\tilde{R}_{1,C})}{D}.$$

The maximum basic production ratio is given by $\tilde{\mathscr{P}}_{1,\max} = \frac{dR^0 + g_1(R^0)}{D}$. The asymptotic behavior of (B_2) is classified into two cases according to the signs of $\phi(0) - D$ and $\tilde{\mathscr{P}}_{1,\max} - 1$.

(i) Cultivated regime: Revisited
Assume that

$$\phi(0) > D.$$

There is always a unique interior equilibrium point, simply denoted by $\tilde{E}_{1+} = \tilde{E}^C_{1+}$. We have the following result for the global stability of $\tilde{E}_{1+}$.

Proposition 4. *[22, Theorem 4] Assume that $\phi(0) > D$. Then $\tilde{E}_{1+}$ is globally asymptotically stable.*

(ii) Self-sufficiency regime: Revisited
Assume that

$$\phi(0) < D.$$

We can show that $\phi(x_1)$ is a one-hump function. If $\tilde{\mathscr{P}}_{1,\max} > 1$, there exist two interior equilibrium points of (B_2) if there exists $\hat{x}_1 > 0$ such that $\hat{x}_1$ satisfies

$$\phi(\hat{x}_1) > D. \tag{SNB$_2$}$$

We have the following result.

Proposition 5. *[22, Theorem 5] Assume that $\phi(0) < D$, $\tilde{\mathscr{P}}_{1,max} > 1$ and (SNB$_2$) holds. Then washout equilibrium point E_0 and persistent equilibrium point $\tilde{E}^C_{1+}$ are locally asymptotically stable. $\tilde{E}^S_{1+}$ is saddle.*

Another subsystem of (E_0) necessary to introduce is given by

$$\begin{cases} R'(t) = (R^0 - R(t))D - dR(t)x_1(t) - g_1(R(t))x_1(t), \\ S'(t) = (S^0 - S(t))D + dR(t)x_1(t), \\ x_1'(t) = x_1(t)(g_1(R(t)) - D) \end{cases} \tag{B$_3$}$$

with initial condition (I_1). Subsystem (B_3) corresponds to the basic chemostat equations coupled with the subjectional dynamics for metabolite S. Let $\tilde{E}_{1+} = (\tilde{R}_1, \tilde{S}_1, \tilde{x}_1)$ denote an interior equilibrium point of (B_3). Note that (B_3) is deduced by letting $f_1 \equiv 0$ in (B_2).

Proposition 6. *[27, Chapter 1] Assume that $R^0 > \tilde{R}_1$. Then $\tilde{E}_{1+}$ is globally asymptotically stable.*

2.2. Population dynamics of strain-2

One of subsystems of (E_0) which describes the population dynamics of strain-2 is given by

$$\begin{cases} R'(t) = (R^0 - R(t))D - g_2(R(t))x_2(t), \\ S'(t) = (S^0 - S(t))D - f_2(S(t))x_2(t), \\ x_2'(t) = x_2(t)(f_2(S(t)) + g_2(R(t)) - D) \end{cases} \tag{B$_4$}$$

with the initial condition

$$(R(0), S(0), x_2(0)) \in \Omega_2, \tag{I$_2$}$$

where

$$\Omega_2 := \{(R,S,x_2) \in \mathbb{R}^3_+ | R \geq 0, S \geq 0, x_2 \geq 0\}. \tag{3}$$

Compared with (B$_2$), we always assume that $d = 0$ for the population dynamics of strain-2. Let $\bar{E}_{2+} := (\bar{R}_2, \bar{S}_2, \bar{x}_2)$ denote an interior equilibrium point of (B$_4$). We have the following result for the global stability of $\bar{E}_{2+}$.

Proposition 7. *[22, Corollary of Theorem 4] Assume that $\bar{S}_2 < S^0$ and $\bar{R}_2 < R^0$. Then $\bar{E}_{2+}$ is globally asymptotically stable.*

Finally, we consider the following subsystem of (E$_0$)

$$\begin{cases} R'(t) = (R^0 - R(t))D - g_2(R(t))x_2(t), \\ S'(t) = (S^0 - S(t))D, \\ x_2'(t) = x_2(t)(g_2(R(t)) - D) \end{cases} \tag{B$_5$}$$

with initial condition (I$_2$). (B$_5$) corresponds to the basic chemostat equations coupled with the independent dynamics for metabolite S. Let $\tilde{E}_{2+} := (\tilde{R}_2, S^0, \tilde{x}_2)$ denote an interior equilibrium point of (B$_5$). We have the following result.

Proposition 8. *[27, Chapter 1] Assume that $R^0 > \tilde{R}_2$. Then $\tilde{E}_{2+}$ is globally asymptotically stable.*

All results are summarized on Table 1.

TABLE 1. Existence and stability conditions for persistent (and washout) equilibrium points of one-strain models.

Subsystem	Stability conditions	Equilibrium point(s)
(B$_1$)	$S^0 > \bar{S}_1$	$\bar{E}^C_{1+}$
	$S^0 < \bar{S}_1$, $\bar{\mathscr{P}}_{1,\max} > 1$ and (SNB$_1$)	E_0 and $\bar{E}^C_{1+}$
(B$_2$)	$\phi(0) > D$	$\tilde{E}_{1+}$
	$\phi(0) < D$, $\tilde{\mathscr{P}}_{1,\max} > 1$ and (SNB$_2$)	E_0 and $\tilde{E}^C_{1+}$
(B$_3$)	$R^0 > \tilde{R}_1$	$\tilde{E}_{1+}$
(B$_4$)	$\bar{S}_2 < S^0$ and $\bar{R}_2 < R^0$	$\bar{E}_{2+}$
(B$_5$)	$R^0 > \tilde{R}_2$	$\tilde{E}_{2+}$

3. COEXISTENCE OF TWO STRAINS: GENERAL CASE

We derive conditions for the existence of an interior equilibrium point of (E$_0$). Coexistence of strains is evaluated by criteria for permanence [17]. A theory on average Lyapunov functions together with criteria for invasibility is exploited to prove the permanence of (E_0). Finally we shall investigate the dynamics of (E_0) by numerical simulations.

3.1. Existence of interior equilibrium point

Let $E_* = (R^*, S^*, x_1^*, x_2^*)$ denote an interior equilibrium point of (E_0). Components of E_* satisfy

$$\begin{cases} (R^0 - R^*)D - dR^*x_1^* - g_1(R^*)x_1^* - g_2(R^*)x_2^* = 0, \\ (S^0 - S^*)D + dR^*x_1^* - f_1(S^*)x_1^* - f_2(S^*)x_2^* = 0, \\ x_1^*(f_1(S^*) + g_1(R^*) - D) = 0, \\ x_2^*(f_2(S^*) + g_2(R^*) - D) = 0. \end{cases} \tag{4}$$

The basic production ratio defined for E_* is given by

$$\mathscr{P}_* := \frac{dR^* + g_1(R^*)}{D}. \tag{5}$$

We need to assume $x_1^* > 0$ and $x_2^* > 0$. It follows from the third and fourth equations of (4) that R^* and S^* are derived as positive roots of the following system of equations

$$\begin{aligned} f_1(S^*) + g_1(R^*) &= D, \\ f_2(S^*) + g_2(R^*) &= D. \end{aligned} \tag{6}$$

There are several possibilities of having positive roots depending on the form of functions f_j and g_j $(j = 1,2)$. Hereafter we assume that there exists a unique set of positive roots of (6). Adding all equations in (4) gives

$$(R^0 - R^*) + (S^0 - S^*) = x_1^* + x_2^*. \tag{7}$$

A system of linear equations with respect to x_1^* and x_2^* can be obtained from the first and second equations of (4) as follows.

$$\begin{aligned} (dR^* + g_1(R^*))x_1^* + g_2(R^*)x_2^* &= (R^0 - R^*)D, \\ (-dR^* + f_1(S^*))x_1^* + f_2(S^*)x_2^* &= (S^0 - S^*)D. \end{aligned}$$

Define Δ by $D\Delta = (dR^* + g_1(R^*))f_2(S^*) - g_2(R^*)(-dR^* + f_1(S^*))$. Substituting $f_1(S^*) = D - g_1(R^*)$ and $f_2(S^*) = D - g_2(R^*)$ into Δ gives

$$\Delta = \mathscr{P}_*D - g_2(R^*). \tag{8}$$

If $\Delta \neq 0$, there exists a unique set of positive roots of (6), denoted by (x_1^*, x_2^*), to which we have the following explicit expressions

$$\begin{aligned} x_1^* &= \frac{f_2(S^*)(R^0 - R^*) - g_2(R^*)(S^0 - S^*)}{\Delta}, \\ x_2^* &= \frac{(\mathscr{P}_* - 1)(R^0 - R^*)D + \mathscr{P}_*(S^0 - S^*)D}{\Delta}. \end{aligned} \tag{9}$$

Note that $E_* \in \text{int}\mathbb{R}^4_+$ if and only if $x_1^* > 0$ and $x_2^* > 0$.

3.2. Criteria for invasibility and permanence

We shall introduce criteria for the invasibility of a strain to another strain population. Let E_2^{Bd} denote an arbitrary boundary equilibrium point of (E_0) which satisfies $x_1 \equiv 0$, or equivalently, an interior equilibrium point on boundary region Ω_2. As summarized in subsection 2.2, E_2^{Bd} corresponds to either $\bar{E}_{2+}$ or $\tilde{E}_{2+}$. The transversal eigenvalue (see Hofbauer and Sigmund [16]) at boundary equilibrium point E_2^{Bd} is defined as follows.

$$\begin{aligned}\Lambda_{1\to 2} :=& \frac{\partial}{\partial x_1}\{x_1(f_1(S)+g_1(R)-D)\}\Big|_{E_2^{\text{Bd}}} \\ =& f_1(S_2^{\text{Bd}})+g_1(R_2^{\text{Bd}})-D. \end{aligned} \tag{10}$$

The sign of $\Lambda_{1\to 2}$ determines the invasibility of strain-1 to the population consisting of strain-2 individuals. If $\Lambda_{1\to 2} > 0$, strain-1 can successfully invade. Otherwise, or $\Lambda_{1\to 2} < 0$, E_2^{Bd} is locally asymptotically stable and hence strain-1 fails to invade. In the same way, the invasibility of strain-2 to strain-1 population is determined by the sign of the transversal eigenvalue at boundary equilibrium point E_1^{Bd} on boundary region Ω_1 as follows.

$$\begin{aligned}\Lambda_{2\to 1} :=& \frac{\partial}{\partial x_2}\{x_2(f_2(S)+g_2(R)-D)\}\Big|_{E_1^{\text{Bd}}} \\ =& f_2(S_1^{\text{Bd}})+g_2(R_1^{\text{Bd}})-D. \end{aligned} \tag{11}$$

As summarized in subsection 2.1, E_1^{Bd} corresponds to either $\bar{E}_{1+}^C$, $\bar{E}_{1+}^S$, $\tilde{E}_{1+}^C$ or $\tilde{E}_{1+}^S$. Two strains are called *mutually invasible* if $\Lambda_{1\to 2} > 0$ and $\Lambda_{2\to 1} > 0$. By applying a theory on average Lyapunov functions developed in [16], [17] to (E_0), we obtain the following result for the permanence of (E_0).

Lemma 1. (permanence) *Assume that there are unique, invasible and globally stable boundary equilibrium points E_1^{Bd} on Ω_1 and E_2^{Bd} on Ω_2, respectively. Then (E_0) is permanent.*

Proof. The dynamical system defined by the solution of (E_0) with initial condition (I_0) is dissipative (see section 1) and leaves the boundaries Ω_1 and Ω_2 invariant. An average Lyapunov function $P_1 : \mathbb{R}^4_+ \to \mathbb{R}$ is defined by $P_1(R,S,x_1,x_2) = x_1$. The derivative of P_1 along the solution of (E_0) evaluated at E_2^{Bd} is given by

$$\frac{\dot{P}_1}{P_1} = \Lambda_{1\to 2} > 0.$$

In the same way, we define $P_2 : \mathbb{R}^4_+ \to \mathbb{R}$ by $P_2(R,S,x_1,x_2) = x_2$. The derivative of P_2 along the solution of (E_0) evaluated at E_1^{Bd} is given by

$$\frac{\dot{P}_2}{P_2} = \Lambda_{2\to 1} > 0.$$

Note that omega limit sets in Ω_1 and Ω_2 are E_1^{Bd} and E_2^{Bd}, respectively. A theorem on average Lyapunov functions implies that (E_0) is permanent. This completes the proof. □

3.3. Numerical simulations

We perform numerical simulations to investigate the dynamics of (E_0). All functional responses f_j & g_j are specified to the Monod kinetics with strain-specific half saturation constants a_{fj} & a_{gj} and maximum growth rates m_{fj} & m_{gj} ($j = 1,2$), respectively. First, we fix the parameters for (E_0) as follows.

$$\begin{aligned} &m_{f1} = 3.0, m_{f2} = 1.0, m_{g1} = 2.0, m_{g2} = 3.0, a_{f1} = 0.3, a_{f2} = 2.0, \\ &a_{g1} = 1.5, a_{g2} = 0.5, R^0 = 2.0, S^0 = 0.5, d = 2.0, D = 1.0. \end{aligned} \tag{P$_1$}$$

The dynamics of (E_0) with (P_1) depicted in the left panel of Figure 1 exhibits the coexistence of two strains in steady state. The parameter value of d, representing the degradation rate of resource R to metabolite S, is increased to 3.0. In this case, strain-1 enables to stay in the chemostat whereas strain-2 is washed out (the right panel of Figure 1). These results suggest that productivity of metabolite influence the coexistence of two strains. Interestingly, high productivity of metabolite results in the exclusion of strain-2.

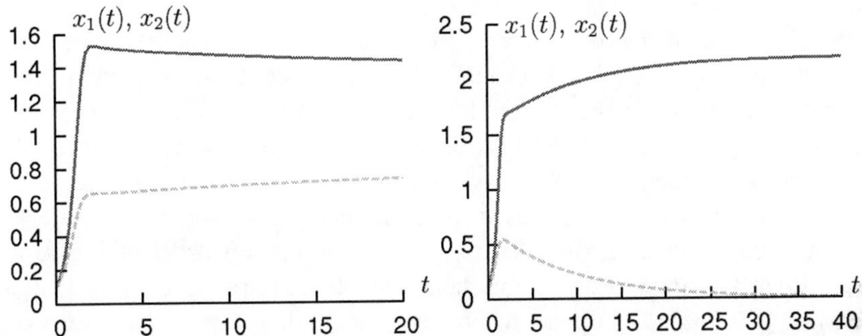

FIGURE 1. Numerical simulations performed for (E_0) with (P_1). The value of d is varied. The trajectories represent the concentration of strain-1 (solid, red lines) and that of strain-2 (dotted, green lines), respectively. Coexistence of two strains (left: $d = 2.0$) and competitive exclusion of strain-2 (right: $d = 3.0$).

Next, the value of m_{g1} is decreased to 0 in order to examine the dependence of strain-1 on resource R. Strain-1 is not able to utilize resource R as an energy source on this setting. However, two strains still coexist (the left panel of Figure 2). By contrast, strain-1 persists whereas strain-2 is washed out if m_{g1} is increased to 3.0 (the right panel of Figure 2), suggesting that it is not necessary for strain-1 to utilize resource R, rather the growth of strain-1 by utilizing resource R can lead to the exclusion of strain-2.

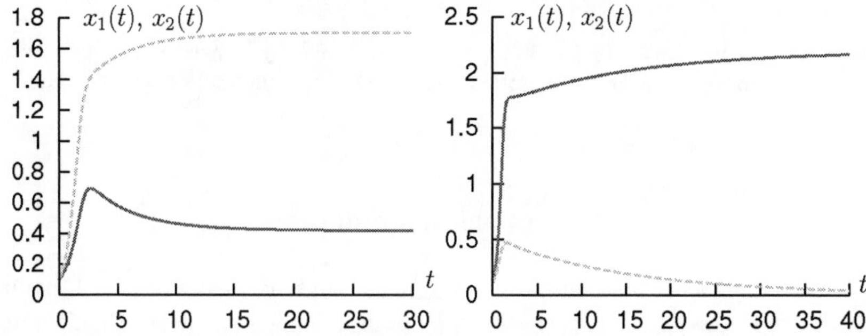

FIGURE 2. Numerical simulations performed for (E_0) with (P_1). The value of m_{g1} is varied. See the caption of Figure 1 for the manner of descriptions for the trajectories (as well as for the following all Figures). Coexistence of two strains (left: $m_{g1} = 0$) and competitive exclusion of strain-2 (right: $m_{g1} = 3.0$).

TABLE 2. Summary of numerical simulations with (P_1).

Figure 1	$d = 2.0$ (default): coexistence	$d = 3.0$ (increased): exclusion
Figure 2	$m_{g1} = 0$ (decreased): coexistence	$m_{g1} = 3.0$ (increased): exclusion

We change the values of parameters as follows.

$$m_{f1} = 1.0, m_{f2} = 2.0, m_{g1} = 3.0, m_{g2} = 2.1, a_{f1} = 2.0, a_{f2} = 0.3,$$
$$a_{g1} = 0.4, a_{g2} = 1.5, R^0 = 2.0, S^0 = 0, d = 1.0, D = 1.0. \quad (P_2)$$

Note that $S^0 = 0$. In other words, there is no external input of metabolite S. The dynamics of (E_0) with (P_2) depicted in the left panel of Figure 3 exhibits the competitive exclusion of strain-2. The value of d is increased from 1.0 to 2.0. In this case, two strains coexist in steady state (the right panel of Figure 3). In contrast to the results depicted on Figure 1, the production of metabolite S operates to facilitate the coexistence of two strains.

Next, the value of m_{f1} is decreased to 0. Strain-1 is not able to utilize metabolite S on this setting. However, two strains still coexist (the left panel of Figure 4). By contrast, strain-1 enables to stay in the chemostat whereas strain-2 is washed out if m_{f1} is increased to 2.0 (the right panel of Figure 4), suggesting that it is not necessary for strain-1 to utilize metabolite S whereas it seems essential for strain-2 to utilize metabolite S. Moreover, high growth of strain-1 by utilizing metabolite S can lead to the competitive exclusion of strain-2.

These numerical simulation results highlight the different roles of metabolite production in determining the outcome. These observations could be explained if the manner of

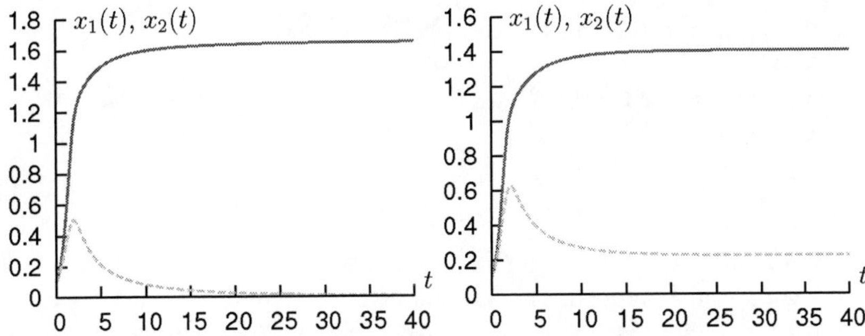

FIGURE 3. Numerical simulations performed for (E_0) with (P_2). The value of d is varied. Competitive exclusion of strain-2 (left: $d = 1.0$) and coexistence of two strains (right: $d = 2.0$).

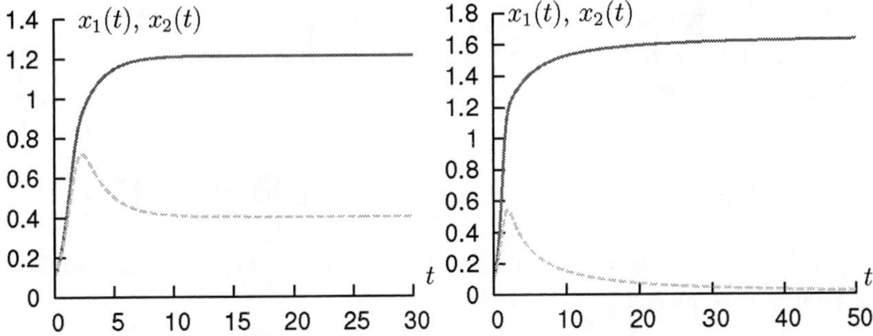

FIGURE 4. Numerical simulations performed for (E_0) with (P_2). The value of m_{f1} is varied. Coexistence of two strains (left: $m_{f1} = 0$) and competitive exclusion of strain-2 (right: $m_{f1} = 2.0$).

interactions between two strains is identified. In the next section, analytically we investigate four specific cases of (E_0) as the typical examples of characterizing the the manner of interactions between two strains.

TABLE 3. Summary of numerical simulations with (P_2) ($S^0 = 0$).

Figure 3	$d = 1.0$ (default): exclusion	$d = 2.0$ (increased): coexistence
Figure 4	$m_{f1} = 0$ (decreased): coexistence	$m_{f1} = 2.0$ (increased): exclusion

4. COEXISTENCE IN SPECIFIC CASES

As a baseline, we assume that $g_1(R) = g_2(R) \equiv 0$. It follows from the third and fourth equations of (E_0) that

$$\begin{cases} x_1'(t) = x_1(t)(f_1(S(t))-D), \\ x_2'(t) = x_2(t)(f_2(S(t))-D). \end{cases} \tag{12}$$

There is generically no interior equilibrium point of (E_0). Hence no utilization of resource R results in the competitive exclusion of one strain. Similarly we see that no utilization of metabolite S also results in the same consequence. In these situations, only one strain with the lowest break-even concentration enables to persist in the chemostat. In this section, analytically we study four specific cases of (E_0). As it is aimed to obtain clear interpretations for the mechanisms underlying the coexistence of two strains, we assume that one of two strains can utilize only resource R or metabolite S as an energy source. These four specific cases are not general but indeed helpful to figure out the mechanisms for coexistence.

4.1. Strain-1 specializing to metabolite utilization ($g_1(R) \equiv 0$)

In this subsection, we assume that strain-1 is incapable of utilizing resource R as an energy source. We consider

$$\begin{cases} R'(t) = (R^0 - R(t))D - dR(t)x_1(t) - g_2(R(t))x_2(t), \\ S'(t) = (S^0 - S(t))D + dR(t)x_1(t) - f_1(S(t))x_1(t) - f_2(S(t))x_2(t), \\ x_1'(t) = x_1(t)(f_1(S(t))-D), \\ x_2'(t) = x_2(t)(f_2(S(t)) + g_2(R(t)) - D) \end{cases} \tag{E_1}$$

with initial condition (I_0). Subsystems of (E_1) on boundary regions Ω_1 and Ω_2 are (B_1) and (B_4), respectively. By Proposition 1, $\bar{E}_{1+} = (\bar{R}_1, \bar{S}_1, \bar{x}_1)$ is a globally stable equilibrium point of subsystem (B_1) if $S^0 > \bar{S}_1$. By Proposition 7, $\bar{E}_{2+} = (\bar{R}_2, \bar{S}_2, \bar{x}_2)$ is a globally stable equilibrium point of subsystem (B_4) if $\bar{S}_2 < S^0$ and $\bar{R}_2 < R^0$. We thereby assume that

$$\bar{S}_1 < S^0, \ \bar{S}_2 < S^0 \text{ and } \bar{R}_2 < R^0 \tag{13}$$

throughout the remainder of this subsection. Under condition (13), there exist three boundary equilibrium points, E_0, $\bar{E}_{1+}$ and $\bar{E}_{2+}$. Note that washout equilibrium E_0 is always unstable. Under condition (13), we will see that it suffices to derive conditions for the existence of an interior equilibrium point of (E_1) and confirm the mutual invasibility conditions (see Lemma 1) in order to obtain the necessary and sufficient conditions for the permanence of (E_1).

Note that $S^* = \bar{S}_1$. Substituting S^* into the fourth equation of (4) gives $g_2(R^*) = D - f_2(S^*)$. If $D > f_2(S^*)$ and the maximum growth rate of g_2 satisfies $m_{g2} > D - f_2(S^*)$, there exists a positive constant R^* such that $R^* = g_2^{-1}(D - f_2(S^*))$. Thus we assume $m_{g2} > D - f_2(S^*) > 0$ throughout the remainder of this subsection. Accordingly, x_1^* and x_2^* are uniquely determined if and only if $\Delta = dR^* - g_2(R^*) \neq 0$. By (9), x_1^* and x_2^* are explicitly given by

$$x_1^* = \frac{f_2(S^*)(R^0 - R^*) - g_2(R^*)(S^0 - S^*)}{\Delta},$$
$$x_2^* = \frac{(\mathscr{P}_* - 1)(R^0 - R^*)D + \mathscr{P}_*(S^0 - S^*)D}{\Delta},$$

where $\mathscr{P}_* = dR^*/D$. The following relations hold among boundary and interior equilibrium points.

$$S^* = \bar{S}_1 = f_1^{-1}(D), \tag{14}$$
$$f_2(\bar{S}_2) + g_2(\bar{R}_2) = D, \tag{15}$$
$$f_2(S^*) + g_2(R^*) = D. \tag{16}$$

By (14) and (16), we need to assume

$$f_2(S^*) < f_1(S^*) = D \tag{17}$$

for the coexistence of two strains. It follows from (15) and (16) that $\bar{S}_2 > S^*$ if and only if $\bar{R}_2 < R^*$.

We shall introduce the *trade-off* condition by

$$f_2(S^*) < f_1(S^*), \;\; \bar{S}_2 > S^*(= \bar{S}_1) \text{ and } \bar{R}_1 > R^*(> \bar{R}_2). \tag{TC$_1$}$$

Note that the better competitor realizes the lower break-even concentration. Thus trade-off condition (TC_1) requires that strain-1 must be superior to strain-2 in exploitative competition for metabolite ($\bar{S}_2 > S^* = \bar{S}_1$ and $f_2(S^*) < f_1(S^*)$) whereas strain-2 must be superior to strain-1 in exploitative competition for resource ($\bar{R}_1 > R^* > \bar{R}_2$). By ($TC_1$), The following result for the positivity of x_1^* and x_2^* is derived by assuming (TC_1).

Proposition 9. *Assume that* $\bar{\mathscr{P}}_1 = \frac{d\bar{R}_1}{D} < 1$. *If* ($TC_1$) *holds,* $f_2(S^*)(R^0 - R^*) - g_2(R^*)(S^0 - S^*) < 0$ *and* $(\mathscr{P}_* - 1)(R^0 - R^*)D + \mathscr{P}_*(S^0 - S^*)D < 0$.

Proof. We claim that

$$f_2(\bar{S}_2)(R^0 - \bar{R}_2) - g_2(\bar{R}_2)(S^0 - \bar{S}_2) = 0. \tag{18}$$

Since now we assume (13), (18) is deduced by adding the first equation of (B_4) multiplied by $f_2(\bar{S}_2)$ to the second equation of (B_4) multiplied by $-g_2(\bar{R}_2)$. It follows from (18) that

$$f_2(S^*)(R^0 - R^*) - g_2(R^*)(S^0 - S^*) =$$
$$f_2(S^*)(R^0 - R^*) - g_2(R^*)(S^0 - S^*) - (f_2(\bar{S}_2)(R^0 - \bar{R}_2) - g_2(\bar{R}_2)(S^0 - \bar{S}_2)).$$

By (TC_1), $S^0 - \bar{S}_2 < S^0 - S^*$ and $R^0 - \bar{R}_2 > R^0 - R^*$. Hence the first assertion is true. Next, we claim that

$$(\bar{\mathscr{P}}_1 - 1)(R^0 - \bar{R}_1)D + \bar{\mathscr{P}}_1(S^0 - \bar{S}_1)D = 0. \tag{19}$$

Since $\bar{\mathscr{P}}_1 < 1$, (19) is deduced by adding the first equation of (B_1) multiplied by $\bar{\mathscr{P}}_1 - 1$ to the second equation of (B_1) multiplied by $\bar{\mathscr{P}}_1$. It follows from (19) that

$$\begin{aligned}&(\mathscr{P}_* - 1)(R^0 - R^*)D + \mathscr{P}_*(S^0 - S^*)D = \\ &(\mathscr{P}_* - 1)(R^0 - R^*)D + \mathscr{P}_*(S^0 - S^*)D - ((\bar{\mathscr{P}}_1 - 1)(R^0 - \bar{R}_1)D + \bar{\mathscr{P}}_1(S^0 - \bar{S}_1)D).\end{aligned}$$

By (TC_1), $1 > \bar{\mathscr{P}}_1 > \mathscr{P}_*$ and $R^0 - \bar{R}_1 < R^0 - R^*$. Moreover $S^* = \bar{S}_1$. Hence the second assertion is also true. This completes the proof. □

We shall introduce the *productivity* condition by

$$\Delta < 0 \Longleftrightarrow \mathscr{P}_* < \frac{g_2(R^*)}{D} = \frac{g_2(R^*)}{f_2(S^*) + g_2(R^*)} (< 1). \tag{PC$_1$}$$

Productivity condition (PC_1) requires that the basic production ratio does not exceed 1. In other words, it requires the low productivity of metabolite. Since now we assume (13), $S^0 > \bar{S}_1$. It follows from (19) that $\bar{\mathscr{P}}_1 < 1$. Assume that the assumptions of Proposition 9 hold. Then $E_* \in \text{int}\mathbb{R}^4_+$ if and only if (TC_1) and (PC_1) hold. We obtain the following result.

Theorem 1. (Type-I coexistence) *Assume that (13) holds. (E_1) is permanent if and only if (TC_1) and (PC_1) hold.*

Proof. (Sufficiency) Straightforward calculation yields that $\Lambda_{1\to 2} = f_1(\bar{S}_2) - f_1(S^*)$, where we used (14). It follows from (15) and (16) that $\Lambda_{2\to 1} = g_2(\bar{R}_1) - g_2(R^*)$. By ($TC_1$), $\Lambda_{1\to 2} > 0$ and $\Lambda_{2\to 1} > 0$. Lemma 1 implies that the assertion is true. (Necessity) By Proposition 9, $E_* \in \text{int}\mathbb{R}^4_+$ if and only if (TC_1) and (PC_1) hold. If (E_1) is permanent, there must be an interior equilibrium point. Hence we can prove the assertion by contradiction. This completes the proof. □

We call the coexistence mediated by low productivity of metabolite, 'Type-I coexistence'. We shall propose biological interpretations to the conditions and results obtained here. Trade-off condition (TC_1) requires that strain-1 must outcompete strain-2 in exploitative competition for metabolite S whereas strain-2 must outcompete strain1 in exploitative competition for resource R. Productivity condition (PC_1) indicates that less production of metabolite by strain-1 (corresponding to the case of $\mathscr{P}_* < 1$) is necessary for the coexistence of two strains. Over-production of metabolite can leads to the competitive exclusion of one strain (see Appendix A.1). We will discuss why over-production of metabolite results in competitive exclusion in section 5.

4.2. Strain-2 specializing to metabolite utilization ($g_2(R) \equiv 0$)

This case has already been studied in [22, Section 4]. We derived conditions for the permanence of (E_0) with $g_2 \equiv 0$ if $S^0 > S^*$ [22, Theorem 6]. There is no boundary equilibrium point on Ω_2 if $S^0 < S^*$. This implies that strain-2 cannot persist in the chemostat by itself. The authors revealed by numerical simulations that strain-2 persists in the chemostat even though $S^0 < S^*$, provided that $\mathscr{P}_* > 1$ and $x_1^* > \frac{S^*-S^0}{\mathscr{P}_*-1}$. This type of coexistence is mediated by the produced metabolites via the degradation of resource R mediated by strain-1. In other words, strain-1 assists the persistence of strain-2 ([22, Section 4] and Figure 5 below). We shall investigate whether this type of coexistence can occur on the other specific cases of (E_0) in the following subsections.

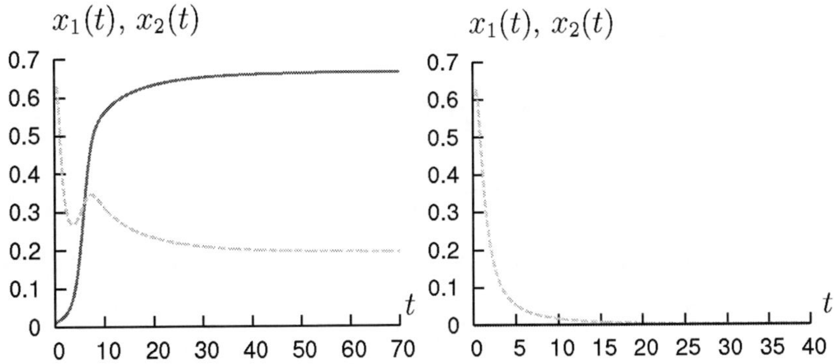

FIGURE 5. [22, Section 4]. $S^0 < S^*$ and $\mathscr{P}_* > 1$. Coexistence in the presence of strain-1 (left: $x_1(0) = 0.1$) and washout of strain-2 in the absence of strain-1 (right: $x_1(0) = 0$).

4.3. Strain-1 specializing to resource utilization ($f_1(S) \equiv 0$)

In this subsection, we assume that strain-1 cannot utilize metabolite S as an energy source. The model is given by

$$\begin{cases} R'(t) = (R^0 - R(t))D - dR(t)x_1(t) - g_1(R(t))x_1(t) - g_2(R(t))x_2(t), \\ S'(t) = (S^0 - S(t))D + dR(t)x_1(t) - f_2(S(t))x_2(t), \\ x_1'(t) = x_1(t)(g_1(R(t)) - D), \\ x_2'(t) = x_2(t)(f_2(S(t)) + g_2(R(t)) - D) \end{cases} \tag{E_2}$$

with initial condition (I_0). Subsystems of (E_2) are (B_3) and (B_4), respectively. By Proposition 6, $\tilde{E}_{1+} = (\tilde{R}_1, \tilde{S}_1, \tilde{x}_1)$ is a globally stable equilibrium point of subsystem (B_3) if $R^0 > \tilde{R}_1$. By Proposition 7, $\bar{E}_{2+} = (\bar{R}_2, \bar{S}_2, \bar{x}_2)$ is a globally stable equilibrium point of subsystem (B_4) if $\bar{S}_2 < S^0$ and $\bar{R}_2 < R^0$. Throughout the remainder of this subsection, we assume that

$$\tilde{R}_1 < R^0, \;\; \bar{S}_2 < S^0 \text{ and } \bar{R}_2 < R^0. \tag{20}$$

Under condition (20), there exist three boundary equilibrium points, E_0, $\tilde{E}_{1+}$ and $\bar{E}_{2+}$.

Note that $R^* = \tilde{R}_1$. Substituting R^* into the fourth equation of (4) gives $f_2(S^*) = D - g_2(R^*)$. If $D > g_2(R^*)$ and the maximum growth rate of f_2 satisfies $m_{f2} > D - g_2(R^*)$, there exists a positive constant S^* such that $S^* = f_2^{-1}(D - g_2(R^*))$. Hence we assume $m_{f2} > D - g_2(R^*) > 0$ throughout the remainder of this subsection. Since $f_1(S) \equiv 0$,

$$\mathscr{P}_* = \frac{dR^* + g_1(R^*)}{D} = 1 + \frac{dR^*}{D},$$
$$\Delta = \mathscr{P}_* D - g_2(R^*) = dR^* + f_2(S^*).$$

Simultaneously $\Delta > 0$. x_1^* and x_2^* are explicitly given by

$$x_1^* = \frac{f_2(S^*)(R^0 - R^*) - g_2(R^*)(S^0 - S^*)}{\Delta},$$
$$x_2^* = \frac{dR^*(R^0 - R^*) + (dR^* + D)(S^0 - S^*)}{\Delta}.$$

The following relations hold among boundary and interior equilibrium points.

$$R^* = \tilde{R}_1 = g_1^{-1}(D), \tag{21}$$
$$f_2(\bar{S}_2) + g_2(\bar{R}_2) = D, \tag{22}$$
$$f_2(S^*) + g_2(R^*) = D. \tag{23}$$

By (21) and (23), we need to assume

$$g_2(R^*) < g_1(R^*) = D \tag{24}$$

for the coexistence of two strains. It follows from (22) and (23) that $\bar{R}_2 > R^*$ if and only if $\bar{S}_2 < S^*$. The trade-off condition in this case is given by

$$g_2(R^*) < g_1(R^*), \;\; \bar{R}_2 > R^*(= \tilde{R}_1) \text{ and } \tilde{S}_1 > S^*(> \bar{S}_2). \tag{TC$_2$}$$

By following the similar proof for Proposition 9, we obtain the following result.

Proposition 10. *Assume that* $\bar{R}_2 > R^*$. *Then* $f_2(S^*)(R^0 - R^*) - g_2(R^*)(S^0 - S^*) > 0$.

By proposition 10, $x_1^* > 0$ if $\bar{R}_2 > R^*$. The productivity condition is given by

$$\mathscr{P}_* \geq 1. \tag{PC$_2$}$$

Note that (PC$_2$) always holds. The equality holds if and only if $d = 0$. We have two different directions according to the sign of $S^0 - S^*$ to determine the sign of x_2^*, or equivalently, the sign of $dR^*(R^0 - R^*) + (dR^* + D)(S^0 - S^*)$.

(i) Cultivated regime
We assume that $S^0 > S^*$. It follows from (B_3) that

$$\tilde{S}_1 = S^0 + \left(1 - \frac{1}{\tilde{\mathscr{P}}_1}\right)(R^0 - \tilde{R}_1), \tag{25}$$

where $\tilde{\mathscr{P}}_1 = \mathscr{P}_* = 1 + d\tilde{R}_1/D$. Thus $\tilde{S}_1 > S^0$. Since $S^0 > S^*$, $\tilde{S}_1 > S^*$ and $x_2^* > 0$. Note that (TC_2) includes the assumption of Proposition 10. We obtain the following result.

Theorem 2. (Type-I coexistence) *Assume that $S^0 > S^*$ and (20) holds. (E_2) is permanent if and only if (TC_2) holds.*

Proof. (Sufficiency) Straightforward calculation yields that $\Lambda_{1\to 2} = g_1(\bar{R}_2) - g_1(R^*)$, where we used (21). It follows from (22) and (23) that $\Lambda_{2\to 1} = f_2(\tilde{S}_1) - f_2(S^*)$. By ($TC_2$), $\Lambda_{1\to 2} > 0$ and $\Lambda_{2\to 1} > 0$. Lemma 1 implies that the assertion holds true. (Necessity) By Proposition 10, $E_* \in \mathrm{int}\mathbb{R}^4_+$ if and only if (TC_2) holds. If (E_2) is permanent, there must be an interior equilibrium point. Hence we can prove the assertion by contradiction. This completes the proof. □

(ii) Self-sufficiency regime
We assume that $S^0 < S^*$. In this case, simultaneously $x_1^* > 0$. $x_2^* > 0$ if and only if

$$S^* < S^0 + \left(1 - \frac{1}{\mathscr{P}_*}\right)(R^0 - R^*) = S^0 + \frac{R^*}{R^* + \frac{D}{d}}(R^0 - R^*).$$

Define $\varphi : [0, R^0] \to [0, \infty)$ by

$$\varphi(r) := \frac{r}{r + \frac{D}{d}}(R^0 - r).$$

The condition $x_2^* > 0$ can be rewritten as $S^* < S^0 + \varphi(R^*)$. By (25), $\tilde{S}_1 = S^0 + \varphi(\tilde{R}_1)$. Since $R^* = \tilde{R}_1$, it is further deduced to $S^* < \tilde{S}_1$. Hence $x_2^* > 0$ if and only if $S^* < \tilde{S}_1$. Note that (TC_2) includes the condition $S^* < \tilde{S}_1$. We obtain the following result.

Theorem 3. (Type-II coexistence) *Assume that $S^0 < S^*$ and (20) holds. (E_2) is permanent if and only if (TC_2) holds.*

We call the type of coexistence mediated by high productivity of metabolite, 'Type-II coexistence'. The amount of the produced metabolites by strain-1 is $S^* - S^0$. Since the condition $\tilde{S}_1 > S^*$ is required for the coexistence of two strains, $\tilde{S}_1$ determines the maximum capability for the coexistence of two strains. Note that $\varphi(R^*) = \tilde{S}_1 - S^0$ corresponds not only to the upper threshold for the invasibility of strain-2 ($\Lambda_{2\to 1} > 0$) but also to the upper threshold for the persistence of strain-2 ($x_2^* > 0$).

We conclude this subsection by investigating the property of $\varphi(r)$. Define $\mathscr{P}_{*,\max}$ by $\mathscr{P}_{*,\max} = 1 + \frac{dR^0}{D}$. φ is maximized at $r = r_{\max}$, where

$$r_{\max} = -\frac{D}{d} + \frac{D}{d}\sqrt{1 + \frac{dR^0}{D}} = -\frac{D}{d} + \frac{D}{d}\sqrt{\mathscr{P}_{*,\max}}.$$

The graph of $\varphi(r) + S^0$ together with the two threshold values S^0 and $S^0 + \varphi(R^*)$ are depicted in Figure 6. Admissible amount of metabolite production for the coexistence of two strains ranges from above the line $y = S^0$ to below the line $y = S^0 + \varphi(R^*)$.

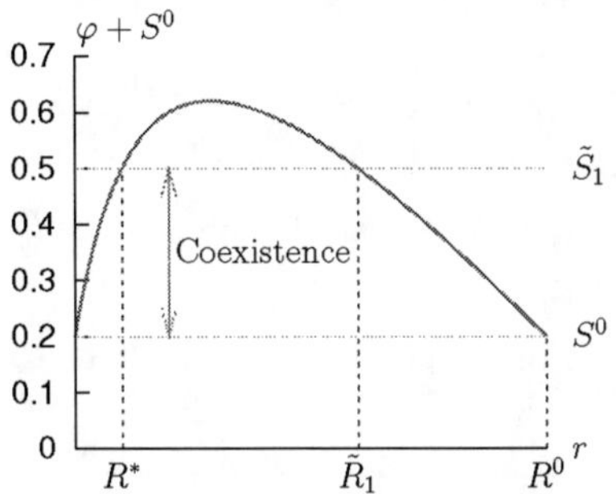

FIGURE 6. Admissible amount of metabolite production for the coexistence of two strains ranges from the above of $y = S^0$ to the below of $y = S^0 + \varphi(R^*)$.

As represented by (TC_2), strain-1 must be superior to strain-2 in exploitative competition for resource R ($g_2(R^*) < g_1(R^*)$ and $\bar{R}_2 > R^* = \tilde{R}_1$), whereas strain-2 must be superior to strain-1 in exploitative competition for metabolite S ($\tilde{S}_1 > S^* > \bar{S}_2$). In this case, productivity condition (PC_2) is always satisfied. Thus two types of coexistence can occur. Both types of coexistence require the trade-off in the availability of resource and metabolite. By contrast, these two types of coexistence differ from each other in terms of the amount of the produced metabolite $S^* - S^0$ by strain-1. If $S^* - S^0 < 0$, the manner of coexistence is the same as that of identified in subsection 4.1. The produced metabolite does not change the situation. In other words, two strains can coexist even if there is no metabolite produced by strain-1 ($d = 0$). Conversely, if $S^* - S^0 > 0$, the manner of coexistence is rather different. The fate of strain-2 now depends on the amount of metabolite produced by strain-1. The produced metabolite indeed changes the situation. Type-II coexistence has been firstly identified in this subsection, although it has already been suggested to occur in subsection 4.2 by numerical simulations (see also [22]).

4.4. Strain-2 specializing to resource utilization ($f_2(S) \equiv 0$)

In this subsection, we assume that strain-2 is incapable of utilizing S as an energy source. The model is given by

$$\begin{cases} R'(t) = (R^0 - R(t))D - dR(t)x_1(t) - g_1(R(t))x_1(t) - g_2(R(t))x_2(t), \\ S'(t) = (S^0 - S(t))D + dR(t)x_1(t) - f_1(S(t))x_1(t), \\ x_1'(t) = x_1(t)(f_1(S(t)) + g_1(R(t)) - D), \\ x_2'(t) = x_2(t)(g_2(R(t)) - D) \end{cases} \tag{E$_3$}$$

with initial condition (I_0). Subsystems of (E_3) on boundary regions Ω_1 and Ω_2 are (B_2) and (B_5), respectively. By Proposition 4, $\tilde{E}_{1+} = (\tilde{R}_1, \tilde{S}_1, \tilde{x}_1)$ is a globally stable equilibrium point of subsystem (B_2) if $\phi(0) > D$. By Proposition 8, $\tilde{E}_{2+} = (\tilde{R}_2, S^0, \tilde{x}_2)$ is a globally stable equilibrium point of subsystem (B_5) if $\tilde{R}_2 < R^0$. Thus we assume that

$$\phi(0) > D \text{ and } \tilde{R}_2 < R^0 \tag{26}$$

throughout the remainder of this subsection. Under condition (26), there exist three boundary equilibrium points, E_0, $\tilde{E}_{1+}$ and $\tilde{E}_{2+}$.

In this case, $R^* = \tilde{R}_2$. Substituting R^* into the third equation of (4) gives $f_1(S^*) = D - g_1(R^*)$. If $D > g_1(R^*)$ and the maximum growth rate of f_1 satisfies $m_{f1} > D - g_1(R^*)$, there exists a positive constant S^* such that $S^* = f_1^{-1}(D - g_1(R^*))$. Hence we assume $m_{f1} > D - g_1(R^*) > 0$ throughout the remainder of this subsection. Since $g_2(R^*) = D$, Δ is given by

$$\Delta = (\mathscr{P}_* - 1)D.$$

If $\mathscr{P}_* \neq 1$, x_1^* and x_2^* are explicitly given as follows.

$$x_1^* = \frac{-(S^0 - S^*)}{\mathscr{P}_* - 1},$$
$$x_2^* = \frac{(\mathscr{P}_* - 1)(R^0 - R^*) + \mathscr{P}_*(S^0 - S^*)}{\mathscr{P}_* - 1}.$$

The following relations hold among boundary and interior equilibrium points.

$$R^* = \tilde{R}_2 = g_2^{-1}(D), \tag{27}$$
$$f_1(\tilde{S}_1) + g_1(\tilde{R}_1) = D, \tag{28}$$
$$f_1(S^*) + g_1(R^*) = D. \tag{29}$$

By (27) and (29), we need to assume

$$g_1(R^*) < g_2(R^*) = D \tag{30}$$

for the coexistence of two strains. It follows from (28) and (29) that $\tilde{S}_1 < S^*$ if and only if $\tilde{R}_1 > R^*$. The trade-off condition in this case is given by

$$g_1(R^*) < g_2(R^*), \ \ (\tilde{R}_2 =)R^* < \tilde{R}_1 \text{ and } (\tilde{S}_1 <)S^* < S^0. \tag{TC$_3$}$$

The productivity condition is given by

$$\Delta < 0 \Longleftrightarrow \mathscr{P}_* < 1. \tag{PC$_3$}$$

Note that $x_1^* > 0$ if and only if $\Delta < 0$. In the same manner to the proof for Proposition 9, we can prove the following result.

Proposition 11. *Assume that $\mathscr{P}_* < 1$ and (TC$_3$) holds. Then $(\mathscr{P}_* - 1)(R^0 - R^*) + \mathscr{P}_*(S^0 - S^*) < 0$.*

Theorem 4. (Type-I coexistence) *Assume that (26) holds. (E_3) is permanent if and only if (TC_3) and (PC_3) hold.*

Proof. (Sufficiency) Straightforward calculation yields that $\Lambda_{1\to2} = f_1(S^0) - f_1(S^*)$, where we used (29). It follows from (28) and (29) that $\Lambda_{2\to1} = g_2(\tilde{R}_1) - g_2(R^*)$. By ($TC_3$), $\Lambda_{1\to2} > 0$ and $\Lambda_{2\to1} > 0$. Lemma 1 implies that the assertion holds true. (Necessity) By proposition 11, $E_* \in \mathrm{int}\mathbb{R}^4_+$ if and only if (TC_3) and (PC_3) hold. If (E_3) is permanent, there must be an interior equilibrium point. Hence we can prove the assertion by contradiction. This completes the proof. □

As represented by (TC_3), strain-2 must be superior to strain-1 in exploitative competition for resource R ($g_1(R^*) < g_2(R^*)$ and $\tilde{R}_2 = R^* < \tilde{R}_1$) whereas strain-1 must be superior to strain-2 in exploitative competition for metabolite S ($\tilde{S}_1 < S^* < S^0$). It is sufficient for strain-1 to be cultivated only by metabolite ($S^0 > S^*$). In this case, over-production of metabolite does not support the coexistence of two strains as suggested in subsection 4.1. We will discuss why over-production of metabolite can result in competitive exclusion in the next section. All results investigated in this section are summarized on Table 4.

TABLE 4. Summary of analytical results.

System	Utilization	Coexistence ($d = 0$)	Coexistence ($S^0 = 0$)
(E_1) ($g_1(R) \equiv 0$)	strain-1: metabolite strain-2: resource (c.f. Figure 2)	Possible Theorem 1 (c.f. Figure 1)	Impossible
[22] ($g_2(R) \equiv 0$)	strain-1: resource strain-2: metabolite	Possible [22, Theorem 6]	Possible (c.f. Figure 5)
(E_2) ($f_1(S) \equiv 0$)	strain-1: resource strain-2: metabolite (c.f. Figure 4)	Possible Theorem 2	Possible Theorem 3 (c.f. Figure 3)
(E_3) ($f_2(S) \equiv 0$)	strain-1: metabolite strain-2: resource	Possible Theorem 4	Impossible

5. DISCUSSION

We considered mathematical models of resource competition between two bacterial strains for nutrient substrate and its metabolite which can be utilized as two types of energy source. In other words, a stage-structure is introduced in nutrient substrate. The metabolic transition of nutrient substrate is mediated by degradative activity of microorganisms mainly achieved by extracellular enzymes bound to microorganisms. In [22], the authors studied several one-strain models and a competition model between two strains. (E_0) includes all models studied in [22]. The competition model studied in [22] was introduced again in subsection 4.2 as one of the specific cases of (E_0).

Most results for the population dynamics of single strain summarized in section 2 are referred from [22]. In section 3, coexistence of two strains was evaluated by criteria for invasibility and permanence (Lemma 1). Numerical simulations were performed to investigate how the coexistence of two strains depends on the availability of resource and metabolite (represented by the change of m_{fj} and f_{gj}), and the productivity of metabolite (represented by the change of d).

Four specific cases of (E_0) were studied in section 4. The necessary and sufficient conditions were derived under which two strains coexist in terms of criteria for permanence. In all cases of coexistence, trade-offs in the availability of resource and metabolite are necessarily required. We further revealed that there exist two different types of coexistence, distinguished by the productivity of metabolite. The productivity of metabolite is determined by the value of basic production ratio $\mathscr{P}_*$ such that high productivity implies $\mathscr{P}_* > 1$ whereas low productivity implies $\mathscr{P}_* < 1$. Type-I coexistence, namely coexistence in cultivated regime, necessitates the low productivity of metabolite. Conversely, type-II coexistence, namely coexistence in self-sufficiency regime, always satisfies high productivity of metabolite.

In subsection 4.1 and Appendix A.1, (E_1) was studied. (E_1) is derived from (E_0) by assuming that strain-1 is specialized to utilize metabolite S. (E_1) is permanent if strain-1 is the better competitor for utilizing metabolite S whereas strain-2 is the better competitor for utilizing resource R. The productivity of metabolite must be low (Theorem 1). Bistability can occur if the productivity of metabolite is high (Appendix A.1). Indeed, we analytically showed that E_* is unstable if $\mathscr{P}_* > 1$ (Appendix A.2). Although a few differences are observed between (E_1) and the model studied by Tang and Wolkowicz in section 6 of their paper [28], the model studied in [28] mostly corresponds to (E_1) with $S^0 = 0$. If $S^0 = 0$, it follows from (9) that we must assume $\mathscr{P}_* > 1$ in order to ensure the existence of interior equilibrium point E_*. Thus the results in subsection 4.1 are consistent with those obtained in [28].

In subsection 4.3, we revealed that both types of coexistence occur. (E_2) is derived from (E_0) by assuming that strain-2 is specialized to utilize metabolite S. (E_2) is permanent if strain-1 is the better competitor for utilizing resource R whereas strain-2 is the better competitor for utilizing metabolite S (Theorems 2 and 3). Note that in both types of coexistence high productivity of metabolite is simultaneously satisfied, as opposed to subsection 4.1 in which low productivity is required for coexistence. Recently, Heßeler et. al. [15] studied a chemostat model based on certain experiments in which two species are competing for one limiting nutrient. They revealed that the metabolic by-product (metabolite) can facilitate the coexistence of species. Although (E_2) and the model considered in [15] are different, the model considered in [15] is similar to (E_2) ($f_1(S) \equiv 0$) with $S^0 = 0$. The results in subsection 4.3 are consistent with those obtained in [15]. In subsection 4.4, (E_3) was studied. (E_3) is derived from (E_0) by assuming that strain-2 is specialized to utilize resource R. (E_3) is permanent if strain-1 is the better competitor for utilizing metabolite S whereas strain-2 is the better competitor for utilizing resource R.

Productivity of metabolite must be low (Theorem 4). The results obtained in subsection 4.4 are similar to those obtained in subsection 4.1.

Suppose that $S^0 = 0$. This assumption implies that no input of metabolite is given to the chemostat. Coexistence of two strains is possible only if conditions of Theorem 3 are satisfied (see subsection 4.3), or situation studied in [22] (see also subsection 4.2). In both cases, strain-1 is superior to strain-2 in exploitative competition for resource R. Hence our study suggests that the exploitation of nutrient substrate which is metabolically in the first stage is important for the strain which is capable of producing metabolite. In subsections 4.1 and 4.4, we found that over-production of metabolite S can result in the competitive exclusion of strain-2. We shall propose possible explanations why over-production of metabolite can lead to the exclusion. In subsections 4.1 and 4.4, we postulated that strain-1, the mediator of metabolic transition from resource R to metabolite S, is superior to strain-2 in exploitative competition for metabolite S. Moreover, the trade-offs suggest that resource R is the primary energy source for strain-2. Thus the large amount of metabolite production is indeed disadvantageous to strain-2, because production of metabolite reduces the amount of resource R as a material for producing metabolite. This may lead to the insufficient amount of resource R available to strain-2.

The importance of trade-offs in the availability of nutrient among competing strains has been implicated in the previous studies [10], [19], [23], [29] in different ways. Type-II coexistence is particularly important for the maintenance of a consortium if metabolite is not provided from outside. We could explain the coexistence observed in the experiment by Christensen *et al.* [6] in which the metabolite (benzoate) produced by strain R1 facilitates the growth of the cross-feeding strain C6. The better competitor, strain R1, for the primary nutrient substrate (benzyl alcohol) provides the metabolite (benzoate) to the cross-feeding strain C6. Thus the growth of strain R1 and strain C6 could be maintained as type-II coexistence, although the development of biofilm is essential so that more precise model formulation for this situation would be necessary to provide theoretical explanations. Type-II coexistence cold be regarded as commensalism in the sense that the metabolic activity of metabolite specialist strain seems no benefit to the metabolite providing strain. Our model would be further extended to the case where strains are in mutualistic relation such as syntrophic associations. The authors revealed on another study that type-II coexistence occurs in xenobiotics degrading bacteria [18] in which mutualism plays an important role. It is interesting to address a question of how mutualism mediates the coexistence of metabolically related two species. One possible extension to deal with mutualism is to consider a situation where the produced metabolite is inhibitory to the growth of the strain involved in the degradation of resource to metabolite. This is often the case observed in real systems that the produced metabolite increases the pH value in a culture by which the growth of organisms are inhibited. Then degradative activity of metabolite-feeding strain could help the growth of metabolite-producing strain. These are left for our future investigations.

REFERENCES

1. R.A. Armstrong and R. McGehee (1980), Competitive exclusion, *Am. Nat.* **115**, 151–170.
2. J.L. Alonso, C. Sabater, M.J. Ibañez, I. Amoros, M.S. Botella and J. Carrasco (1997), Fenitrothion and 3-methyl-4-nitrophenol degradation by two bacteria in natural waters under laboratory conditions, *J Environ Sci Health* **A32** 799–812.
3. M.M. Ballyk and G.S.K. Wolkowicz (1993), Exploitative competition in the chemostat for two perfectly substitutable resources, *Math Biosci* **118**, 127–180.
4. G.J. Butler and G.S.K. Wolkowicz (1987), Exploitative competition in a chemostat for two complementary and possibly inhibitory resources, *Math Biosci* **83**, 1–48.
5. P. Chesson (2000), Mechanisms of maintenance of species diversity. *Annu Rev Ecol Syst* **31** 343–366.
6. B.B. Christensen, J.A.J. Haagensen, A. Heydorn, S. Molin (2002), Metabolic commensalism and competition in a two-species microbial consortium, *Applied and Environmental Microbiology* **68** 2495–2502.
7. E. Costa, J. Pérez, J-U. Kreft (2006), Why is metabolic labour divided in nitrification, *TRENDS in Microbiology* **14** 213–219.
8. M.K. Cowan and K.P. Talaro (2005), "Microbiology: A Systems Approach", McGraw-Hill Science.
9. W. Dejonghe E. Berteloot, J. Goris, N. Boon, K. Crul, S. Maertens, M. Hofte, P.D. Vos, W. Verstraete, E.M. Top (2003), Synergistic degradation of linuron by a bacterial consortium and isolation of a single linuron-degrading Variovorax strain, *Appl Environ Microbiol* **69** 1532–1541.
10. M. Doebeli (2002), A model for the evolutionary dynamics of cross-feeding polymorphisms in microorganisms, *Population Ecology* **44** 59–70.
11. S.A.H. Geritz, E. Kisdi, G. Meszéna, J.A.J. Metz (1998), Evolutionary singular strategies and the adaptive growth and brancing of the evolutionary tree, *Evol Ecol* **12** 35–57.
12. J.P. Grover (1997), "Resource Competition", Population and Community Biology Series, 19, Chapman and Hall, New York.
13. J. Hale (1988), "Asymptotic Behavior of Dissipative Systems", Providence, RI: American Mathematical Society.
14. S.R. Hansen and S.P. Hubbell (1980), Single-nutrient microbial competition: quantitative agreement between experimental and theoretical forecast outcomes, *Science* **207**, 1491–1493.
15. J. Heßeler, J.K. Schmidt, U. Reichl and D. Flockerzi (2006), Coexistence in the chemostat as a result of metabolic by-products, *J Math Biol* **53**, 556–584.
16. J. Hofbauer and K. Sigmund (1998), Evolutionary Games and Population Dynamics, Cambridge University Press.
17. V. Hutson (1984), A theorem on average Lyapunov functions *Monash Math* **98** 267–275.
18. C. Katsuyama, S. Nakaoka, Y. Takeuchi, K. Tago, M. Hayatsu, K. Katoh, Mathematical analysis of an experimental study of syntrophic association in pesticide degradation, *submitted.*
19. J-U. Kreft and S. Bonhoeffer (2005), The evolution of groups of cooperating bacteria and the growth rate versus yield trade-off, *Microbiology* **151** 637–641.
20. B. Li, G.S.K. Wolkowicz, and Y. Kuang (2000), Global asymptotic behavior of a chemostat model with two perfectly complementary resources and distributed delay, *SIAM J Appl Math*, **60**, 2058–2086.
21. B. Li and H.L. Smith (2001), How many species can two essential resources support?, *SIAM J Appl Math* **62**, 336-66.
22. S. Nakaoka and Y. Takeuchi, Microbial survival and coexistence by self-produced resource, *submitted.*
23. T. Pfeiffer and S. Bonhoeffer (2004), Evolution of cross-feeding in microbial populations, *Am Nat* **163** E126–E135.
24. A.A. Raghoebarsing, A. Pol, K.T. Pas-Schoonen, A.J.P. Smolders, K.F. Ettwig , W.I.C. Rijpstra, S. Schouten, J.S.S. Damsté, H.J.M.O. Camp, M.S.M. Jetten, M. Strous (2006), A microbial consortium couples anaerobic methane oxidation to denitrification, *Nature* **440** 918–921.

25. D.E. Rozen and R.E. Lenski (2000), Long-term experimental evolution in *Escherichia coli.* VIII. Dynamics of a balanced polymorphism, *Am Nat* **155** 24–35.
26. D. Smith, S. Alvey, D.E. Crowley (2005), Cooperative catabolic pathways within an atrazine-degrading enrichment culture isolated from soil, *FEMS Microbiol Ecol* **53** 265–273.
27. H.L. Smith and P. Waltman (1994), "The Theory of the Chemostat: Dynamics of Microbial Competition", Cambridge University Press.
28. B. Tang and G.S.K. Wolkowicz (1992), Mathematical models of microbial growth and competition in the chemostat regulated by cell-bound extracellular enzymes," *J Math Biol* **31**, 1–23.
29. T.L.S. Vincent, J.S. Scheel, J.S. Brown and T.L. Vincent (1996), Trade-offs and coexistence in consumer-resource models: it all depends on what and where you eat, *Am Nat* **148** 1038–1058.
30. U. Zissi and G. Lyberatos (2001), Partial degradation of *p*-Aminoazobenzene by a defined mixed culture of Bacillus subtilis and Stenotrophomonas maltophilia, *Biotechnology and Bioengineering* **72** 49–54.

A. SUPPLEMENTARY ANALYSES

In subsection 4.1, we derived the necessary and sufficient conditions for the permanence of (E_1) under which (13) holds. One of assumptions ensuring the coexistence of two strains is violated. Bistability occurs if high amount of metabolite is produced.

A.1. Bistability and competitive exclusion

We assume that (13) still holds for (E_1) to ensure the existence of interior equilibrium point E_*. Bistability is expected if two strains are not mutually invasible. Two strains may not be mutually invasible if the conditions opposite to trade-off condition (TC_1) and productivity condition (PC_1) are satisfied. These observations suggest us to assume

$$\bar{S}_2 < S^*(=\bar{S}_1) \text{ and } \bar{R}_1 < R^*(<\bar{R}_2). \tag{RTC$_1$}$$

and

$$\Delta > 0 \Longleftrightarrow \mathscr{P}_* > \frac{g_2(R^*)}{D}. \tag{RPC$_1$}$$

Condition (RTC_1) describes a relationship between two strains opposite to trade-off condition (TC_1). Under condition (RTC_1), strain-1 is superior to strain-2 in exploitative competition for resource ($\bar{R}_1 < R^* < \bar{R}_2$) whereas strain-2 is superior to strain-1 in exploitative competition for metabolite ($\bar{S}_2 < S^* = \bar{S}_1$). Similarly, condition (RPC_1) requires that the productivity of metabolite must be high in opposite to condition (PC_1). We claim that the following result holds. The statement can be proved by the similar way for Proposition 9.

Proposition 12. *Assume that (RTC_1) holds. Then* $f_2(S^*)(R^0-R^*)-g_2(R^*)(S^0-S^*)>0$ *and* $(\mathscr{P}_*-1)(R^0-R^*)D+\mathscr{P}_*(S^0-S^*)D>0$.

Assume that the assumption of Proposition 12 holds. Then $E_* \in \mathrm{int}\mathbb{R}^4_+$ if and only if $\Delta > 0$, that is, (RPC_1) holds. It immediately follows from (RTC_1) that $\Lambda_{1\to 2} < 0$

and $\Lambda_{2\to 1} < 0$, implying that there exists interior equilibrium point E_* but two strains are not mutually invasible. Numerical simulations are performed to confirm whether bistability between two strains occurs. The parameters are fixed as $R^0 = 2.0$, $S^0 = 0.5$, $D = 1.0$, $d = 10.0$, $m_{f1} = 2.0$, $m_{f2} = 1.0$, $m_{g2} = 1.6$, $a_{f1} = 0.3$, $a_{f2} = 1.0$ and $a_{g2} = 0.4$. With these parameter values, we have $\bar{S}_2 = 0.195 < S^* = 0.3$, $\bar{R}_1 = 0.09 < R^* = 0.370$ and $\mathscr{P}_* = 0.905 > g_2(R^*)/D = 0.769$. Hence (RTC$_1$) and (RPC$_1$) hold. The trajectory of (E$_1$) starting at $(2.0, 0.5, 0.1, 1.0)$ is depicted in the left panel of Figure 7. In this case, strain-1 persists whereas strain-2 is washed out. By contrast, the trajectory of (E$_1$) starting at $(2.0, 0.5, 0.05, 1.0)$ is depicted in the right panel of Figure 7. Note that the initial concentration of strain-1 is decreased from 0.1 to 0.05. In this case, strain-2 persists whereas strain-1 is washed out. Hence bistability between two strains indeed occurs.

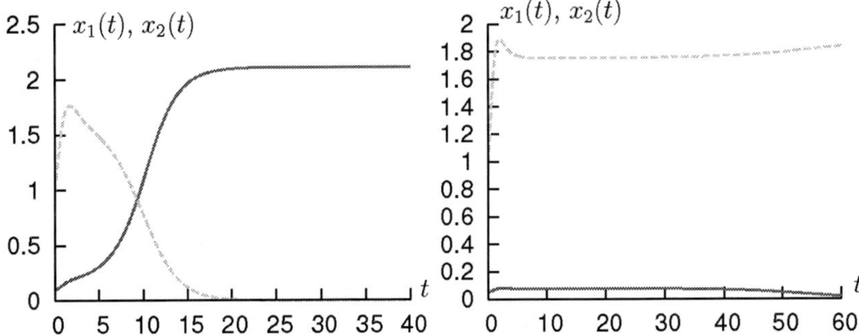

FIGURE 7. Bistability. The trajectories represent the concentration of strain-1 (solid, red lines) and that of strain-2 (dotted, green lines), respectively. Strain-1 outcompetes strain-2 if $x_1(0) = 0.1$ (left) whereas strain-2 outcompetes strain-1 if $x_1(0) = 0.05$ (right).

A.2. Local stability of E_* ($g_1(R) \equiv 0$ or $f_2(S) \equiv 0$)

We analyze the local asymptotic stability of E_*. The characteristic equation defined for the linearized equations of (E$_0$) around interior equilibrium point E_* is given by

$$\begin{aligned}&(\lambda + D)[\lambda^3 + (dx_1^* + D + f_1'x_1^* + f_2'x_2^* + g_1'x_1^* + g_2'x_2^*)\lambda^2\\&+ [\{(dx_1^* + D) - D\mathscr{P}_*\}f_1'x_1^* + (dx_1^* + D - g_2)f_2'x_2^* + (D\mathscr{P}_* + f_1'x_1^* + f_2'x_2^*)g_1'x_1^*\\&+ (g_2 + f_1'x_1^* + f_2'x_2^*)g_2'x_2^*]\lambda + (D\mathscr{P}_* - g_2)(g_1'f_2' - g_2'f_1')x_1^*x_2^*] = 0.\end{aligned} \tag{31}$$

Here we abused to write subscripts in (31). If $g_1(R) \equiv 0$, (31) is reduced to

$$\begin{aligned}&\lambda^3 + (dx_1^* + D + f_1'x_1^* + f_2'x_2^* + g_2'x_2^*)\lambda^2 + [\{(dx_1^* + D) - dR^*\}f_1'x_1^*\\&+ (dx_1^* + D - g_2)f_2'x_2^* + (g_2 + f_1'x_1^* + f_2'x_2^*)g_2'x_2^*]\lambda - D\left(\mathscr{P}_* - 1 + \frac{f_2}{D}\right)g_2'f_1'x_1^*x_2^* = 0.\end{aligned} \tag{32}$$

It immediately follows from the constant term of (32) that (32) has at least one positive real root if $\mathscr{P}_* > 1$, implying that E_* is unstable whenever $\mathscr{P}_* > 1$. If $f_2(S) \equiv 0$, (31) is reduced to

$$\begin{aligned}\lambda^3 &+ (dx_1^* + D + f_1'x_1^* + g_1'x_1^* + g_2'x_2^*)\lambda^2 + [\{(dx_1^* + D) - D\mathscr{P}_*\}f_1'x_1^* \\ &+ (D\mathscr{P}_* + f_1'x_1^*)g_1'x_1^* + (D + f_1'x_1^*)g_2'x_2^*]\lambda - D(\mathscr{P}_* - 1)g_2'f_1'x_1^*x_2^* = 0.\end{aligned} \tag{33}$$

The consequence is the same as that is obtained for the case of $g_1(R) \equiv 0$. Thus the coexistence of two strains in steady state never occurs both in the cases of $g_1(R) \equiv 0$ and $f_2(S) \equiv 0$ whenever $\mathscr{P}_* > 1$. This supports the conclusion obtained in Appendix A.1 in which we have shown that bistability between $\bar{E}_{1+}^C$ and $\bar{E}_{2+}$ occurs whenever $\mathscr{P}_* > 1$.

ACKNOWLEDGMENTS

This work was partly supported by the Sasakawa Scientific Research Grant from The Japan Science Society.

Optimization Techniques in Computing Good Quality Solutions to Sequence Alignment Problems

Paola Festa

Dipartimento di Matematica e Applicazioni, Università di Napoli Federico II, Via Cintia, I-80126 Napoli, Italy, E-mail: paola.festa@unina.it

Abstract. In the past few years a large number of molecular biology problems have been formulated as combinatorial optimization problems, including sequence alignment problems, genome rearrangement problems, string selection and comparison problems, and protein structure prediction and recognition. This paper describes the combinatorial formulation of some among the most interesting molecular biology problems and surveys the most efficient state-of-the-art techniques and algorithms to exactly or approximately solve them.

Keywords: Computational biology, molecular structure prediction, protein and sequences alignment, combinatorial optimization, metaheuristics.
PACS: 02.10.Ox, 87.10.Vg, 87.18.Wd, 87.55.de, 87.55.kd

1. INTRODUCTION

A fundamental remark made by researchers in molecular biology regards the abstraction of the real three-dimensional structure of DNA and its representation as a unidimensional sequence of characters from an alphabet of four symbols. The same type of assumption involves also the protein represented as a sequence of characters from an alphabet of twenty symbols. As a result of the linear coding of DNA and proteins, many molecular biology problems have been formulated as computational and optimization problems involving strings and sequences, such as, for example, to rebuild long DNA sequences starting from overlapping fragments (fragment assembly), to compare two or more sequences looking for their similarities (strings coding the same function), to look for patterns that occur with a certain frequency in DNA and/or protein sequences. In particular, optimization and mathematical programming techniques have been already applied to solve several molecular biology problems, including sequence alignment problems, genome rearrangement problems, string selection and comparison problems, and protein structure prediction and recognition. An exhaustive survey can be found in a recent paper of Greenberg, Hart, and Lancia [1].

The scope of this paper is to describe the combinatorial formulation of some interesting biological system problems and the mathematical programming methods that can be applied to solve them.

The remainder of this paper is organized as follows. In Section 2 sequence alignment problems are described. Properties and solution approaches for the pairwise and the multiple alignment problem are discussed in subsection 2.1 and subsection 2.2, respectively. In Section 3 concluding remarks and some open questions are discussed.

CP1028, *Collective Dynamics: Topics on Competition and Cooperation in the Biosciences*
edited by L. M. Ricciardi, A. Buonocore, and E. Pirozzi

2. SEQUENCE ALIGNMENT PROBLEMS

In the field of molecular biology, DNA and RNA sequences have an immediate and intrinsic interpretation as sequences of symbols. In analyzing them one could be interested in finding the specific activity of some of their pieces. Since a fundamental remark made by researchers in molecular biology claims that similar biological primary structures correspond to similar activities, techniques that compare sequences are used to get information about an unknown sequence from the knowledge of two or more already studied sequences. Comparing genomic sequences drawn from individuals of different species consists in determining their *similarity* or *distance* and is one of the fundamental problems in molecular biology. In fact, such comparisons are needed to identify not only functionally relevant DNA regions, but also spot fatal mutations and evolutionary relationships. Unfortunately, the concept of similarity has no unique definition. In fact, for example proteins and DNA segments can be considered similar in several different aspects: they can have a similar structure or can have similar amino acid primary sequence. Usually, given two linearly ordered sequences, one evaluates their similarity through a measure that associates to a pair of sequences a number related to the cardinality of subsets of corresponding elements in each (*alignment*). The correspondence must be order-preserving, i.e. if the i-th element of the first sequence s corresponds to the k-th element of the second sequence t, no element following i in s can correspond to an element preceding k in t.

The difficult task of providing a unique definition of similarity and distance among sequences has been nicely described in a paper appeared in 1994 by Gusflied et al. [2]. Since in DNA and amino acid sequences there is considerable disagreement about how to weight matches, mismatches, insertions/deletions, the authors propose to study the so called *parametric sequence alignment problem* which consists of computing the optimal valued alignment between two sequences as a function of variable weights for matches, mismatches, spaces, and gaps. The goal is to partition the parameter space into regions (which are necessarily convex) such that in each region one alignment is optimal throughout and such that the regions are maximal for this property. In their paper, Gusflied et al. studied first the structure of this convex decomposition, and then the complexity of computing the decomposition. They have shown that for the special case where only matches, mismatches, and spaces are counted, and where spaces are counted throughout the alignment, the decomposition is simple and leads to at most $n^{\frac{2}{3}}$. Moreover, the computational complexity needed by their algorithm to compute the decomposition is $O(knm)$, where k is the actual number of regions, and $n < m$ are the lengths of the two strings[1].

Stated that a genomic sequence can be represented as a string over an alphabet Σ

---

[1] Denoting with n the size of the input, an algorithm $\mathscr{A}$ has computational complexity $O(g(n))$ if the number $f(n)$ of elementary operations that it performs is such that $f(n) \in O(g(n))$, where $O(g(n)) = \{h(n) \mid \text{there exist positive constants } c \text{ and } n_0 \text{ s.t. } 0 \leq f(n) \leq cg(n)\ \forall\, n \geq n_0\}$.

If $g(n)$ is a polynomial in n, then the computational complexity of $\mathscr{A}$ is said *polynomial* and the problem that it solves is said *computationally tractable*. A computationally intractable mathematical programming problem $\mathscr{P}$ is an optimization problem such that no polynomial time method able to find an optimal solution for $\mathscr{P}$ is known.

whose elements are either the four nucleotide letters or the twenty letters corresponding to the amino acids, the most studied alignment problems involve either a pair of sequences or a set of more than two sequences, where it is assumed that sequences are of the same length[2]. In the first case, the resulting problem is known as *pairwise alignment problem* and it is computationally tractable. In the case of a set of more than two sequences the problem is known as *multiple alignment problem* and it is generally computationally intractable.

2.1. Pairwise alignment problem

It is the simplest alignment problem and it is computationally tractable. Given two sequences s and t, the pairwise alignment problem consists in aligning s and t either *locally* or *globally*. When looking for a global alignment, the problem consists in aligning s and t over their entire length. In local alignment, instead one discards portions of the sequences that do not share any homology. Most alignment methods are global, leaving it up to the user to decide on the portion of sequences to be incorporated in the alignment. Formally, the global pairwise alignment problem can be stated as follows: Given symmetric costs $\gamma(a,b)$ for replacing the character a with b and costs $\gamma(a,-)$ for deleting or inserting the character a, find a minimum cost set of character operations that transform s into t. In case of genomic strings, the costs are specified by the so called *substitution matrices* which score the likelihood of specific letter mutations, deletions, and insertions. An alignment A of two sequences s and t is a bi-dimensional array containing the gapped input sequences in its rows. The total cost of A is computed by summing up the costs for the pairs of characters in corresponding positions. An optimal alignment A^* of s and t is an alignment corresponding to the minimum total cost denoted by $d_A^*(s,t)$ and also called *edit distance* of s and t.

As underlined, the pairwise alignment problem is computationally tractable, i.e. an optimal solution can be found by a polynomial algorithm. One of the most efficient exact algorithm is a Dynamic Programming method proposed in 1981 by Smith and Waterman [3] and whose computational complexity is $O(n^2)$, where n denotes the length of the input sequences.

2.2. Multiple alignment problem

The problem of aligning a set $\{s_1, s_2, \ldots, s_k\}$ of k sequences is a generalization of the pairwise alignment problem and is called the *multiple alignment problem*. For the multiple sequences case, an alignment is a matrix A having each gapped sequence in a row. Columns of A in a multiple alignment should share a target biological feature that depends on the application of the alignment. For example, the multiple alignment can be used to illustrate evolutionary relationships among the sequences and in this

[2] This is not a restrictive assumption, since it is always possible to make two or more sequences of the same length possibly inserting gaps, represented by the '-' character.

case the residues in each column should have a shared evolutionary history. Another target of the multiple alignment can be to get information about sequences structure and in this case the residues in each column should have a shared structural or functional role. Clearly, it is very difficult to choose an appropriate objective function that exactly takes into account this kind of feature extraction. One among the most used objective functions is the so called *Sum-Of-Pairs* (SP) score. An optimal solution is an alignment A^* corresponding to the minimum *Sum-Of-Pairs* (SP) score, denoted by $SP(A^*)$, in which the cost is computed by summing up costs of symbols matched up at the same positions, over all the pairs in the sequences. Formally, the SP-score of an alignment A is given by

$$SP(A) = \sum_{1 \leq i < j \leq k} d_A(s_i, s_j) = \sum_{1 \leq i < j \leq k} \sum_{l=1}^{|A|} \gamma(A_{il}, A_{jl}),$$

where $|A|$ is the alignment length, i.e. its number of columns. Contrary to the case of only two sequences, the multiple alignment problem is a computationally intractable optimization problem. Its intractability was proved in 1994 by Wang and Jiang [4].

A generalization of Smith and Waterman's dynamic programming algorithm [5, 6] has exponential computational complexity given by $O(2^k l^k)$ and therefore it cannot be used to solve even small size instances of the problem. Particularly interesting are some observation reported in 1988 by Carrillo and Lipman in [7], leading to further constraints of the problem that the authors proved useful in reducing computation in the dynamic programming method. In more detail, they showed how to establish a correspondence between a set of k sequences $\{s_1, s_2, \ldots, s_k\}$ and a *lattice* $\mathscr{L}(s_1, s_2, \ldots, s_k)$ in the k-dimensional space that consists of the k-dimensional hypercubes obtained from the Cartesian product of k strings of squares. Each of these strings is associated to a particular sequence and has as many squares as characters in the corresponding sequence. All hypercubes in the lattice form a k-dimensional parallelepiped, whose corner corresponding to the first character of each sequence is called original corner, while the corner furthest from the original corner is called end corner. Then, a path $\gamma(s_1, s_2, \ldots, s_k)$ between the sequences $\{s_1, s_2, \ldots, s_k\}$ is a connected broken line joining the original corner to the end corner, where the segments of this broken line join vertices of the lattice belonging to the same hypercube.

In their paper, the authors proposed a method to efficiently determine an estimated path γ. Even if their approach is not guaranteed to always find paths close in measure to an optimal one, the idea behind it is elegant and very original.

In 2000, Kececioglu et al. [8] proposed an alternative and more effective method based on the *maximum weight trace problem*, which has been introduced by Kececioglu [9] as a different optimization problem that generalizes the SP-score objective function. A *trace* has a graph theoretic definition, but in the following it will be specialized for alignments problems. Let us suppose that k sequences have to be aligned having $o_1, o_2, \ldots, o_k$ elements, respectively. Then, a complete k-partite graph[3] $G = (V, E)$ can be

[3] A graph $G = (V, E)$ is said *complete* if E contains an edge connecting each pair of vertices in V. A k-partite graph $G = (V, E)$ is a graph whose vertices set V can be partitioned into k sets, $V_1, V_2, \ldots, V_k$, and whose edges set $E \subseteq \cup_{1 \leq i < j \leq k} V_i \times V_j$. If a k-partite graph $G = (V, E)$ is complete, then

defined, where $V = \cup_{i=1}^{k} V_i$ (for each $i = 1, 2, \ldots, k$, $V_i = \{v_{i1}, v_{i2}, \ldots, v_{io_k}\}$ is the subset of vertices of level i) and $E = \cup_{1 \le i < j \le k} V_i \times V_j$ is the set of edges called *lines*. A line $[i, j] \in E$ is realized by an alignment A if i and j are put in the same column of A and a trace T is a set of lines such that there exists an alignment A_T that realizes all the lines in T. Given weights w_{ij} associated with the lines $[i, j] \in E$, the maximum weight trace problem (MWT) consists in finding a trace T^* corresponding to the maximum total weight. Let $\overline{E} = \{(v_{ih}, v_{ir}) \mid i = 1, \ldots, k,\ 1 \le h < r \le o_i\}$ be the set of directed arcs connecting each element v_{ir} ($i = 1, \ldots, k$) to the elements that follow it in the sequence s_i, then Kececioglu et al. [10] have shown that a trace T must satisfy the following proposition:

Proposition 1 *T is a trace if and only if there is no directed mixed cycle*[4] *$\mathscr{C}$ in the graph $\overline{G} = (V, T \cup \overline{E})$.*

The property of a trace T stated by Proposition 1 is important since it has been used to mathematically formulate the maximum weight trace problem. In fact, by defining a Boolean variable x_{ij} for each line $[i, j] \in E$, the problem admits the following 0/1 integer programming formulation:

$$\max \sum_{[i,j] \in E} w_{ij} x_{ij}$$

s.t.

$$\sum_{[i,j] \in \mathscr{C} \cap E} x_{ij} \le |\mathscr{C} \cap \overline{E}| - 1 \quad \forall \mathscr{C} \text{ in } \overline{G}, \quad \text{(a)}$$

$$x_{ij} \in \{0, 1\}, \quad \forall [i, j] \in E. \quad \text{(b)}$$

Mixed-cycle inequality constraints (a) have been used by Kececioglu et al. who in [8] proposed a branch-and-cut algorithm[5] that was the first to find an optimal alignment for a set of 15 proteins of about 300 amino acids. However, it has been shown experimentally that it is not suitable for aligning sequences longer than a few hundred characters. Given the high computational complexity of multiple sequence alignment problems, to deal with long sequences several polynomial time approximation and heuristic methods have been designed and proposed in the literature. In the operations research community, an approximation method finds a suboptimal solution providing an approximation-guarantee on the quality of the solution found. An algorithm $\mathscr{A}lg$ for a minimization problem is said to be an approximation algorithm if the worst solution returned by $\mathscr{A}lg$

$E = \cup_{1 \le i < j \le k} V_i \times V_j$.

[4] A cycle in a graph $G = (V, E)$ is a sequence of vertices $[v_1, v_2, \ldots, v_m]$ such that $m \ge 1$, $[v_i, v_{i+1}] \in E$ for $i = 1, \ldots, m-1$, $v_1 = v_m$, and except for the first and the last vertex there are no repeated vertices. A directed mixed cycle is a cycle containing both directed arcs (elements of $\overline{E}$) and undirected edges (lines of T), in which the arcs are crossed according to their orientation, while the edges can be crossed in either direction.

[5] A branch-and-cut algorithm belongs to the branch-and-bound algorithms family. A branch-and-bound algorithm is based on the idea of intelligently enumerating all the feasible solutions of a combinatorial optimization problem by successive partitioning of the solution space. *Branch* refers to this partitioning process, while *bound* refers to bounds on the objective function value that are used to construct a proof of optimality without exhaustive search.

is not greater than ε times the best solution, for $\varepsilon > 1$. In this case, ε is called the approximation guarantee of the proposed algorithm. Such approximation algorithms are important for intractable combinatorial optimization problems, since they are in a sense the best that can be done to give a guarantee of quality for the returned solution. A vast literature on approximation algorithms has been developed in the last decade, and good starting points are the books [11] and [12]. Due to Gusfield [13] is one among the most effective approximation methods for multiple alignment problems with the minimum Sum-Of-Pairs (SP) score as objective cost function satisfying triangle inequality. In 1993 he proposed an iterative technique that progressively builds an alignment by considering one sequence at time. It starts by building a star[6] having a node for each of the k sequences. One of these nodes is selected as center. The procedure then uses the tree as a guide for aligning the input sequences. In more detail, denoting by s_c, $c \in \{1,\ldots,k\}$, the sequence selected as the center of the star, Gusfield's algorithm first computes $k-1$ pairwise alignments A_i, one for each sequence s_i, $i \neq c$, aligned with s_c such that each A_i corresponds to the edit distance between the pair (s_i,s_c). The $k-1$ alignments can be then merged into a single alignment $A(c)$, by simply putting in the same column two characters if they are aligned to the same character of s_c. Note that, in $A(c)$ each sequence s_i remains optimally aligned with s_c, i.e.

$$d_{A(c)}(s_i,s_c) = d_{A_i}(s_i,s_c), \qquad \forall\, i = 1,\ldots,k,\ i \neq c. \tag{1}$$

Therefore, it is possible to write

$$d_{A_i}(s_i,s_c) = d(s_i,s_c), \qquad \forall\, i = 1,\ldots,k,\ i \neq c, \tag{2}$$

neglecting the specific alignment the edit distance d is referred to. Moreover, by assuming that the cost function γ is a metric over the alphabet Σ, the distance is a metric defined over the set of sequences and therefore triangle inequality holds, i.e.

$$d_{A(c)}(s_i,s_j) \leq d(s_i,s_c) + d(s_j,s_c), \qquad \forall\, i,j = 1,\ldots,k,\ i \neq j \neq c. \tag{3}$$

Let I be a generic multiple sequence alignment problem instance and let SP_I^* and $\hat{SP}_I$ be the optimal objective function value and the objective function value corresponding to the suboptimal solution found by Gusfield's algorithm, respectively. Then, in [13] Gusfield proved that $\hat{SP}_I \leq (2 - \frac{2}{k})SP_I^*$, for any instance I. Basically, he achieved this approximation ratio by assembling an alignment of k sequences from optimal alignments of pairs of sequences. His algorithm produced meaningful biological results in the sense that he obtained biologically plausible alignments: a computational experiment with an alignment of 19 sequences gave a suboptimal solution only 2% worse than the optimal one. Since then a natural and obvious research direction for improvement has been to use optimal alignments of $l > 2$ sequences ($l < k$), and then assemble them to approximately align k sequences. However, devising an efficient assembling procedure is not an easy task that for an arbitrary l remained an open problem until 1997, when Bafna et al. [14]

[6] A tree is a connected and acyclic graph and a star is a tree with at most one node (center of the star) which is not a leaf.

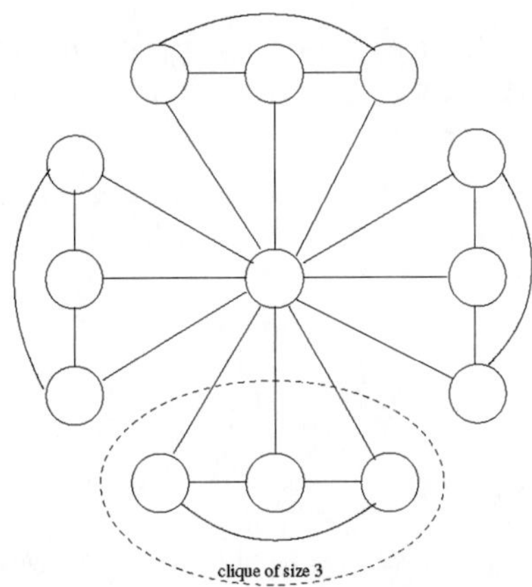

FIGURE 1. Example of a 5-star on 13 nodes.

showed that for arbitrary $l < k$ it is possible to obtain a performance guarantee $2 - \frac{l}{k}$ with an algorithm polynomial in n and k. They extended the star data structure used by Gusflied to a l-star, defined as a graph $G = (V, E)$ having $r = \frac{k-1}{l-1}$ cliques[7] of size l whose vertex sets intersect in only one center vertex (Figure 1) and showed that this representation preserves compatibility[8] through the following Lemma.

Lemma 2 *Let $A_1, A_2, \ldots, A_r$ denote alignments for the r cliques, with each A_i aligning l sequences.*

For any l-star and any specified alignments $A_1, A_2, \ldots, A_r$ for its cliques, there is an alignment A for the k sequences that is compatible with each of the alignments $A_1, A_2, \ldots, A_r$.

Bafna et al. introduced weights on the edges of an l-star G having node center c, as follows.

$$\forall\, i, j \quad w_{ij} = \begin{cases} k - (l-1), & \text{if } i = c \text{ or } j = c, \\ 1, & \text{if } i, j \neq c,\ i \text{ and } j \text{ in the same clique}, \\ 0, & \text{otherwise}. \end{cases}$$

Let $W(G) = [w_{ij}]$ be the resulting $k \times k$ matrix of weights and let $W_1, W_2, \ldots, W_r$ denote the submatrices of weights for the r cliques of an l-star G. Let $A_1^*, A_2^*, \ldots, A_r^*$ be optimal weighted sum-of-pairs alignments for the r cliques.

---

[7] Let $G = (V, E)$ be an undirected weighted graph, where $V = \{v_1, v_2, \ldots, v_n\}$ is the set of n nodes and $E \subseteq V \times V$ is the set of edges. Given a node subset $S \subseteq V$, the subgraph induced by S is denoted by $G(S) = (S, E \cap S \times S)$. A graph $G = (V, E)$ is *complete* if and only if its nodes are pairwise adjacent, i.e. for each pair of nodes $v_i, v_j \in V$, $(v_i, v_j) \in E$.

A *clique* $\mathscr{C}$ is a subset of V such that the induced graph $G(C)$ is complete.

[8] Given an alignment A on the sequences $\{s_1, s_2, \ldots, s_k\}$ and an alignment A' on some subset of the sequences, A is said *compatible* with A' if A aligns the characters of the sequences aligned by A' in the same way as A'.

Bafna et al. proved that for any l-star G it is possible to compute an alignment A_G optimal with respect to the weight matrix $W(G)$ by computing optimal weighted alignments for each clique of G. In fact, from Lemma 2 it holds the following result.

Lemma 3 *Given an l-star G, there is an optimal (weighted with respect to $W(G)$) alignment A_G for the k sequences that is compatible with each of the alignments $A_1^*, A_2^*, \ldots, A_r^*$.*

Moreover, it holds that

$$W(G) \cdot SP(A_G) = C_1 \cdot SP(A_1^*) + C_2 \cdot SP(A_2^*) + \cdots + C_r \cdot SP(A_r^*). \tag{4}$$

The additive property of equation (4) states the correctness of Bafna et al.'s assembly operation that can be performed in $O(k \cdot g(l,n))$, where $g(l,n)$ denote the computational complexity required to obtain an optimal solution to the weighted SP score for l sequences of length n. Since $g(l,n)$ can be clearly prohibitively high, the authors reduced their approximation problem to that of finding an optimal alignment for each l-star in a *balanced set*, defined as follows.

Definition 1 *Let $\mathcal{G}$ be a collection of l-stars, and let $W(G)$ denote the weight matrix for star G. The collection $\mathcal{G}$ is said balanced if for some scalar $p > 1$ it results that*

$$\sum_{G \in \mathcal{G}} W(G) = p \cdot E.$$

In fact, in the case of balanced set, the SP score of an optimal alignment A^* can be easily lower bounded by the minimum value for $W(G) \cdot SP(A_G)$ over all possible stars G, as stated by the following lemma.

Lemma 4 *Let $\mathcal{G}$ be a balanced set of l-stars, then*

$$\min_{G \in \mathcal{G}} W(G) \cdot SP(A_G) \leq \frac{p}{|\mathcal{G}|} \min_{A} E \cdot SP(A). \tag{5}$$

Since inequality (5) holds for an arbitrary alignment A, it also holds for A^*. A trivial balanced set $\mathcal{G}$ that could be used is the set of all l-stars, which is obviously balanced by symmetry. Note that, Gusfield's algorithm [13] worked for $l = 2$ and there are only k 2-stars. For $l > 2$ the number of l-stars grows exponentially with k making *naive* Bafna et al.'s algorithm with high running time. Fortunately, Bafna et al. have been able to show that it is not necessary to exhaustively compute alignments for all possible l-stars by applying a dynamic programming algorithm.

Another interesting paper appeared in 1999 due to Wu et al. [15], who generalized Gusfield's method [13] to use any tree and not only stars, preserving an approximation guarantee of 2.

The Sum-Of-Pairs is a widely used score function, even if it has no direct reasonable biological justification. A more biologically significant score function should take into account the evolutionary relationships between the sequences. An attempt in this direction has been made by Hein [16] who proposed a heuristic that at the same time infers and uses the evolutionary relationships among sequences. In 1994 Krogh et al. [17]

described a heuristic, which use hidden Markov models to describe the relationships between sequences.

To solve real instances of the multiple sequence alignment problem, any method requires a huge amount of computer memory. Recently, in [18] Korf and Zhang presented a new algorithm that reduces the space complexity of heuristic search compared to state-of-the-art algorithms. This target is achieved by improving a classical technique to find an optimal global alignment of several DNA or amino-acid sequences that consists in finding a lowest-cost corner-to-corner path in a k-dimensional grid. In fact, in [18] the authors proposed an original way to save computer memory by storing only a subset of the used data structures designing an algorithm that reduces the memory requirement from $O(n^k)$ to $O(n^{k-1})$, where n is the length of the sequences. In the case of aligning two sequences of length n, if the 2-dimensional grid is small enough to fit into memory, the single source single destination shortest path algorithm proposed in 1959 by Dijkstra [19] can solve the problem in $O(n^2)$ occupying $O(n^2)$ memory space. Shortest path problems are classical combinatorial problems that arise as subproblems when solving many optimization problems. In fact, they have been widely studied leading to a great number of algorithms adapted to find an optimal solution in various special conditions and/or constraint formulations [20, 21, 22, 23, 24]. Since 1959 several algorithms [25] have been proposed efficiently implementing Dijkstra's algorithm, but using special data structures and so obtaining a better computational complexity bound. Particularly suitable for computing the shortest path in the case of aligning two sequences are the so called A^* *search techniques*. An A^* search algorithm for the single source-single destination shortest path problem with nonnegative edge lengths (see among others [26]) focuses an *informed search* of a shortest path from a given source node s to a given destination node d through the use of a heuristic function e that expresses an estimate of the shortest distance from any node to d. An A^* algorithm is guaranteed to produce an optimal solution path (whenever a path from the source node to the destination node exists) if the heuristic function e is monotone.

A further and elegant solution approach is due to Korostensky and Gonnet [27] who found near optimal multiple sequence alignments by solving a related Symmetric Traveling Salesman Problem (STSP)[9]. In particular, the authors established a correspondence between the cities and the sequences and defined the distances as the scores of the pairwise alignments. Once reduced the original problem into a STSP instance over a graph G, they look for the longest Hamiltonian cycle in G, since scores represent probabilities and we are interested in the maximum probability. Then, to use any state-of-the-art algorithm for the STSP, they simply subtract the scores from a large number and then compute the shortest cycle. The authors proved that the proposed algorithm calculates a near optimal solution to multiple alignment problems and has a performance guarantee of $\frac{n-1}{n} \cdot score(A^*)$, where $score(A^*)$ is the optimal score. The computational complexity

[9] Given is a distance matrix M that contains a distance for each pair among the n cities, the STSP consists in finding the shortest Hamiltonian cycle, i.e. a cycle involving each city exactly once. The underlying graph G with n nodes is directed. In the asymmetric version of the problem the matrix M is not symmetric and the underlining graph G is undirected.

```
algorithm iter-impr (c, x, 𝒩)
    NoLocal := true;
    while (NoLocal) do
        /* exploration of the neighborhood N(x) */
        if (∃ x̄ ∈ N(x) : c'x̄ < c'x,  c'x̄ ≤ c'y,  ∀ y ∈ N(x)) then
            x := x̄;
        else NoLocal := false;
        endif
    endwhile
    return (x);
end iter-impr
```

FIGURE 2. Local search procedure iter-impr.

of the algorithm is pseudo-polynomial[10] and is equal to $O(k^2 n^2)$, where k is the length of the longest input sequence.

As already underlined, an approximation method finds a suboptimal solution providing an approximation-guarantee on the quality of the solution found. Sometimes it is preferable to approach and solve the problem heuristically, especially in the presence of large scale problem instances. A heuristic approach finds a suboptimal solution whose quality can be only experimentally verified and therefore in general it is meaningful to apply it for solving optimization problems that are hard even to approximate. The main ingredients of any heuristic method are the criterion of constructing a feasible solution x and the definition of the so called *local search phase*, that starting from x explores a suitable defined *neighborhood* of x (a set of feasible solutions "close" to x) until a local optimum is found. More formally,

Definition 2 *Let Pr be a combinatorial optimization problem and let P be its (finite) set of feasible solutions. A neighborhood function is any function $N : P \to 2^P$ that to a solution $x \in P$ associates a subset $N(x) \subset 2^P$ of feasible solutions "close" to x, accordingly to some defined metric.*

The subset $N(x)$ is said neighborhood of x and each $y \in N(x)$ is said neighbor of x.

Any local search procedure takes as input an initial feasible solution x and explores a suitable defined neighborhood of x looking for better solutions in terms of objective function value. The simplest local search procedure is known as *iterative improvement* and its pseudo-code for a minimization problem is shown in Figure 2. Given a cost vector c and once defined a neighborhood function N suitable for the problem to be solved, starting from an initial solution x the iterative improvement procedure explores $N(x)$ looking for a better solution $\bar{x}$. If such a solution exists, then the search continues from $\bar{x}$; otherwise, the procedure provides as output the current solution x which is locally optimal respect to the defined neighborhood. In the last two decades special attention of combinatorial optimization research community has been put on the design of *metaheuristics*, a new family of algorithms that tries to combine basic heuristic

[10] Denoting with n the size of the input and with k a parameter characterizing the input problem instance, an algorithm $\mathscr{A}$ has pseudo-polynomial computational complexity if it runs in $O(g(n) \cdot h(k))$, where $g(n)$ is a polynomial in n and $h(k)$ is a polynomial in k.

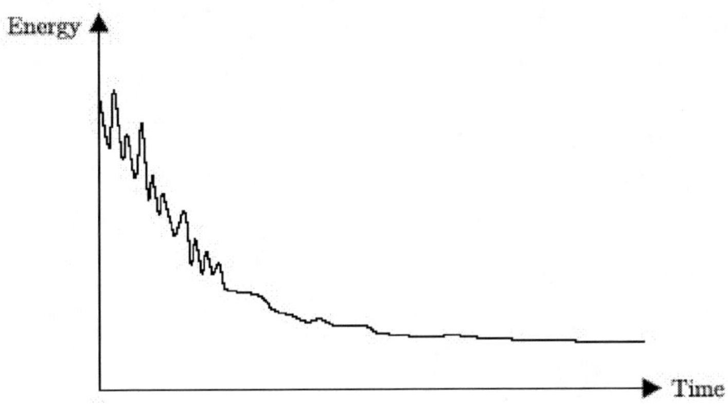

FIGURE 3. Annealing process of simulated annealing.

methods in higher level frameworks aimed at efficiently exploring the set of feasible solution of a given combinatorial problem. They include but are not restricted to Tabu Search, Genetic Algorithm, GRASP (Greedy Randomized Adaptive Search Procedure), Simulated Annealing, and Iterated Local Search. For the multiple alignment problem, a Simulated Annealing (SA) and Genetic Algorithms (GAs) have been proposed.

SA is a generalization of the Monte Carlo method for examining the equations of state and frozen states of n-body systems [28]. The idea is based on the process of annealing of a system. During the annealing state, process will melt and initially it will start at high temperature T and it will slowly cooled down so that the system will always be at thermodynamic equilibrium. As cooling proceeds, the system becomes more ordered and approaches a frozen ground state at $T = 0$. Hence, the process can be thought of as an adiabatic approach to the lowest energy state (see Figure 3). If the initial temperature of the system is too low or cooling is done insufficiently, slowly the system may become quenched forming defects or freezing out in meta stable states (i.e. trapped in a local minimum energy state).

In the original scheme of the method, an initial state x of a thermodynamic system having energy $E(x)$ and temperature T is chosen. Holding T constant, the initial configuration is perturbed and the energy variation $\Delta E = E(x) - E(\bar{x})$ in the perturbed configuration $\bar{x}$ is computed. If $\Delta E < 0$, the new perturbed configuration $\bar{x}$ is accepted, while if $\Delta E \geq 0$, $\bar{x}$ is accepted with a probability given by the Boltzmann factor $e^{-\frac{\Delta E}{KT}}$. These processes are repeated for a number of times to obtain good sampling statistics for the current temperature. The temperature is decremented and the entire process is repeated until a frozen state is achieved at $T = 0$.

In any SA-like algorithm for combinatorial problems, the analogy with the generalization of this Monte Carlo is straight forward. Given the following generic optimization problem

$$(Pr) \quad \min\{c'x : \; x \in P\},$$

where c is a given cost vector, x is the vector of decision variables, and P the finite set of

```
algorithm simulated-annealing (c,T,𝒩)
1   Build a feasible solution x;
2   x_b := x;
3   while (T > 0 and stopping criterion not met^a) do
4       Select x̄ ∈ 𝒩(x); /* local search: select x̄ with probability 1/|𝒩(x)| */
5       ΔE := c'x̄ − c'x;
6       if (ΔE < 0) then
7           x := x̄;
8           if (c'x < c'x_b) then x_b := x̄;
9       else x := x̄ with probability e^(−ΔE/KT);
10      endif
11      Decrease T;
12  endwhile
13  return (x_b);
end simulated-annealing
```

[a] Two possible stopping criteria: a maximum number of iterations to be performed; a maximum number of performed iterations without improvements.

FIGURE 4. A generic SA algorithm for a minimization problem.

feasible solution, then the current state x of the thermodynamic system is analogous to a current feasible solution x for Pr, the energy equation $E(x)$ for the thermodynamic system is analogous to the objective function value $c'x$, and ground state is analogous to the global minimum. Figure 4 depicts the pseudo-code of a generic SA for a minimization problem.

Starting from an initial feasible solution x, a SA selects at random a neighbor solution $\bar{x}$ and computes a possible improvement in terms of objective function value. If a lower objective function value corresponds to state $\bar{x}$, then the system goes from state x to state $\bar{x}$, the incumbent solution x_b is possibly updated, and the search continues starting from $\bar{x}$. Otherwise, the system goes from state x to state $\bar{x}$ with probability $e^{-\frac{c'\bar{x}-c'x}{KT}}$. For multiple alignment problems, given the input set of sequences, a first alignment x is randomly generate (line 1). A perturbation is then applied to x (line 4) by shifting of an existing gap or by introducing a new gap so obtaining a neighbor alignment $\bar{x}$.

As for any metaheuristic, the major problem in implementing this type of method is the right tuning of the free parameters to avoid local minima. In the case of SA, attention must be paid for the choice of the initial temperature T, the number of iterations to be performed at each temperature, and the amount of decreasing of the temperature at each step as cooling proceeds. About the tuning of parameter T, it strongly affects the probability of accepting not improving solutions. Usually, a SA starts with a high temperature that is then progressively decreased in order to not let acceptance of not improving solutions. Note that, for $T = 0$ (line 9) the process looses any stochastic characteristics and corresponds to a simple iterative improvement.

Other metaheuristic frameworks applied for multiple alignment problems and producing promising results have been genetic algorithms (GAs). Since the 1970s [29] has began the interest in genetic heuristic search algorithms, later theoretically and empirically proven to be a robust search method [30] having a high probability of locating the

```
algorithm GA
    t := 0;
    Initialize population P(0);
    Evaluate P(0);
    while not termination-criteria do
        t := t + 1;
        Select P(t) from P(t − 1);
        Alter P(t);
        Evaluate P(t);
    end;
end GA;
```

FIGURE 5. Pseudo-code of a generic genetic algorithm.

global solution optimally in a multimodal search landscape. They derive from Darwin's evolutionary theory and the principles of natural selection, where stronger individuals are likely the winners in a competing environments. In complete analogy with the nature, once encoded each possible point in the search space of the optimization problem into a suitable representation, a GA transforms a population of individual solutions, each with an associated *fitness* (or objective function value), into a new generation of the population. The new population is generated by applying genetic operators, such as *crossover* and *mutation* [31]. Crossover is the exchange of genetic material among chromosomes possibly generating good combination of genes for better individuals, while mutation causes regenerating lost genetic material. In Figure 5 a simple genetic algorithm is described by the pseudo-code, where $P(t)$ is the population at iteration t. GA is powerful and broadly applicable stochastic search and optimization technique, generally consisting of the following basic components:

1. Randomly create an initial population $P(0)$ of individuals.
2. Iteratively perform the following substeps on the current generation of the population until the termination criterion has been satisfied. As for a SA, two possible stopping criteria are a maximum number of iterations to be performed or a maximum number of performed iterations without improvements.
   (a) Assign fitness value to each individual using the fitness function.
   (b) Select parents to mate.
   (c) Create children from selected parents by crossover and mutation.
   (d) Identify the best-so-far individual for this iteration of the GA.

To successfully apply GA on multiple alignment problems, one needs to encode each possible point in the search space (alignment) into a suitable representation/data structure that could be easily manipulated (in terms of recombination and mutation). Moreover, a proper evaluation function or objective function should be defined, which in this case is related to the overall alignment score of the sequences.

Attempts to apply GAs to the multiple sequence alignment problem started in 1993 when Ishikawa et al. [32] published a hybrid GA with dynamic programming. The resulting algorithm does not try to directly optimize the alignment but rather the order in which the sequences should be aligned using dynamic programming. Ishikawa et al.'s method obtained results satisfactory enough to prompt the development of the use of GAs in sequence analysis. In fact, in 1996 and 1997 two GAs appeared able to deal with sequences in a more general manner [33, 34]. In particular, in [33] Notredame and Higgins described SAGA (Sequence Alignment by Genetic Algorithm), where the population is made of complete multiple sequence alignments and the operators manipulate the aligned sequences, by inserting and/or shifting gaps in a random or semi-random manner. Each individual is a multiple alignment. The data structure chosen for the individual representation a two-dimensional array where each line represents an aligned sequence and each cell is either a residue or a gap. The population has a constant size ($|P(t)| = k$, k given constant for each iteration t) and does not contain any duplicate (i.e. identical individuals). To generate an initial population $P(0)$ as diverse as possible in terms of genotype and as uniform as possible in terms of scores (objective function values), SAGA generates at random 100 multiple alignments that only contain terminal gaps and that have the same length. Initial generation can also be carried out by generating sub-optimal alignments using an implementation of dynamic programming that incorporates some randomness. The crossover operator is meant to generate a new alignment by combining two existing ones. In SAGA two types of crossovers are applied: the *one point crossover* that combines two parents through a single point of exchange and the *uniform crossover* that promotes multiple exchanges between two parents by swapping blocks between consistent bits of their representation. Only the fittest child between the two children produced by the crossovers is kept and inserted into the new population (if it is not a duplicate).

About mutation, in SAGA several different operators have been described and the most promising one has been the gap insertion operator, since it allows evolution memory to be incorporated in the GA. In fact, by applying this operator it is possible to reconstitute backward some of the events of insertion/deletions through which a set of sequences might have evolved.

Recently, to improve the efficiency and the accuracy of the algorithms several hybrid GAs with SA have been proposed [35, 36]. In 2004, Omar et al. [35] represent the population as an array of sequences over an alphabet of 20 characters. As always, the symbol '-' refers to the gap in the alignment and represents an insertion or a deletion of an amino acid residue. The initial population $P(0)$ is generated at random and all positions that are not associated with amino acid are filled with a gap. The size of each population is constant and can be decided by the user. Selection of parents to mate is performed at random with a probability proportional to the fitness function value: the fitter the individual, the more likely it will be chosen at random to mate. To crossover SAGA operators [33] are used and as in SAGA only the fittest child is kept and inserted into the new population, i.e. the fittest offspring will survive in the next GA iteration. The mutation operator picks a random amino acid from a randomly chosen sequence in the alignment and checks whether one of its neighbors has a gap. If this is the case, the algorithms swaps the selected amino acid with a gap neighbor. To generate the population of the next iteration, old chromosome are replaced with new offspring by using fitness function. Unlike a pure GA, after this step Omar et al.'s hybrid algorithm [35] performs a further step, called *Aligning Improver* that tries to improve alignment quality from a single solution produced from GA. In more detail, if the fitness value is less than a threshold value (e.g. $< 70\%$ fitness), the algorithm proceeds to a simulated annealing phase to align a solution produced from the GA phase. The invoked SA uses the same representation and the same fitness function used in GA phase and defines a *swap* neighborhood. Initially, acid amino with gaps are picked at random from a random sequence and the temperature should be high enough to melt the system completely (the authors set $T = 10$). Then, T should be reduced towards its freezing point as the algorithm proceeds in the search in the feasible solution space. Initially, the authors defined a parameter $COOLING - RATE = 0.9$ and at the end of each iteration, T is cooled down by setting $T = COOLING - RATE * T$. In their paper, the authors discuss how the integration of SA into a GA framework can help to escape from local minima problem compared to dynamic programming.

3. CONCLUSIONS AND OPEN QUESTIONS

A large number of problems in molecular biology can be formulated as combinatorial optimization problems. In fact, the applicability of operations research models and methods in molecular biology is a rapidly growing and fruitful area in the cross section of several different scientific disciplines.

Still many topics of interest remain open for further research in this area. Two main basic open questions regard multiple alignment problems. First of all, it would be interesting to investigate whether there exists any algorithm with further better approximation guarantee. Secondly, to approach large scale instances of this family of molecular biology problems it would be fruitful to study the applicability of heuristic and metaheuristic techniques other than SA and GA.

REFERENCES

1. H. Greenberg, W. Hart, and G. Lancia, *INFORMS Journal of Computing* **16**, 211–231 (2004).
2. D. Gusfield, K. Balasubramanian, and D. Naor, *Journal of Algorithmica* **12**, 312–326 (1994).
3. T. Smith, and M. Waterman, *Journal of Molecular Biology* **147**, 195–197 (1981).
4. L. Wang, and T. Jiang, *Journal of Computational Biology* **1**, 337–348 (1994).
5. D. Sankoff, and J. B. K. (eds), *Time Warps, String Edits and Macromolecules: The Theory and Practice of Sequence Comparison*, Addison-Wesley, 1983.
6. D. Sankoff, C. Morel, and R. J. Cedergren, *Nature New Biology* **245**, 232–234 (1973).
7. H. Carrillo, and D. Lipman, *SIAM Journal of Applied Mathematics* **48**, 1073–1082 (1988).
8. J. Kececioglu, H.-P. Lenhof, K. Mehlhorn, P. Mutzel, and K. Reinelt, *Discrete Applied Mathematics* **104**, 143–186 (2000).
9. J. Kececioglu, "The maximum weight trace problem in multiple sequence alignment," in *Proceedings of the Annual Symposium on Combinatorial Pattern matching (CPM)*, 1993, vol. 684 of Lecture Notes in Computer Science, pp. 106–119.
10. K. Reinelt, H.-P. Lenhof, P. Mutzel, K. Mehlhorn, and J. Kececioglu, "A branch-and-cut algorithm for multiple sequence alignment," in *Proceedings of the Annual International Conference of Computational Molecular Biology (RECOMB)*, 1997, pp. 241–249.
11. D. H. (editor), *Approximation algorithms for NP-hard problems*, PWS Publishing, 1996.
12. V. Vazirani, *Approximation algorithms*, Springer-Verlag, 2001.
13. D. Gusfield, *Bulletin of Mathematical Biology* **55**, 141–154 (1993).
14. V. Bafna, E. Lawler, and P. Pevzner, *Theoretical Computer Science* **182**, 233–244 (1997).

15. B. Wu, G. Lancia, V. Bafna, K. Chao, R. Ravi, and C. Tang, *SIAM Journal on Computing* **29**, 761–778 (1999).
16. J. Hein, *Methods in Enzymology* **183**, 626–645 (1990).
17. A. Krogh, M. Brown, I. Mian, K. Sjölander, and D. Haussler, *Journal of Molecular Biology* **235**, 1501–1531 (1994).
18. R. Korf, and W. Zhang, "Divide-and-Conquer Frontier Search Applied to Optimal Sequence Alignment," in *Proceedings of the National Conference on Artificial Intelligence*, 2000, vol. 17, pp. 910–916.
19. E. Dijkstra, *Numer. Math.* **1**, 269–271 (1959).
20. R. Ahuja, T. Magnanti, and J. Orlin, *Network Flows: Theory, Algorithms and Applications*, Prentice-Hall Englewood Cliffs, 1993.
21. G. Gallo, and S. Pallottino, *Math. Programming Study* **26**, 38–64 (1986).
22. G. Gallo, and S. Pallottino, *Ann. Oper. Res.* **7**, 3–79 (1988).
23. P. Festa, "Shortest path algorithms," in *Handbook of Optimization in Telecommunications*, edited by P. Pardalos, and M. Resende, Springer, 2006, pp. 185–210.
24. P. Festa, *Shortest path algorithms*, Aracne Editrice, 2007, iSBN: 978-88-548-1482-0.
25. B. Cherkassky, A. Goldberg, and T. Radzik, *Mathematical Programming* **73**, 129–174 (1996).
26. G. Lugar, and W. Stubblefield, *Artificial Intelligence Structures and Strategies for Complex Problem Solving*, Addison Wesley Longman, 1998.
27. C. Korostensky, and G. Gonnet, "Near Optimal Multiple Sequence Alignments Using a Traveling Salesman Problem Approach," in *SPIRE '99: Proceedings of the String Processing and Information Retrieval Symposium & International Workshop on Groupware*, IEEE Computer Society, Washington, DC, USA, 1999, p. 105, ISBN 0-7695-0268-7.
28. S. Kirkpatrick, C. Gellat, and M. Vecchi, *Science* **220**, 671–680 (1983).
29. J. Holland, *Adaptation in Natural and Artificial Systems*, Univ. of Michigan Press, Ann Arbor, Mich., USA, 1975.
30. D. Goldberg, *Genetic Algorithms in Search, Optimization, and Machine Learning*, Addison-Wesley, 1989.
31. J. Koza, F. B. III, D. Andre, and M. Keane, *Genetic Programming III, Darwinian Invention and Problem Solving*, Morgan Kaufmann Publishers, 1999.
32. M. Ishikawa, T. Toya, and Y. Tokoti, "Parallel Iterative Aligner with Genetic Algorithm," in *13th International Conference on Artificial Intelligence*, 1993, pp. 13–22.
33. C. Notredame, and D. Higgins, *Nucleic Acids Res.* **24**, 1515–1524 (1996).
34. C. Zhang, and A. Wong, *Comput. Appl. Biosci.* **13**, 565–581 (1997).
35. M. F. Omar, R. Salam, R. Abdullah, and N. Rashid, *Int. Journ. of Computational Intelligence* **1**, 81–89 (2004).
36. S.-M. Chen, and C.-H. Lin, *Information and Management Sciences* **18**, 97–111 (2007).

Combining Experimental Evidences from Replicates and Nearby Species Data for Annotating Novel Genomes

Claudia Angelini*, Luisa Cutillo†, Italia De Feis*, Pietro Liò** and Richard van der Wath**

*Istituto per le Applicazioni del Calcolo "Mauro Picone" (CNR), Napoli, Italy
†Telethon Institute of Genetics and Medicine, Napoli, Italy
**Computer Laboratory, University of Cambridge, Cambridge, United Kingdom

Abstract. For several years now, there has been an exponential growth of the amount of life science data (e.g., sequenced complete genomes, 3D structures, DNA chips, Mass spectroscopy data) generated by high throughput experiments. Carrying out analyses of complex, voluminous, and heterogeneous data and guiding the analysis of data using a statistical and mathematical sound methodology is thus of paramount importance. Here we make and justify the observation that experimental replicates and phylogenetic data may be combined to strength the evidences on identifying transcriptional motifs, which seems to be quite difficult using other currently used methods. We present a case study considering sequences and microarray data from fungi species. Although we show that our methodology can result of immediate practical utility to bioinformaticians and biologists for annotating new genomes, here the focus is also on discussing the dependent interesting mathematical problems that high throughput data integration poses.

Keywords: Bayesian variable selection, DNA regulatory motifs, gene expression, phylogenetic tree.
PACS: 81.10.Rt, 87.18.Vf, 87.18.Wd

1. INTRODUCTION

Good practice in science is to replicate experiments in that reproducibility allows statistical support of the results. This goes back to the Galileo's principle of '*Provare e riprovare*' (test and test again)[11]. When experiments are costly, particularly in high throughput biology, replicates are in the smallest number to assure 'above' the threshold statistical reliability for disseminating and publishing results; funding constraints sometimes result in seriously hampering statistical robustness. Each organism has homeostatic mechanisms that maintain the genetic information and its expression which allow to replicate experiments if the conditions remain the same. In molecular biology it is often difficult to retain the exact experimental setup but close conditions are widely accepted. In a field where technology changes constantly and at great pace, results that are few months distant may be based on slightly different manufacturing of few of a dozen of components and show different accuracy. The biological samples used in the experiments, i.e. culture cell lines, for examples bacterial or yeast cells which have short generation times, although kept in the same medium, may change because of mutations and contamination during their periodical proliferation. Cells may have minimal different concentration of constituents such as nucleic acid, proteic, lipid or sugar factors, ions,

CP1028, *Collective Dynamics: Topics on Competition and Cooperation in the Biosciences*
edited by L. M. Ricciardi, A. Buonocore, and E. Pirozzi

giving them different fitness with respect other colonies. However, a large body of experimental evidences in comparative genomics is showing that recently diverged species may retain similarities in gene sequence, expression and genome organization. Examples on using information from different species comes from comparative genomics. The current methodology in annotating genomes uses pair hidden Markov approach [7, 8] to take advantage of the already sequenced genomes. These considerations suggest that, in absence of experimental replicates, or in addition to these, statistical support to experimental evidences may also be searched by analyzing close variants of the species under examination or phylogenetic nearby species, i.e. species which have recently diverged. Of course the validity of the assumption of replicas through phylogenetical relatedness should be proved by copious experimental evidences or literature. Such useful exploitation of species richness is hampered by the lack of any theoretical statistical framework leading to combine the knowledge from true replicas and the nearby species replicas.

This paper is attempting to remedy to this problem. A meaningful example of the difficulty of retaining the exact experimental conditions and of the potential availability of several quasi replicas in nearby species is the analysis of gene expression at genome level. Cells have evolved a highly interconnected transcriptional network composed of signaling molecules, transcription factors (TFs), and their DNA target [18]. The mRNA expression level of a gene, which is measured using microarray experiments, is typically determined by several input signals, through the regulatory logic encoded in its noncoding regulatory DNA sequences embedded in the upstream regions of the genes [2]. An important and fortuitous aspect of transcriptional network organization is that large sets of genes tend to be coexpressed at the mRNA level [9], consistent with the notion that many cellular processes require the simultaneous participation of many gene products. Comparative analysis has shown that the coexpression of many of these genes is conserved across diverse species [9, 14], but little is known about the underlying mechanisms by which these genes are regulated.

Here we describe a new systematic genome-wide statistical approach for identifying overrepresented DNA sequence elements, or motifs, of newly sequenced species, from gene expression data of nearby species. In particular we make use of the phylogenetic tree to assess the distance among the species, then we use the Bayesian variable selection to combine the information on the DNA sequences of the species under analysis with the genome expression information of other sufficiently closed species from which several experimental results are available. Pooling information across studies can help to more accurately identify the true target genes, as pointed out in [5], allowing both to share the final cost of the analysis and to use already available data which are contained in classical repositories.

The paper is organized as follows: in Section 2 we first present a case study which focuses on yeast species; the phylogenetic relationship between these species is analyzed and used to identify those species that are enough close to potentially share similarity in gene expression and motifs locations; in Section 3 we describes the statistical methodology, based on the variable selection, whose results will be presented and discussed in Section 4. Conclusions will be drawn in Section 5.

2. CASE STUDY

A first attempt of annotation crossing from one species to another one was proposed in [1], considering the fungi *S. Pombe* and *S. Japonicus* organisms. Fungi are rather simple organisms with respect to human genome. Although experimentally characterized and computationally predicted binding sites evolve slower than surrounding sequence [16, 23], their short generation time make them accumulate lot of mutations and diverge very fast [28]. In [1] we focused our attention on the ENG1 cluster, a set of very strongly cell cycle-regulated genes, which in *S. Pombe* contains eight genes, involved in cell separation [24]. The genes are `adg1` and `adg2` (cell surface glycoproteins), `adg3` (glucosidase), `agn1` and `eng1` (glycosyl hydrolases), `cfh4` (chitin synthase regulatory factor), `mid2` (an anillin needed for cell division and septin organization), and `SPCC306.11`, a sequence orphan of unknown function. Motif searches showed that each gene of the cluster has at least one binding site for the `Ace2` transcription factor (consensus CCAGCC).

We first assessed the distance between *S. Pombe* and *S. Japonicus* by means of a phylogenetic tree. A maximum likelihood tree, based on the ENG1 protein showing the position of *S. Japonicus* with respect other fungi is shown in the Figure 1 where we used the Jones, Taylor and Thornton amino acid substitution model (JTT model) of evolution (see [15, 34]) due to the fact that the ENG1 protein family are globular cytoplasmic proteins. The phylogeny shows the short distance between *S. Pombe* ENG1 protein and *S. Japonicus* (branch length is in 10 amino acid replacement units) with respect to the other fungi genomes. Note that *S. Japonicus S. Pombe* ENG1 and ENG2 proteins have clearly diverged although they retain a high sequence similarity. The divergence is larger for *S. Cerevisiae* ACF1 and ACF2 proteins, pointing to the difficulty of predicting the similarity among genes in similar species.

Then we proposed a statistical methodology for finding DNA binding sites in *S. Japonicus* based on bayesian variable selection, which is an extension of that from [31]. While the authors of [31] showed that variable selection is more effective than the linear regression used by [4], in [1] we extended their procedure to the use of gene expression from different species/strains using a single experiment. Here we further extend the methodology pooling together both phylogenetically closely related species and several microarray studies carried out in different laboratories and with multiple sources of replications available. Indeed biologists often conduct multiple but different studies that all target the same biological system or pathway and within each study, replicate slides within repeated measurements under the same experimental conditions are often produced which can be used for improving and validating the statistical inference.

The statistical procedure for annotation crossing from *S. Pombe* to *S. Japonicus* will be described in the next section.

3. METHODOLOGY

Our methodology goes through the following steps: 1) selection of a group of co-regulated genes in *S. Pombe* which represents the species for which biological results are already available and determination of the species to infer by means of a phylo-

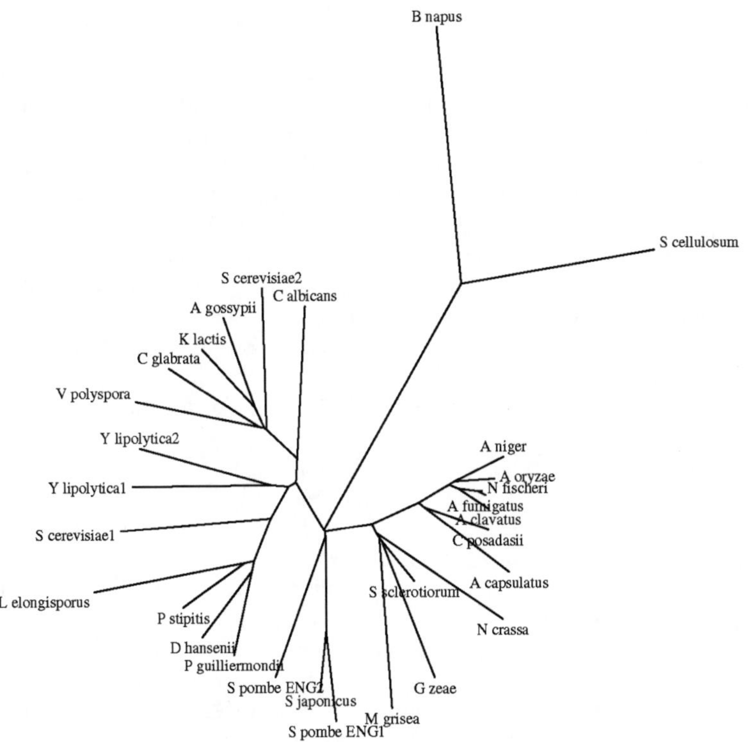

FIGURE 1. Maximum Likelihood tree inferred using ENG1 protein sequence from the following species: *S. Japonicus, S. Cerevisiae, S. Pombe, Kluyveromyces lactis, Debaryomyces hansenii, Candida albicans, Yarrowia lipolytica, Aspergillus oryzae, Phaeosphaeria nodorum, Neurospora crassa, Vanderwaltozyma polyspora Neosartorya fischeri Pichia guilliermondii,Coccidioides posadasii, Gibberella zeae, Ashbya gossypii, Sclerotinia sclerotiorum, Magnaporthe grisea, Ajellomyces capsulatus, Aspergillus clavatus, Aspergillus niger, Pichia stipitis, Lodderomyces elongisporus, Candida glabrata*; *Brassica napus* and *Sorangium cellulosum* are plant sequences used as outgroups, i.e. to facilitate the rooting of fungi phylogeny; we also include *S. Japonicus* ENG1 and ENG2 proteins and *S. Cerevisiae* ACF1 and ACF2 proteins.

genetic tree built on the selected genes; 2) identification of the homologous genes in *S. Japonicus*, i.e., the specie currently under investigation, using BLAST [10]; 3) choice of a set of about 200 biologically independent genes of *S. Japonicus* after a phylogenetic analysis; 4) determination and scoring a set of about 150 candidate motifs using the software MDSCAN proposed in [20]; 5) selection of few "best motifs" using Bayesian variable selection, a procedure which allows the user to account for the different sources of replications available. The details for steps 1)-5) will be given in the next subsections.

3.1. Standard methodologies for Likelihood Calculations, statistical test for Model Comparison and motif selection

As regard point 1), the likelihood of a tree (the hypothesis) is equal to the probability of observing the data if that hypothesis were correct. The observed data is usually taken to be the sequence alignment, although it would of course be more reasonable to say that the sequences are what have been observed and the alignment should then be inferred along with the phylogeny. The statistical method of maximum likelihood chooses amongst hypotheses by selecting the one which maximizes the likelihood; that is, which renders the data the most plausible. In the context of molecular phylogenetics, a model of nucleotide or amino acid replacement across each column of the alignment, permits the calculation of the likelihood for any possible combination of tree topology and branch lengths. All likelihood calculations were based on standard methods described, for example, by [30] for models of evolution using a replacement matrix for all sites. Likelihood maximization and maximum likelihood parameter estimation were performed by numerical optimization routines. Likelihood ratio tests, see [13, 19, 33], were used to compare competing pairs of evolutionary models. Our LRTs were based on calculation of $\Delta l = l_1 - l_0$, where l_1 and l_0 are the maximum log-likelihoods for the null and alternative hypotheses, respectively. Good performance in phylogenetic estimation is expected to follow from accurate models of sequence evolution, particularly when the phylogenetic history has involved biologically complex patterns of evolutionary dynamics [21]. Thus, the fitting of the evolutionary model to the data is also a test of hypotheses on the biological assumptions.

Point 2) is based on a standard and widely accepted tool in bioinformatics, see for example [10].

The gene selection in point 3) is guided from the fact that extensive comparative genomic analysis has revealed that all the eukaryotic genomes contain families of duplicated genes which have recently diverged. In many cases these families have retained large par of the upstream regulatory sequences. In particular the remnants of whole genome duplications have been identified in different yeast strains [17, 29, 35] as well as in other species. The redundancy of yeast genome suggest we should select a meaningful non redundant ensemble of genes that contains all the relevant statistical characteristics of the genome. In order to select this set of genes we have performed a phylogenetic analysis of a manually cured (by one of the authors) data set of genes. The phylogenetic analysis, based on neighbor joining which is a distance method, allows to identify the genes which have very large sequence similarities and therefore may derive from a common ancestor. All members except one of these groups of very similar genes were pooled out. Eventually we end up with a number of approximately 200 genes. In [6] it is showed that a 3rd order Markov chain based on similar number of genes and promoter sequences (about 300 kb) is quite accurate in estimating oligonucleotide frequencies. Also in [12] and [22] the transcriptional regulatory code was analyzed using a similar number of yeast proteins.

As concern point 4) of the procedure, we used MDSCAN described in [20] to search for nucleotide patterns which appears in the upstream sequences of the genes of interest. The algorithm starts by enumerating each segment of width w (seed) in the top t

sequences. For each seed, it looks for w-mers with at least n base pair matches in the t sequences. These are used to form a motif matrix and the highest scoring seeds are retained. The updating step is done iteratively by scanning all w-mers in the remaining sequences and adding in or removing from the weight matrix segments that increase the score. This is repeated until the alignment stabilizes. The score for each gene and each candidate motif was calculated as in [1, 4, 31] using the following equation

$$X_{mg} = \log_2 \left[\sum_{x \in X_{wg}} \Pr(x \,\text{from}\, \theta_m) / \Pr(x \,\text{from}\, \theta_0) \right]$$

where $m = 1, \ldots, M$, M number of motifs; $g = 1, \ldots, G$, G number of genes; θ_m is the probability matrix of motif m of width w, and the background model θ_0 was computed using a Markov chain of the third order either from the intergenic regions of *S. Japonicus* or its complete genome.

The main novelty of the present paper is on point 5) where we fit a linear regression model relating gene expression levels ($Y^e \in R^{G \times S_e}$), where S_e indicates the number of technical replicates for each experiment e, $e = 1, \ldots, E$, performed on G genes of *S. Pombe*, to the pattern scores (X) of dimension $G \times M$, M being the number of candidate motifs, evaluated on *S. Japonicus*. The proposed method is described in the following section.

3.2. Bayesian Statistical model

In this section we describe a Bayesian variable selection method that takes into account the different and multiple sources information available, pools together results of several experiments and allow the users to select the motifs that best explain and predict the changes in expression level in a group of co-regulated genes.

Suppose that E similar microarray experiments have been carried out on a set of G genes in a given species in order to investigate the same biological pathway and that for each experiment e ($e = 1, \ldots, E$) S_e technical replicates are available.

We assume the following model

$$y_{ges} \mid \mu_{ge}, \sigma_e^2 \sim \mathrm{N}\left(\mu_{ge}, \sigma_e^2\right) \quad g = 1, \ldots, G;\ s = 1, \ldots, S_e;\ e = 1, \ldots, E \qquad (1)$$

where y_{ges} represents the observed gene expression value of the gene g in the eth experiment and the sth replicate and μ_{ge} is the true expression of gene g under the experimental condition e.

A latent vector, $\gamma \in R^M$, with binary entries, is introduced to identify variables included in the model; γ_m takes on value 1 if the mth variable (motif) is included and 0 otherwise. Let $m\gamma = \sum_{i=1}^{M} \gamma_i$.

The true gene expression value μ_{ge} is connected to a specific subset of the M candidate motifs identified by the latent vector γ by the following relation

$$\mu_{ge}|\gamma = \sum_{\{m:\gamma_m=1\}} x_{gm}\beta_{me} = X_{g.(\gamma)}\beta_{e(\gamma)},$$

with $X_{g.(\gamma)}$ row of $X\gamma$ referring to gene g for all the included motifs.

For the coefficients and the variance the following priors have been elicitated

$$\beta_{e(\gamma)} \quad | \quad \sigma_e^2, \gamma \sim \mathrm{N}_{m\gamma}\left(0, \sigma_e^2 H_{(\gamma)}\right) \tag{2}$$

$$\sigma_e^2 \quad \sim \quad \mathrm{IG}(\nu, S) \tag{3}$$

where $H_{(\gamma)}, \nu$ and S need to be assessed through a sensitivity analysis. We specify Bernoulli priors for the elements of γ:

$$p(\gamma) = \prod_{j=1}^{M} \theta^{\gamma_j}(1-\theta)^{1-\gamma_j} \tag{4}$$

where $\theta = m_{\text{prior}}/M$ and m_{prior} is the number of covariates expected *a priori* to be included in the model.

Model (1) means that

$$Y_{.s}^e \sim \mathrm{N_G}\left(X_{(\gamma)}\beta_{e(\gamma)}, \sigma_e^2 I\right)$$

where $Y_{.s}^e = (y_{1s}^e, \ldots, y_{Gs}^e)'$, $s = 1, \ldots, S_e$ and $e = 1, \ldots, E$. Note that without loss of generality we further assume that the columns of X and $Y_{.s}^e$ are mean-centered.

Having set the prior distributions, a Bayesian analysis proceeds by updating the prior beliefs with information that comes from the data. After some standard algebra, the posterior distribution of the vector γ given the data, i.e., $f(\gamma|X, Y^1, \ldots, Y^E)$, can be obtained:

$$f(\gamma|X, Y^1, \ldots, Y^E) \propto \prod_{e=1}^{E} a_e \frac{|H_{(\gamma)}|^{-1/2} |K_{(\gamma)}^e|^{-1/2}}{\left(c_e - M_e' \left(K_{(\gamma)}^e\right)^{-1} M_e + S\right)^{(GS_e+\nu)/2}} \tag{5}$$

with

$$\begin{aligned}
K_{(\gamma)}^e &= \left(S_e X_{(\gamma)}' X_{(\gamma)} + H_{(\gamma)}^{-1}\right) \\
c_e &= \sum_{s=1}^{S_e} (Y_{.s}^e)'(Y_{.s}^e) \\
M_e &= X_{(\gamma)}' \left(\sum_{s=1}^{S_e} Y_{.s}^e\right) \\
a_e &= (1/2\pi)^{GS_e/2} S^{\nu/2} \left[\Gamma((GS_e+\nu)/2) 2^{(GS_e+\nu)/2}\right] / \left[\Gamma(\nu/2) 2^{\nu/2}\right]
\end{aligned}$$

and the analysis can them be carried out as showed in Section 3.2.1.

We observe that model (1)-(2)-(3)-(4) can be generalized in order to handle the presence of missing data which are typically encountered when analyzing real data experiment. For each fixed experiment $e = 1, \ldots, E$ and each fixed technical replicate

$s = 1,\ldots,S_e$, let $G_{e,s}$ the number of genes whose values are measured on the array. In this case the posterior becomes

$$f(\gamma|X_{e,s},Y^e_{.s},s=1,\ldots,S_e;e=1,\ldots,E) \propto \prod_{e=1}^{E} a_e \frac{|H_{(\gamma)}|^{-1/2}|K^e_{(\gamma)}|^{-1/2}}{\left(c_e - M_e'\left(K^e_{(\gamma)}\right)^{-1} M_e + S\right)^{(G_e+\nu)/2}} \tag{6}$$

with

$$\begin{aligned}
G_e &= \sum_{s=1}^{S_e} G_{e,s} \\
K^e_{(\gamma)} &= \left(\sum_{s=1}^{S_e} X'_{e,s,(\gamma)} X_{e,s,(\gamma)} + H^{-1}_{(\gamma)}\right), \quad X_{e,s} \text{ of dimension } G_{e,s} \times M \\
c_e &= \sum_{s=1}^{S_e} (Y^e_{.s})'(Y^e_{.s}), \quad Y^e_{.s} \text{ of dimension } G_{e,s} \times 1 \\
M_e &= \sum_{s=1}^{S_e} X'_{e,s,(\gamma)} Y^e_{.s} \\
a_e &= (1/2\pi)^{G_e/2} S^{\nu/2} \left[\Gamma((G_e+\nu)/2) 2^{(G_e+\nu)/2}\right] / \left[\Gamma(\nu/2) 2^{\nu/2}\right].
\end{aligned}$$

Vector values with high posterior probability (5)-(6) identify the most promising sets of candidate motifs, and the sampling procedure to determine them will be presented in the next section.

3.2.1. Stochastic search

Given the large number of possible vector values (2^M possibilities with M covariates), we use a stochastic search Markov Chain Monte Carlo (MCMC) technique to search for sets with high posterior probabilities.

The method visits a sequence of models that differ successively in one or two variables. At each iteration, a candidate model, γ^{new}, is generated by randomly choosing one of these two transition moves:

(i) Add or delete one variable from γ^{old}.

(ii) Swap the inclusion status of two variables in γ^{old}.

The proposed γ^{new} is accepted with a probability that depends on the ratio of the relative posterior probabilities of the new versus the previously visited models:

$$\min\left\{\frac{f(\gamma^{\text{new}}|X,Y^1,\ldots,Y^E)}{f(\gamma^{\text{old}}|X,Y^1,\ldots,Y^E)},1\right\}, \tag{7}$$

which leads to the retention of the more probable set of patterns, see [3, 31]. An analogous formula is obtained considering the posterior probability given by formula (6).

The stochastic search results in a list of visited sets (i. e. combination of candidate motifs) and the corresponding relative posterior probabilities, then the selection of few "best motifs" can be done either using the global MAP principle or by selecting the covariates on the basis of their marginal probability to be included. Indeed the marginal posterior probability of inclusion for a single motif j, $P(\gamma_j = 1|X,Y^1,\ldots,Y^E)$, can be computed from the posterior probabilities of the visited models:

$$\begin{aligned} f(\gamma_j = 1|X,Y^1,\ldots,Y^E) &= \int f\left(\gamma_j = 1, \gamma_{(-j)}|X,Y^1,\ldots,Y^E\right) d\gamma_{(-j)} \\ &\propto \int f\left(Y^1,\ldots,Y^E|X,\gamma_j = 1,\gamma_{(-j)}\right) \cdot f(\gamma) d\gamma_{(-j)} \\ &\approx \sum_{t=1}^{NI} f\left(Y^1,\ldots,Y^E|X,\gamma_j = 1,\gamma_{(-j)}^{(t)}\right) \cdot f\left(\gamma_j = 1,\gamma_{(-j)}^{(t)}\right), \end{aligned} \tag{8}$$

where $\gamma_{(-j)}^{(t)}$ is the vector γ at the tth iteration without the jth motif.

4. RESULTS

Five experiments have been chosen for the analysis of the regulatory mechanism of cluster ENG1: ELUTRIATION A described in [24] (`http://www.redgreengene.com/oliva$\_$plos$\_$2005/index.html`) and ELUTRIATION 1, ELUTRIATION 2, ELUTRIATION 3 and CDC25 BLOCK RELEASE 1 described in [27] (`http://www.sanger.ac.uk/PostGenomics/S$\_$pombe/projects/cellcycle/`). All the experiments explore the transcriptional responses of the fission yeast *S. Pombe* to cell cycle and measure gene expression as a function of time in cells synchronized through different approaches: centrifugal elutriation and the use of temperature sensitive cell cycle mutants. All these experiments have no technical replicates, so formulas (5)-(6) simplify because $S_e = 1$ for $e = 1,\ldots,5$. The sequence data consists of the *S. Japonicus*' genome (`http://www.broad.mit.edu`).

From the *S. Japonicus* genome 174 genes have been selected in order to represent most of the biological functions in the cell and to result almost biologically independent; and their homologues in *S. Pombe* have been determined. We added 7 of the 8 genes of the ENG1 cluster to this selection (`SPCC306.11` has been removed since the observed values were missing in 4 out of 5 experiments). Hence a total of 181 genes represented the response Y^e in the variable selection procedure. The value y_{ge}, for gene g, $g = 1,\ldots,181$, and experiment e, $e = 1,\ldots,5$, represents the average of the genes expression values measured on the microarray in the interval where the ENG1 genes show their common activation peak, approximately 30 minutes-90 minutes.

The prior parameters have been chosen in the following way: $H_{(\gamma)} = c\,[\text{diag}(X'X)^+]_{(\gamma)}$, with $c = 0.05$ to be comparable in size with the variability of the regression coefficients of the full model averaged over the experiments; $\nu = 3$, to give weak prior knowledge, since it corresponds to the smallest integer such that the expectation of σ_e^2, $\text{E}(\sigma_e^2) = \nu/(\nu-2)S$, exists, and the scaling value $S = 0.16$ to be comparable in size with the variability of the data averaged over the experiments. We repeated the analysis with several values of $m_{\text{prior}} = 1,3,5$ to study the effects of the sparsity request on the models selection. We have chosen these values for m_{prior} because fungi are rather simple organisms, and it is known that the regulation mechanism is based on few motifs.

As regards the evaluation of the score matrix X, *S. Japonicus* sequences up to 1000 bp upstream have been extracted, shortening them, if necessary, to avoid any overlap with adjacent ORF's. For genes with negative orientation, this was done taking the reverse complement of the sequences. Indeed the motif finding algorithms are sensitive to noise, which increases with the size of upstream sequences examined and, as reported in [32], the vast majority of the yeast regulatory sites from the TRANSFAC database are located within 800 bp from the translation start site. Moreover nucleotide patterns of length 5 to 12 bp have been searched and up to 30 distinct candidates for each width have been considered. This has originated 146 candidate motifs to be used in the regression.

We have run 10 parallel MCMC chains of length 100.000. The searches were started with a randomly selected γ^{start}. We pooled together the sets of patterns visited by the MCMC chains and computed the normalized posterior probabilities of each distinct visited set. We also computed the marginal posterior probabilities for the inclusion of single nucleotide patterns.

Figures 2, 3, 4 show the marginal posterior probability computed from *S. Japonicus* for $m_{\text{prior}} = 1,3,5$. In the figure the x-axes correspond to the pattern indices and the y -axes correspond to the marginal posterior probability. The spikes indicate patterns included in the model with high probability. Table 1 shows the motifs whose marginal posterior probability is greater than 0.3 for $m_{\text{prior}} = 1,3,5$, respectively. We can notice that the results are robust with respect to the choice of prior parameters. Indeed five motifs (**CCAGCC**TCGT, CTTCCGATCCG, GA**CCAGCC**TAG, CAGATTTCGTGC, TACGTTTAAGCT) show a high marginal posterior probability (greater than 0.9) and all of them are contained in the *best* model according to the MAP principle, for all the choice of m_{prior}.

In order to study the robustness of the proposed methodology and the influence of each single experiment on the model selection, we have run the leave one out cross validation strategy over the experiments. For each choice of m_{prior} we have carried out the variable selection analysis only on 4 experiments, out of 5, in all possible combinations. Table 2 shows, for each of the 11 motifs listed in Table 1, the interval whose bounds are the minimum and the maximum value of the marginal probability over cross validation experiments for each fixed m_{prior}. As we expected, the motifs with less interval variability are the ones with the higher marginal posterior probability, as shown in Table 1. Note that long patterns are more often selected than short ones. This corresponds to the background model not so efficiently rejecting association among nearby DNA bases. Eukaryotic DNA is highly heterogeneous, patchy and repetitious [25] and state of art background models cannot adequately take into account for the variations in base association. Note also the high values of the marginal probabilities that become much higher (about 3 fold) than those obtained using single replicates and one species [1], stressing the importance of considering new mathematical and statistical techniques able to pool all the available information about one species.

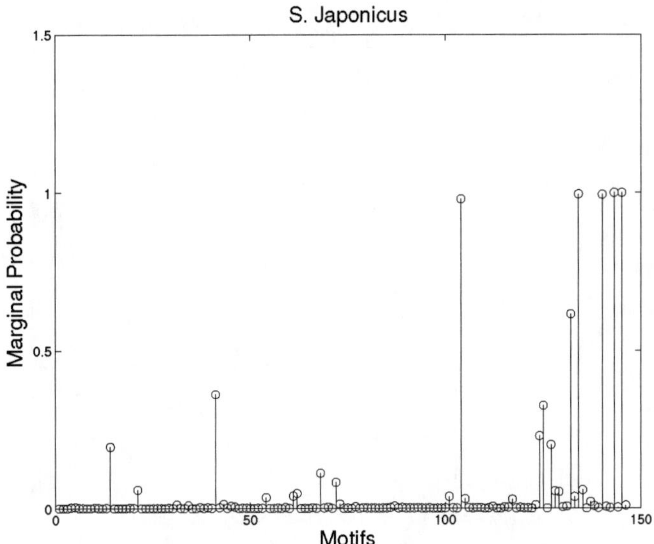

FIGURE 2. Marginal posterior probabilities for $m_{\text{prior}} = 1$.

TABLE 1. Motifs and related marginal posterior probabilities > 0.3 for $m_{\text{prior}} = 1, 3, 5$.

Motifs	$m_{\text{prior}} = 1$	$m_{\text{prior}} = 3$	$m_{\text{prior}} = 5$
GGAAA	-	0.6615	0.9129
ACCAGCC	0.3611	0.4840	0.5345
CCAGCCTCGT	0.9801	0.9975	0.9984
AGACAAAGGAA	-	0.3890	0.5393
ATTCCCTTTCT	0.6153	0.9194	0.9779
CTTCCGATCCG	0.9951	0.9992	0.9999
GA**CCAGCC**TAG	0.9937	0.9995	0.9995
GGCTGGCACAC	0.3251	0.4748	0.5621
GTCTATATACG	-	0.3796	0.5630
CAGATTTCGTGC	0.9999	0.9999	1.0000
TACGTTTAAGCT	0.9999	1.0000	1.0000

5. DISCUSSION AND CONCLUSIONS

DNA sequences contain the information of the phylogenetic relationships among species and of the evolutionary processes that have caused the sequences to divergence. However the cell has developed sophisticated machines to assure a certain degree of fidelity of the conservation of the genetic information and the homeostasis of the physiological conditions. The mathematical and statistical methods try to capture these two types of information to determine how and why DNA and protein molecules work the way they do. Given the high cost of performing a large number of high throughput experimental

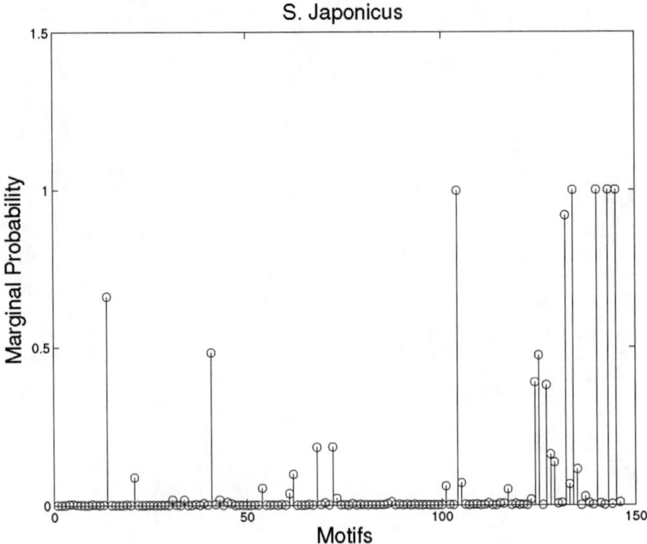

FIGURE 3. Marginal posterior probabilities for $m_{\text{prior}} = 3$.

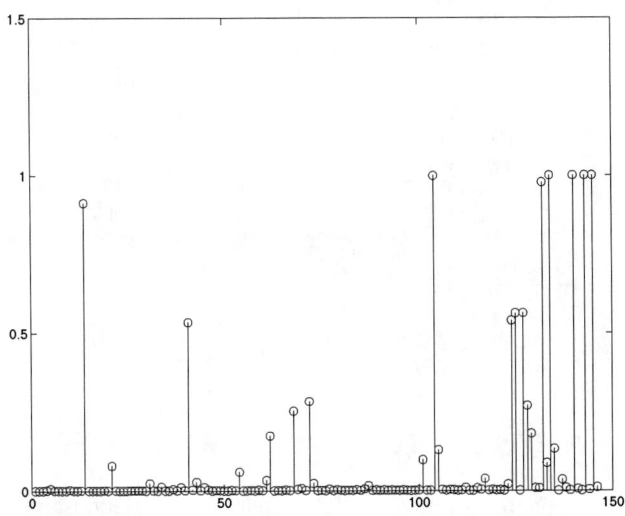

FIGURE 4. Marginal posterior probabilities for $m_{\text{prior}} = 5$.

TABLE 2. Motifs and related minimum and maximum marginal posterior probabilities over cross validation experiments for $m_{\text{prior}} = 1, 3, 5$.

Motifs	$m_{\text{prior}} = 1$	$m_{\text{prior}} = 3$	$m_{\text{prior}} = 5$
GGAAA	-	[0.0722, 0.5680]	[0.1568, 0.8422]
A**CCAGCC**	[0.1816, 0.3571]	[0.2484, 0.5123]	[0.2741, 0.6218]
CCAGCCTCGT	[0.7980, 0.9737]	[0.9853, 0.9984]	[0.9847, 0.9992]
AGACAAAGGAA	-	[0.1357, 0.4471]	[0.1628, 0.6001]
ATTCCCTTTCT	[0.2221, 0.4603]	[0.4503, 0.8492]	[0.6421, 0.9749]
CTTCCGATCCG	[0.9300, 0.9908]	[0.9853, 0.9984]	[0.9928, 0.9995]
GA**CCAGCC**TAG	[0.9258, 0.9848]	[0.9702, 0.9984]	[0.9875, 0.9996]
GGCTGGCACAC	[0.0906, 0.3051]	[0.1596, 0.4755]	[0.2140, 0.5473]
GTCTATATACG	-	[0.0947, 0.4636]	[0.1224, 0.5739]
CAGATTTCGTGC	[0.9993, 0.9999]	[0.9998, 0.9999]	[0.9999, 1.0000]
TACGTTTAAGCT	[0.9999, 1.0000]	[0.9999, 1.0000]	[0.9999, 1.0000]

replicates, we make the hypothesis that experimental evidences from species similar to that under analysis, may provide additional statistical support. Obviously the closer the species to the one under investigation, the better; if the two samples are just simple variants, the difference between experimental results may be very small and we may define the second species as a 'quasi replica'. With the aim of combining experimental replicates and phylogenetic data to identifying transcriptional regulatory motifs, we present a fungi case study for which both types of information is available. Although we show that this approach can result of immediate practical utility to bioinformaticians and biologists for annotating new genomes, we find important to also highlight the interesting mathematical problems raised by high throughput data integration problems.

5.1. Mathematical problems

The most important results we have obtained is to show that both considering more replicates or data from more species, the marginal probabilities become much higher (about 3 fold) than those obtained using single replicates and one species [1]. Would be better to add a species or a not-so-great quality replicate? Obviously we need to estimate the quality of an experimental replicate (the quality may be even different for different genes) and to estimate the quality of the replicate from a nearby species. Therefore, a first non trivial statistical problem is which phylogenetically close species to choose and how much support they may give. Note that species variants may be certainly better than another species. For instance schizosaccharomyces has two close variants (*pombe* and *japonicus*), *aspergillum* has three variants (*clavatus*, *niger*, *oryzae*). Species variants are much closer on the tree, reflecting that they have recently diverged. A criterion is to identify a distance threshold on the phylogenetic tree that would give some confidence on combining data. This would require generating the tree using phylogenetic advanced methodology, considering the same genes under investigation. Note that regulatory regions, although under purifying selection, are nevertheless diverging at higher rate than coding regions. In [26] the authors have demonstrated an inverse correlation between the rate of evolution of transcription factors and the number of genes that they regulate. Therefore, for small gene networks, distant species may not provide adequate support and in general the distance may depend on the size of the genetic network. Work in progress focuses on generating an empirical Bayes approach which would to determine such distance threshold. Finally we would like to devise a methodology to put replicates and other available species data in order of goodness of statistical support, such that a practitioner may datamine sequences and microarray resources to increase the evidences.

5.2. Biological problems

As a summary of results, we got both confirmation of known results and new findings (motifs) which have high marginal probability values. There are two important aspects of using replicates and other species data. Replicates allow to investigate the strength of a regulatory network in a species, i.e. small variations in gene expression may mean that the gene network is finely regulated and does not allow for so much natural variation. The inclusion of phylogenetic data tells us about how much that network has changed in different species in terms of number of genes involved, motif patterns, changes in expression. Indeed *S. Japonicus* show some changes in number of genes and motif patterns. Therefore an important result of including additional species is the insight that this analysis gives on the evolution the gene regulatory network under investigation.

ACKNOWLEDGMENTS

We acknowledge the Schizosaccharomyces Japonicus Sequencing Project at Broad Institute of Harvard and MIT (`http://www.broad.mit.edu`) and the CNR DG.RSTL.004.002 project.

REFERENCES

1. C. Angelini, L. Cutillo, I. De Feis, R. van der Wath, and P.Liò, "Identifying regulatory sites using neighborhood species", in *Evolutionary Computation, Machine Learning and Data Mining in Bioinformatics: 5th European Conference, EvoBIO 2007, Valencia, Spain, April 11-13, 2007, Proceedings*, edited by E. M. et al., Series: Lecture Notes in Computer Science 4447, 2007, pp. 1–10.
2. M. A. Beer, and S. Tavazoie, *Cell* **117**, 185–198 (2004).
3. P. J. Brown, M. Vannucci, and T. Fearn, *J. R. Stat. Soc. Ser. B* **60**, 627–641 (1998).
4. E. M. Conlon, X. S. Liu, J. D. Lieb, and J. S. Liu, *Proc. Natl. Acad. Sci. USA* **100**, 3339–3344 (2003).
5. E. M. Conlon, J. J. Song, and J. S. Liu, *BMC Bioinformatics* **7**, 247–250 (2006).
6. A. J. Cuticchia, D. A. Newsome, W. W. Jennings 3rd, and R. Ivarie, *Nucleic Acids Res.* **16**, 7145–7158 (1988).
7. I. M. Meyer, and R. Durbin, *Bioinformatics* **18**, 1309–1318 (2002).
8. I. M. Meyer, and R. Durbin, *Nucleic Acids Res.* **32**, 776–783 (2004).
9. M. B. Eisen, P. T. Spellman, P. O. Brown, and D. Botstein, *Proc. Natl. Acad. Sci. USA* **95**, 14863–14868 (1998).
10. W. J. Ewens, and G. R. Grant,"BLAST", in *Statistical Methods in Bioinformatics*, edited by K. D. et al., Springer-Verlag, New York, 2001, pp. 269–300.

11. G. Galilei, "Discorsi e dimostrazioni matematiche, intorno a due nuove scienze attenenti alla meccanica & i movimenti locali" (1638).
12. C. T. Harbison, D. B. Gordon, T. I. Lee, N. J. Rinaldi, K. D. Macisaac, T. W. Danford, N. M. Hannett, J. B. Tagne, D. B. Reynolds, J. Yoo, E. G. Jennings, J. Zeitlinger, D. K. Pokholok, M. Kellis, P. A. Rolfe, K. T. Takusagawa, E. S. Lander, D. K. Gifford, E. Fraenke, and R. A. Young, *Nature* **431**, 99–104 (2004).
13. J. P. Huelsenbeck, and B. Rannala, *Science* **276**, 227–232 (1997).
14. J. D. Hughes, P. W. Estep, S. Tavazoie, and G. M. Church, *J. Mol. Biol.* **296**, 1205–1214 (2000).
15. D. T. Jones, W. R. Taylor, and J. M. Thornton, *Comput. Appl. Biosci* **8**, 275–282 (1992).
16. M. Kellis, N. Patterson, B. Birren, B. Berger, and E. S. Lander, *J. Comput. Biol.* **11(2-3)**, 319–55 (2004).
17. M. Kellis, B.W. Birren, and E.S. Lander, *Nature* **428**, 617–624 (2004).
18. M. Levine, and R. Tjian, *Nature* **424**, 147–151 (2003).
19. P. Liò, and N. Goldman, *Genome Res.* **8**, 1233–1244 (1998).
20. X. S. Liu, D. L. Brutlag, and J. S. Liu, *Nat. Biotechnol.* **20**, 835–839 (2002).
21. P. J. Lockhart, D. Huson, U. Maier, M. J. Fraunholz, Y. Van de Peer, A. C. Barbrook, C. J. Howe, and M. A. Steel, *Mol. Biol. Evol.* **17**, 835–838 (2000).
22. K. D. MacIsaac, T. Wang, D. B. Gordon, D. K. Gifford, G. D. Stormo, and E. Fraenkel, *BMC Bioinformatics* **7:113** (2006).
23. A. M. Moses, D. Y. Chiang, M. Kellis, E. S. Lander, and M. B. Eisen, *BMC Evol. Biol.*, **3:19** (2003).
24. A. Oliva, A. Rosebrock, F. Ferrezuelo, S. Pyne, H. Chen, S. Skiena, B. Futcher, and J. Leatherwood, *PLoS Biol.* **3(7):e225**, 1239–1260 (2005).
25. F. Piazza, P. Liò, *Physica A* **347**, 472–488 (2005).
26. N. Rajewsky, N. D. Socci, M. Zapotocky, and E. D. Siggia, *Genome Res.* **12**, 298–308 (2002).
27. G. Rustici, J. Mata, K. Kivinen, P. Liò, C. J. Penkett, G. Burns, J. Hayles, A. Brazma, P. Nurse, and J. Bähler, *Nature Genet.* **36**, 809–817 (2004).
28. G. Rustici, H. van Bakel, D. H. Lackner, F. C. Holstege, C. Wijmenga, J. Bähler, and A. Brazma, *Genome Biol.* **8(5):R73** (2007).
29. C. Seoighe, and K. H. Wolfe, *Gene* **238**, 253–261 (1999).
30. D. L. Swofford, G. J. Olsen, P. J. Waddell, and D. M. Hillis, "Phylogenetic inference", in *Molecular systematics*, edited by D. M. H. et al., Sinauer, Sunderland, MA, 1996, pp. 407–514.
31. M. G. Tadesse, M. Vannucci, and P. Liò, *Bioinformatics* **20**, 2553–2561 (2004).
32. J. van Helden, B. Andr, and J. Collado-Vides, *J. Mol. Biol.* **281(5)**, 827–842 (1998).
33. Z. Yang, *J. Mol. Evol.* **39**, 306–314 (1994).
34. S. Whelan, P. Li, and N. Goldman, *Trends Genet.* **17(5)**, 262–272 (2001).
35. K. H. Wolfe, and D. C. Shields, *Nature* **387**, 708–713 (1997).

Analysis and Comparison of Information Theory-based Distances for Genomic Strings

Walter Balzano*, Ferdinando Cicalese†, Maria Rosaria Del Sorbo** and Ugo Vaccaro‡

*Dipartimento di Scienze Fisiche, Università di Napoli Federico II, Complesso Universitario di Monte Sant'Angelo Via Cintia, 80126 Napoli, Italy
†AG Genominformatik, Technische Fakultaet, Bielefeld Univeristaet, Germany
**Dipartimento di Matematica e Applicazioni, Università di Napoli Federico II, Complesso Universitario di Monte Sant'Angelo Via Cintia, 80126 Napoli, Italy
‡Dipartimento di Informatica ed Applicazioni, Università di Salerno, Via Ponte Don Melillo, 84084 Fisciano (SA), Italy

Abstract. Genomic string comparison via alignment are widely applied for mining and retrieval of information in biological databases. In some situation, the effectiveness of such alignment based comparison is still unclear, e.g., for sequences with non-uniform length and with significant shuffling of identical substrings. An alternative approach is the one based on information theory distances. Biological data information content is stored in very long strings of only four characters. In last ten years, several entropic measures have been proposed for genomic string analysis. Notwithstanding their individual merit and experimental validation, to the nest of our knowledge, there is no direct comparison of these different metrics. We shall present four of the most representative alignment-free distance measures, based on mutual information. Each one has a different origin and expression. Our comparison involves a sort of arrangement, to reduce different concepts to a unique formalism, so as it has been possible to construct a phylogenetic tree for each of them. The trees produced via these metrics are compared to the ones widely accepted as biologically validated. In general the results provided more evidence of the reliability of the alignment-free distance models. Also, we observe that one of the metrics appeared to be more robust than the other three. We believe that this result can be object of further researches and observations. Many of the results of experimentation, the graphics and the table are available at the following URL: http://people.na.infn.it/~wbalzano/BIO

Keywords: Alignment-free genomic string distance, information, entropy.
PACS: 87.10.-e, 89.70.Cf, 89.70.-a

1. INTRODUCTION

The total amount of life sciences data has an average doubling rate of six months. Just to give an idea of the orders of magnitude, the number of human genes is more than 3×10^4 involving more than 3×10^9 chemical letters; and it is estimated that at least 5×10^4 proteins and their average length is about 6×10^2 residues. This, only regarding some of the data related to humans. If one projects this numbers over the enormous variety of organisms under investigation, the above doubling rate appear as deemed to increase rather than diminish. The peculiar features of such data, their size and the necessity of highly reliable methodologies to extract significant information from them has given rise to a new branch in data mining, the genomic mining, exclusively devoted to the management of information hidden in genomic databases. In fact, the digital char-

CP1028, *Collective Dynamics: Topics on Competition and Cooperation in the Biosciences*
edited by L. M. Ricciardi, A. Buonocore, and E. Pirozzi

acter of biological sequence data, long chains of symbols from a quaternary alphabet, has a deep influence on the nature of algorithms proposed and applied for computational analysis of living materials [1]. The "comparison" between two or more strings is a widely explored branch in computational biology, since it plays a primary role in analysis, phylogeny and classification of organic entities [2] [3]. Discovering analogies or clear differences in biological sequences via comparison is also a basic tools in genome mining. Many sequence alignment methods, which assume conservation of contiguity between homologous segments, have been applied to deal with genomic strings. The underlying idea is to generate a distance matrix, based on models of nucleotide substitution or amino acid replacement (see, e.g., Jukes-Cantor [32], Kimura [33],Lake [34]. These methods need the alignment of the sequences, assume a kind of evolutionary model and suffer by high computational complexity. Moreover alignment cost criteria and evolutionary models are not univocal, do not scale well with the data size and for entire genome comparison might be far from efficient as the loss of contiguity of the homologous region might be beyond the possibility of alignment [4], [5], [6]. In phylogenetic analysis, whole genome-based methods are of primary interest because single gene sequences haven't enough information to outline the evolutionary history of organisms. In fact, non-uniform rates of evolution and horizontal gene transfer make phylogenetic analysis of species by single gene sequences complicated. The approaches we analyze in this paper share a common basic assumption: biological sequences codify information. Therefore evaluation of differences and similarity between sequences can be measured in terms of the information shared by sequences. It's likely that after evolutional phenomena, as inversions, insertion, deletion and swapping, two monophyletic organisms genes don't share any more information localized in the regions where these mutations took place. Thus, it makes sense to accept alignment-free algorithms that do not consider single and punctual discrepancies, but global parameters referred to the whole amount of information hidden in biological data. Dawy *et al.* [23]gives a detailed account of alignment-free methods in biological applications. Two big categories are presented: methods based on word frequencies, relying on statistical distances between frequency vectors; and information-theory based measures, using Kolmogorov complexity theory [27]and scale-independent representation of sequences by iterative maps. The latter methodologies have a more recent history. Most of them have been developed in the last decade. Although this new area of investigation is considered rich of promises, the literature is rather heterogeneous. In fact despite the variety of proposals, reciprocal comparisons and evaluations are still lacking.. Our idea is to perform a survey of the most representative work in the area. We shall present four information theory-based metrics recently proposed, and compare them. The test bed will be the construction of a phylogenetic tree for of 35 Eukaryotes mtDNA. The paper is organized as follows: in Section 2 we introduce the four algorithms, object of our examination; in Section 3 we discuss string compression, as it plays a critical role in information-theory based algorithms implementation; our test bed procedure is detailed in Section 4; in Section 5 we provide several tables illustrating the experimental results. Together with the distance data produced by the different algorithms and the corresponding phylogenetic tree, we show the believed-true phylogenetic tree over the set of mtDNA's used for the experiments. This tree is taken from Reyes *et al.* [28]. In the last section we summarize our findings and discuss some open questions that deserve further investigation.

2. METRICS IN COMPARISON

In this section we will schematically list and explain the four alignment-free models for measuring genomic information discrepancy, which are the object of our comparative study. We refer to [31] for undefined information theoretic quantities. Here we just remind that, given a probabilistic source S, its entropy $H(S)$ is defined as

$$H(S) = -\sum_{s} P(s) \log P(s)$$

where $(P(s)$ is the probability that the source output is s. Given two sources S and T, the conditional entropy of S given T is defined by

$$H(S/T) = -\sum_{s,t} P(st) \log P(s/t)$$

where $P(s/t)$ is the conditional probability that S emits s, given that the source output of T is t. Finally, the mutual information between S and T is defined as

$$I(S;T) = H(S) - H(S/T).$$

2.1. First metric in synthesis

Proposing Author: Pavol Hanus et al. [23, 24, 25];
Year of first presentation: 2005;
Information theory supports: Shannon fundamental theorem on data compression [29];
Preliminary hypothesis:

(a) If s is a message generated by a source S, the entropy rate is approximable by a compression ratio

$$H(S) \approx \frac{|\texttt{comp}(s)|}{|s|}$$

according to the idea that DNA compression can be conceived as a measure of the information stored in it. The symbol $|\cdot|$ indicates the size in bits or symbols, $\texttt{comp}(s)$ denotes the compressed version of string s, where some fixed but otherwise arbitrary *optimal* compression algorithm is used.

(b) The conditional entropy $H(S/T)$ of two different sources S and T is approximable as the compression ratio achieved for the message s when the compressor is conditioned (i..e, the compressor already has processed) on the message t:

$$H(S/T) \approx \frac{|\texttt{comp}(st)| - |\texttt{comp}(t)}{|s|}$$

Distance definition: If S and T are two sources, their mutual information describes how much related the two sources are. By normalizing w.r.t. the maximum mutual information between the two sources, we have

$$d_1(S,T) = 1 - \frac{H(S) - H(S/T)}{\max\{H(S), H(T)\}}.$$

Justification: The distance measure d_1 is zero only for identical sources $S = T$, because only identical sources share the maximum possible information and have identical entropies at the same time $H(S) = H(T)$. This distance additionally reflects whether the sources are identical or not. The symmetry property directly follows by the symmetry of the mutual information between the two sources, since $I(S,T) = H(S) - H(S/T) = H(T) - H(T/S)$. The distance is also proved to satisfy the triangular inequality [23]. Therefore d_1 is a normalized metric.

Notes and warnings: The authors of [23] use different compressor to test their classification metric on different high level animals mtDNA: Lempel-Ziv (LZ), Context Tree Weighting (CTW), Burrows Wheeler Transform (BWT), Prediction by Partial Matching (PPM) and DNACompress. This distance measure based on a trained compression is said to be specifically suitable for phylogenetic trees construction.

2.2. Second metric in synthesis

Proposing Author: Ming Li et al. [6], [7], [26];
Year of presentation: 2001;
Information theory bases: Kolmogorov complexity [26];

Distance definition: given two sequences S and T, this distance is defined as follows:

$$d_2(S,T) = 1 - \frac{K(S) - K(S/T)}{K(ST)}$$

where $K(S/T)$ is the conditional Kolmogorov complexity, or algorithmic entropy, meaning the length of the shortest program causing a standard universal computer to output S on input T. $K(S)$ is equal to $K(S/\varepsilon)$, where ε is the empty string.

Descriptions: $K(S/T)$ is a measure of the randomness of S given T [26]. $K(S) - K(S/T)$ is the information T knows about S and in [26] it is proved to be equal to $K(S) - K(S/T) \approx K(T) - K(T/S)$, so this measure is said to be "almost" symmetric. The term $K(ST)$ is the information in the string S concatenated with T and it is a normalization factor. d_2 is a universal similarity metric for strings, where universality means it is a lower bound for all the computable distance and similarity functions, including the ones considered for biological applications. There is a relationship between Kolmogorov complexity of sequences and Shannon information theory: the expected Kolmogorov complexity of a sequence S is asymptotically close to the entropy of the information source emitting S.

Notes and warnings: Since the measure d_2 is based on Kolmogorov complexity [26], it is not computable in the Turing sense. However, one can still get a practical tool from such a beautiful theoretical finding since the Kolmogorov complexity of a string can be approximated via data compression [8]. This leaves open the problem of how to best approximate the universal measure, which can be regarded more as a methodology than a formula quantifying how similar two objects are. The heuristic approximations of $K(S)$ and $K(S/T)$ are respectively $\texttt{comp}(S)$ and $\texttt{comp}(S/T)$, even if it is evident that these two approximations are absolute and not relative measures of information. The compressor exploited is the one designed and presented in 1999 by the authors themselves [9], that is GenCompress, explicitly aimed to better shrink biological data than all other compressor now existing on the information theory scene.

2.3. Third metric in synthesis

Proposing Author: Hasan H. Otu et al. [20];
Year of presentation: 2003; *Information theory bases: Lempel-Ziv complexity* [30];

Distance definition: Given two sequences S and T, we define the function $d_3(S,T)$ as follows:

$$d_3(S,T) = \frac{c(S/T)+c(T/S)}{c(ST)}.$$

where the function $c(\cdot)$ is in turn defined as follows. Let $c_H(S)$ be the number of components in a *history* of S. Then the LZ complexity of S is $c(S) = min c_H(S)$ [30] over all histories of S. It can be shown that $c(S) = c_E(S)$ where $c_E(S)$ is the number of components in the exhaustive history of S.

Descriptions: this distance measure, based on the relative complexity between sequences, implies the evolutionary distance between organisms and is proved to be a metric. The distance between sequences S and T was obtained using the exhaustive histories of the sequences S, T, ST and TS. These exhaustive histories were obtained by parsing the sequences using the production rules described in [30]. The number of components in the exhaustive histories were then used as described above to compute the various distance measures.

Notes: This distance measure naturally follows from the idea of building sequence T using S and it is the 'sum distance'. We use this term in the sense that it accounts for the total number of steps it takes to build T from S and vice versa. The one proposed above is a normalized version of the sum distance. In our application the function $c(\cdot)$ is approximated as the compression function, evaluated by Gencompress.

2.4. Fourth metric in synthesis

Proposing Author: Hasan H. Otu et al. [20];
Year of presentation: 2003; *Information theory bases: Lempel-Ziv complexity* [30];

Distance definition: Given two sequences S and T, we define the function $d_4(S,T)$ as follows:

$$d_4(S,T) = \frac{\max\{c(S/T), c(T/S)\}}{\max\{c(S), C(T)\}}.$$

Descriptions: We can formulate the number of steps it takes to generate a sequence T from a sequence S by $c(ST) - c(S)$. Thus, if S is closer to T than R then we would expect $c(ST) - c(S)$ to be smaller than $c(RT)c(R)$ as is the case in the above example.

Notes: This normalized distance is demonstrated to be a metric and makes use of the max function to operate the most robust selection for a distance definition. The denominator, taking into account the triangular inequality, assures the normalization.

3. COMPLEX FUNCTIONS APPROXIMATION

The idea of using compression for phylogenetic classification of whole genomes was first introduced in 2001 [6]. It is based on the idea that it is expectable that a better compression is achieved for the concatenations of two similar sequences than for the concatenation of two dissimilar ones. This observation is very useful in the definition of an effective distance from the theoretic setting outlined above. Let comp be a compression algorithm and $\texttt{comp}(s)$ denoted the compressed version of the string s. The compressed string can be assumed to be a binary string. Let $|\texttt{comp}(s)|/|s|$ be the compression rate, namely the ratio of the lengths of the two strings. The compression rate of a good compressor usually comes close to the entropy of the information source emitting s [31]. Entropy sets up a lower bound on compression rates, but measuring the entropy is generally computationally hard. An empirical way to circumvent this problem is to estimate the entropy by means of the compression rate of a good compressor on a sufficiently long representative string. That is, the compression rate becomes a *compressive estimate of entropy*.

For the second metric, we must be satisfied to approximating $K(S)$ by the length of $\texttt{comp}(s)$. Moreover, since $K(S,T) = K(ST)$ up to additive logarithmic precision [27], $K(S,T)$ can be also approximated by the length $|comp(ST)|$of the compression of the concatenation of S and T.

The choice of the real compressor wasn't critical, because we relied on the results of the recent broad and full documented study conducted by Ferragina *et al.* [5]. In a comparison of 25 different compression algorithms, selected among the most common and used, GenCompress come out to be the best algorithm currently available to compress DNA sequences. Presented just before 2000 in [7] by Ming Li *et al.*, GenCompress is based on an approximate matching method, specifically designed to compress genomic strings, which present very strong discrepancies from other kinds of text: there are only four characters, representing the four nucleotide bases, storable by only two bits. Furthermore, strings are formed by very frequent repeats, palindromes and so on. Based on these and other similar considerations, GenCompress exploits approximate occurrences of substrings in DNA sequences to achieve the best compression to date for genomic strings.

4. THE COMPARISON MODEL

Our aim is to provide indications about the reliability of the some of the most relevant distance concepts for genomic applications, which have been recently proposed in biology, and inspired by information theoretic arguments. These distance concepts seem to be valid substitutes of traditional alignment-based distances, as they are able to solve some ordinary problems afflicting the traditional metrics.

Our idea is to use these four distance models to produce distance data on the mtDNA of 35 different species of Eukaryotes. Such distance measurement are used to produce 4 different phylogenetic tree, one for each distance model. These tree are then evaluated by comparing them with the most commonly accepted phylogenetic tree on the same set of 35 taxa [28]. We choose to experiment on mitochondrial DNA sequences, because they do not recombine. This lack of recombination lets sequences to be traced through only one genetic line and all polymorphisms are supposed to be caused by mutations. Mitochondrial DNA in mammals has a faster mutation rate than nuclear DNA sequences. This faster rate of mutation generates a higher variance between sequences and this is an advantage when studying closely related species. The mitochondrial control region (Displacement or D-loop) is one of the fastest mutating sequence regions in animal DNA.

Phylogenetic analysis were carried out on the complete mammalian mtDNA sequences available in the CoreNucleotide subsection of NCBI database [19]:

human (*Homo sapiens*, V00662)	cow (*Bos taurus*, V00654)
chimpanzee (*Pan troglodytes*, D38116)	sheep (*Ovis aries*, AF010406)
common, pigmy chimpanzee (*Pan paniscus*, D38113)	pig (Sus scrofa, AJ002189)
gorilla (*Gorilla gorilla*, D38114)	hippopotamus (*Hippopotamus amphibius*, AJ010957)
orangutan (*Pongo pygmaeus*, D38115)	Neotropical fruit bat (*Artibeus jamaicensis*, AF061340)
gibbon (*Hylobates lar*, X99256)	african elephant (*Loxodonta africana*, AJ224821)
baboon (*Papio hamadryas*, Y18001)	aardvark (*Orycteropus afer*, Y18475)
horse (*Equus caballus*, X79547)	armadillo (*Dasypus novemcintus*, Y11832)
donkey (*Equus asinus*, X97337)	rabbit (*Oryctolagus cuniculus*, AJ001588)
indian rhinoceros (*Rhinoceros unicornis*, X97336)	guinea pig (*Cavia porcellus*, AJ222767)
white rhinoceros (*Ceratotherium simum*, Y07726)	fat dormouse (*Glis glis*, AJ001562)
harbor seal (*Phoca vitulina*, X63726)	rat (*Rattus norvegicus*, X14848)
gray seal (*Halichoerus grypus*, X72004)	mouse (Mus musculus, V00711)
cat (*Felis catus*, U20753)	hedgehog (*Erinaceus europaeus*, X88898)
dog (*Canis familiaris*, U96639)	opossum (*Didelphis virginiana*, Z29573)
fin whale (*Balenoptera physalus*, X61145)	wallaroo (*Macropus robustus*, Y10524)
blue whale (*Balenoptera musculus*, X72204)	platypus (*Ornithorhyncus anatinus*, X83427)
european red squirrel (*Sciurus vulgaris*, AJ238588)	

These strings of nucleotide bases can be encoded by binary strings (two bits for each character) and can be considered as the input of our evaluation model, as shown in Figure 1 below.

The scheme above describes the sequence of steps necessary to achieve the complete transformation of our data, from genomic strings to a Phylogenetic tree.

We have implemented an automatic tool in a Unix shell suited to manage the considerable amount of input data, 35 strings of about 1.6×10^4 characters, performing the

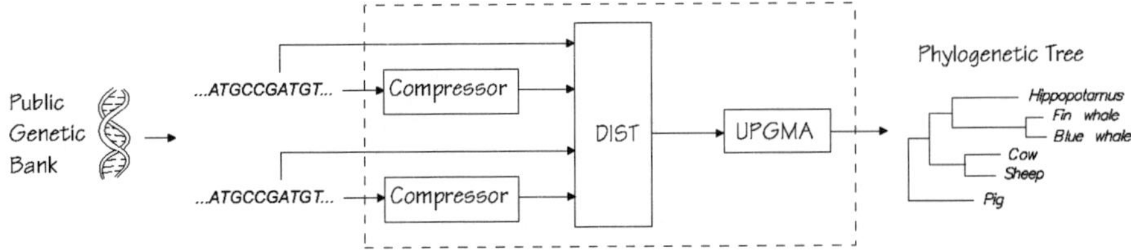

FIGURE 1. Evaluation and comparison model scheme

computation of 35×35 distance matrix, almost symmetric, which stores the pairwise "relatedness" between strings, evaluated by the 4 models presented in Section 2. The elements of this matrix are the input data for a MATLAB® (Bioinformatics toolbox) procedure to draw phylogenetic trees, on the base of the distances provided in input. The first stage block in the construction of the distance functions is the compression performed by GenCompress, allegedly, the best compression algorithm existing at the moment for genomic strings. The compressor evaluate also the conditional compression, implemented in GenCompress, which performs the compression of a string tuning the compression parameters on another string. The second stage block receives as input the compressed and the original strings, transforming them in numerical data, calculating the distance matrix according to the distance definition presented above. As stressed by Chen *et al.* [7], the distance matrix may not be exactly symmetric, because not all the distance algorithms are symmetric. Our resolution of this weak dissymmetry was based on the consideration of the mean and the variance of the difference between two symmetric elements (further details are presented at URL: http://people.na.infn.it/~wbalzano/BIO). The remarkable exiguity of these values has determined our decision to symmetrize the matrix, substituting every element with the average value between itself and its symmetric element: $d(S,T)+d(T,S)/2$. The fourth metric is an exception, being it intrinsically symmetric. Finally the block based on Unweighted Pair Group Method Average (UPGMA) method develops the phylogenetic expressions of the "relatedness" before calculated and turns the data in a plain diagram, visually interpretable and comparable to other phylogenetic expressions.

5. RESULTS

In this section we present the four distance matrix with respective phylogenetic trees. We start by presenting the phylogenetic tree taken from [28], that we shall use as a benchmark for our findings.

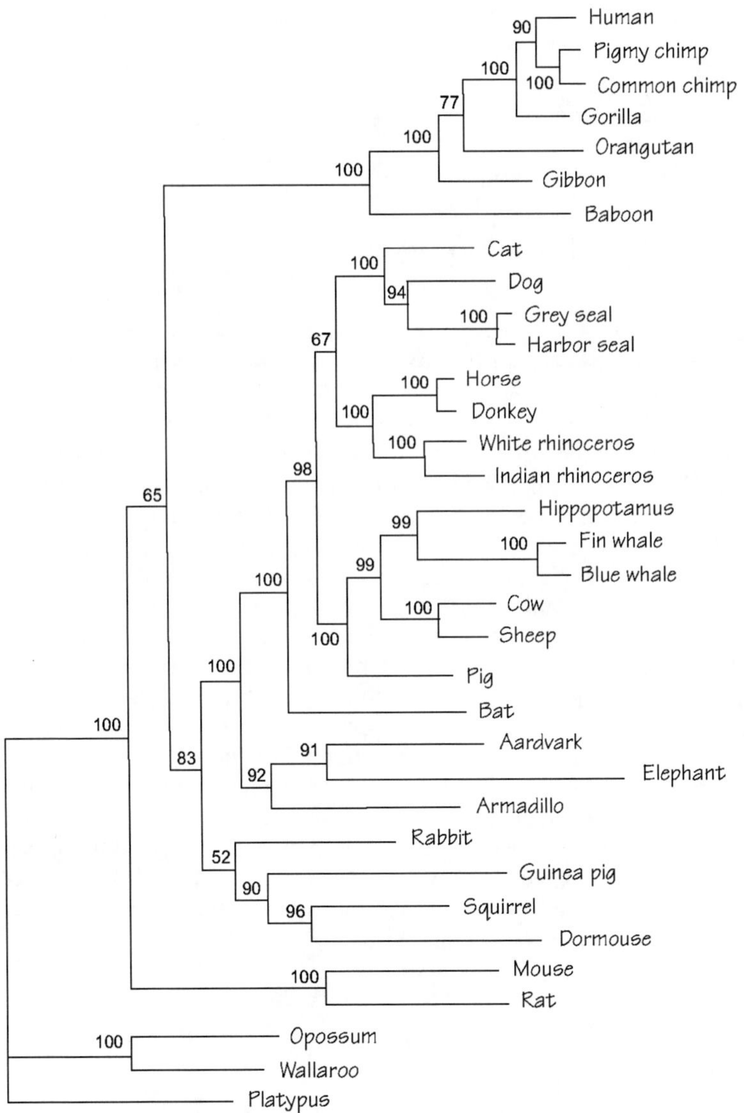

FIGURE 2. Reference phylogenetic tree by Reyes [28]

We present below the Phylogenetic tree obtained by applying the first metric d_1

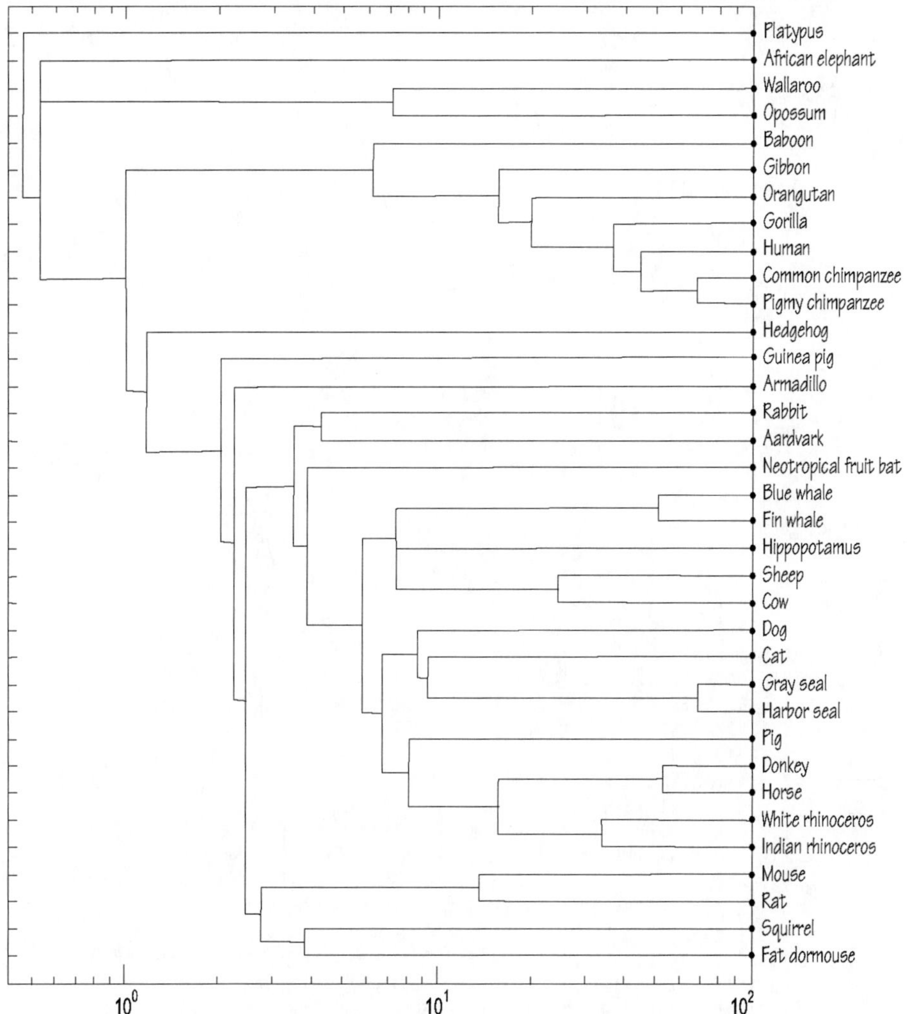

FIGURE 3. Phylogenetic tree for first metric, Hanus approximation

and the relative distance matrix

	Human	Pigmy chimpanzee	Common chimpanzee	Gorilla	Orangutan	Gibbon	Baboon	Horse	Donkey	Indian rhinoceros	White rhinoceros	Harbor seal	Gray seal	Cat	Dog	Fin whale	Blue whale	Cow	Sheep	Pig	Hippopotamus	Neotropical fruit bat	African elephant	Aardvark	Armadillo	Rabbit	Guinea pig	Fat dormouse	Rat	Mouse	Squirrel	Hedgehog	Opossum	Wallaroo	Platypus
Human	-																																		
Pigmy chimpanzee	52,0	-																																	
Common chimpanzee	52,8	29,2	-																																
Gorilla	61,4	60,1	61,2	-																															
Orangutan	77,5	76,7	77,6	78,0	-																														
Gibbon	82,0	80,7	80,7	81,5	83,5	-																													
Baboon	91,3	90,7	91,5	91,4	91,2	90,7	-																												
Horse	95,2	95,6	95,6	96,2	96,5	95,3	96,3	-																											
Donkey	95,8	96,2	96,1	95,9	96,4	95,4	96,7	44,4	-																										
Indian rhinoceros	95,9	96,2	95,7	95,5	95,9	95,7	96,1	80,9	82,2	-																									
White rhinoceros	95,5	96,1	95,6	95,3	96,7	95,4	96,4	81,2	82,1	63,8	-																								
Harbor seal	96,1	96,1	95,4	96,1	96,5	96,5	96,7	90,2	91,1	89,8	89,8	-																							
Gray seal	96,3	96,4	95,7	95,9	96,8	96,4	96,9	90,3	91,3	90,5	89,5	29,0	-																						
Cat	96,2	96,4	95,9	96,0	96,8	96,1	96,8	89,5	90,3	90,4	89,9	87,9	88,2	-																					
Dog	96,8	96,1	96,1	96,4	97,2	96,4	96,7	90,9	91,2	92,8	91,8	88,7	88,2	89,2	-																				
Fin whale	96,5	96,5	96,2	96,8	96,6	97,0	96,2	91,8	91,7	92,6	91,8	93,3	93,5	93,2	93,5	-																			
Blue whale	96,2	95,8	95,1	96,4	96,4	96,6	96,3	92,0	90,6	92,2	91,3	92,5	92,3	93,0	93,6	46,2	-																		
Cow	95,8	95,6	95,3	95,5	96,0	95,6	96,3	90,1	90,8	90,0	90,1	91,1	90,8	91,5	92,9	90,1	89,4	-																	
Sheep	96,5	95,8	95,3	95,9	96,1	95,3	96,7	91,0	90,8	90,5	91,0	91,7	91,6	92,5	93,5	90,7	89,7	72,9	-																
Pig	96,1	95,5	95,2	95,7	95,9	96,0	96,3	89,7	89,4	89,1	88,8	90,1	91,0	90,9	92,9	90,0	89,6	89,7	89,1	-															
Hippopotamus	95,9	95,4	95,1	95,8	96,3	96,4	96,0	91,0	90,8	90,6	90,9	91,8	92,0	92,9	93,8	90,9	89,2	89,4	90,5	90,1	-														
Neotropical fruit bat	96,4	96,1	96,0	96,0	97,0	95,8	97,4	93,4	93,4	93,9	93,3	93,7	94,0	93,9	93,3	94,0	93,6	93,3	92,2	92,4	94,4	-													
African elephant	97,6	97,8	97,9	97,6	97,8	97,5	98,2	96,6	96,4	96,2	96,0	95,7	96,0	96,5	96,6	97,5	97,5	95,9	95,9	96,4	96,0	96,6	-												
Aardvark	96,2	96,6	96,0	96,4	97,1	96,8	96,4	93,9	94,2	92,3	93,0	93,6	93,2	93,8	93,8	94,8	94,3	93,6	93,9	93,6	94,5	94,7	95,4	-											
Armadillo	96,9	96,7	96,5	96,3	97,5	96,6	97,5	95,1	94,4	94,2	94,5	94,1	94,7	94,8	96,4	95,9	95,3	93,7	94,4	93,5	95,2	96,7	96,7	94,9	-										
Rabbit	95,7	95,8	95,8	95,4	96,2	96,2	96,3	94,1	94,1	94,0	93,6	94,0	93,7	93,5	93,9	94,6	95,0	93,8	93,2	92,9	94,7	93,8	96,1	93,1	95,6	-									
Guinea pig	97,2	96,7	96,9	97,0	96,8	96,5	97,5	95,6	94,8	94,5	94,8	95,2	95,4	95,1	95,0	95,6	95,1	95,3	95,3	94,9	94,9	95,8	97,2	95,7	96,7	95,3	-								
Fat dormouse	96,6	95,9	96,2	96,8	97,0	96,6	97,0	95,0	94,6	94,1	94,6	94,7	94,8	94,3	94,8	96,3	95,5	94,4	94,8	95,5	95,5	95,2	97,5	95,6	97,4	94,7	94,8	-							
Rat	97,3	96,5	96,6	97,2	97,0	96,6	97,6	94,6	94,8	94,3	93,9	95,1	95,7	94,0	94,6	95,7	95,7	95,4	95,3	94,6	95,4	95,1	97,8	95,3	95,7	95,6	96,1	94,2	-						
Mouse	97,3	97,1	97,2	97,5	97,7	97,1	97,3	95,8	95,4	94,9	94,7	94,5	95,0	94,3	94,9	96,0	95,3	94,7	95,0	94,9	95,8	95,4	97,5	95,4	95,4	95,0	96,0	95,2	83,7	-					
Squirrel	96,0	95,7	95,3	96,1	96,6	95,8	97,1	93,9	94,2	94,1	94,9	94,2	94,6	93,8	94,7	95,5	95,0	93,7	94,0	93,8	94,4	94,8	96,9	95,3	94,9	93,7	95,0	93,6	94,4	94,7	-				
Hedgehog	97,4	97,4	97,3	97,4	97,7	97,7	97,7	96,4	96,5	96,2	96,2	96,2	96,3	96,4	96,5	96,9	96,4	95,4	95,5	95,4	95,7	96,1	97,7	96,2	96,6	95,6	96,9	96,2	96,8	96,0	95,5	-			
Opossum	97,6	97,7	97,6	97,7	98,1	97,3	97,8	96,1	96,5	96,3	96,4	96,8	96,9	96,8	96,4	97,0	97,2	96,2	96,9	96,7	96,9	96,8	97,6	96,1	96,5	96,5	97,1	96,6	96,4	95,8	96,1	97,1	-		
Wallaroo	97,8	97,9	97,7	98,1	98,0	97,7	98,3	96,8	96,8	96,8	95,6	96,3	96,4	96,6	96,5	97,3	97,2	96,3	96,2	96,5	96,9	97,1	97,8	96,7	96,7	95,8	96,3	96,7	95,8	96,2	96,8	97,3	90,2	-	
Platypus	98,2	98,0	97,9	97,9	98,2	98,3	98,6	97,1	97,3	96,3	97,8	96,9	96,7	97,0	96,8	97,3	97,1	96,4	97,0	96,7	97,3	97,1	98,6	96,9	97,8	97,1	97,3	96,3	97,1	97,1	97,3	97,9	97,6	96,9	-

FIGURE 4. Distance matrix for first metric, Hanus approximation

We present below the Phylogenetic tree obtained by applying the second metric d_2

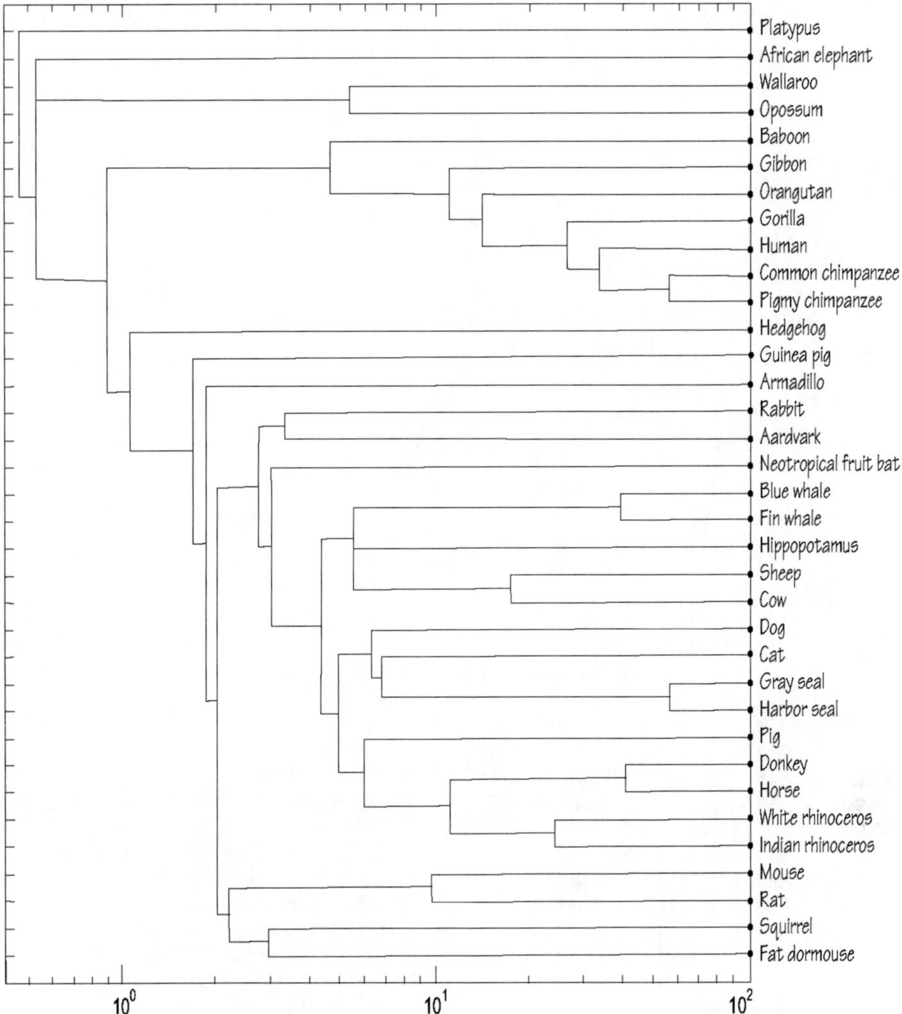

FIGURE 5. Phylogenetic tree for second metric, Ming Li metric approximation

and the relative distance matrix

	Human	Pigmy chimpanzee	Common chimpanzee	Gorilla	Orangutan	Gibbon	Baboon	Horse	Donkey	Indian rhinoceros	White rhinoceros	Harbor seal	Gray seal	Cat	Dog	Fin whale	Blue whale	Cow	Sheep	Pig	Hippopotamus	Neotropical fruit bat	African elephant	Aardvark	Armadillo	Rabbit	Guinea pig	Fat dormouse	Rat	Mouse	Squirrel	Hedgehog	Opossum	Wallaroo	Platypus
Human	-																																		
Pigmy chimpanzee	68,3	-																																	
Common chimpanzee	68,9	45,0	-																																
Gorilla	76,1	75,0	75,8	-																															
Orangutan	87,4	86,8	87,4	87,5	-																														
Gibbon	90,1	89,3	89,3	89,9	91,0	-																													
Baboon	95,5	95,1	95,5	95,5	95,4	95,1	-																												
Horse	97,5	97,7	97,7	98,1	98,2	97,6	98,1	-																											
Donkey	97,8	98,0	98,0	97,9	98,1	97,6	98,3	61,4	-																										
Indian rhinoceros	97,9	98,0	97,8	97,7	97,9	97,8	98,0	89,5	90,3	-																									
White rhinoceros	97,7	98,0	97,7	97,6	98,3	97,6	98,1	89,6	90,2	77,9	-																								
Harbor seal	98,0	98,0	97,6	98,0	98,2	98,2	98,3	94,9	95,3	94,6	94,6	-																							
Gray seal	98,1	98,1	97,8	97,9	98,4	98,1	98,4	94,9	95,4	95,0	94,4	44,8	-																						
Cat	98,0	98,2	97,9	98,0	98,3	98,0	98,3	94,4	94,8	94,9	94,7	93,5	93,7	-																					
Dog	98,3	98,0	98,0	98,2	98,6	98,1	98,3	95,2	95,3	96,2	95,7	94,0	93,7	94,3	-																				
Fin whale	98,2	98,2	98,1	98,4	98,3	98,5	98,1	95,7	95,7	96,1	95,7	96,5	96,6	96,4	96,6	-																			
Blue whale	98,1	97,9	97,5	98,2	98,2	98,3	98,1	95,8	95,0	95,9	95,4	96,1	96,0	96,4	96,7	63,0	-																		
Cow	97,9	97,7	97,6	97,7	97,9	97,7	98,1	94,8	95,1	94,7	94,7	95,3	95,2	95,5	96,3	94,8	94,4	-																	
Sheep	98,2	97,8	97,6	97,9	98,0	97,6	98,3	95,3	95,2	95,0	95,3	95,7	95,6	96,1	96,6	95,1	94,5	84,3	-																
Pig	98,0	97,7	97,5	97,8	97,9	97,9	98,1	94,6	94,4	94,2	94,0	94,8	95,3	95,2	96,3	94,7	94,5	94,5	94,2	-															
Hippopotamus	97,9	97,6	97,5	97,8	98,1	98,2	97,9	95,2	95,1	95,0	95,1	95,7	95,8	96,2	96,8	95,2	94,3	94,4	95,0	94,8	-														
Neotropical fruit bat	98,2	98,0	98,0	97,9	98,4	97,8	98,7	96,6	96,6	96,8	96,5	96,8	96,9	96,8	96,5	96,9	96,7	96,5	95,9	96,1	97,1	-													
African elephant	98,8	98,9	98,9	98,8	98,9	98,7	99,1	98,3	98,1	98,1	97,9	97,8	98,0	98,2	98,3	98,7	98,7	97,9	97,9	98,1	97,9	98,3	-												
Aardvark	98,1	98,2	97,9	98,1	98,5	98,3	98,2	96,8	97,0	96,0	96,4	96,7	96,5	96,8	96,8	97,3	97,0	96,6	96,8	96,7	97,1	97,3	97,6	-											
Armadillo	98,4	98,3	98,2	98,1	98,7	98,3	98,7	97,5	97,1	97,0	97,2	97,0	97,3	97,3	98,1	97,9	97,6	96,7	97,1	96,6	97,5	98,3	98,3	97,3	-										
Rabbit	97,8	97,8	97,8	97,6	98,0	98,0	98,1	96,9	96,9	96,8	96,6	96,9	96,7	96,6	96,8	97,2	97,4	96,7	96,4	96,3	97,2	96,8	98,0	96,4	97,7	-									
Guinea pig	98,5	98,3	98,4	98,5	98,4	98,2	98,7	97,8	97,3	97,1	97,3	97,5	97,6	97,5	97,4	97,8	97,5	97,6	97,6	97,4	97,3	97,8	98,6	97,8	98,3	97,6	-								
Fat dormouse	98,2	97,9	98,0	98,4	98,5	98,2	98,5	97,4	97,2	96,9	97,2	97,3	97,3	97,1	97,3	98,1	97,7	97,1	97,3	97,7	97,7	97,5	98,7	97,7	98,7	97,3	97,3	-							
Rat	98,6	98,2	98,2	98,5	98,5	98,3	98,8	97,2	97,3	97,0	96,8	97,5	97,8	96,9	97,2	97,8	97,8	97,6	97,6	97,2	97,6	97,5	98,9	97,6	97,8	97,7	98,0	97,0	-						
Mouse	98,6	98,5	98,6	98,7	98,8	98,5	98,6	97,9	97,6	97,4	97,2	97,2	97,4	97,0	97,4	97,9	97,6	97,2	97,4	97,4	97,8	97,6	98,7	97,7	97,6	97,4	97,9	97,5	91,1	-					
Squirrel	97,9	97,8	97,6	98,0	98,3	97,8	98,5	96,8	97,0	96,9	97,4	97,0	97,2	96,8	97,3	97,7	97,4	96,7	96,9	96,8	97,1	97,3	98,4	97,6	97,3	96,7	97,4	96,6	97,1	97,3	-				
Hedgehog	98,6	98,6	98,6	98,7	98,8	98,8	98,8	98,1	98,2	98,0	98,0	98,0	98,1	98,2	98,2	98,4	98,1	97,6	97,6	97,6	97,8	98,0	98,8	98,0	98,3	97,7	98,4	98,0	98,3	97,9	97,6	-			
Opossum	98,8	98,8	98,8	98,8	99,0	98,6	98,9	98,0	98,2	98,1	98,2	98,4	98,4	98,4	98,1	98,5	98,5	98,0	98,4	98,3	98,4	98,3	98,7	98,0	98,2	98,2	98,5	98,2	98,1	97,8	98,0	98,5	-		
Wallaroo	98,9	99,0	98,8	99,0	99,0	98,8	99,1	98,3	98,4	98,3	97,7	98,1	98,2	98,2	98,2	98,6	98,5	98,1	98,0	98,2	98,4	98,5	98,9	98,3	98,3	97,8	98,1	98,3	97,8	98,1	98,4	98,6	94,8	-	
Platypus	99,1	99,0	98,9	98,9	99,1	99,1	99,3	98,5	98,6	98,1	98,9	98,4	98,3	98,5	98,4	98,6	98,5	98,1	98,5	98,3	98,6	98,5	99,3	98,4	98,9	98,5	98,6	98,1	98,5	98,5	98,6	98,9	98,8	98,4	-

FIGURE 6. Distance matrix for second metric, Ming Li approximation

We present below the Phylogenetic tree obtained by applying the third metric d_3

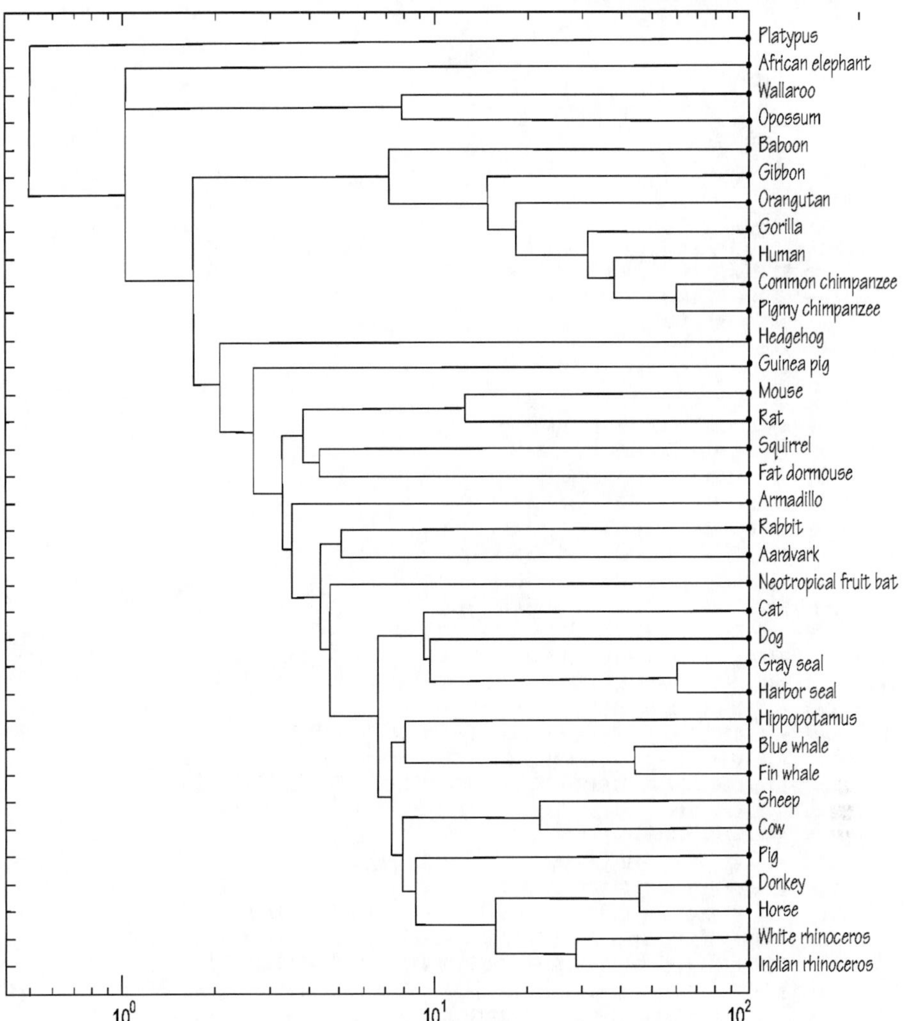

FIGURE 7. Pylogenomic tree for the third metric, Otu approximation

and the relative distance matrix

	Human	Pigmy chimpanzee	Common chimpanzee	Gorilla	Orangutan	Gibbon	Baboon	Horse	Donkey	Indian rhinoceros	White rhinoceros	Harbor seal	Gray seal	Cat	Dog	Fin whale	Blue whale	Cow	Sheep	Pig	Hippopotamus	Neotropical fruit bat	African elephant	Aardvark	Armadillo	Rabbit	Guinea pig	Fat dormouse	Rat	Mouse	Squirrel	Hedgehog	Opossum	Wallaroo	Platypus
Human	-																																		
Pigmy chimpanzee	68,6	-																																	
Common chimpanzee	69,4	45,2	-																																
Gorilla	75,9	75,2	76,1	-																															
Orangutan	86,9	86,4	87,1	88,1	-																														
Gibbon	90,0	89,3	89,4	89,4	90,9	-																													
Baboon	95,4	94,6	95,6	95,7	95,6	95,4	-																												
Horse	97,8	98,2	98,3	98,5	98,5	97,8	98,4	-																											
Donkey	98,2	98,5	98,6	98,3	98,5	98,0	98,6	61,6	-																										
Indian rhinoceros	98,4	98,6	98,2	97,9	98,2	98,2	98,4	88,4	89,8	-																									
White rhinoceros	98,1	98,5	98,2	98,0	98,9	97,9	98,7	89,1	89,7	78,0	-																								
Harbor seal	98,5	98,4	97,9	98,4	98,6	98,7	98,8	94,8	95,4	94,7	94,7	-																							
Gray seal	98,6	98,7	98,0	98,2	98,9	98,4	98,9	95,0	95,8	95,4	94,2	45,0	-																						
Cat	98,4	98,6	98,2	98,4	98,8	98,4	98,8	94,6	95,3	95,2	94,7	93,8	94,1	-																					
Dog	98,9	98,2	98,4	98,5	99,1	98,3	98,8	95,5	95,7	96,5	95,8	93,9	93,6	94,3	-																				
Fin whale	98,7	98,8	98,6	99,0	98,7	99,1	98,6	95,9	95,9	96,4	95,7	96,6	96,8	96,7	96,8	-																			
Blue whale	98,5	98,1	97,8	98,8	98,4	98,8	98,6	96,2	95,0	96,0	95,4	96,5	96,3	96,6	96,9	63,2	-																		
Cow	98,2	98,1	97,9	98,3	98,5	98,0	98,4	94,7	95,3	94,5	94,9	95,7	95,4	95,7	96,9	95,4	95,0	-																	
Sheep	98,7	98,4	97,9	98,1	98,3	97,8	98,7	95,2	95,3	95,0	95,3	95,8	95,8	96,3	96,9	95,1	94,8	84,0	-																
Pig	98,5	98,1	97,8	98,2	98,0	98,3	98,5	94,8	94,6	94,2	94,1	94,6	95,2	95,5	96,6	94,8	94,7	94,5	94,4	-															
Hippopotamus	98,4	98,2	98,1	98,4	98,6	98,7	98,3	95,6	95,2	95,0	95,1	95,9	95,9	96,7	97,2	95,5	94,2	95,0	95,0	94,8	-														
Neotropical fruit bat	98,6	98,5	98,4	98,2	98,8	98,0	99,2	96,8	96,9	97,4	96,9	97,1	97,2	97,3	96,8	97,1	97,1	97,0	96,2	96,0	97,5	-													
African elephant	99,3	99,4	99,6	99,4	99,4	99,2	99,7	98,8	98,8	98,5	98,5	98,4	98,5	98,9	98,7	99,3	99,4	98,3	98,5	98,8	98,6	98,7	-												
Aardvark	98,4	98,7	98,2	98,6	98,9	98,7	98,5	97,0	97,4	96,1	96,5	96,8	96,8	97,0	97,2	97,7	97,3	96,8	97,2	97,1	97,4	97,8	98,2	-											
Armadillo	99,0	98,7	98,7	98,4	99,2	98,6	99,3	97,7	97,2	97,3	97,4	97,0	97,5	97,3	98,7	98,0	97,7	96,8	97,5	96,5	97,9	98,8	99,0	97,7	-										
Rabbit	98,0	98,2	98,2	98,1	98,5	98,5	98,5	97,4	97,4	97,3	97,0	97,3	97,1	97,2	97,3	97,6	98,0	97,0	96,7	96,4	97,7	96,8	98,6	96,7	98,1	-									
Guinea pig	99,0	98,6	98,8	98,9	98,8	98,5	99,3	98,1	97,8	97,5	97,8	98,0	98,1	98,0	98,0	98,1	98,0	98,1	98,0	97,9	97,7	98,3	99,2	98,4	98,7	98,2	-								
Fat dormouse	98,6	98,1	98,4	98,8	98,6	98,4	98,8	97,7	97,4	97,2	97,5	97,5	97,7	97,3	97,8	98,6	98,0	97,4	97,8	97,9	97,9	98,0	99,2	98,2	99,1	97,6	97,7	-							
Rat	99,1	98,5	98,6	99,2	98,9	98,4	99,2	97,5	97,7	97,1	97,2	97,7	98,0	97,0	97,5	98,5	98,5	98,2	97,8	97,5	98,3	97,8	99,4	97,8	98,1	98,2	98,3	97,2	-						
Mouse	99,1	98,7	99,0	99,2	99,3	98,9	98,9	98,1	97,9	97,6	97,7	97,3	97,6	97,5	97,9	98,6	98,2	97,8	97,8	97,8	98,4	98,0	99,5	98,2	98,1	97,7	98,6	97,9	91,7	-					
Squirrel	98,2	97,9	97,7	98,3	98,6	98,0	98,7	97,1	97,3	97,4	97,8	97,2	97,6	97,2	97,7	98,3	98,0	97,0	97,1	97,1	97,1	97,7	99,0	97,9	97,7	97,0	97,7	97,2	97,1	97,6	-				
Hedgehog	99,0	98,9	98,9	99,0	99,1	98,9	99,1	98,3	98,2	98,3	98,4	98,3	98,4	98,5	98,7	98,6	98,5	98,0	98,0	98,0	97,9	98,5	99,4	98,4	98,6	98,1	98,9	98,4	98,5	98,3	98,0	-			
Opossum	99,0	99,1	99,1	99,1	99,3	98,8	99,3	98,3	98,4	98,5	98,7	98,7	98,9	98,7	98,5	98,7	98,9	98,3	98,9	98,9	98,8	98,8	99,3	98,3	98,3	98,6	99,0	98,6	98,5	98,1	98,4	99,0	-		
Wallaroo	99,4	99,5	99,3	99,6	99,5	99,4	99,7	98,8	99,0	98,8	98,2	98,6	98,7	98,7	98,8	99,0	99,1	98,5	98,2	98,6	98,8	98,9	99,6	98,5	98,9	98,3	98,6	98,8	98,3	98,6	98,8	98,9	95,0	-	
Platypus	99,4	99,3	99,3	99,2	99,4	99,4	99,8	99,0	99,1	98,4	99,4	98,9	98,8	99,1	99,0	99,1	99,1	98,6	98,8	98,7	99,0	99,1	100	98,9	99,4	99,0	99,2	98,6	98,9	99,1	99,2	99,5	99,3	98,7	-

FIGURE 8. Distance matrix for the third metric, Otu approximation

We present below the Phylogenetic tree obtained by applying the fourth metric d_4

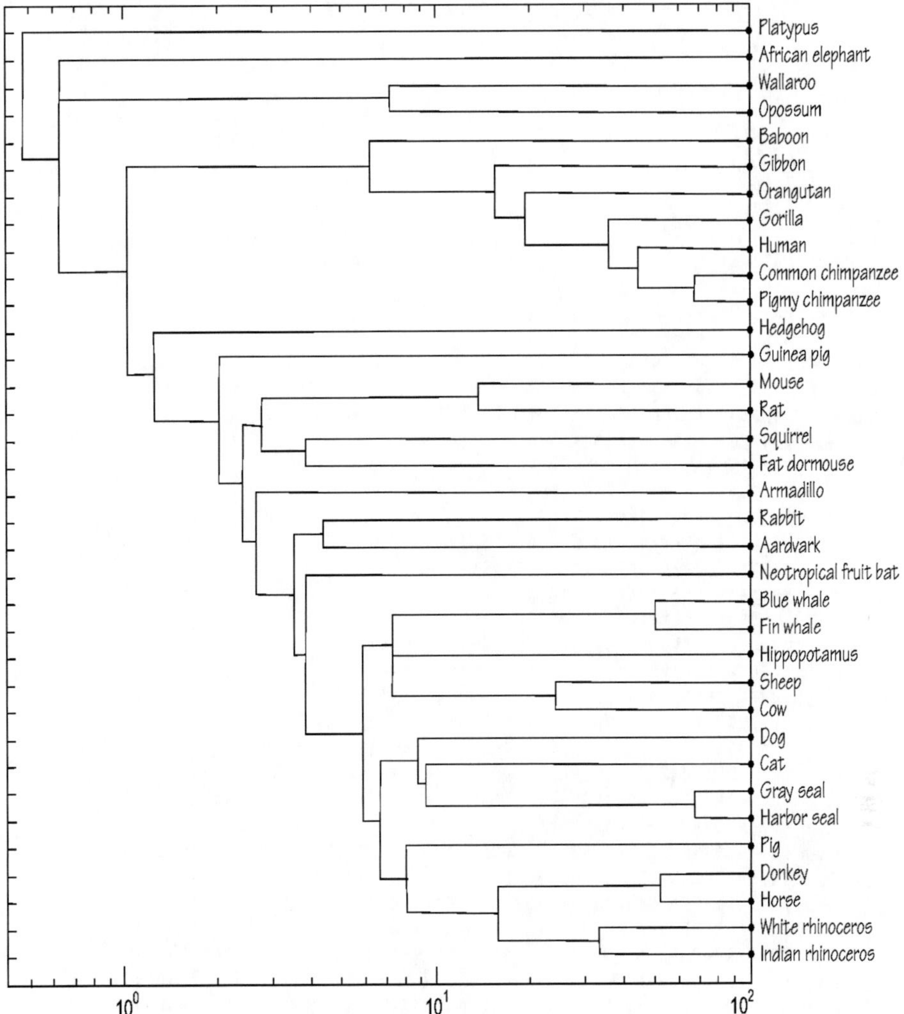

FIGURE 9. Pylogenomic tree for fourth metric, Otu approximation

and the relative distance matrix

	Human	Pigmy chimpanzee	Common chimpanzee	Gorilla	Orangutan	Gibbon	Baboon	Horse	Donkey	Indian rhinoceros	White rhinoceros	Harbor seal	Gray seal	Cat	Dog	Fin whale	Blue whale	Cow	Sheep	Pig	Hippopotamus	Neotropical fruit bat	African elephant	Aardvark	Armadillo	Rabbit	Guinea pig	Fat dormouse	Rat	Mouse	Squirrel	Hedgehog	Opossum	Wallaroo	Platypus
Human	-																																		
Pigmy chimpanzee	51,94	-																																	
Common chimpanzee	52,78	29,15	-																																
Gorilla	61,52	60,08	61,13	-																															
Orangutan	77,69	76,81	77,73	78,31	-																														
Gibbon	81,98	80,61	80,66	81,55	83,53	-																													
Baboon	91,35	90,69	91,41	91,41	91,21	90,68	-																												
Horse	95,18	95,64	95,57	96,33	96,50	95,27	96,34	-																											
Donkey	95,85	96,19	96,14	95,93	96,33	95,39	96,69	44,55	-																										
Indian rhinoceros	95,88	96,14	95,67	95,48	95,90	95,69	96,09	80,70	82,09	-																									
White rhinoceros	95,43	96,09	95,60	95,36	96,73	95,24	96,37	81,02	81,94	63,89	-																								
Harbor seal	96,08	96,09	95,40	96,26	96,43	96,49	96,71	90,32	91,21	89,87	89,85	-																							
Gray seal	96,30	96,34	95,62	95,93	96,73	96,34	96,86	90,34	91,27	90,44	89,45	29,10	-																						
Cat	96,15	96,39	95,84	96,06	96,68	96,04	96,81	89,41	90,16	90,47	89,93	87,87	88,14	-																					
Dog	96,62	96,07	96,14	96,49	97,16	96,41	96,71	91,09	91,16	92,74	91,70	88,54	88,12	89,11	-																				
Fin whale	96,52	96,47	96,19	96,77	96,64	97,03	96,22	91,71	91,71	92,54	91,61	93,16	93,37	93,14	93,44	-																			
Blue whale	96,26	95,81	95,03	96,51	96,44	96,64	96,34	91,97	90,52	92,05	91,19	92,40	92,17	92,95	93,48	46,11	-																		
Cow	95,90	95,70	95,48	95,53	96,08	95,67	96,41	90,12	90,95	90,02	90,15	91,10	90,80	91,50	92,89	90,16	89,57	-																	
Sheep	96,62	95,79	95,30	96,01	96,10	95,32	96,69	90,95	90,85	90,40	90,95	91,72	91,60	92,35	93,46	90,66	89,69	72,80	-																
Pig	96,08	95,52	95,20	95,78	95,82	95,96	96,34	89,74	89,30	89,01	88,79	90,22	91,01	90,98	92,74	89,96	89,54	89,69	89,00	-															
Hippopotamus	95,88	95,43	95,16	95,73	96,33	96,38	95,98	90,91	90,74	90,54	90,79	91,66	91,94	92,73	93,81	91,01	89,16	89,31	90,64	90,09	-														
Neotropical fruit bat	96,42	96,12	96,04	95,96	96,96	95,76	97,46	93,41	93,36	93,88	93,26	93,90	94,06	93,73	93,36	93,97	93,58	93,34	92,18	92,41	94,31	-													
African elephant	97,62	97,76	97,91	97,61	97,76	97,46	98,21	96,63	96,36	96,22	95,97	95,90	96,09	96,59	96,59	97,47	97,49	96,10	95,90	96,34	96,03	96,61	-												
Aardvark	96,22	96,54	95,94	96,36	97,11	96,71	96,42	93,83	94,19	92,22	92,99	93,68	93,29	93,82	93,78	94,77	94,25	93,84	93,90	93,59	94,51	94,66	95,38	-											
Armadillo	96,90	96,66	96,52	96,28	97,51	96,59	97,46	95,01	94,39	94,24	94,54	94,17	94,72	94,85	96,37	95,79	95,21	93,84	94,30	93,37	95,11	96,63	96,71	94,85	-										
Rabbit	95,68	95,74	95,77	95,28	96,07	96,19	96,27	94,03	94,04	93,89	93,51	93,95	93,61	93,50	93,85	94,47	94,93	93,69	93,08	92,82	94,63	93,66	96,05	92,99	95,59	-									
Guinea pig	97,17	96,69	96,84	97,11	96,73	96,39	97,46	95,72	94,91	94,54	94,83	95,23	95,45	95,16	94,91	95,57	95,06	95,25	95,38	94,91	94,78	95,82	97,26	95,72	96,76	95,30	-								
Fat dormouse	96,57	95,89	96,17	96,86	96,98	96,54	97,01	95,03	94,66	94,06	94,66	94,61	94,78	94,28	94,84	96,17	95,51	94,60	94,83	95,51	95,43	95,36	97,53	95,56	97,38	94,73	94,78	-							
Rat	97,24	96,46	96,59	97,14	96,88	96,41	97,59	94,63	95,07	94,28	93,97	95,12	95,83	93,97	94,59	95,72	95,81	95,50	95,42	94,63	95,33	95,14	97,81	95,35	95,65	95,57	96,08	94,18	-						
Mouse	97,34	97,09	97,19	97,46	97,74	97,13	97,26	95,97	95,57	94,93	94,72	94,67	95,03	94,32	94,86	95,97	95,21	94,65	95,08	94,90	95,71	95,46	97,50	95,49	95,44	95,13	95,97	95,18	83,65	-					
Squirrel	96,00	95,64	95,22	96,11	96,58	95,79	97,11	93,88	94,16	94,09	94,89	94,16	94,54	93,73	94,64	95,44	95,03	93,74	94,01	93,76	94,41	94,81	96,84	95,29	94,86	93,61	94,94	93,51	94,31	94,54	-				
Hedgehog	97,32	97,34	97,24	97,41	97,66	97,68	97,61	96,28	96,38	96,14	96,16	96,22	96,31	96,40	96,40	96,87	96,34	95,30	95,30	95,34	95,61	95,98	97,65	96,12	96,55	95,56	96,86	96,18	96,67	95,97	95,34	-			
Opossum	97,62	97,71	97,64	97,67	98,14	97,26	97,81	96,08	96,48	96,26	96,43	96,86	96,90	96,81	96,37	96,97	97,14	96,23	96,90	96,75	96,90	96,73	97,53	96,05	96,55	96,48	97,13	96,58	96,39	95,82	96,04	97,03	-		
Wallaroo	97,74	97,93	97,66	98,09	97,99	97,66	98,26	96,80	96,82	96,80	95,65	96,36	96,48	96,58	96,51	97,25	97,14	96,25	96,21	96,51	96,85	97,17	97,80	96,75	96,73	95,75	96,34	96,70	95,82	96,24	96,87	97,26	90,16		
Platypus	98716	98,03	97,86	97,89	98,24	98,31	98,58	97,15	97,33	96,31	97,76	96,89	96,71	97,01	96,84	97,25	97,09	96,35	97,03	96,65	97,25	97,08	98,57	96,89	97,83	97,07	97,33	96,23	97,15	97,17	97,34	97,91	97,54	96	

FIGURE 10. Distance matrix for fourth metric, Otu approximation

6. CONCLUSIONS

The direct comparison of genetic trees with the one proposed by Reyes *et al.* [28] can be object of reflection about the reliability of the models inspected. Many evidences confirm that the distance estimation provided by the four algorithms alignment-free inspired by the ones of Hanus *et al.* [23], [24], [25], Ming Li *et al.* [6], [7], [26], and Otu *et al.* [20] can produce very interesting results for a new point of view in evolution of Eukaryotes interpretations. Many branches of the trees can testify the suitability of information theory models to data mining in genomic and proteomic databases. In particular the third metric shows particular robustness and precision in the branch containing as leafs the dog, the cat and the seals. Here the other three metrics misplace these hierarchical relations, even if the interpretation is still a challenge to which all interested observers are called. As a last proposal, we foresee, among many other developments, the possibility to introduce multiple comparison for alignment-free models, based on the most robust and promising information theory-based metric.

ACKNOWLEDGMENTS

The work of the second author was supported by the Sofja Kovalevskaja Award of the Alexander von Humboldt Foundation and the German Federal Ministry of Research and Education.

REFERENCES

1. P. Baldi and S. Brunak, *Bioinformatics, the machine learning approach, Second edition*, The MIT Press, Cambridge (USA) - London (UK), 2001.
2. M. Waterman, *Introduction to Computational Biology. Maps, Sequences and Genomes*, Chapman Hall, 1995.
3. D. Gusfield D, *Algorithms on Strings, Trees and Sequences: Computer Science and computational Biology*, Cambridge University Press, 1997.
4. S. Vinga and J. Almeida, Alignment-Free Sequence Comparison: A Review, *Bioinformatics* **19**(3) 513–523 (2003).
5. G. Ferragina, R. Giancarlo *et al.*, Compression-based classification of biological sequences and structures via the Universal Similarity Metric: experimental assessment, *BMC Bioinformatics* **8**:252 (2007).
6. M. Li, H.J. Badger *et al.*, An information sequence distance and its application to whole mitochondrial genome phylogeny, *Bioinformatics* **17**(2)149–154 (2001).
7. X. Chen,S. Kwong S., and M. Li, A Compression Algorithm for DNA Sequences and Its Applications in Genome Comparison, *Genome Informatics (GIW'99)*, Tokyo, Japan, 51–61, 1999.
8. H.M. Aktulga, I. Kontoyiannis, L.A. Lyznik, L. Szpankowski, A.Y. Grama, and Szpankowski W., Identifying Statistical Dependence in Genomic Sequences via Mutual Information Estimates, *EURASIP Journal on Bioinformatics and Systems Biology*, Article ID 14741, 11 pages, 2007. doi:10.1155/2007/14741.
9. X. Chen, S. Kwong, M. Li, A compression algorithm for DNA sequences based on approximate matching, *Proceedings of the Tenth Workshop on Genome Informatics (GIW)*, number 10 in Genome Informatics Series, Tokyo, December 14-15 1999.
10. J. K. Lanctot, M. Li, and E. Yang, Estimating DNA sequence entropy, *Proceedings of the eleventh annual ACM-SIAM symposium on Discrete algorithms*, 409-418, January 09-11, 2000, San Francisco, California, United States
11. G. Korodi and I. Tabus, An efficient normalized maximum likelihood algorithm for DNA sequence compression, *ACM Transactions on Information Systems (TOIS)*, **23**(1), 3-34, (2005).
12. K.G. Srinivasa, M. Jagadish , K.R. Venugopal, and Patnaik L. M., Non-repetitive DNA Sequence Compression Using Memoization, *Lectures Notes in Computer Science* **4345** 402–412, (2006) Springer Berlin Heidelberg.
13. G. Manzini and M Rastero, A simple and fast DNA compressor, *Software-Practice & Experience*, **35**(4), 1397–1411, (2005).
14. T. Matsumoto, K. Sadakane, and H. Imai, Biological sequence compression algorithms, in: *Proc. Genome Informatics Workshop*, 43–52. Universal Academy Press, Tokyo, (2000).

15. K. Sadakane, T. Okazaki, T. Matsumoto, H. Imai, Implementing the context tree weighting method by using conditional probabilities. *Proc. of 22th Symposium on Information Theory and its Applications*, 673–676, December 1999, (in Japanese).
16. X. Chen, M. Li, B. Ma, and J. Tromp, DNACompress: Fast and effective DNA sequence compression, *Bioinformatics*, **18**(12), 1696-1698, (2002).
17. J.G. Cleary I.H. and Witten, Data compression using adaptive coding and partial string matching, *IEEE Trans. on Commun.*, **COM-32**(4), 396-402, (1984).
18. S. Grumbach and F. Tahi, A new challenge for compression algorithms: genetic sequences, *Information Processing & Management*, **30** 875-886, (1994).
19. National Center for Biotechnology Information, http://www.ncbi.nlm.nih.gov.
20. H.H. Otu and K. Sayood, K., A new sequence distance measure for phylogenetic tree construction, *Bioinformatics*, **19**(16) (2003).
21. A. Milosavjevic, Discovering by minimal length encoding: A case study in molecular evolution, *Machine Learning*, **12**, 68-87, (1993).
22. F.M.J. Willems, Y.M. Shtarkov, and T.J. Tjalkens, The context tree weighting method: basic properties, *IEEE Trans. Inform. Theory*, **IT-41**(3) 653-664, (1995).
23. Z. Dawy, J. Hagenauer, P. Hanus, J.C. Mueller, Mutual Information Based Distance Measures for Classification and Content Recognition with Applications to Genetics, *IEEE International Conference on Communications (ICC 2005)*, Seoul, South Korea, **2**, 820–824 (2005).
24. P. Hanus, Z. Dawy, J. Hagenauer,J.C. Mueller, DNA Classification Using Mutual Information Based Compression Distance Measures, in: *14th International Conference of Medical Physics*, Nürnberg, Germany, September 2005, Biomedizinische Technik, vol. 50, supp. vol. 1, part 2, S. 1434-1435.
25. P. Hanus, J. Dingel, J. Zech, J. Hagenauer, J.C. Mueller, Information Theoretic Distance Measures in Phylogenomics, in: *International Workshop on Information Theory and Applications (ITA)*, San Diego, CA, USA, January 2007.
26. M. Li, X. Chen, X. Li, B. Ma, P. M. B. Vitanyi, The Similarity Metric, *IEEE Transactions On Information Theory*, **50**(12), 3250–3264, (2004).
27. M. Li, P. Vitanyi, *An Introduction to Kolmogorov Complexity and its Applications, 2nd ed.* New York, Springer-Verlag, 1997.
28. A. Reyes *et al.*, Where Do Rodents Fit? Evidence from the Complete Mitochondrial Genome of Sciurus vulgaris, *Molecular Biology and Evolution* **17** 979–983 (2000).
29. C. Shannon, *A Mathematical Theory of Communication*, Reprinted with corrections from The Bell System Technical Journal, **27**, pp. 379-423, 623-656, July, October, (1948).
30. A. Lempel and J. Ziv, On the complexity of finite sequences, *IEEE T. Inform. Theory* **22**, 75–81 (1976).
31. T.M. Cover and J.A. Thomas, *Elements of Information Theory*, Wiley, 1990.
32. T. H. Jukes and C. Cantor, *Mammalian Protein Metabolism*, chapter: Evolution of protein molecules, 21–132, Academic Press, New York, (1969).
33. M. Kimura, A simple method for estimating evolutionary rates of base substitutions through comparative studies of nucleotide sequences. *Molecular Evolution*, **16**, 111–120, (1980).
34. J.A. Lake, A rate-independent technique for analysis of nucleic acid sequences: Evolutionary parsimony, *Molecular Biology and Evolution*, **4** 167–191, (1987).

Modelling of 3D Early Blood Vessel Formation: Simulations and Morphological Analysis

F. Cavalli*, A. Gamba†, G. Naldi*, S. Oriboni*, M. Semplice*, D. Valdembri** and G. Serini**

*Dipartimento di Matematica, Università degli Studi di Milano, via Saldini 50, 20133 Milano, Italia.
†Dipartimento di Matematica, Politecnico di Torino, Corso Duca degli Abruzzi 24, 10129, Torino, and INFN, unità di Torino, via P. Giura 1, 10125 Torino, Italia.
**Department of Oncological Sciences and Division of Molecular Angiogenesis, Institute for Cancer Research and Treatment, University of Torino, Strada Provinciale 142 Km 3.95, 10060 Candiolo (TO), Italia.

Abstract. Vascular networks form by a self-aggregation process of individual endothelial cells that differentiate at seemingly random sites in the embryo and collectively migrate toward each other forming a preliminary vascular plexus (*vasculogenesis*), followed by functional remodelling that gives rise to the final hierarchical system (*angiogenesis*).

The study of this phenomenon is performed by biologists using *in vitro* and *in vivo* assays, both in two and three dimensional settings. The lack of direct biological evidence of the chemotactic autocrine loop that is thought to be the main responsible for the early aggregation, called for the development of mathematical models of this process, in order to study the possible effects of such a loop. After successful two-dimensional studies, the model was recently extended to a three dimensional setting and a suitably efficient approximation scheme for the numerical simulations has been developed, while three-dimensional images of embryo vascular networks are becoming available through confocal microscopy.

This paper is concerned with the comparison of experimental and simulated data on embryo vascular plexi. Critical exponents of percolation, Euler-Poincaré characteristic, fractal dimension, power spectrum decay and maximum distance from a vessel are considered and compared.

Keywords: Chemotaxis, pattern formation, scaling laws, numerical simulations, structures and organization in complex systems
PACS: 87.17.Jj, 87.18.Hf, 89.75.Da, 89.75.Fb

1. INTRODUCTION

Vasculogenesis is the process of blood vessel network formation by the spontaneous aggregation of individual cells that, in the early stages of embryo development, migrate towards vascularization sites [1].

The study of this phenomenon is performed by biologists using *in vitro* and *in vivo* assays, both in two and three dimensional settings. A chemotactic autocrine loop mediated by VEGF signalling among endothelial cells is thought to be the main responsible at least for the early organization of the network. However no direct biological experimental evidence of this loop is available. On the other hand, a successful theoretical model of two dimensional experimental vasculogenesis has been recently proposed, showing the relevance of percolation concepts and of cell cross-talk (chemotactic autocrine loop) to the understanding of this self-aggregation process [2]. A natural 3D extension of the

CP1028, *Collective Dynamics: Topics on Competition and Cooperation in the Biosciences*
edited by L. M. Ricciardi, A. Buonocore, and E. Pirozzi

earlier proposed computational model, which is relevant for the investigation of the genuinely three dimensional process of vasculogenesis in vertebrate embryos was proposed in [3].

In order to evaluate the accuracy of the theoretical models one may perform numerical simulations based on them and compare the results with experimental data. Due to the large biological variability of the samples, a quantitative comparison of experimental observations with computational results must be based on the measurement of robust geometrical features. Moreover, observable quantities must have a principally statistical meaning. A lesson on this regard comes from the study of two dimensional vasculogenesis, where experimental assays are realized by scattering a controlled cell density n on an appropriate gel surface. By varying the cell density n one observes a percolative transition, studied in [2].

Up to now, the comparison of the computational model with experimental observations has been successful but limited to two dimensional settings. This is related to the fact that most experiments in experimental vasculogenesis are actually realized by seeding cells on a two dimensional gel matrix. It is however worth noticing that a complete understanding of the vascularization process is possible only if it is considered in its natural three dimensional setting ([4, 5]). Since recently confocal microscopy is providing three dimensional images, *in vivo* 3D experiments will be soon possible. In this paper we give preliminary results on the computation and comparison of the following geometrical and topological properties of three-dimensional vascular networks: percolation, fractal dimension, Euler-Poincaré characteristic and maximum distance from a vessel.

On a lattice whose bonds may be in open and closed states, percolation consists in the presence of a continuous path connecting opposite sides of the sample. Analogously, on digitized images of vascular networks, percolation is the existence of a cluster of occupied pixels (a vessel) connecting opposite sides of the sample: this is clearly fundamental since the vascular system must reach all parts of the body. *In vitro* two-dimensional experiments showed a transition whereby for low cell densities percolating clusters do not exist, while they appear at a well-defined critical density n_c. This behaviour is confirmed by the two-dimensional simulations of [2]. In the vicinity of the critical density n_c the geometric properties of clusters show a peculiar scaling behaviour, described by *critical exponents* that are observables characterizing the aggregation dynamics. We have thus two reasons to study percolation in relation to vascular network formation: (*i*) percolation is a fundamental property for vascular networks since blood should have the possibility to travel across the whole tissue sample along the network to carry nutrients to tissues; (*ii*) critical exponents are robust observables characterizing the aggregation dynamics.

The fractal dimension is an indicator of the self-similarity of the network structure at different scales. The Euler-Poincaré characteristic is a topological feature of the capillary network, proportional to the number of loops present in the structure. Finally we study the maximum distance from a vessel, checking what portion of tissue (in the experimental images) and of computational grid (in the simulations) is within 200μm of a vessel, this being the average diffusion length of oxygen in tissues.

The paper is organized as follows. Section 2 summarizes some background knowledge on the biological problem of vascular network formation and contains a description of the experimental observations. Section 3 describes the three-dimensional model and its numerical approximation. In Section 4 we present the computation of the various geometrical properties of the simulated and the experimental capillary structures.

2. BIOLOGICAL BACKGROUND

Capillaries supply tissues with nutrients in an optimal way: they are made of endothelial cells and develop in a hierarchical vascular system with intercapillary distances ranging from 50 to 300 μm. Their growth is essentially driven by two processes: vasculogenesis and angiogenesis [1]. Vasculogenesis consists of local differentiation of precursor cells to endothelial ones, that assemble into a vascular network by directed migration and cohesion. Angiogenesis is essentially characterized by sprouting and merging of vessels, and remodelling of an immature structure. In two dimensional assays, capillary network formation proceeds along different stages. Microvascular endothelial cells are randomly plated on Matrigel, a natural basal membrane matrix, and tracked by video microscopy [2]. They migrate and form an early network then, after a network remodelling stage, some tubular structures appear.

The dynamics and an accurate statistics of individual cells trajectories has been presented in [2], showing that, in the first stage the cell motion has marked directional persistence, pointing toward zones of higher cell concentration. This suggests that cells communicate emitting some soluble chemical factors that diffuse (and degrade) in the surrounding medium and moving along the gradients of this chemical field. Cells behave like indirectly interacting particles, the interaction being mediated by the release of soluble chemotactic factors.

In the embryo, endothelial cells are produced and migrate in a three dimensional scaffold, the extracellular matrix. Migration is actually performed through a series of biochemical processes, such as sensing of chemotactic gradients, and of mechanical operations, such as extensions, contractions, and degradation of the extracellular matrix along the way. During the first phase cells migrate over distances which are an order of magnitude larger than their radius and aggregate when they adhere with one of their neighbours.

Today several major developments in three dimensional cell culture and in cell and tissue imaging allow the observation of the mechanisms of cell migration and aggregation in three dimensional settings in real time [6, 7]. Observation of *in vitro* vasculogenesis can be performed in transparent tissues, such as in chick embryos, with a confocal microscope. As a sample of typical vascular structures that are observed in a three dimensional setting in the early stages of development of a living being, we provide here $(750\,\mu\text{m})^2$ images of chick embryo brain at different development stages (Fig. 1). Embryo development stages are usually classified according to Hamilton and Hamburger (HH classification). Wheat germ agglutinin (WGA) marked with the fluorescent traces Alexa-488 was injected in the embryos. After fixing, the embryo brains were imaged at the confocal microscopy facility at IRCC (Candiolo, Italy) obtaining images as a series of 1024×1024 pictures of sections taken at distances varying from 4 to 10 μm. Observed development stages were HH stages 17, 20 and 26. At an early stage (about 52-64 hours) one observes a typical immature vascular network formed by vasculogenesis and characterized by a high density of similar blood vessels. At the next stage (70-72

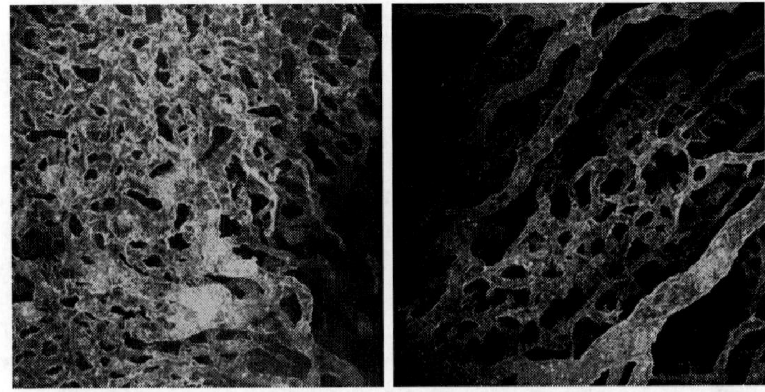

FIGURE 1. Vascular networks formed by vasculogenesis in chick embryo brain, at various stages of development, classified according to Hamilton and Hamburger (HH). These Free Maximum Projections have been constructed by the microscope software by superimposing the different sections and assigning higher luminosity intensities to higher standing sections, thus obtaining a pseudo 3D image giving an overall idea of the imaged sample. Left: HH stage 17, corresponding to 52-64 hours; right: HH stage 26 (5 days).

hours) we observe initial remodelling of the vascular network. Remodelling becomes more evident when the embryo is 5 days old, when blood vessels are organized in a mature, hierarchically organized vascular tree. Free Maximum Projections of two samples at different stages of development are shown in Figure 1.

From the study of in vitro vasculogenesis we can argue that the formation of experimentally observed structures can be explained as the consequence of cell motility and of cell cross-talk mediated by the exchange of soluble chemical factors (chemotactic autocrine loop). The theoretical model also shows that the main factors determining the qualitative properties of the observed vascular structures are the available cell density and the diffusivity and half-life of the soluble chemical exchanged. It seems that only the dynamical rules followed by the individual cell are actually encoded in the genes. The interplay of these simple dynamical rules with the geometrical and physical properties of the environment produces the highly structured final result. At the moment, no direct observation of the chemotactic autocrine loop regulating vascular network formation is available, although several indirect biochemical observations point to it [8], so, the main evidence in this sense still comes from the theoretical analysis of computational models.

The mathematical model described in this paper is about the formation of an early network (like the one in Fig. 1A). However, a good understanding and prediction of the formation of the early vascular network is essential for the subsequent remodelling. Since we are interested in developing a quantitative comparison between experimental data and theoretical model, we select a set of observable quantities providing robust quantitative information on the network geometry and topology. Images corresponding to the different sections will be reconstructed to provide 3D models to be compared with the results of numerical simulations of the theoretical model.

3. COMPUTATIONAL MODEL OF BLOOD VESSEL GROWTH

The multidimensional Burgers' equation gives a coarse hydrodynamic description of the motion of independent agents performing rectilinear motion and interacting at short ranges (see e.g. [9, 10]). A particularly interesting case is that of weakly interacting sticky matter. Each particle starts moving with a constant velocity given by a random statistical distribution. This dynamics gives rise to intersection of trajectories and formation of shock waves. After the birth of these local singularities, regions of high density grow and form a peculiar network-like structure. The main feature of this structure is the existence of comparatively thin layers and filaments of high density that separate large low-density regions.

For the blood vessel formation one has to take into account evidence suggesting that cells do not behave as independent agents, but rather exchange information in the form of soluble chemical factors. This lead to the model proposed by Gamba et al. in [2] and Serini et al. in [11]. The model describes the motion of a fluid of randomly seeded independent particles which communicate through emission of a soluble factor and move toward its concentration gradients, and allows to well reproduce both the observed percolative transition and the typical scale of two dimensional vascular networks.

3.1. Model equations

We assume that the cell population can be described by a continuous distribution of density $n(\vec{x},t)$, where $\vec{x} \in \mathbb{R}^3$ is the space variable, and $t \geq 0$ is the time variable. The population density moves with velocities $\vec{v}(\vec{x},t)$, that are accelerated by chemical gradients of a soluble factor. The chemoattractant soluble factor is released by the cells, diffuse and degrade in time and it is described by a scalar chemical concentration field $c(\vec{x},t)$.

Thus our model for the vasculogenesis process combines the mass conservation law for cell matter, a momentum balance law which links the acting forces and the cell acceleration, and a reaction-diffusion equation for the production, degradation and diffusion of the concentration of the chemotactic factor. The first equation expresses the conservation of the number of cells. The second one takes into account the phenomenological chemotactic force, the dissipation by interaction with the substrate, the phenomenon of cell directional persistency along their trajectories and a term implementing an excluded volume constraint [2, 12]. The third equation is a reaction diffusion equation for the field $c(\vec{x},t)$, that includes a source term proportional to the cell concentration $n(\vec{x},t)$, thus closing the description of the autocrine loop. Explicitly, we consider the following system:

$$\frac{\partial n}{\partial t} + \nabla \cdot (n\mathbf{v}) = 0 \tag{1a}$$

$$\frac{\partial \mathbf{v}}{\partial t} + \mathbf{v} \cdot \nabla \mathbf{v} = \mu(c)\nabla c - \nabla \phi(n) - \beta(c)\mathbf{v} \tag{1b}$$

$$\frac{\partial c}{\partial t} = D\Delta c + \alpha(c)n - \frac{c}{\tau} \tag{1c}$$

where μ measures the cell response to the chemotactic factor, while D and τ are respectively the diffusion coefficient and the characteristic degradation time of the soluble chemoattractant. The function α determines the rate of release of the chemical factor. The friction term $-\beta\mathbf{v}$ mimics the dissipative interaction of the cells with the extracellular matrix.

The term $\nabla\phi(n)$ is a density dependent pressure term, where $\phi(n)$ is zero for low densities, and increases for densities above a suitable threshold. This is a phenomenological term which models short range interaction between cells and the fact that cells do not interpenetrate and have some degree of rigidity. For simplicity we choose the following functional form

$$\phi(n) = \begin{cases} B_p(n-n_0)^{C_p} & n > n_0 \\ 0 & n \leq n_0 \, . \end{cases} \tag{2}$$

A simple model can be obtained by assuming that the cell sensitivity μ, the rate of release of the chemoattractant α and the friction coefficient β are constant. A more realistic description may be obtained including saturation effects as functional dependencies of the aforementioned coefficients on the concentration c. We point out that μ determines the time scale of the evolutionary process, i.e. the time elapsed before the stationary state is reached.

However, biological observations suggest that the dynamics of cell changes when they establish cell-cell contacts. It is reasonable to suppose that a different genetic program is activated at this moment, disabling cell motility. We therefore switch off cell motility as soon as the cell concentration, signalled by chemoattractant emission, reaches a given threshold. In this way the computational system is guaranteed to reach a stationary state.

These effects can be taken into account using a non-constant sensitivity $\mu(c)$, a non-linear emission rate $\alpha(c)$, or a variable friction coefficient $\beta(c)$. We choose a threshold c_0 and functions of the form

$$\mu(c) = \mu_0[1 - \tanh(c - c_0)] \tag{3a}$$

$$\alpha(c) = \alpha_0[1 - \tanh(c - c_0)] \tag{3b}$$

$$\beta(c) = \beta_0[1 + \tanh(c - c_0)] \tag{3c}$$

The effect of the first two terms is that the sensitivity of the cells and their chemoattractant production is strongly damped when the concentration c reaches the threshold c_0. We did not observe a significant dependence on the exact form of the damping function, provided that it approximates a step function that is nonzero only when $c < c_0$.

$\beta(c)$, on the other hand has the effect of turning on a strong friction term at locations of high chemoattractant concentration. In [13] we performed several tests and observed that the different choices (3) are approximately equivalent in freezing the system into a network-like stationary state.

3.2. Numerical simulations

>From available experimental results [14] it is known that the major angiogenic growth factors have diffusion constant and half-life approximately equal to $D = 10^{-7}\text{cm}^2\text{s}^{-1}$ and $\tau = 4000$ s. We fix the other constant parameters by dimensional analysis and fitting to the characteristic scales of the biological system. In particular, we choose: $\mu_0 = 10^{-11}\text{mm}^4/\text{s}^3$, $\alpha = 1\text{s}^{-1}$, $\beta = 10^{-3}\text{s}^{-1}$. For the coefficients in the expression (2) of the pressure function ϕ we take $n_0 = 1.0, C_p = 3$ and $B_p = 10^{-3}$. Finally, in the absence of biological motivations in favour of any of the mechanisms (3), we take $\mu(c)$ as in (3a), leaving β and α constants.

Since in the early stages of development almost all intraembryonic mesodermal tissues contain migrating endothelial precursors, we use initial conditions representing a randomly scattered distribution of cells, *i.e.*, we throw an assigned number of cells in random positions inside the cubic box, with zero initial velocities and zero initial concentration of the soluble factor, with a single cell given initially by a Gaussian bump of width σ of the order of the average cell radius ($\simeq 15\mu$m) and unitary weight in the integrated cell density field n. The initial velocity is set to zero.

Our numerical scheme is described in [3] and it is based on a second order finite difference method. Space and time discretization are performed by using TVD schemes and, respectively, IMEX schemes. Very fine grids have to be used in order to resolve the details of the $n(\mathbf{x},t)$ field, which may contain hundreds of small bumps, each representing a single cell. Since each cell has radius $\sigma = 15\mu$m, one needs a grid spacing such that $\Delta x < 10\mu$m and therefore we employed grids of 128^3 cells for a cubic domain of 1mm side. Due to the computational costs of realistic simulations, we have implemented the numerical algorithm on the high performance cluster for parallel computation at the Department of Mathematics of the University of Milano (*http://cluster.mat.unimi.it/*). In particular, to avoid restrictions on the time step, time advancement for the linear diffusion equation 1c is performed with the implicit Euler scheme. The linear system arising from the approximation of equation 1c is solved with the conjugate gradient algorithm, with block-Jacobi global preconditioner and Jacobi preconditioner on each block.

We performed numerical simulations with varying initial average cell density $\bar{n}$. We observed that the initially randomly distributed cells (Fig. 2) coalesce forming elongated structures (Fig. 3) and evolve towards a stationary state (Fig. 4) mimicking the geometry of a blood vessel network in the early stages of formation. The figures show an isosurface of the density variable u, together with two cross-section density plots.

We assigned $\bar{n}$ in the range $2100 - 3500\,\text{cells}/\text{mm}^3$ and performed 10 to 15 runs for each density value with a $128 \times 128 \times 128$ grid on a biological system of 1 mm^3. The characteristic lengths and geometric properties of the stationary state depend on $\bar{n}$ and we observed a percolative phase transition similar to the one described in [2] for the twodimensional case.

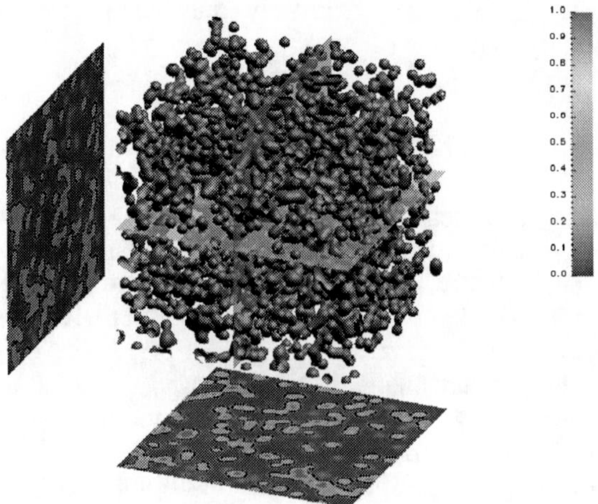

FIGURE 2. Initial state of a numerical simulation with 2500cells/mm^3. The colorbar on the right is referred to the coloring of the cross sections. The red threedimensional isosurface corresponds to the black contour lines in the cross sections ($t = 0$).

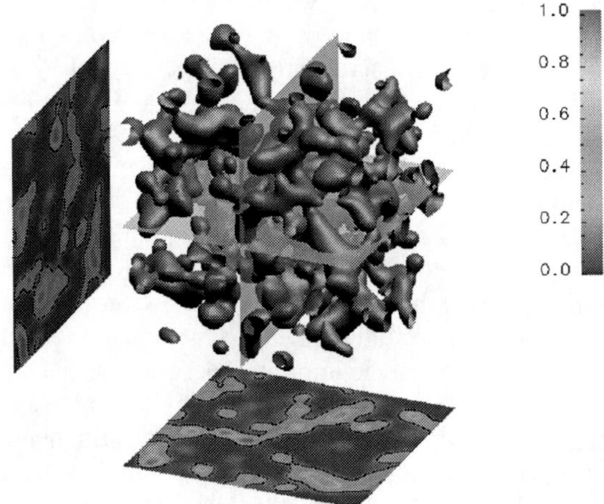

FIGURE 3. Transient state of the evolution of the initial state depicted in Fig. 2 according to model (1). The initial formation of network-like structures is observed ($t = 40$).

4. MORPHOLOGICAL AND GEOMETRICAL ANALYSIS

4.1. Analysis of the percolative phase transition

In experimental blood vessel formation a percolative transition is observed by varying the initial cell density [2]. For low cell densities only isolated clusters of endothelial

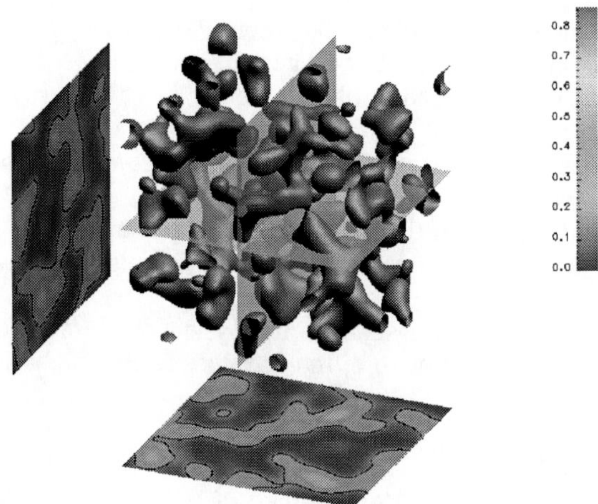

FIGURE 4. Stationary state of the evolution of the states depicted in Figs. 2 and 3 according to model (1). Well developed threedimensional network-like structures are observed($t = 60$).

cells are observed, while for very high densities cells fill the whole available space. In between these two extreme behaviours, close to a critical cell density n_c, one observes the formation of critical percolating clusters connecting opposite sides of the domain, characterized by well defined scaling laws and exponents. These exponents are known to independent of the microscopic details of the process while their values characterize different classes of aggregation dynamics [15, 2].

The purely geometric problem of percolation is one of the simplest phase transitions occurring in nature. Many percolative models show a second order phase transition at a critical value n_c of the average density $\bar{n}$, *i.e.* the probability Π of observing an infinite, percolating cluster is 0 for $\bar{n} < n_c$ and 1 for $\bar{n} > n_c$ [15]. The phase transition can be studied by focusing on the values of an order parameter, i.e. an observable quantity that is zero before the transition and takes on values of order 1 after it. In a percolation problem the natural order parameter is the probability P that a randomly chosen site belongs to the infinite cluster (on finite grids, the infinite cluster is substituted by the largest one).

In the vicinity of the critical density n_c the geometric properties of clusters show a peculiar scaling behaviour. For instance, in a system of finite linear size L, the probability of percolation $\Pi(n,L)$, defined empirically as the fraction of computational experiments that produce a percolating cluster, is actually a function of the combination $(n-n_c)L^{1/\nu}$, where ν is a universal exponent [15].

In a neighbourhood of the critical point and on a system of finite size L, the following finite size scaling relations is also observed:

$$\Pi(\bar{n},L) \sim \widehat{\Pi}[(\bar{n}-n_c)L^{1/\nu}] \tag{4}$$

There are two main reasons to study percolation in relation to vascular network formation: (*i*) percolation is a fundamental property for vascular networks, since blood should have the possibility to travel across the whole vascular network to carry nutrients to tissues; (*ii*) critical exponents are robust observables characterizing the aggregation dynamics.

A rather complete characterization of percolative exponents in the twodimensional case has been provided in [2].

As a first step in the study of the more realistic three dimensional case, we compute the exponent ν characterizing the structures produced by the model dynamics (1) with varying initial cell density.

To this aim, extensive numerical simulation of system (1) were performed using lattice sizes $L = 1, 0.78, 0.62, 0.5\,\mathrm{mm}$, with different values of the initial density $\bar{n}$. For each point 10 to 15 realizations of the system of size 1mm were computed, depending on the proximity to the critical point.

The continuous density at final time $n(\mathbf{x})$ was then mapped to a set of occupied and empty sites by choosing a threshold n_0. Each region of adjacent occupied sites (cluster) was marked with a different index. The percolation probability Π for each set of realizations was then measured. In Fig. 5 we show clusters obtained in a box with $L = 0.5\,\mathrm{mm}$ with $\bar{n} = 3100$. The largest percolating cluster is shown in red, together with some other smaller clusters shown in different colours.

Using relation (4), we estimate the position of the critical point n_c and the value of the critical exponent ν. The data for different box side length and initial density should lie on a single curve after rescaling the densities as $\hat{n} = (\bar{n} - n_c)L^{1/\nu}$. For fixed n_c and ν we rescale $\bar{n}$ and fit the data with a logistic curve, then compute the distance of the data from the curve. The squared distance is minimized to obtain estimates for n_c and ν.

Using $n_0 = 0.35$ we obtain $n_c = 2658$ and $\nu = 0.84$ (the data collapse in shown in Fig. 6). This latter value is compatible with the known value 0.88 for random percolation in three dimensions [15].

4.2. Euler-Poincaré characteristic

Recently, new insights in the study of the critical properties of clusters in percolation theory have emerged based on ideas coming from mathematical morphology [16] and integral geometry [17]. One of these measures is the Euler-Poincaré characteristic, which is a well-known descriptor of the topological features of geometric patterns. It belongs to the finite set of Minkowski functionals whose origin lies in the mathematical study of convex bodies and integral geometry (see [17]). Minkowski functionals encompass standard parameters such as volume, area, length, as well as Euler-Poincaré characteristic. These measures share the following remarkable property: any homogeneous, additive, isometry-invariant and conditionally continuous functional on a compact subset of the $d-$dimensional Euclidean space can be expressed as a linear combination of the Minkowski functionals (Hadwiger's theorem of integral geometry [17]).

The estimation of the Euler-Poincaré characteristic of a spatial set is of current interest in stereology as well as in image analysis. In stereology the estimation is based mostly on observations on parallel planar sections (see e.g [18] and references therein).

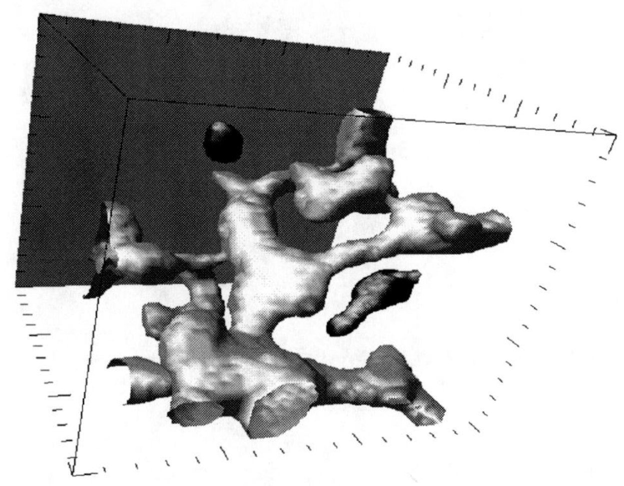

FIGURE 5. Cluster percolation with cell density $n = 2900\text{cells/mm}^3$. We show three connected clusters in a realization of model (1). The largest cluster shown in light gray percolates. The other two are plot in dark gray.

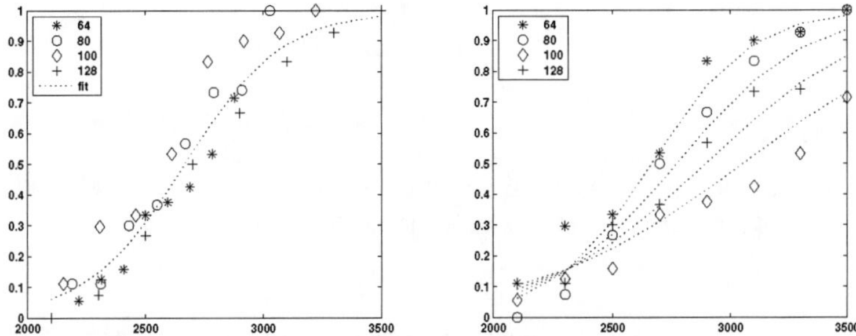

FIGURE 6. Collapse of percolation probability data using formula (4). The original data are shown in the right panel.

Observations on lattices of points are treated mainly in the context of image analysis, and there the estimation of the Euler number is based on the Euler formula of graph theory, see [16].

Here we start from the discretized three dimensional image I_N of a spatial structure which is obtained from the numerical simulation of model 1. Then we compute the isosurface geometry for data I_N of final density at some isosurface. The shape of the isosurface geometry is described by a 3D triangulation. The triangulation data structure can be seen as a container for faces and vertices maintaining incidence and adjacency relations among them. By using the 3D triangulation we can consider the Euler characteristic χ

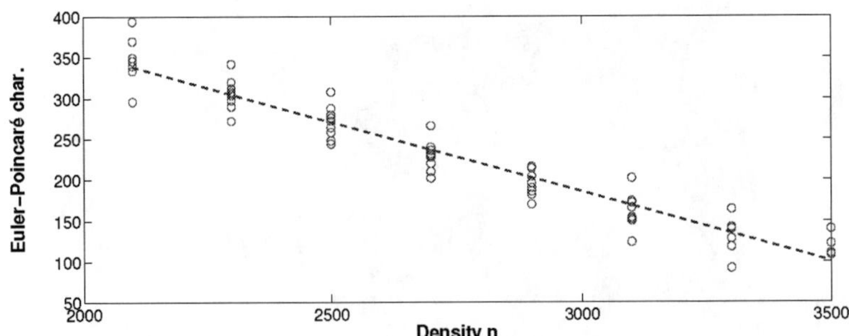

FIGURE 7. Euler-Poincaré characteristic of spatial images obtained as a stationary state of the computational model for different initial density.

as classically defined for polyhedra P, according to the formula

$$\chi(P) = V - E + F$$

where V, E, and F are respectively the numbers of vertices (corners), edges and faces in the given polyhedron. The vascular structures are composed by many disjoint clusters C_j, then the Euler characteristic of their disjoint union is the sum of their Euler characteristics [19]:

$$\chi(\bigsqcup_j C_j) = \sum_j \chi(C_j).$$

In Fig. 7 we have plotted the behaviour of the Euler characteristic χ with respect to different initial densities. The figure shows that the values of the characteristic χ decrease (almost linearly) when the initial density increases. We recall that the Euler characteristic of a closed orientable surface is related to its genus g as

$$\chi = 2 - 2g.$$

Moreover for a surface with c connect component, as for the vessel networks, we have

$$\chi = 2(c - g).$$

Then from our computational results appear that the genus of the spatial structure increases with respect to the initial density. This means that for low initial density many small disjoint spherical cluster form. While, for big values of the initial density the number of cluster decreases and some of them involve a higher number of sites. Then, for high initial density more "holes" and loops appear.

4.3. Fractal dimension

Many anatomic structures display a fractal-like geometry. The self-similarity of these structures allows fast and efficient transport over complex, spatially distributed networks. Examples may be found in the nervous system (information dissemination), in

TABLE 1. Average of the D_0 fractal dimension computed on the simulated vascular networks. We list the initial cell density (left), the values of D_0 computed with the box-counting algorithm (centre) and the values of D_0 computed with the sandbox algorithm (right).

$\bar{n}$	$\widehat{D_0}$	$\widetilde{D_0}$
2100	2.3444 ± 0.0047	2.2746 ± 0.0110
2300	2.3724 ± 0.0036	2.3050 ± 0.0091
2500	2.4020 ± 0.0027	2.3407 ± 0.0100
2700	2.4292 ± 0.0028	2.3713 ± 0.0093
2900	2.4570 ± 0.0026	2.4043 ± 0.0061
3100	2.4852 ± 0.0026	2.4320 ± 0.0065
3300	2.5062 ± 0.0031	2.4547 ± 0.0046
3500	2.5185 ± 0.0028	2.4687 ± 0.0049

the bowel (nutrient absorption), as well as in the vascular system, where the branched arterial and venous trees allow for efficient oxygen transport to the tissues.

We are here interested in studying the fractal dimension of the early vascular plexus that we obtain as stationary state of our simulations. First we turn the density variable $n(x)$ of the stationary state of a simulation into a three-dimensional matrix of zeros and ones, representing empty and occupied sites on the computational grid. Next, let $\vec{y}$ be a list of all the positions of the occupied sites and define

$$C(q,r) = \frac{1}{M}\sum_{k=1}^{M}\left[\frac{1}{M}\sum_{j=1}^{M}\theta(r-|y(j)-y(k)|)\right]^{q-1}, \tag{5}$$

where the sums range over all occupied sites and θ is the Heaviside function. The inner sum thus computes the (normalized) number of occupied sites within distance r of the k-th occupied site. The outer sum performs a weighted average over all the possible centres $y(k)$.

Following the theory started by [20], in the $r \to 0$ limit,

$$C(q,r) \sim r^{(q-1)D_q}$$

where D_q is the fractal dimension. Note that for $q = 0$ one recovers the box-counting dimension and for $q = 2$ the correlation dimension.

The computational cost of (5) can be reduced approximating the fractal dimension D_0 of our simulated vascular networks using both the box-counting and the sandbox algorithm. The former gives an estimation of D_0 by successively splitting the numerical grid into cubes of side $L/2^k$ and recording the fraction of subsamples that contain at least one pixel of the network. This gives the points $C(0, L/2^k)$ of the curve $C(0,r)$ and D_0 can be estimated by looking at the slope of the approximately linear log-log graph of $C(0,r)$.

The sand-box algorithm follows more closely the definition (5) except that, in order to save on computational cost, the range of the outer sum is reduced. In practice, $K < M$ pixel are randomly selected among the full pixels, cubes of side r centred at these sites are investigated for the density of the fractal restricted to each of them (inner sum of (5)) and the results are averaged over these K pixels (outer sum of (5)). We observed that, for our images, using $K = 300$ is a good compromise between stability of the estimated values and the computational cost. The function $C(0,r)$ was approximated as above for $r = 3,5,9,17,33,65$ pixels and then D_0 was estimated as described above for the box-counting algorithm.

In Table 1 we list the results obtained with the box-counting ($\widehat{D_0}$) and the sand-box ($\widetilde{D_0}$). The results are averaged over all the simulations with the same initial density. The small standard deviations and the very small discrepancies between the two estimated values at each density show the robustness of this procedure. Both the series of values show a slight increase in the fractal dimension when the initial cell density is increased. This is a three-dimensional realisation of the swiss-cheese phenomenon observed in two dimensional in vitro experiments: when the cell density is increased too much above the threshold for the network formation, the cells do not aggregate in filaments (1D structures) but in large blobs of matter (2D-structures) [2]. In the present three dimensional setting, these results on D_0, show that when the initial cell density is increased, the fractal dimension of the vascular networks is driven towards 3. We conjecture that this is a footprint of the presence of larger three dimensional blobs in the fractal.

4.4. Power spectrum exponent

The behaviour of the power spectrum of biological signals and structures has been investigated, originally in the context of the heartbeat periodicity [21] and later on in many other situations [22]. Up to now, this technique is typically employed to reveal the presence of long-range correlations and complexity in temporal time-series. In the present context we perform a power spectrum analysis of the real and simulated vascular networks, looking for spatial correlations.

For a one-dimensional signal, denoting with $S(f)$ the absolute value of the FFT coefficient at frequency f, one looks for the exponent β that best fits the data with the power law $S(f) \sim (1/f)^{\beta}$. In our three-dimensional setting, denoting $S(i,j,k)$ the FFT coefficients, we set

$$S(f) = \langle |S(i,j,k)| \rangle_{f=i+j+k}$$

where $\langle . \rangle$ denotes averaging. We can now proceed to estimate β from the slope of the log-log graph of $S(f)$ against f, as usual in the one-dimensional setting.

When analyzing the experimental images, a source of major difficulties was their z-axis resolution. Recall that each of the 3-D images is composed of a series of 1024×1024 pixel slices of a portion of chick brain $750 \times 750\mu$m wide. The slices however are interspaced by 4 up to 10μm, depending on the experiment. Thus the z-axis resolution (perpendicular to the plane of the slices) is much lower than the resolution in the other two directions. The procedure followed was to first select a 256×256 pixels wide, significant portion of the images; these were cleared of *salt-and-pepper* noise via median

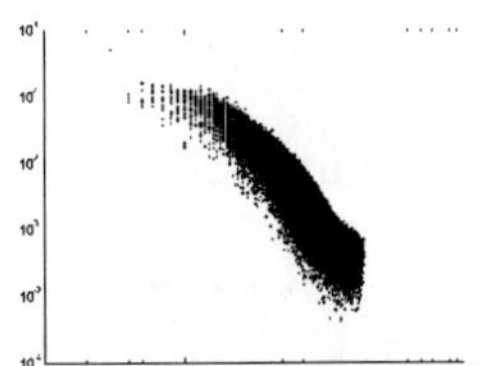

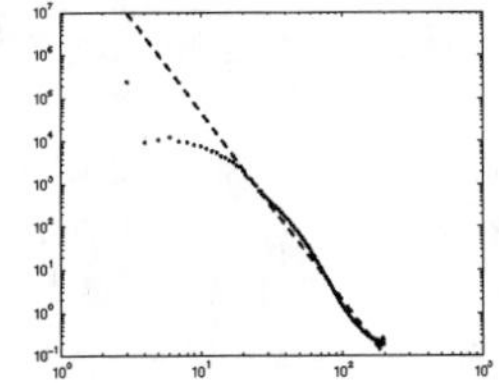

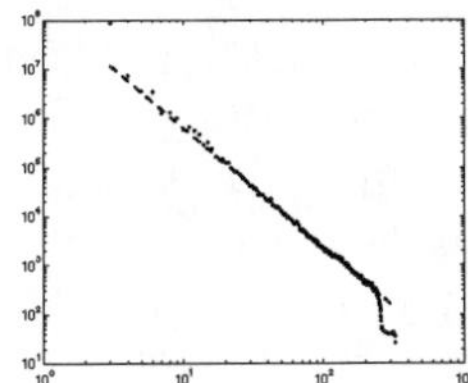

FIGURE 8. Power spectrum exponent for experimental and simulated networks, logarithmic scales are used for the x-axes of frequencies.

TABLE 2. Power spectrum exponent for experimental and simulated networks.

$\bar{n}$	average β	standard deviation
2100	4.3079	0.0899
2300	4.2456	0.0993
2500	4.3246	0.1162
2700	4.3878	0.0676
2900	4.4514	0.0600
3100	4.5353	0.0616
3300	4.6410	0.0597
3500	4.6959	0.0239

embryo stage	average β	standard deviation
HH17	2.23	0.51
HH20	1.63	0.22
HH26	1.96	0.50

filtering and then the z resolution was increased via cubic spline interpolation in order to have the same resolution in all the three cartesian directions. The exponent β for experimental images was estimated on this preprocessed data.

Figure 8 shows our results on a typical experimental and simulated network. On the left we show the graphs of $S(i,j,k)$ against $f = i+j+k$, for a simulation with initial average density $\bar{n} = 2500$, which is close to the percolation threshold. In the centre we show the $S(f)$ graph with the best fit for $(1/f)^{\beta}$. On the right hand side, we show the $S(f)$ graph for an experimental image on a chick brain from an embryo at HH17 stage.

As shown in Table 2, the estimated values for β are around 2 for the simulations and above 4 for the experimental images. Even taking into account the higher variability of the latter estimations, as clearly shown by the standard deviations, this difference is noticeable. The reason might be partly sought in the interpolation procedure used to get the β estimate. However this could also be the footprint of the angiogenic remodelling process that starts in the embryo at around stage HH17, but is not taken into account by our mathematical model.

4.5. Oxygen perfusion

We describe the vascular network as a matrix of zeros and ones, marking places where a capillary is present. These are obtained from the simulations through simple thresholding and from the experimental images via median filtering, thresholding and further refinement by hand.

Convoluting the vascular network with a sphere of radius 200μm, tells us whether each pixel of the embryo can be reached by oxygen perfusing from one of the capillary. Strikingly, results show that real vascular networks are able to oxygenate the whole embryo, while the simulated networks do not possess this property. This underlines the importance of the remodelling of the network, that is not yet taken into account by our model.

5. CONCLUSIONS

We have exposed results on the comparison, in a three dimensional setting, of numerical simulations of vascular network formation with *in vivo* images.

Understanding the dynamical process of vascularization is an important challenge for contemporary biology, which has clinical implications, like the possibility of testing medical treatments *in silico*. Recent biological research suggests that morphogenetic processes should be studied in their intrinsic three dimensional setting, where peculiar effects are observable, which are lost in lower dimensional models. Progress in visualization and experimental techniques is on the other hand making possible to get real three dimensional data of this process.

Taking into account directed cell motility and an autocrine loop of chemoattractant signalling, a differential model can be formulated. It is relevant to check whether this model is able to reproduce the main features of the natural process. In [3] we described a scheme for the efficient numerical approximation of the mathematical model, that allowed to run several test runs on a cluster for parallel computation.

Images of vascular network formation in chick embryo at different stages of development have been obtained by confocal microscopy, showing different stages of development of a vascular network in chick embryo. At an early stage of development one observes typical immature vascular networks formed by vasculogenesis. At subsequent stages one observes initial remodelling, which becomes more evident at later stages, where blood vessels become organized in a mature and hierarchically organized vascular network.

As a starting point towards a quantitative comparison between experimental data and the theoretical model we need to select a set of observables which provide robust quantitative information on the network geometry. The lesson learned from the study of two dimensional vasculogenesis is that percolative exponents are an interesting set of such observables, so we tested the computation of percolative exponents on simulated network structures.

Other properties of the topology (Euler characteristic), self-similarity (fractal dimension, power spectrum) and functionality (oxygen perfusion) of the simulated and real networks have been computed and, whenever possible, compared. Despite the difficulties posed by the computation of these quantities of the available images, this study provides a satisfactory validation of the theoretical model. It also enlightens possible shortcomings of the model, in that it does not account for the remodelling phase, and shows that images with a more uniform resolution in every direction should be acquired for a thorough comparison to be carried out.

REFERENCES

1. P. Carmeliet, *Nature Medicine* **6**, 389–395 (2000).
2. A. Gamba, D. Ambrosi, A. Coniglio, A. de Candia, S. Di Talia, E. Giraudo, G. Serini, L. Preziosi, and F. Bussolino, *Phys. Rev. Lett.* **90**, 118101 (2003).
3. F. Cavalli, A. Gamba, G. Naldi, M. Semplice, D. Valdembri, and G. Serini, *J. Comput. Phys.* **225**, 2283–2300 (2007).
4. A. Abbott, *Nature* **424**, 870–2 (2003).
5. E. Cukierman, R. Pankov, D. Stevens, and K. Yamada, *Science* **294**, 1708–12 (2001).
6. P. Friedl, *Curr Opin Immunol* **16**, 389–93 (2004).
7. P.J. Keller, F. Pampaloni, and E.H.K. Stelzer, *Curr Opin Cell Biol* **18** (2006).
8. L. Coultas, K. Chawengsaksophak, and J. Rossant, *Nature* **438**, 937–945 (2005).
9. S. Shandarin, and Y. Zeldovich, *Rev. Mod. Phys.* **61**, 185–220 (1989).
10. M. Kardar, G. Parisi, and Y. Zhang, *Phys. Rev. Lett.* **56**, 889–892 (1986).
11. G. Serini, D. Ambrosi, E. Giraudo, A. Gamba, L. Preziosi, and F. Bussolino, *EMBO J* **22**, 1771–9 (2003).
12. D. Ambrosi, A. Gamba, and G. Serini, *Bull. Math. Biol.* **66**, 1851–73 (2004).
13. F. Cavalli, A. Gamba, G. Naldi, and M. Semplice, "APPROXIMATION OF 2D AND 3D MODELS OF CHEMOTACTIC CELL MOVEMENT IN VASCULOGENESIS," in *MATH EVERYWHERE. Deterministic and Stochastic Modelling in Biomedicine, Economics and Industry. Dedicated to the 60th birthday of V. Capasso.*, edited by G. Aletti, M. Burger, A. Micheletti, and D. Morale, Springer, Heidelberg, 2006, pp. 179–191.
14. A. Pluen, P. A. Netti, R. K. Jain, and D. A. Berk, *Biophysical Journal* **77**, 542–552 (1999).
15. D. Stauffer, and A. Aharony, *Introduction to Percolation Theory*, Taylor & Francis, London, 1994.
16. J. Serra, *Mathematical Morphology*, Academic Press, 1982.
17. R. Schneider, *Convex bodies: the Brunn-Minkowski theory*, vol. 44 of *Encyclopedia of Mathematics and its Applications*, Cambridge University Press, Cambridge, 1993, ISBN 0-521-35220-7.
18. J. Ohser, and W. Nagel, *J. Microsc.* **184**, 117–126 (1996).
19. E. H. Spanier, *Algebraic topology*, Springer-Verlag, 1966.
20. P. Grassberger, and I. Procaccia, *Phys. Rev. Lett.* **50**, 346–349 (1983), ISSN 0031-9007.
21. C. K. Peng, S. Havlin, J. M. Hausdorff, J. E. Mietus, H. E. Stanley, and A. L. Goldberger, *J Electrocardiol* **28 Suppl**, 59–65 (1995).
22. C. K. Peng, J. E. Mietus, Y. Liu, C. Lee, J. M. Hausdorff, H. E. Stanley, A. L. Goldberger, and L. A. Lipsitz, *Ann Biomed Eng* **30**, 683–92 (2002).

Learning Decision Trees over Erasing Pattern Languages

Yasuhito Mukouchi and Masako Sato

Faculty of Liberal Arts and Sciences, Osaka Prefecture University, Sakai, Osaka 599-8531, Japan

Abstract. In this paper, we consider a learning problem of decision trees over *erasing* patterns from positive examples in the framework of identification in the limit due to Gold and Angluin. An *erasing* pattern is a string pattern with constant symbols and *erasable* variables. A decision tree over erasing patterns can be applied to identify or express transmembrane domains of amino acid sequences, and gives intuitive knowledge expressions.

We first show that the ordinary decision trees with height 1 over erasing regular patterns are learnable but those with height at most 2 are not learnable from positive examples. Then we introduce a co-pattern p^c for an erasing pattern p, and we redefine the language of a decision tree over erasing patterns as a language obtainable by finitely many applications of union operations and intersection operations to the languages of erasing patterns and co-patterns. Under the new definition of decision trees, we show that these decision trees with height at most n are learnable from positive examples. Moreover, we investigate efficient learning algorithms for decision trees with height 1. Terada et al. discussed the same problem for decision trees over *nonerasing* patterns, and the results obtained in the present work are natural extensions of Terada's results.

Keywords: Decision Tree, pattern language, inductive inference, formal language.
PACS: 01.40.Ha

1. INTRODUCTION

Finding a motif from functional domains of DNA sequences is an important problem in molecular biology. Arikawa et al. [3], Miyano [9] and so on have developed and investigated learning/discovery systems of a motif of DNA sequences in the framework of *PAC learning paradigm from positive and negative examples.* In their studies, decision trees over regular patterns are applied to as a knowledge representation model. In this paper, we deal with the problem of learning decision trees over erasing patterns in the framework of *identification in the limit from positive examples* due to Gold [7] and Angluin [1].

A *pattern* over an alphabet Σ is a string consisting of constant symbols in Σ and variables. For example, $p = axby$ and $q = axbx$ are patterns over an alphabet $\{a,b\}$. An *erasing pattern* is a pattern with erasable variables. A *semantic meaning* $L(p)$ of a pattern p is a language of all constant strings over Σ obtained by substituting *nonempty* constant strings for all variables of p, and that of an *erasing pattern* p is defined by allowing to substitute the empty string for erasable variables. A language L is called a *pattern language* if there is a pattern p such that $L = L(p)$. In this case, if p is an *erasing* pattern, we call it an *erasing pattern language*. A pattern or an erasing pattern p is called *regular*, if every (erasable) variable in p appears at most once.

A *decision tree T over (erasing) patterns* is a binary tree whose root node and internal

CP1028, *Collective Dynamics: Topics on Competition and Cooperation in the Biosciences*
edited by L. M. Ricciardi, A. Buonocore, and E. Pirozzi

nodes are labeled with (erasing) patterns and whose leaves are labeled with 0 or 1. For a given constant string w, we repeat the following process until it reaches a leaf: (1) visit an internal node (initially the root node), (2) if w matches p, i.e., $w \in L(p)$, then move to the left child, otherwise move to the right child. Finally if it reaches a leaf with label 1, then the decision tree T accepts the word w, otherwise T does not accept w. The *language* $L(T)$ *generated by a decision tree* T is the set of constant strings accepted by T. Then this language $L(T)$ can be obtained by performing finitely many times union operations and intersection operations for pattern languages and their complements.

A class of pattern languages is introduced by Angluin [2] as a language class that is inferable from positive examples only. A language learning from positive examples is a process to find general rules expressed by e.g. patterns or decision trees from their strings. In the present paper, we consider language learning in the framework of *identification in the limit from positive examples* due to Gold [7] and Angluin [1], and discuss learning problems of decision trees over erasing patterns.

Wright [20] introduced a notion of *finite elasticity* for a language class, and showed that a class of languages with finite elasticity is inferable from positive examples and that this property is closed under a union operation. Using this property, the class of unions of at most n pattern languages is shown to be inferable from positive examples. Furthermore Moriyama and Sato [10] showed that the property of finite elasticity is also closed under an intersection operation, a concatenation operation and so on. Using these results on closedness, it is shown that the class of languages generated by decision trees over erasing patterns is inferable from positive examples, if the class of erasing pattern languages and their complements has finite elasticity.

From this point of view, we introduce a string p^c called a *co-pattern* of an erasing pattern p, and define its semantic meaning $L(p^c)$ as the language of constant strings each of which does not belong to $L(p)$ and is not shorter than $c(p)$, where $c(p)$ is the constant string obtained from p by deleting all variables in p. A language L is called a *co-pattern language* if there is a co-pattern p^c such that $L = L(p^c)$. Then we show that the class $\mathscr{J}\mathscr{E}\mathscr{R}\mathscr{P}\mathscr{L}$ of erasing regular pattern languages and their co-pattern languages has finite elasticity, and thus the class $\mathscr{J}\mathscr{E}\mathscr{R}\mathscr{P}\mathscr{L}(n)$ obtained by performing union operations and intersection operations at most n times for $\mathscr{J}\mathscr{E}\mathscr{R}\mathscr{P}\mathscr{L}$ also has finite elasticity. By using this result, we show that the class $\mathscr{T}\mathscr{E}\mathscr{R}\mathscr{P}\mathscr{L}_n$ of languages generated by decision trees over erasing regular patterns with height at most n is inferable in the limit from positive examples.

Finally, we also present an efficient learning algorithm for the class $\mathscr{J}\mathscr{E}\mathscr{R}\mathscr{P}\mathscr{L}$ ($= \mathscr{T}\mathscr{E}\mathscr{R}\mathscr{P}\mathscr{L}_1$). Terada et al. [19] discussed the same problem for decision trees over *nonerasing* patterns, and the results obtained in the present work are natural extensions of Terada's results.

2. LANGUAGE LEARNING FROM POSITIVE EXAMPLES

Firstly, we introduce the notion of *identification in the limit from positive examples* (cf. Gold [7], Angluin [2]).

Let Σ be a finite alphabet of *constant* symbols, and we denote by Σ^+ the set of nonempty constant strings over Σ and by Σ^* the set of possibly empty constant strings over Σ. Let us denote the empty string by ε, and thus $\Sigma^* = \Sigma^+ \cup \{\varepsilon\}$. For nonnegative integer n, let us denote by Σ^n the set of constant strings with length just n and by $\Sigma^{\geq n}$ the set of those with length at least n. A subset L of Σ^* will be called a *language* over Σ.

In what follows, N denotes the set of nonnegative integers, for a finite set S, $\sharp S$ denotes the cardinality of S, and for a constant string w, $|w|$ denotes the length of w.

A *positive presentation*, or a *text*, of a nonempty language L is an infinite sequence of strings $w_1, w_2, \cdots$ in Σ^* such that $\{w_n \mid n \geq 1\} = L$. By the symbol σ, we denotes a positive presentation of a language and by $\sigma[n]$ the σ's initial segment of length $n \in N$.

A class $\mathscr{L} = L_0, L_1, \cdots$ of languages over Σ is an *indexed class of recursive languages*, if there is a computable function $f : N \times \Sigma^* \to \{0,1\}$ such that $f(i,w) = 1$ iff $w \in L_i$. In what follows, we assume that a class of languages is an indexed family of recursive languages without any notice.

An *inductive inference machine* (IIM, for short) M is an effective procedure that runs in stages $1, 2, \cdots$ and, at stage n, requests the n-th example w_n and produces nonnegative integer h_n as the n-th *guess* based on examples received so far. By $M(\sigma[n])$ we denote the n-th guess h_n of an IIM M which is successively presented $\sigma[n]$ on its input requests.

An IIM M *converges* to $h \in N$ for a positive presentation σ, if there is an $n \in N$ such that for any $m \geq n$, $M(\sigma[m]) = h$. An IIM M *identifies* or *infers* a language $L \in \mathscr{L}$ *in the limit from positive examples*, if M converges to an index j for σ such that $L_j = L$ for any positive presentation σ of L. An IIM M *infers* a class $\mathscr{L}$ of languages *in the limit from positive examples*, if for any language $L_i \in \mathscr{L}$, M infers L_i in the limit from positive examples. A class $\mathscr{L}$ is *inferable in the limit from positive examples*, if there is an IIM which infers $\mathscr{L}$ in the limit from positive examples.

In the above definitions, we omit the phrase "in the limit", if it is clear from the context.

An IIM M is *polynomial time updating*, if for each stage, M produces the n-th guess h_n after receiving the n-th example w_n within time polynomial in the total length $|w_1| + |w_2| + \cdots + |w_n|$ of examples received so far. We consider an IIM efficient, if it is a polynomial time updating IIM.

Inferability of a class has been characterized as follows: Let $\mathscr{L} = L_0, L_1, \cdots$ be a class of languages. A set $T \subseteq \Sigma^*$ is called a *finite tell-tale set of L within* $\mathscr{L}$, if T is a finite subset of L and for any $L_j \in \mathscr{L}$, $T \subseteq L_j$ implies $L_j \not\subsetneq L$.

Theorem 1 (Angluin [2]) *A class $\mathscr{L}$ is inferable in the limit from positive examples, if and only if there is an effective procedure that, on input an index i, enumerates a finite tell-tale set of $L_i \in \mathscr{L}$ within $\mathscr{L}$.*

A notion of *finite thickness* for a class of languages due to Angluin [2] is a very useful sufficient condition for inferability and is defined as follows: A class $\mathscr{L}$ has *finite thickness*, if for any nonempty subset $S \subseteq \Sigma^*$, $\{L \in \mathscr{L} \mid S \subseteq L\}$ is a finite set.

Theorem 2 (Angluin [2]) *If a class $\mathscr{L}$ has finite thickness, then $\mathscr{L}$ is inferable in the limit from positive examples.*

On the other hand, a notion of *finite elasticity* for a class of languages due to Wright [20] is defined as follows: A class $\mathscr{L}$ has *finite elasticity*, if there does not exist an infinite sequence $w_0, w_1, w_2, \cdots \in \Sigma^*$ and an infinite sequence $L_1, L_2, \cdots \in \mathscr{L}$ such that

$$\{w_0, w_1, \cdots, w_{k-1}\} \subseteq L_k, \quad \text{but} \quad w_k \notin L_k \quad (k \geq 1).$$

Wright [20] showed that finite elasticity is a sufficient condition for a class to be inferable in the limit from positive examples and that this property is closed under a union operation $\tilde{\cup}$, defined below, for classes of languages. Furthermore, Moriyama and Sato [10] also showed that this property is closed under an intersection operation $\tilde{\cap}$ defined below, a concatenation operation and some other operations for classes of languages.

In the present paper, we focus the following two operations for classes of languages:

$$\mathscr{L}_1 \tilde{\cup} \mathscr{L}_2 = \{L_1 \cup L_2 \mid L_1 \in \mathscr{L}_1,\ L_2 \in \mathscr{L}_2\},$$
$$\mathscr{L}_1 \tilde{\cap} \mathscr{L}_2 = \{L_1 \cap L_2 \mid L_1 \in \mathscr{L}_1,\ L_2 \in \mathscr{L}_2\}.$$

For a class $\mathscr{L}$, we denote by $\mathscr{L}(n)$ the class obtained by applying a union operation $\tilde{\cup}$ or an intersection operation $\tilde{\cap}$ at most n times for the class $\mathscr{L}$ $(n \geq 1)$.

To sum up, the following theorem holds:

Theorem 3 (Wright [20], Moriyama and Sato [10]) *On the property of finite elasticity, the following propositions hold:*

(1) If a class $\mathscr{L}$ has finite thickness, then the class $\mathscr{L}$ also has finite elasticity.

(2) If a class $\mathscr{L}$ has finite elasticity, then the class $\mathscr{L}$ is inferable in the limit from positive examples.

(3) If classes $\mathscr{L}_1$ and $\mathscr{L}_2$ have finite elasticity, then the classes $\mathscr{L}_1 \tilde{\cup} \mathscr{L}_2$ and $\mathscr{L}_1 \tilde{\cap} \mathscr{L}_2$ also have finite elasticity.

(4) If a class $\mathscr{L}$ has finite elasticity, then the class $\mathscr{L}(n)$ also has finite elasticity $(n \geq 1)$.

We note that it is easily shown by definition that for classes $\mathscr{L}_1$ and $\mathscr{L}_2$ with finite elasticity, the ordinary union class $\mathscr{L}_1 \cup \mathscr{L}_2 = \{L \mid L \in \mathscr{L}_1 \text{ or } L \in \mathscr{L}_2\}$ also has finite elasticity.

Moriyama and Sato [10] also showed that even if a class $\mathscr{L}$ has finite elasticity, it does not follow that the class $\{L^c \mid L \in \mathscr{L}\}$ has finite elasticity, where for a language $L \subseteq \Sigma^*$, $L^c = \Sigma^* \setminus L$ represents the complement of L.

3. LEARNING LANGUAGES GENERATED BY DECISION TREES OVER ERASING PATTERNS

3.1. Erasing patterns and their languages

Let Σ be a finite alphabet with at least two *constant symbols* and X be a set of *variables* with $\Sigma \cap X = \phi$. A *pattern* is a nonempty string over $\Sigma \cup X$.

A *substitution* is a homomorphism from strings to strings which maps each constant symbol $a \in \Sigma$ to itself and each variable $x \in X$ to a nonempty string in $(\Sigma \cup X)^+$. By θ, we denote a substitution. Then, $p\theta$ denotes the image of a pattern p by a substitution θ.

The language $L(p)$ generated by a pattern p is the set of all constant strings w such that $w = p\theta$ for some substitution θ. For example, let $\Sigma = \{a,b\}$. Then $p = axbxay$ is a pattern, and the set $L(p) = \{aabaaa, abbbaa, aabaab, abbbab, \cdots\}$ is the language of p. A language L is a *pattern language*, if there is a pattern p such that $L = L(p)$.

In the definitions above, a mapping of variables to the empty string ε is not allowed. A substitution which allows mapping of a variable to the empty string ε is called an *erasing substitution*. In applying patterns to real problems, there are many cases that erasing substitutions are more suitable (cf. Arimura et al. [5], Shinohara [17] and so on). In case we use erasing substitutions, variables, patterns and pattern languages are called *erasable variables*, *erasing patterns* and *erasing pattern languages*, respectively. For example, we consider the above $p = axbxay$ as an erasing pattern. Then the language of p is $L(p) = \{aba, aabaa, abbba, abaa, abab, \cdots\}$. Let us denote by $\mathscr{P}$ and $\mathscr{E}\mathscr{P}$ the class of patterns and that of erasing patterns, respectively, and by $\mathscr{P}\mathscr{L}$ and $\mathscr{E}\mathscr{P}\mathscr{L}$ the classes of their languages.

For a pattern or an erasing pattern p, the length of p is denoted by $|p|$. For an erasing pattern p, the constant string obtained by deleting all erasable variables from p is denoted by $c(p)$. Then, by definition, for a pattern p and a string w, $w \in L(p)$ implies $|p| \leq |w|$, and thus the length of shortest strings in $L(p)$ is $|p|$. On the other hand, for an erasing pattern p and a string w, $w \in L(p)$ implies $|c(p)| \leq |w|$, and thus the length of shortest strings in $L(p)$ is $|c(p)|$.

For patterns p and q, p is an *instance* of q, denoted by $p \preceq q$, if there is a substitution θ such that $p = q\theta$. In this case, we also say that q is a *generalization* of p and denote it as $q \succeq p$. In case $p \preceq q$ but $q \not\preceq p$, we say that p is a *proper instance* of q or q is a *proper generalization* of p, and denote it as $p \precneq q$ or $q \succneq p$. Similarly, we also define these notions for erasing patterns p and q. Clearly, for patterns p and q, $p \preceq q$ implies $|q| \leq |p|$, and for erasing patterns p and q, $p \preceq q$ implies $|c(q)| \leq |c(p)|$.

For patterns p and q, $p \preceq q$ and $q \preceq p$ hold if and only if they are identical except for renaming of variables, and thus we do not distinguish one from the other. However, for erasing patterns p and q, $p \preceq q$ and $q \preceq p$ imply $c(p) = c(q)$ but do not imply that they are identical except for renaming of erasable variables. For example, let us consider erasing patterns $p = axxyb$ and $q = azb$. By substituting xxy for z in q, we obtain p from q, and thus $p \preceq q$ holds. On the other hand, by substituting ε and z for x and y in p, we obtain q from p, and thus $q \preceq p$ holds. For (erasing) patterns p and q, we write $p \simeq q$ if $p \preceq$ and $q \preceq p$ hold.

The *membership problem* for a language L is a problem of deciding whether or not $w \in L$ for a given constant string w. The membership problem for (erasing) pattern languages is computable but NP-complete (cf. Angluin [1]).

For (erasing) patterns p and q, $p \preceq q$ implies $L(p) \subseteq L(q)$, but the converse is not valid (cf. Angluin [1]).

On inferability of pattern languages, the following result holds:

Theorem 4 (Angluin [2]) *The class $\mathscr{P}\mathscr{L}$ of pattern languages has finite thickness, and thus it is inferable in the limit from positive examples.*

To the contrary, on erasing pattern languages, the following negative result has been shown:

Theorem 5 (Reidenbach [13, 14]) *Let Σ be a finite alphabet with $2 \leq \sharp\Sigma \leq 4$.*

The class $\mathscr{EPL}$ of erasing pattern languages is not *inferable in the limit from positive examples.*

We note that in case $\sharp\Sigma \geq 5$, the inferability of the class $\mathscr{EPL}$ above is still unknown.

In the present paper, we mainly consider erasing regular patterns and their languages. A pattern or an erasing pattern p is *regular*, if every (erasable) variable in p appears at most once. For example, $p = axybxa$ is not regular, but $q = axybza$ is regular. For (erasing) regular pattern languages, the membership problem is computable in linear time (cf. Shinohara [16]). Furthermore, on inferability of (erasing) regular pattern languages, the class is *efficiently* inferable in the limit from positive examples, that is, there is a polynomial time updating IIM for the class (cf. Shinohara [16]).

For (erasing) regular patterns p and q, the semantic inclusion $L(p) \subseteq L(q)$ can be reduced to the syntactic inclusion $p \preceq q$.

Theorem 6 (Shinohara [16], Mukouchi [12]) *Let $\sharp\Sigma \geq 3$.*

For (erasing) regular patterns p and q, $L(p) \subseteq L(q)$ if and only if $p \preceq q$.

Let us denote by $\mathscr{RP}$ and $\mathscr{ERP}$ the class of regular patterns and that of erasing regular patterns, respectively, and by $\mathscr{RPL}$ and $\mathscr{ERPL}$ the classes of their languages.

3.2. Decision trees over patterns

A *decision tree T over patterns* is a binary tree whose root node and internal nodes are labeled with patterns and whose leaves are labeled with 0 or 1.

For a given constant string w, we repeat the following process until it reaches a leaf: (1) visit an internal node (initially the root node), (2) for the pattern p associated with the node, if $w \in L(p)$, then move to the left child, otherwise move to the right child. Finally if it reaches a leaf with label 1, then the decision tree T accepts w, otherwise T does not accept w. The *language $L(T)$ generated by a decision tree T* is the set of constant strings accepted by T.

Let us denote by $\mathscr{TP}_n$ the class of decision trees over patterns with height at most n and by $\mathscr{TPL}_n$ the class of their languages.

The decision tree as in Figure 1 is a decision tree with height 2.

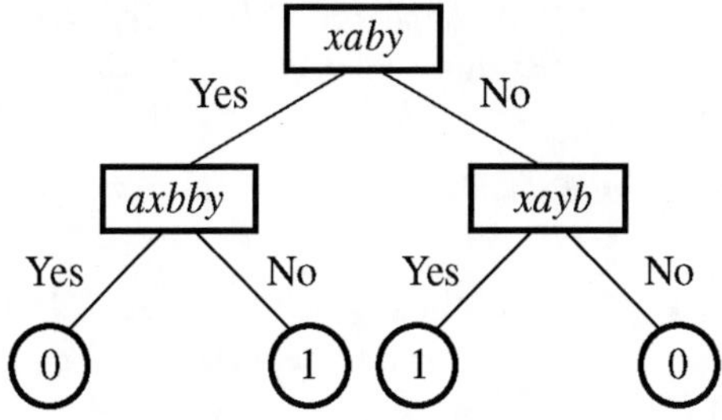

FIGURE 1. A decision tree over patterns

The language $L(T)$ of the decision tree T above can be expressed as follows:

$$L(T) = (L(xaby) \cap L(axbby)^c) \cup (L(xaby)^c \cap L(xayb))$$

Let n be the height of a decision tree T over patterns. As one can see from this example, the language of constant strings each of which is accepted at the leaf labeled with 1 via internal nodes labeled with patterns $p_1, p_2, \cdots, p_k$ $(k \leq n)$ is expressed as an intersection of $L(p_i)$s or it's complement $L(p_i)^c$s with $1 \leq i \leq k$, where p_1 is the pattern associated with the root node. The language of a decision tree T is expressed as a union of such languages.

Thus if we put $\mathscr{L} = \mathscr{PL} \cup \{L(p)^c \mid p \in \mathscr{P}\}$, then the class $\mathscr{TPL}_n$ is a subclass of $\mathscr{L}(m)$ with $m = n \times 2^{n-1}$. If $\mathscr{L}(m)$ is inferable in the limit from positive examples, then so is the subclass $\mathscr{TPL}_n$.

We also define the classes $\mathscr{TRP}_n$, $\mathscr{TEP}_n$ and $\mathscr{TERP}_n$ by using "regular patterns", "erasing patterns" and "erasing regular patterns", respectively, instead of "patterns" in the definition of $\mathscr{TP}_n$, and we denote their languages by $\mathscr{TRPL}_n$, $\mathscr{TEPL}_n$ and $\mathscr{TERPL}_n$, respectively. Similar discussions are also valid for these classes.

3.3. Inferability of decision trees over pattern

In this section, we consider inferability of classes introduced. Firstly, we consider the class of complements of pattern languages. As easily seen, for any pattern p, the complement $L(p)^c$ of $L(p)$ is not a pattern language.

Here, for a pattern p, we introduce a special string p^c called a *co-pattern* of p, and we denote by *co-*$\mathscr{P}$ the class of co-patterns. We can consider several semantic meanings for a co-pattern p^c of a pattern p. Firstly, we define the semantic meaning or the language of the co-pattern p^c as the complement $L(p)^c$ of $L(p)$, that is, we put $L(p^c) = L(p)^c$. Under this definition, we denote by *co-*$\mathscr{PL}$ the class of co-pattern languages. Then, unfortunately, this class *co-*$\mathscr{PL}$ does not have finite thickness. This can be confirmed as follows: Let w be a string with length 1. Then, $w \in L(p^c)$ holds for every pattern p with length at least 2, and thus there are infinitely many languages $L \in$ *co-*$\mathscr{PL}$ with $\{w\} \subseteq L$. Furthermore, this class *co-*$\mathscr{PL}$ does not have finite elasticity. In fact, a sequence of strings $a, aa, aaa, \cdots$ and a sequence of co-pattern languages $L(aa^c), L(aaa^c), L(aaaa^c), \cdots$ shall witness to it. However, Shinohara [18] showed that the class $\mathscr{PL}$ is inferable in the limit from *negative examples*, and this directly implies the following theorem:

Theorem 7 (Based on Shinohara [18]) *The class co-*$\mathscr{PL}$ *is inferable in the limit from positive examples.*

Now, we put $\mathscr{JP} = \mathscr{P} \cup co\text{-}\mathscr{P}$ and $\mathscr{JPL} = \mathscr{PL} \cup co\text{-}\mathscr{PL}$. As easily seen from Figure 2, the class $\mathscr{JPL}$ is equal to the class $\mathscr{TPL}_1$ of languages generated by decision trees over patterns with height 1.

Theorem 8 (Terada et al. [19]) *The class* $\mathscr{JPL}$ $(= \mathscr{TPL}_1)$ *is inferable in the limit from positive examples.*

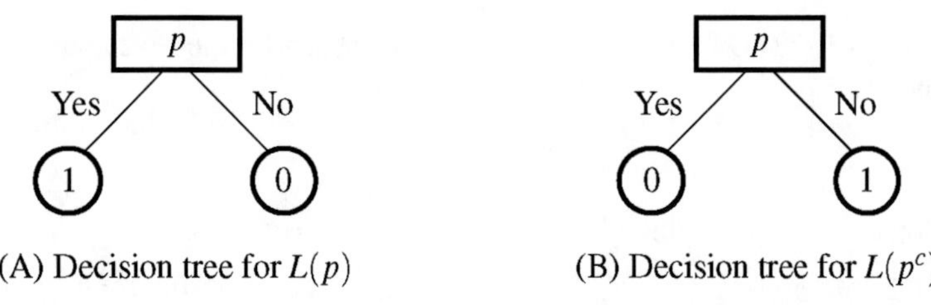

(A) Decision tree for $L(p)$ (B) Decision tree for $L(p^c)$

FIGURE 2. Decision trees with height 1

On the other hand, the following negative result has been shown:

Theorem 9 (Terada et al. [19]) *The class* $\mathcal{TPL}_2$ *is not inferable from positive examples.*

By this theorem, under the former semantic meanings of decision trees and co-patterns, the class $\mathcal{TPL}_n$ is not inferable in the limit from positive examples for $n \geq 2$.

In what follows, for a class $\mathcal{SP}$ of (erasing) patterns, let us denote by *co-*$\mathcal{SP}$ the class of co-patterns of (erasing) patterns in $\mathcal{SP}$ and by *co-*$\mathcal{SPL}$ the class of their languages. Then we put $\mathcal{JSP} = \mathcal{SP} \cup$ *co-*$\mathcal{SP}$ and $\mathcal{JSP} = \mathcal{SPL} \cup$ *co-*$\mathcal{SPL}$. For example, *co-*$\mathcal{RP}$ represents the class of co-patterns of regular patterns and *co-*$\mathcal{RPL}$ represents the class of their languages.

Concerning the class $\mathcal{TRPL}_n$ of languages generated by decision trees over regular patterns, the same results hold, that is, the following corollary hold:

Corollary 1 *The class* $\mathcal{JRPL}$ $(= \mathcal{TRPL}_1)$ *is inferable in the limit from positive examples.*

For $n \geq 2$*, the class* $\mathcal{TRPL}_n$ *is not inferable from positive examples.*

Next, we consider inferability of the class of languages generated by decision trees over erasing (regular) patterns.

By Theorem 5, the class $\mathcal{EPL}$ is not inferable in the limit from positive examples if $2 \leq \sharp\Sigma \leq 4$. Since the class $\mathcal{EPL}$ is a subclass of $\mathcal{JEPL}$, the following theorem holds:

Theorem 10 *Let* $2 \leq \sharp\Sigma \leq 4$.

The class $\mathcal{JEPL}$ $(= \mathcal{TEPL}_1)$ *is not inferable in the limit from positive examples.*

In case $\sharp\Sigma \geq 5$, the inferability of the class $\mathcal{JEPL}$ $(= \mathcal{TEPL}_1)$ is still unknown.

Theorem 11 *Let* $\sharp\Sigma \geq 3$.

The class $\mathcal{JERPL}$ $(= \mathcal{TERPL}_1)$ *is inferable in the limit from positive examples.*

Proof. For an erasing regular pattern p, we put $S_p = \{w \in L(p) \mid |w| \leq |c(p)| + 1\}$, $S_{p^c} = \{w \in L(p^c) \mid |w| \leq |c(p)| + 1\}$. We can show that S_p is a finite tell-tale set of

$L(p)$ within $\mathscr{JERPL}$ and S_{p^c} is that of $L(p^c)$. Clearly, these sets are recursively enumerable.

Thus, by Theorem 1, we see that the class $\mathscr{JERPL}$ is inferable in the limit from positive examples. □

On the other hand, the following negative result has been shown:

Theorem 12 *The classes $\mathscr{TEPL}_n$ and $\mathscr{TERPL}_n$ are not inferable from positive examples for $n \geq 2$.*

Proof. Since the class $\mathscr{TERPL}_2$ is a subclass of $\mathscr{TEPL}_n$ as well as $\mathscr{TERPL}_n$ for $n \geq 2$, it suffices for us to show that the class $\mathscr{TERPL}_2$ is not inferable from positive examples.

Suppose the converse, that is, suppose that the class $\mathscr{TERPL}_2$ is inferable from positive examples. Then, by Theorem 1, every language in the class has finite tell-tale set with in the class. Let $w \in \Sigma^*$ be an arbitrary constant string. We can regard w as an erasing regular pattern. Then the language $L(w^c)$ is generated by the decision tree as in Figure 3 (A), and thus $L(w^c) \in \mathscr{TERPL}_2$ hold.

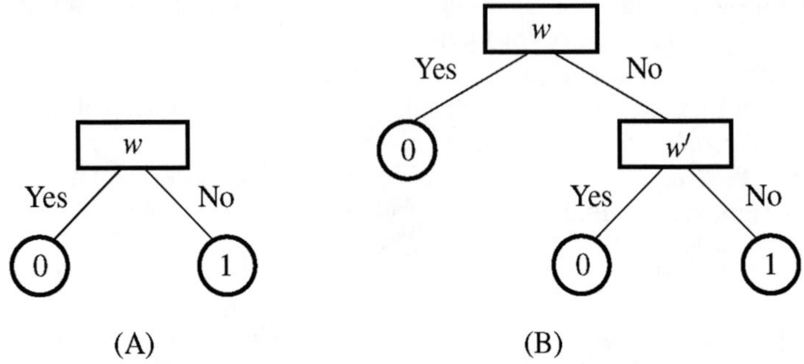

FIGURE 3. Decision trees for $L(w^c)$ and $L(w^c) \cap L(w'^c)$

Now let $S \subseteq L(w^c)$ be a finite tell-tale set of $L(w^c)$ within $\mathscr{TERPL}_2$. Since S is a finite set, there is a constant string $w' \in \Sigma^*$ such that $w' \notin S$ and $w' \neq w$. Here, we consider the language $L = L(w^c) \cap L(w'^c)$. Then L is generated by the decision tree with height 2 as in Figure 3 (B), and thus $L \in \mathscr{TERPL}_2$ hold.

On the other hand, since $S \subseteq L(w'^c)$, $w' \notin L$ and $w' \in L(w^c)$ hold, it follows that $S \subseteq L \subsetneq L(w^c)$ holds. This contradicts that S is a finite tell-tale set of $L(w^c)$ within $\mathscr{TERPL}_2$. Therefore $\mathscr{TERPL}_2$ is not inferable from positive examples. □

3.4. Another semantic meaning of co-patterns

Here, as another semantic meaning of the co-pattern p^c of a pattern p, we consider the following language $L(p^c)$ which is obtained from the complement $L(p)^c$ of $L(p)$ by deleting finitely many constant strings shorter than p:

$$L(p^c) = \{w \in L(p)^c \mid |w| \geq |p|\} = L(p)^c \cap \Sigma^{\geq |p|}.$$

Under this semantic meaning of co-patterns, $w \in L(p^c)$ implies $|p| \leq |w|$, and by this property, the following result is obtained:

Theorem 13 (Terada et al. [19]) *The class $\mathscr{J}\mathscr{P}\mathscr{L}$ has finite thickness, and thus $\mathscr{J}\mathscr{P}\mathscr{L}$ is inferable in the limit from positive examples.*

In correspondence with the semantic meaning $L(p^c)$ of a co-pattern p^c above, we reconsider the semantic meaning of decision tree T. In the former definition, for a given constant string w, we repeat the following process until it reaches a leaf: (1) visit an internal node (initially the root node), (2) for the pattern p associated with the node, if $w \in L(p)$, then move to the left child, otherwise move to the right child. In this definition, at each internal node, there is a choice of $w \in L(p)$ or $w \notin L(p)$, i.e., $w \in L(p)^c$. Under the above semantic meaning of co-patterns, there is a choice of $w \in L(p)$, $w \in L(p^c)$ or $w \notin L(p) \cup L(p^c)$, and thus we change the process above to the following one: (1) visit an internal node (initially the root node), (2) if $w \in L(p)$ for the pattern p associated with the node, then move to the left child, (3) if $w \in L(p^c)$, then move to the right child, (4) otherwise halt the process. Finally if it reaches a leaf with label 1, then the decision tree T accepts w, otherwise T does not accept w. We note the decision tree does not accept w, if the process halts on the way. For a given constant string w, the process halts at the node associated with a pattern p if and only if $|w| < |p|$ holds. Therefore, there are at most finitely many constant strings w for which the process halts. Now we redefine the language $L(T)$ generated by a decision tree T as the set of constant strings accepted by T in the way redefined above. Then, for example, the language $L(T)$ of the decision tree T as in Figure 1 can be expressed as follows:

$$L(T) = (L(xaby) \cap L(axbby^c)) \cup (L(xaby^c) \cap L(xayb))$$

Under the above semantic meaning of decision trees, the following theorem holds:

Theorem 14 (Terada et al. [19]) *The class $\mathscr{J}\mathscr{P}\mathscr{L}(n)$ has finite elasticity for $n \geq 1$. Therefore, for $n \geq 1$, the classes $\mathscr{T}\mathscr{P}\mathscr{L}_n$ and $\mathscr{T}\mathscr{R}\mathscr{P}\mathscr{L}_n$ are inferable in the limit from positive examples.*

Next, we consider another semantic meaning of the co-pattern p^c of an *erasing* pattern p. For an erasing pattern p, let us put

$$L(p^c) = \{w \in L(p)^c \mid |w| \geq |c(p)|\} = L(p)^c \cap \Sigma^{\geq |c(p)|}.$$

We note that under this semantic meaning of co-patterns, $w \in L(p^c)$ implies $|c(p)| \leq |w|$.

An erasing regular pattern p is *canonical*, if p is of the form $w_0 x_1 w_1 x_2 \cdots x_n w_n$ for some $n \geq 0$, $w_0, w_n \in \Sigma^*$ and $w_1, \cdots, w_{n-1} \in \Sigma^+$. We note that a canonical erasing regular pattern p does not contain consecutive erasable variables in it. Shinohara [17] showed that for an erasing regular pattern p, there is a canonical erasing regular pattern q such that $p \simeq q$, i.e., $p \preceq q$ and $q \preceq p$.

Theorem 15 *The class $\mathscr{J}\mathscr{E}\mathscr{R}\mathscr{P}\mathscr{L}$ has finite thickness, and thus $\mathscr{J}\mathscr{E}\mathscr{R}\mathscr{P}\mathscr{L}$ is inferable in the limit from positive examples.*

Proof. Let $w \in \Sigma^*$ be a constant string.

For an erasing regular pattern p, if $w \in L(p)$, then $|c(p)| \leq |w|$. By this relation, we can show that there are at most finitely many canonical erasing regular patterns p such that $w \in L(p)$.

On the other hand, for an erasing regular pattern p, if $w \in L(p^c)$, then $|c(p)| \leq |w|$. Again, by this relation, we can show that there are at most finitely many canonical erasing regular patterns p such that $w \in L(p^c)$.

Therefore, $\mathscr{J}\mathscr{E}\mathscr{R}\mathscr{P}\mathscr{L}$ has finite thickness. □

We note that the class $\mathscr{J}\mathscr{E}\mathscr{P}\mathscr{L}$ does not have finite thickness.

By this theorem, under the new semantic meaning of decision trees, the following theorem holds:

Theorem 16 *The class $\mathscr{J}\mathscr{E}\mathscr{R}\mathscr{P}\mathscr{L}(n)$ has finite elasticity for $n \geq 1$.*

Therefore, for $n \geq 1$, the class $\mathscr{T}\mathscr{E}\mathscr{R}\mathscr{P}\mathscr{L}_n$ is inferable in the limit from positive examples.

In the rest of this section, we give some basic properties on the classes $\mathscr{E}\mathscr{R}\mathscr{P}\mathscr{L}$ and co-$\mathscr{E}\mathscr{R}\mathscr{P}\mathscr{L}$.

Lemma 1 *For an erasing regular pattern p, the following propositions hold:*

(1) $|c(p)| = \min\{n \mid \Sigma^n \not\subseteq L(p^c)\}$ holds.

(2) $\Sigma^n \subseteq L(p^c)$ for some $n \geq 0$, if and only if p is a constant string, i.e., p does not contain any erasable variable.

Lemma 2 *For two erasing patterns p and q with $|c(p)| = |c(q)|$, $L(p) \subseteq L(q)$ if and only if $L(q^c) \subseteq L(p^c)$.*

We note that for two erasing patterns p and q, $L(p) \subseteq L(q)$ does not imply $L(q^c) \subseteq L(p^c)$. In fact, for two patterns e.g. $p = xaay$ and $q = xay$, $L(p) \subseteq L(q)$ and $L(q^c) \not\subseteq L(p^c)$ hold. Furthermore, $L(q^c) \subseteq L(p^c)$ also does not imply $L(p) \subseteq L(q)$. In fact, for two patterns e.g. $p = xay$ and $q = xaay$, $L(p) \not\subseteq L(q)$ and $L(q^c) \subseteq L(p^c)$ hold.

4. EFFICIENT LEARNING OF DECISION TREES OVER ERASING REGULAR PATTERNS WITH HEIGHT 1

In the present section, we consider efficient learning of classes of languages generated by decision trees with height 1. Firstly, for the class $\mathscr{T}\mathscr{R}\mathscr{P}\mathscr{L}_1$, Terada et al. [19] has obtained the following result:

Theorem 17 (Terada et al. [19]) *Let $\sharp\Sigma \geq 3$.*

The class $\mathscr{T}\mathscr{R}\mathscr{P}\mathscr{L}_1$ is inferable in the limit from positive examples by a polynomial time updating IIM.

In this paper, we consider efficient inferability of the class $\mathscr{T}\mathscr{E}\mathscr{R}\mathscr{P}\mathscr{L}_1$ of languages generated by decision trees over *erasing regular patterns*. As stated above, the class $\mathscr{T}\mathscr{E}\mathscr{R}\mathscr{P}\mathscr{L}_1$ just consists of erasing regular pattern languages and co-pattern languages of erasing regular patterns, that is, $\mathscr{T}\mathscr{E}\mathscr{R}\mathscr{P}\mathscr{L}_1 = \mathscr{J}\mathscr{E}\mathscr{R}\mathscr{P}\mathscr{L} = \mathscr{E}\mathscr{R}\mathscr{P}\mathscr{L} \cup$

co-$\mathscr{ERPL}$. Concerning the class $\mathscr{ERPL}$ only, Shinohara [17] has obtained an polynomial time updating IIM. Here we give polynomial time updating IIM for the class $\mathscr{TERPL}_1$.

In general, for a class $\mathscr{L}$ of languages with finite elasticity, if the MINL problem for $\mathscr{L}$ is computable, then the IIM below is shown to infer $\mathscr{L}$ in the limit from positive examples (cf. Arimura et al. [4]). A language L is a *minimal language* of a set S of constant strings within a class $\mathscr{L}$ of languages, if $S \subseteq L$ holds and there is no language $L' \in \mathscr{L}$ such that $S \subseteq L' \subsetneq L$ holds. The MINL problem for a class $\mathscr{L}$ of languages is a problem of finding a minimal language $L \in \mathscr{L}$ of S for a given finite set S of constant strings.

Algorithm IIM
Input: an infinite sequence $w_1, w_2, \cdots$ of constant strings;
Output: an infinite sequence $g_1, g_2, \cdots$ of indices;
begin
  $S := \phi$; $g_0 := -1$; $n := 1$;
  repeat
    read the next data w_n; $S := S \cup \{w_n\}$;
    if $w_n \notin L_{g_{n-1}}$ **then** $g_n :=$ MINL$_{\mathscr{L}}(S)$
    else $g_n := g_{n-1}$;
    output g_n; $n := n+1$
  forever
end.

In the algorithm above, MINL$_{\mathscr{L}}(S)$ is a procedure that solves the MINL problem for the class $\mathscr{L}$, i.e., that returns an index i of a minimal language $L_i \in \mathscr{L}$ of S within $\mathscr{L}$.

Concerning the MINL problem for the class $\mathscr{ERPL}$, Shinohara [17] gives a procedure MINL(S) which runs in time $O(l^2 n)$, where l is the maximum length of constant strings in S and $n = \sharp S$. We note that the Shinohara's procedure needs one of *maximal common subsequences* of S, which is shown to be computable in time $O(ln)$ (cf. Fraser et al. [6]).

Now we consider the following procedure co-MINL(S):

Procedure co-MINL
Input: a nonempty finite set $S \subseteq \Sigma^*$ of constant strings;
Output: a co-pattern p^c of an erasing regular pattern;
begin
  let n be the length of shortest strings in S;
  if $\Sigma^n \not\subseteq S$ **then begin**
    let $a_1 a_2 \cdots a_n$ be an arbitrary constant string in $\Sigma^n - S$;
    $p_0 := a_1 a_2 \cdots a_n$;
    for $i := 1$ to $n+1$ **do begin**
      $q_i := p_{i-1}\{x_i a_i \leftarrow a_i\}$;
      if $S \subseteq L(q_i^c)$ **then** $p_i := q_i$ **else** $p_i := p_{i-1}$
    end;
    $q_{n+1} := p_n\{a_n x_{n+1} \leftarrow a_n\}$;
    if $S \subseteq L(q_{n+1}^c)$ **then** $p_{n+1} := q_{n+1}$ **else** $p_{n+1} := p_n$;
    $p := p_{n+1}$
  end

else begin

let $b_1 b_2 \cdots b_{n-1}$ be an arbitrary constant string in Σ^{n-1};

$p := b_1 b_2 \cdots b_{n-1}$

end;

output p^c

end.

In the procedure above, the statement " $q_i := p_{i-1}\{x_i a_i \leftarrow a_i\}$ " represents that let q_i be the pattern obtained from p_{i-1} by replacing the constant symbol a_i with the string $x_i a_i$.

Lemma 3 *Let* $\sharp\Sigma \geq 3$.

For a nonempty finite set $S \subseteq \Sigma^*$ *of constant strings, the procedure* co-MINL(S) *outputs a co-pattern* p^c *of an erasing regular pattern* p *such that* $L(p^c)$ *is a minimal language of* S *within co-*$\mathscr{ERPL}$*.*

Proof. By the definition of p in the procedure, clearly, $S \subseteq L(p^c)$ holds. Suppose that there is an erasing regular pattern q such that $S \subseteq L(q^c) \subsetneq L(p^c)$. Without loss of generality, we assume that p and q are canonical.

(1) In case $\Sigma^n \not\subseteq S$. Then $p = p_{n+1}$ and $c(p) = a_1 a_2 \cdots a_n$ hold.

By the assumption of $S \subseteq L(q^c) \subsetneq L(p^c)$, we see that $|c(p)| \leq |c(q)| \leq n$ holds, and thus $|c(p)| = |c(q)| = n$ holds. Then, by Lemma 2 and Theorem 6, we see that $L(p) \subsetneq L(q)$ and $p \precnsim q$ holds. Again, by $|c(p)| = |c(q)| = n$, we see that $c(q) = a_1 a_2 \cdots a_n$ holds.

Let us put $p = Y_1 a_1 Y_2 a_2 \cdots a_n Y_{n+1}$ and $q = Z_1 a_1 Z_2 a_2 \cdots a_n Z_{n+1}$, where Y_i and Z_i are the erasable variable x_i or the empty string ε $(1 \leq i \leq n+1)$.

Then, we can represents p_i, q_i $(1 \leq i \leq n)$, p_{n+1} and q_{n+1} in the procedure as follows:

$$p_i = Y_1 a_1 Y_2 \cdots Y_{i-1} a_{i-1} (Y_i a_i) a_{i+1} \cdots a_n, \qquad p_{n+1} = Y_1 a_1 Y_2 \cdots Y_n (a_n Y_{n+1}),$$
$$q_i = Y_1 a_1 Y_2 \cdots Y_{i-1} a_{i-1} (x_i a_i) a_{i+1} \cdots a_n, \qquad q_{n+1} = Y_1 a_1 Y_2 \cdots Y_n (a_n x_{n+1}),$$

where, by the construction of the procedure, $Y_i = \varepsilon$ implies $S \not\subseteq L(q_i^c)$.

Since $p \precnsim q$ holds, there is an index i with $1 \leq i \leq n+1$ such that Y_i is the empty string ε, Z_i is an erasable variable and $Y_j = Z_j$ for j with $1 \leq j \leq i-1$. For this index i, we put $p = p' a_i p''$ and $q = p'(Z_i a_i) q''$, where

$$p' = Y_1 a_1 Y_2 a_2 \cdots Y_{i-1} a_{i-1} = Z_1 a_1 Z_2 a_2 \cdots Z_{i-1} a_{i-1},$$
$$p'' = Y_{i+1} a_{i+1} \cdots a_n Y_{n+1},$$
$$q'' = Z_{i+1} a_{i+1} \cdots a_n Z_{n+1}.$$

Since Y_i is the empty string ε, $S \not\subseteq L(q_i^c)$ holds. On the other hand, since $q_i = p'(x_i a_i) a_{i+1} \cdots a_n \preceq p'(Z_i a_i) q'' = q$, it follows that $L(q_i) \subseteq L(q)$ holds. By $|c(q_i)| = |c(q)| = n$ and Lemma 2, we see that $L(q^c) \subseteq L(q_i^c)$, and thus $S \not\subseteq L(q^c)$ holds, which contradicts the assumption of $S \subseteq L(q^c)$.

(2) In case $\Sigma^n \subseteq S$. Then $p = b_1 b_2 \cdots b_{n-1}$ holds.

By the assumption of $S \subseteq L(q^c) \subsetneq L(p^c)$ and Lemma 1, we see that q is a constant string with length $n-1$. Therefore $L(q^c) = \Sigma^{\geq n-1} - \{q\}$ and $L(p^c) = \Sigma^{\geq n-1} - \{p\}$ hold, which contradicts the assumption of $L(q^c) \subsetneq L(p^c)$.

By (1) and (2), we see that there is no erasing regular pattern q such that $S \subseteq L(q^c) \subsetneq L(p^c)$, and thus $L(p^c)$ is a minimal language of S within *co-$\mathscr{ERPL}$*. □

Lemma 4 *For a nonempty finite set $S \subseteq \Sigma^*$ of constant strings, the procedure* co-MINL(S) *runs in time $O(l^2m)$, where $l = \max\{|w| \mid w \in S\}$ and $m = \sharp S$.*

Proof. The length of q_i appearing in the procedure is at most $2l+1$. Thus, for a constant string w, whether or not $w \in L(q_i)$ is decidable in time $O(l+|w|)$ (cf. Shinohara [16]). Therefore, whether or not $S \subseteq L(q^c)$ is decidable in time $O(lm)$. In the procedure, the for-loop is executed at most $l+1$ times, and thus the procedure totally runs in time $O(l^2m)$. □

The procedure *co*-MINL(S) outputs a co-pattern p^c which generates a minimal language of S within *co-$\mathscr{ERPL}$*. The following procedure JMINL(S) outputs an erasing regular pattern p or a co-pattern p^c which generates a minimal language of S within $\mathscr{JERPL}$.

Procedure JMINL
Input: a nonempty finite set $S \subseteq \Sigma^*$ of constant strings;
Output: an erasing regular pattern or a co-pattern of an erasing regular pattern;
begin
  $p :=$ MINL(S);
  if $c(p) = \varepsilon$ **then** output co-MINL(S)
  else output p
end.

Theorem 18 *Let $\sharp\Sigma \geq 3$.*

For a nonempty finite set $S \subseteq \Sigma^*$ of constant strings, the procedure JMINL outputs π such that $L(\pi)$ is a minimal language of S within $\mathscr{JERPL}$, where π represents either an erasing regular pattern or a co-pattern of an erasing regular pattern.

Proof. As in the procedure, let us put $p =$ MINL(S).
(1) In case $c(p) \neq \varepsilon$, i.e., p contains at least one constant symbol.

Clearly, $L(p)$ is a minimal language of S within $\mathscr{ERPL}$. Here, we show by contradiction that there is no co-pattern q^c of an erasing regular pattern q such that $S \subseteq L(q^c) \subsetneq L(p)$.

Suppose that there is a co-pattern q^c of an erasing regular pattern q such that $S \subseteq L(q^c) \subsetneq L(p)$.

Since $L(q^c) \subsetneq L(p)$ holds, it follows that $1 \leq |c(p)| \leq |c(q)|$ holds, and thus q also contains at least one constant symbol. Assume that p and q contain constant symbols a and b, respectively.

Since $L(q^c) = \Sigma^{\geq |c(q)|} \cap L(q)^c = \Sigma^{\geq |c(q)|} - L(q)$ holds, $L(q^c) \subsetneq L(p)$ implies $\Sigma^{\geq |c(q)|} \subseteq L(p) \cup L(q)$. Let c be a constant symbol in Σ other than a and b. By the assumption of $\sharp\Sigma \geq 3$, we can take such c.

Then we consider a constant string $w = c^{|c(q)|}$. Clearly $w \in \Sigma^{|c(q)|}$, but $w \notin L(p) \cup L(q)$ because every constant string in $L(p) \cup L(q)$ contains either a or b. This means $\Sigma^{\geq |c(q)|} \not\subseteq L(p) \cup L(q)$, which is a contradiction. Thus there is no co-pattern q^c of an erasing regular pattern q such that $S \subseteq L(q^c) \subsetneq L(p)$.

Therefore $L(p)$ is a minimal language of S within $\mathscr{JERPL}$.

(2) In case $c(p) = \varepsilon$, i.e., p contains no constant symbol, and thus we assume $p = x$ without loss of generality.

By the definition of a minimal language, we see that for every erasing regular pattern q with $c(q) \neq \varepsilon$, $S \not\subseteq L(q)$ holds, because $L(q) \subsetneq L(p) = \Sigma^*$ holds.

Let us put r^c = co-MINL(S). Clearly, $L(r^c)$ is a minimal language of S within co-$\mathscr{ERPL}$. Here, we show by contradiction that there is no erasing regular pattern q such that $S \subseteq L(q) \subsetneq L(r^c)$.

Suppose that there is an erasing regular pattern q such that $S \subseteq L(q) \subsetneq L(r^c)$. Since $L(q) \subsetneq L(r^c) \subseteq \Sigma^*$, it follows that $L(q) \subsetneq \Sigma^*$, and thus q contains at least one constant symbol, i.e., $c(q) \neq \varepsilon$. As stated above, this means $S \not\subseteq L(q)$, which is a contradiction. Thus there is no erasing regular pattern q such that $S \subseteq L(q) \subsetneq L(r^c)$.

Therefore $L(r^c)$ is a minimal language of S within $\mathscr{JERPL}$.

By (1) and (2), we see that the procedure JMINL outputs π such that $L(\pi)$ is a minimal language of S within $\mathscr{JERPL}$. □

Since the class $\mathscr{JERPL}$ $(= \mathscr{TERPL}_1)$ has finite elasticity, we see by Theorem 18 that the following theorem holds:

Theorem 19 *Let* $\sharp\Sigma \geq 3$.

The class $\mathscr{TERPL}_1$ *is inferable in the limit from positive examples by a polynomial time updating IIM.*

REFERENCES

1. D. Angluin, "Finding patterns common to a set of strings," in *Proceedings of the 11th Annual Symposium on Theory of Computing*, 1979, pp. 130–141
2. D. Angluin, "Inductive inference of formal languages from positive data," in *Information and Control* **45**, 1980, pp. 117–135.
3. S. Arikawa, S. Kuhara, S. Miyano, Y. Mukouchi, A. Shinohara, and T. Shinohara, "A machine discovery from amino acid sequences by decision trees over regular patterns," in *New Generation Computing* **11**(3,4), 1993, pp. 361–375.
4. H. Arimura, T. Shinohara and S. Otsuki, "Finding minimal generalizations for unions of pattern languages and its application to inductive inference from positive data," in *Proceedings of the 11th Symposium on Theoretical Aspects Computer Science, Lecture Notes in Computer Science* **775**, 1994, pp. 646–660.
5. H. Arimura, R. Fujino, T. Shinohara and S. Arikawa, "Protein motif discovery from positive examples by minimal multiple generalization over regular patterns," in *Genome Informatics* **5**, 1994, pp. 39–48.
6. C.B. Fraser, R.W. Irving and M. Middendorf, "Maximal common subsequences and minimal common supersequences," in *Information and Computation* **124**, 1996, pp. 145–153.
7. E.M. Gold, "Language identification in the limit," in *Information and Control* **10**, 1967, pp. 447–474.
8. S. Lange and R. Wiehagen, "Polynomial-time inference of arbitrary pattern languages," in *New Generation Computing* **8**, 1991, pp. 361–370.
9. S. Miyano, "Learning theory towards Genome Informatics," in *Proceedings of the 4th Workshop on Algorithmic Learning Theory, Lecture Notes in Artificial Intelligence* **744**, 1993, pp. 19–36.
10. T. Moriyama and M. Sato, "Properties of language classes with finite elasticity," in *IEICE Transactions on Information and Systems* **E78-D**(5), 1995, pp. 532–538.
11. T. Motoki, T. Shinohara and K. Wright, "The correct definition of finite elasticity: Corrigendum to identification of unions," in *Proceedings of the 4th Annual Workshop on Computational Learning Theory* 1991, pp. 375–375.

12. Y. Mukouchi, "Containment Problems for Pattern Languages," in *IEICE Transactions on Information and Systems* **E75-D**(4), 1992, pp. 420–425.
13. D. Reidenbach, "A Negative Result on Inductive Inference of Extended Pattern Languages," in *Proceedings of the 13th International Conference on Algorithmic Learning Theory, Lecture Notes in Computer Science* **2533**, 2002, pp. 308–320.
14. D. Reidenbach, "On the Learnability of E-pattern Languages over Small Alphabets," in *Proceedings of the 17th Annual Conference on Learning Theory, Lecture Notes in Computer Science* **3120**, 2004, pp. 140–154.
15. M. Sato, "Inductive inference of formal languages," in *Bulletin of Informatics and Cybernetics* **27**(1), 1995, pp. 85–106.
16. T. Shinohara, "Polynomial time inference of pattern languages and its applications," in *Proceedings of the 7th IBM Symposium on Mathematical Foundations of Computer Science* 1982, pp. 191–209.
17. T. Shinohara, "Polynomial time inference of extended regular pattern languages," in *Proceedings of RIMS Symposium on Software Science and Engineering, Lecture Notes in Computer Science* **147**, 1982, pp. 115–127.
18. T. Shinohara, "Inductive inference from negative data," in *Bulletin of Informatics and Cybernetics* **21**(3, 4), 1985, pp. 67–70.
19. M. Terada, Y. Mukouchi and M. Sato, "Inductive Inference of Decision Trees over Patterns from Positive Examples" (in Japanese), in *Journal of the IEICE* **J83-DI**(1), 2000, pp. 60–67.
20. K. Wright, "Identification of unions of languages drawn from an identifiable class," in *Proceedings of the 2nd Annual Workshop on Computational Learning Theory* 1989, pp. 328–333.

Some Applications of Hida Distributions to Biology

Si Si

Faculty of Information Science and Technology, Aichi Prefectural University, Nagakute, Aichi 480-1198, Japan, E-mail: sisi@ist.aichi-pu.ac.jp

Abstract. We discuss significant applications of white noise analysis to biological science. Having introduced the space of Hida distributions, we can define partial differential operators ∂_t acting on the space rigorously, and we apply them to typical biological problems.

Keywords: Brownian motion, white noise, Hida distribution, Gaussian, annihilation operator.
PACS: 01.30.Rr, 02.50.Cw, 02.50.Fz

1. INTRODUCTION

In the modern biological science, we often meet random phenomena interfered with by a noise. We are interested in such phenomena in a microscopic level. Important results often occur being influenced by random noise, the idealized version of which is the Gaussian white noise. We are therefore suggested to discuss mathematical models of random complex systems, which are functions of the Gaussian white noise. They are assumed to be evolutional that is they are developing as the time it goes by.

Our first aim is therefore discuss the analysis of functions of Gaussian white noise, a mathematical model of which is the time derivative $\dot{B}(t)$ of a Brownian motion $B(t)$. Then, we are naturally led to discuss the case where the white noise is used to be an input of a non-linear device and where we observe the output to identify the unknown device. We know an interesting and significant case of identifying the retina. This has been done successfully, where we appeal to the white noise calculus.

For the second aim we have to provide tools for the analysis of random complex systems which are expressed as functions of white noise. We shall be able to use differential and integral calculus, which have been established recently.

2. WHITE NOISE

We will review the white noise theory and at the same time we should like to confirm that the theory is fitting for our purpose.

In the communication theory, white noise is viewed as a random signal (or process) with a flat spectral density. This means that the signal's power spectral density has equal power in any band. White noise is considered analogous to white light which contains all frequencies. Gaussian noise (i.e., noise with a Gaussian amplitude distribution) is necessarily white noise. However, while the term 'white' refers to the shape of the flat power spectral density, not only Gaussian but also Poisson, Cauchy, etc., white noises.

CP1028, *Collective Dynamics: Topics on Competition and Cooperation in the Biosciences*
edited by L. M. Ricciardi, A. Buonocore, and E. Pirozzi

Thus, the two words "Gaussian" and "white" are often both specified in mathematical models of systems. Gaussian white noise is a good approximation of many real-world situations and generates mathematically tractable models.

What we are going to discuss is the analysis of generalized white noise functionals which are called **Hida distributions**, thus, we are concerned with the case only of Gaussian white noise. We will just write white noise instead of Gaussian white noise in what follows. White noise is realized as the time derivative of a Brownian motion, denoted by $\dot{B}(t)$. To be able to understand the white noise rigorously in probabilistic sense we will discuss some facts in the following.

2.1. Give a status of $\dot{B}(t)$

1. In a classical stochastic analysis (See, for instance, [1].), Brownian motion is denoted by $B(t)$ and white noise is formally written as $\frac{d}{dt}B(t) = \dot{B}(t)$. But, a sample function of white noise is not ordinary function since $B(t)$ does not have derivative. However, we understand that a sample function of $\dot{B}(t)$ is a generalized function. With a test function $\xi(t)$, we can define a bilinear form
$$\langle \dot{B}, \xi \rangle = \int \xi(t)\dot{B}(t)dt) = \dot{B}(\xi), \tag{1}$$
it may be understood as $\int \xi(t)dB(t)$, where $dB(t)$ is taken as a random measure. We can see that it is subject to a Gaussian distribution with mean 0 and variance $\|\xi^2\|$. If $f \in L^2(R)$, $\langle \dot{B}(t), \xi \rangle$ extends to $\langle \dot{B}(t), f \rangle$. The function f can be approximated by test functions $\{\xi_n\}$, then the limit of $\langle \dot{B}(t), \xi_n \rangle$ can be defined in the topology, namely, in the norm on the space $L^2(\Omega, P)$, where (Ω, P) is a probability space on which the Brownian motion $B(t)$ is defined. The limit is expressed as $\langle \dot{B}, f \rangle$ and we call it a stochastic bilinear form.
2. In 1955 Gel'fand defined a generalized stochastic process for the case that sample function is a generalized function. White noise $\dot{B}$ is such a generalized stochastic process. That is, take a test function ξ and obtain $\langle \dot{B}, \xi \rangle$. An ordinary stochastic process $\{X(t)\}$ is a system of random variables with parameter t, however, for a generalized stochastic process $\{X(\xi)\}$, the parameter is $\xi (\in E)$, where E is a nuclear space.
Generalizing a concept of characteristic function, we can define a characteristic functional such as
$$C(\xi) = E\left(e^{iX(\xi)}\right), \ \xi \in E. \tag{2}$$
In this sense, $\dot{B}(t)$ is a generalized stochastic process and its characteristic functional is
$$C(\xi) = \exp[-\frac{1}{2}\|\xi\|^2]. \tag{3}$$
According to the Bochner-Minlos theorem, there exists a random measure μ in E^*, a dual space of E. The measure space (E^*, μ) is now called white noise. A member x of E^* can be thought as a sample function of $\dot{B}(t)$ almost surely μ.

3. $\dot{B}(t)$ as a generalized functional.

Let H_1 be a Hilbert space spanned by linear functions of Brownian motion :

$$H_1 = \left\{ \int f(t)\dot{B}(t)dt = \langle \dot{B}, f \rangle, \ f \in L^2(R) \right\}, \tag{4}$$

then H_1 is isomorphic to $L^2(R^1)$:

$$H_1 \cong L^2(R^1)$$

with an isomorphic correspondence

$$f \longleftrightarrow \dot{B}(f) \ (= \langle \dot{B}, f \rangle).$$

Here if it were possible to take the delta function δ_t instead of f, it corresponds to $\dot{B}(t)$. It means that we need to expand $L^2(R^1)$ to a space which contains delta function, that is $K^{-1}(R^1)$, the Sobolev space of order -1. It is the dual space of Sobolev space $K^1(R^1)$ of order 1. Then, we have the Gel'fand triple

$$K^1(R^1) \subset L^2(R^1) \subset K^{-1}(R^1), \tag{5}$$

where the inclusions are continuous injection from left to right.

To establish the triple for the generalized white noise functionals, it can be done as follows. Let $H_1^{(1)}$ be a subspace of H_1 with a strong topology such that $H_1^{(1)}$ is isomorphic to $K^1(R^1)$:

$$H_1^{(1)} \cong K^1(R^1).$$

Then, the dual space $H_1^{(-1)}$of $H_1^{(1)}$ is isomorphic to $K^{(-1)}(R^1)$:

$$H_1^{(-1)} \cong K^{-1}(R^1).$$

We now have a Gel'fand triple

$$H_1^{(1)} \subset H_1 \subset H_1^{(-1)}, \tag{6}$$

where $H_1^{(1)}$ is a space of test functionals and $H_1^{(-1)}$ is a space of generalized functionals. It is seen that the white noise $\dot{B}(t)$ exists as a member of $H_1^{(-1)}$. In other words, the *identity* of white noise $\dot{B}(t)$ is given as a member of $H_1^{(-1)}$.

In addition, $\{\dot{B}(t), t \in R\}$ is a system of variables of white noise generalized functionals.

Note. Although finite numbers of $\dot{B}(t)$'s in $H_1^{(-1)}$ are linearly independent, $H_1^{(-1)}$ is not a continuously infinite dimensional space unlike the case where $\{\dot{B}(t), t \in R\}$ is taken to be an orthogonal base.

2.2. Non-linear functionals of white noise

Since the white noise system $\{\dot{B}(t), t \in R\}$ is already defined, we can now discuss a general function. All the linear functionals of white noise are in $H_1^{(-1)}$ and so they are already well defined. Basic nonlinear functionals are polynomials, but we have defined the functionals of only degree one. It is now to consider the case for the degree $n(\geq 2)$. However, here we meet some problems.

i) Naturally there can not be rigorous definition for polynomials in $\dot{B}(t)$ since they are not in a classical space (L^2) (i.e. Fock space $\bigoplus H_n$, where H_n is a multiple Wiener integral of degree n) of nonlinear functions of $B(t)$'s with finite variance. We now have to expand the space (L^2) so as to contain all the polynomials in $\dot{B}(t)$'s.
ii) Each member H_n of the Fock space is to expand to the space which contains the Hermite polynomials in $\dot{B}(t)$ of degree n. Observe the case of degree 1 and determine a suitable degree of the Sobolev space on R^n.

It is favorable to choose the degree m of the *Sobolev space* to be $\frac{n+1}{2}$. The reason is that the trace theorem of the Sobolev spaces tells us that members in $K^{(n+1)/2}(R^n)$ are continuous and the restriction of them to an $(n-1)$-dimensional hyperplane belongs to the space $K^{n/2}\widehat{(R^{n-1})}$, namely the degree decreases by 1/2 when the restriction is made to a subspace of one dimension lower. These properties are convenient when the analysis on H_n is carried out, by taking the isomorphism between H_n and $\widehat{L^2(R^n)}$ into account.

$$K^{\widehat{(n+1)/2}(R^n)} \subset \widehat{L^2(R^n)} \subset K^{-\widehat{(n+1)/2}(R^n)}, \tag{7}$$

where $\widehat{L^2(R^n)}$ denotes the subset containing all symmetric functions of $L^2(R^n)$ ([5], [9]).

As in the case of $n = 1$, we have

$$H^{(n)} \subset H_n \subset H_n^{(-n)}. \tag{8}$$

The definitions of test functionals and generalized functionals are the same as for the case of $n = 1$.

The polynomials in $H_n^{(-n)}$ are renormalized polynomial and they do not contain in classical space H_n. Namely $H_n^{(n)}$ is isomorphic to $K^{\widehat{(n+1)/2}(R^n)}$ and $H_n^{(-n)}$ is the dual space of $H_n^{(n)}$.

Example. Consider the simplest non-linear function $\dot{B}(t)^2$. We will compare with the classical calculation where $dB(t)$ is defined as a random measure. Non-linear calculation can be done, e.g. according to the Itô table :

$$(dB(t))^2 = dt.$$

Hence

$$\left(\frac{dB(t)}{dt}\right)^2 = \frac{1}{dt}.$$

This can not be controlled. On the other hand, it can be seen that the difference

$$(dB(t))^2 - dt$$

is random although it is infinitesimal. It means that there has to be something which can not be visualized. There is no meaning to let them independent each other. However, if dt is approximated to the infinitesimal interval Δ containing t, the approximation of $\left(\frac{dB(t)}{dt}\right)^2 - \frac{1}{dt}$ is

$$\left(\frac{\Delta B}{\Delta}\right)^2 - \frac{1}{\Delta}, \tag{9}$$

the limit of which is in $H_2^{(-2)}$, but not in H_2. This limit is called *renormalized square* of $\dot{B}(t)$ and is denoted by $:\dot{B}(t)^2$. The renormalization can be applied to polynomials in $\dot{B}(t)$'s of higher degree and then we have $H_n^{(-n)}$, $n \geq 1$.

Finally, take a positive non-increasing sequence$\{c_n\}$ and let

$$(L^2)^- = \bigoplus c_n H_n^{(-n)}, \tag{10}$$

which is a generalized white noise functional space. A member of $(L^2)^-$ is called *Hida distribution.*

Once again we emphasize the basic idea of our analysis that we introduce the space $(L^2)^-$ of generalized white noise functionals since $\dot{B}(t)$ (or $x(t)$) is a variable of white noise functionals.

Now it is quite reasonable to define partial derivatives in the variables $\dot{B}(t)$'s. For this purpose we introduce the S-transform.

S-transform
For $\varphi(x) \in (L^2)$, the $(S\varphi)(\xi)$ is given by

$$(S\varphi)(\xi) = C(\xi) \int_{E^*} \exp[\langle x, \xi \rangle] \varphi(x) d\mu(x). \tag{11}$$

It may also be defined in such a manner that

$$(S\varphi)(\xi) = \int_{E^*} \varphi(x + \xi) d\mu(x).$$

Remark 2.1 *The S-transform may be thought of as an analogue of the Laplace transform applied to functionals on R^n. However, it has more important properties in our case and plays important and different roles from the finite dimensional case. For details we refer to the monograph [4].*

Annihilation operator
We wish to introduce a partial differential operators expressed symbolically in the form

$$\partial_t = \frac{\partial}{\partial \dot{B}(t)}.$$

Since the variable $\dot{B}(t)$ is a generalized functional in $H_1^{(-1)}$, the partial differential operator has to be defined in a generalized sense .

Incidentally, we claim that the operator ∂_t should be extended to an *annihilation operator* acting on the space of generalized white noise functionals. Rigorous definition is given as follows.

Definition 2.1 *Let* φ *be a generalized white noise functional and let* $U(\xi)$ *be its S-transform. Then, the annihilation operator is defined by*

$$\partial_t \varphi = S^{-1} \frac{\delta}{\delta \xi(t)} U(\xi), \tag{12}$$

if $\frac{\delta}{\delta \xi(t)} U(\xi)$ *exists and is a U-functional. The notation* $\frac{\delta}{\delta \xi(t)}$ *stands for the Fréchet derivative.*

Intuitively speaking, the Fréchet derivative (See, for instance, [2].) may be viewed as a *continuous* analogue of a differential of a function $u = u(x_1, x_2, \cdots, x_n)$ on R^n, such that

$$du = \sum_1^n \frac{\partial u}{\partial x_i} dx_i,$$

where each coordinate (or direction) contributes equally in $i, 1 \leq i \leq n$. Similarly each $\dot{B}(t)$, where $t \in R$, contributes equally in t in our analysis.

3. APPLICATIONS TO BIOLOGY

The white noise analysis can be applied to the study of biological, in fact, organic bodies. There are two typical cases.

In the first case, the body in question admits white noise input and output can be observed. A good example can be seen in Naka's theory ([7], [8]). In general, we assume that there is an evolutional system for which input signal is taken to be $\dot{B}(t)$ and output $X(t)$ is observed. Obvious additional assumptions are that the system is non-anticipaticing and the output $X(t)$ is in $(L^2)^-$ and is continuous in t.

Under the assumption stated above $X(t), t \geq 0$, is expressed in the form

$$X(t) = \sum_{n=0}^{\infty} X_n(t), \quad X_n(t) \in H_n^{(-n)}. \tag{13}$$

Using the isomorphism described in the previous section $X_n(t)$ has the representation F_n in $\widehat{K^{(-n)}(R^n)}$:

$$X_n(t) \sim F_n(u_1, \cdots, u_n) \in \widehat{K^{(-n)}(R^n)} \tag{14}$$

Theorem 3.1 *Assume that* $X_n(t)$ *is in the domain of* D_n*, where*

$$D^n = \prod_{j=1}^{n} \partial_{t_j}, \ t_j\text{'s are different.}$$

Then

$$F_n(u_1,\cdots,u_n) = D^n X_n(t). \tag{15}$$

Remark 3.1 *If $X_n(t)$ is, so called regular, i.e. F is smooth then the proof is obvious. However, if $X_n(t)$ is not regular we need to modify the theorem. For example, if*

$$X_2(t) = \int_0^t f(u) : \dot{B}(u)^2 : du \tag{16}$$

then

$$\partial_s^2 X_2(t) = f(s)\frac{1}{ds},\ s < t. \tag{17}$$

In the second case $X(t)$ is Gaussian and no assumption on input. We, however assume that $X(t)$ is continuous in t and $X(0) = 0$. Then we can appeal the Hida-Crammér theorem which asserts that

$$X(t) = \sum_{n=0}^{N} X_n(t), \quad (N \text{ may be } \infty), \tag{18}$$

and that $X_n(t)$ is expressed in the canonical form ([3]) such that

$$X_n(t) = \int_0^t F_n(t,u)\dot{B}_n(u)du, \tag{19}$$

where $\dot{B}_n(u)du$ is a Gaussian random measure.

Theorem 3.2 *Under the above assumptions and with the assumption that $B_n(u)$'s are Brownian motions, the operators $\partial_t^n = \frac{\partial}{\partial \dot{B}_n(t)}, n \geq 1$, are defined and*

$$F_n(t,u) = \partial_u^n X(t),\ u < t. \tag{20}$$

The proof is easy and the system is identified by the kernel $F_n(t,u)$.

Note that there are many generalizations of this theorem and are useful in the identification of a Gaussian system (See [6] for related topics.).

ACKNOWLEDGMENTS

The author wishes to express her deep gratitude to Professor Luigi Ricciardi, the organizer of BIOCOMP, who invited her paper to the proceedings. Thanks are also due to HIBI foundation for the financial support.

REFERENCES

1. P. Lévy, *Processus stochastiques et mouvement brownien*, Gauthier-Villars, 1948 [2éme ed. 1965].
2. P. Lévy, *Problèmes concret d'analyse fonctionnelle*, Gauthier-Villars, 1951.
3. T. Hida, *Mwm. Coll. Sci.Univ. Kyoto*, **33**, 258–351 (1960).

4. T. Hida, *Analysis of Brownian functionals*, Carleton Math. Notes, no. 13, 1975.
5. T. Hida and Si Si, *Lectures on white noise functionals*, World Scientific Pub. Co., 2006.
6. T. Hida, T. Shimizu and Si Si, *Gaussian systems*, Volterra Center Notes, 2006.
7. K. Naka and V. Bhanot, "White-Noise analysis in retinal physiology", in *Advanced Mathematical Approach to Biology*, edited by T. Hida, World Scientific, 1997, pp. 109–267.
8. M. Schetzen, *The Volterra & Wiener Theory of Nonlinear Systems*, Kreiger Publ. Com., 2006.
9. Si Si, *Infinite Dimensional Analysis and Quantum Probability and Related Topics*, **11** (1), (2008) (to appear).

Remembering a Friend

Luigi M. Ricciardi

Dipartimento di Matematica e Applicazioni, Università di Napoli Federico II, Via Cintia, I-80126 Napoli, Italy, E-mail: lmr@unina.it

My long-term friendship with Professor Ei Teramoto has left an indelible mark upon my feelings, my private life, and my academic activity. To condense his impact on me within a few pages is difficult, if not completely impossible. Hence, I shall refrain from commenting on Ei's numerous and invaluable scientific achievements, which can be handled ably and thoroughly by others. For readers who would like the great pleasure of knowing him better, however, I hope to contribute some lesser-known facts about the man himself, and share some unique remembrances of our friendship.

I first ran across the name "Teramoto" in the late seventies. I was glancing through the pages of an issue of the Journal of Theoretical Biology when my attention was captured by the title of an article dealing with switching effects in prey-predator systems. It was authored by a name I had never heard before, a Mr. or a Ms. Tansky. Looking more carefully, I found a footnote that disclosed the arcana: the author's name was actually an acronym for a group of people consisting of Ei Teramoto, Hiroshi Ashida, Hisao Nakajima, Nanako Shigesada, Kohkichi Kawasaki, and Norio Yamamura.

I was immediately struck by the name Teramoto because its pronunciation is very close to the Italian word "terremoto", which means "earthquake." In retrospect, this word association turned out to be prophetic, considering the impact this benevolent Teramoto would have on me and many others when we met a few years later.

The occasion was the Symposium on Cybernetics in Biology and Medicine that I organized at the University of Vienna in April 1980. Almost immediately following this meeting, Professor Alwyn C. Scott and I had also planned a second workshop entitled "Biomathematics: Current Status and Future Perspectives" at the University of Salerno. Ei had accepted my invitation to give talks both in Vienna and in Salerno. As the time for the conferences grew nearer, however, I doubted he would really show up, as I heard nothing from him during the several months following his acceptance, and received no manuscript from him for the proceedings. Had I been acquainted with Ei, I would not have harbored these doubts: he was one of those rare individuals of an almost extinct species for whom "Yes" means yes, and "No" means no, without any "buts", "ifs," or other explicit or implicit conditions.

As it happened, Japan and Italy met for the first time in the neutral territory of the Austrian republic, and by a vigorous handshake initiated a friendship that would last for over fifteen years. Unfortunately no written trace remains of his talk in Vienna, but a summary of the elegant lecture he delivered at the Salerno workshop can be found in [1].

I paid my first visit to Ei's laboratory at Kyoto University in the fall of 1981. There I found a community of staff and graduate students diligently and effectively operating under Ei's unquestioned authority. This was augmented by the systematic guidance of

CP1028, *Collective Dynamics: Topics on Competition and Cooperation in the Biosciences*
edited by L. M. Ricciardi, A. Buonocore, and E. Pirozzi

an intelligent, multifaceted lady whose charm still matches her remarkable perseverance, endurance, and scientific ability.

After that first visit, my meetings with Ei became too numerous to be recalled. However, a few facts should be mentioned here as they may be valuable to those involved.

Ei and I had soon realized that through our joint efforts something useful to the scientific communities of both Italy and Japan could be devised. After discussing various possibilities, we concluded that the most effective tool would be a scientific cooperation agreement under the sponsorship of the Italian National Research Council (CNR) and its counterpart, the Japanese Society for the Promotion of Science (JSPS). As a result of our discussions, an ambitious joint proposal entitled "Competition and Cooperation in Complex Dynamical Systems" was organized and submitted to the two agencies. Japanese participants consisted of Professors Shunichi Amari (Tokyo), Takeyuki Hida (Nagoya), Shunsuke Sato (Osaka), Ei Teramoto (the coordinator of the Japanese team), and Masaya Yamaguti (Kyoto). Members of the Italian team included Professors Franco Brezzi (Pavia), Renato Capocelli (Salerno), Piero De Mottoni (L'Aquila), Alberto Tesei (Rome), and myself acting as coordinator. The project that resulted has proved successful over several years. It has led to numerous valuable scientific results and fostered fruitful, long-lasting cooperation among members of the two teams. This has been accomplished through joint research, seminars, conferences, workshops and, particularly, through a systematic exchange of visitors between the two countries.

Looking through my files, I recently found Ei's letter of February 23, 1983, in which he informed me that the JSPS would provide financial support of the project. He also asked when I would be able to join him again in Kyoto, and added: "I can serve you sake with different Kutani cups that Sachiko (that is, Mrs. Teramoto) recently found in a special market at Tooji Temple."

I can still taste the sake that Ei offered me on so many occasions and in so many different types of choko!

In addition to the cooperative agreement between CNR and JSPS, our shared sake led to another significant step in scientific collaboration: the setting up of an official cooperative agreement between the science faculties of Kyoto University and the University of Naples on the topic of "Applied Mathematics and Mathematical Biology." Cooperative activities here included the exchange of staff for research and teaching, the exchange of graduate students for study, and the joint organization of seminars and workshops. The agreement between the Japanese and Italian universities, signed by the academic authorities on May 18, 1984, has proved to be valuable in many respects. For example, under this agreement Dr. H. Seno was able to complete his thesis work in two years of graduate study in Naples, and several exchanges of staff took place as well. I myself made use of the agreement to spend one month in Kyoto in the fall of 1985 and to participate in the International Symposium on Mathematical Biology co-organized by Ei [2].

It would be irrilevant to list here the numerous periods of time that I spent at Ei's laboratory. I was often accompanied by my wife, who shared with Ei many friends and experiences from the world of physics, and by my daughter, at that time a student of the Naples Music Conservatory, who was kindly invited by Mrs. Teramoto to practice daily in her own home, on her own piano.

The International Workshop on Biomathematics and Related Computational Problems that took place in Naples and Capri in May 1987 was also organized as part of the above-mentioned agreement [3]. Ei joined me there on May 24, escorted by Drs. Kohkichi Kawasaki and Youichi Kobuchi. After the workshop, he spent ten additional days at the Mathematics Department of the University of Naples, in the city's historic district. Here he occupied an artistically paneled seventeenth century office, which pleased him immensely.

Disregarding unessential chronology, I would like to spend the remainder of this brief essay sketching a few images of Ei as they now emerge in my memory. Perhaps I can express some of the feelings I experienced in his company, and so offer the reader a human perspective on a scientist who seemed ever to transcend it.

I see the two of us sitting silently on a rock, overwhelmed by the mystic, ineffable beauty of the Katsura-Rikyu gardens, where time and space seem to relinquish their comfortable limitations, merging into indefinite new dimensions that invite the risk of immersion. I recall Ei's admiration of art (he himself was a very gifted painter) wherever he found it: in the graceful movements of a Flamenco lady dancer in Las Palmas de Gran Canaria in the occasion of the International Conference of Mathematics at the Service of Man; or in an incredibly bright firmament over the terrace of the Caesar Augustus Hotel in Capri, after his own stellar presentation at that workshop. I see myself telling Ei about Miramare Castle near the International Center for Theoretical Physics in Trieste, and reporting to him the fate of Archduke Ferdinand Maximilian Joseph of Austria as relived in the poetry of Giosuè Carducci. I recall that from time to time we unexpectedly discovered that we shared experiences or acquaintances: once I happened to mention that some of my work had been inspired by discussions with a physicist named Arnold Siegert and asked Ei if he had ever run across that name, never suspecting that the very same person had been Ei's host and co-worker at Evanston several years before!

I cannot count the times Ei and I discussed perspectives on mathematical biology. He was sensitive to the political fragility of this topic in certain academic circles still firmly anchored in more traditional or fashionable disciplines, and he was much concerned about the careers of his collaborators and his students. I remember a quiet chat one evening over the rise and fall of Mathematical Biology at a certain midwestern university in the United States, and the sudden, vehement intervention of a former professor of that university, which elicited a great deal of surprise in both of us. And can I ever forget Ei's deep consideration over a painful backache that almost immobilized me during a stay in Kyoto, and his prompt arrangements for my care and therapy at a specialized medical center? Or his invitation to me and to my family to admire Daimonjiyaki from his legendary Japanese-style office in the Biophysics Department of his university?

I recall that on this last occasion, at the insistence of several friends, he had agreed to sing a Japanese folk song for us. In the general silence that followed, Ei suddenly turned his witty expression into an intense, deep, painful grimace, and then his voice filled the air. I could not understand the words, but still vivid in my memory are the strong sensations produced by that sound originating from the very roots of Earth: a primordial, pervasive, and at times almost obsessive shakuhachi crying, alternating with soft, sadly modulated melodies …

A man of profound depth, Ei never hesitated to manifest his sense of humor. One day in Naples, while visiting the former Carthusian monastery known as the Certosa di San Martino, I showed him the old graveyard of the monks: a small square piece of land invaded by a profusion of weeds and wild bushes. Suddenly Ei whispered, "This is the way my grave is going to look."

I wonder now if it was merely a facetious remark, or if he truly saw more than his astonished friend. At the time, I hid my surprise, and we laughed.

REFERENCES

1. L.M. Ricciardi and A.C. Scott, Eds. (1982) Biomathematics in 1980. Papers presented at a workshop on Biomathematics: Current Status and Future Perspectives, Salerno, April 1980. Mathematics Studies Vol. 58. North Holland, Amsterdam.
2. E. Teramoto and M. Yamaguti, Eds. (1987) Mathematical Topics in Population Biology, Morphogenesis and Neurosciences. Proceedings of an International Symposium held in Kyoto, November 10–15, 1985. Lecture Notes in Biomathematics Vol. 71, Springer-Verlag, Heidelberg.
3. L.M. Ricciardi, Ed. (1988) Biomathematics and Related Computational Problems (Dedicated to Ei Teramoto). Proceedings of an International Workshop held in Naples and Capri, May 25–30, 1987. Kluwer Academic Publisher, Dordrecht.

BIOCOMP2007

Registered Participants

ABUNDO MARIO
Dipartimento di Matematica, Università Tor Vergata
Via della Ricerca Scientifica, 00133 Roma, Italy
E-mail: abundo@mat.uniroma2.it

AIELLO MARCO
Dipartimento di Matematica e Applicazioni, Università di Napoli Federico II
Via Cintia, 80126 Napoli, Italy
E-mail:

AIHARA IKKYU
Department of Applied Analysis and Complex Dynamical Systems, Graduate school of Informatics, Kyoto University
Kyoto 606-8501, Japan
E-mail: aihara.ikkyu@t02.mbox.media.kyoto-u.ac.jp

AIHARA KAZUYUKI
Complexity Modelling Project, ERATO, JST and Institute of Industrial Science, The University of Tokyo
Tokyo 153-8505, Japan
E-mail: aihara@sat.t.u-tokyo.ac.jp

ALBANO GIUSEPPINA
Università degli Studi di Salerno
Via Ponte Don Melillo, 84084 Fisciano (SA), Italy
E-mail: pialbano@unisa.it

AMATO UMBERTO
Istituto per le Applicazioni del Calcolo "Mauro Picone" CNR
Via P. Castellino 111, 80131 Napoli, Italy
E-mail: u.amato@iac.cnr.it

ANGELINI CLAUDIA
Istituto per le Applicazioni del Calcolo "Mauro Picone" CNR
Via P. Castellino 111, 80131 Napoli, Italy
E-mail: c.angelini@iac.cnr.it

ARIGA TAKAYUKI
Graduate School of Frontier Biosciences, Osaka University
1-3 Yamadaoka, Suita, Osaka, 565-0871 Japan
E-mail: ariga@phys1.med.osaka-u.ac.jp

ASTUMIAN DEAN
Dept. of Physics, University of Maine
5709 Bennet Hall Orono, ME 04473 U.S.A.
E-mail: astumian@maine.edu

BALZANO WALTER
Dipartimento di Scienze Fisiche, Università di Napoli Federico II
Via Cinthia, 4, 8012 Napoli - Italy
E-mail: walter.balzano@gmail.com

BARBIERI RICCARDO
Massachusetts General Hospital -Harvard Medical School, Massachusetts Institute of Technology Clinics 3, 55 Fruit Street
Boston, MA 02114, U.S.A.
E-mail: Barbieri@neurostat.mgh.harvard.edu

BARTUMEUS FREDERIC
Princeton University 106 Guyot Hall
08544 Princeton, NJ, U.S.A.
E-mail: fbartu@princeton.edu

BAZHENOV MAXIM
Salk Institute for Biological Studies
10010 North Torrey Pines Rd La Jolla, CA 92037
E-mail: bazhenov@salk.edu

BENES VIKTOR
Charles University in Prague, Faculty of Mathematics and Physics, Department of Probability and Mathematical Statistics
Sokolovská 83, 18675 Praha 8, Czech Republic
E-mail: benesv@karlin.mff.cuni.cz

BIBBONA ENRICO
INRiM
Strada delle Cacce, 91 - 10135 Torino
E-mail: enrico.bibbona@unito.it

BIER MARTIN
Department of Physics
East Carolina University Greenville, NC 27858 U.S.A.
E-mail: bierm@ecu.edu

BIMONTE GIUSEPPE
Dipartimento di Fisica, Università degli Studi di Napoli Federico II
Via Cintia, 80126 Napoli, Italy
E-mail:

BORISYUK ROMAN
University of Plymouth
A224 Portland Sq., Plymouth, PL4 8AA, UK
E-mail: rborisyuk@plymouth.ac.uk

BRAUN HANS ALBERT
Institute of Physiology, University of Marburg
Deutschhausstr. 2, D-35037 Marburg, Germany
E-mail: braun@staff.uni-marburg.de

BUFFA GIADA
Università di Torino
Via Carlo Alberto 10, 10123 Torino, Italy
E-mail: 253602@studenti.unito.it

BUONOCORE ANIELLO
Dipartimento di Matematica e Applicazioni, Università di Napoli Federico II
Via Cintia, 80126 Napoli, Italy
E-mail: aniello.buonocore@unina.it

CAPASSO VINCENZO
Dipartimento di Matematica
Via Saldini 50, 20133 Milano, Italy
E-mail: vincenzo.capasso@unimi.it

CAPUTO LUIGIA
Dipartimento di Matematica, Università di Torino
Via Carlo Alberto 10, 10123 Torino, Italy
E-mail: luigia.caputo@unito.it

CARFORA MARIA FRANCESCA
Istituto per le Applicazioni del Calcolo "Mauro Picone" CNR
Via P. Castellino 111, 80131 Napoli, Italy
E-mail: f.carfora@iac.cnr.it

CAUSIN PAOLA
Dipartimento di Matematica, Università degli Studi di Milano
Via Saldini 50, Milano, Italy
E-mail: causin@mat.unimi.it

CHIALVO DANTE
Department of Physiology, Northwestern University
303 E. Chicago Ave, Chicago, IL 60611, U.S.A.
E-mail: d-chialvo@northwestern.edu

COLAPS GENNARO
Dipartimento di Matematica e Applicazioni,Università di Napoli Federico II
Via Cintia, 80126 Napoli, Italy
E-mail: colaps@unina.it

COLLI FRANZONE PIERO
Dipartimento di Matematica, Università di Pavia
Via Ferrata 1, 27100 Pavia, Italy
E-mail: colli@imati.cnr.it

COUZIN IAN D.
Department of Ecology and Evolutionary Biology, Princeton University, Princeton, 08544 NJ, U.S.A.
Department of Zoology, University of Oxford, Oxford, OX1 3PS, UK.
E-mail: icouzin@princeton.edu

CULL PAUL
Computer Science, Oregon State University
Corvallis, Oregon, 97331, U.S.A.
E-mail: pc@cs.orst.edu

DE FEIS ITALIA
Istituto per le Applicazioni del Calcolo "Mauro Picone" CNR
Via P. Castellino 111, 80131 Napoli, Italy
E-mail: i.defeis@iac.cnr.it

DEL SORBO MARIA ROSARIA
Dipartimento di Matematica ed Applicazioni, Università di Napoli Federico II
Via Cinthia, 4 - 80126 Napoli - Italy
E-mail: marodel@gmail.com

DERÉNYI IMRE
Dept. of Biological Physics, Eötvös University
Pázmány P. stny. 1A, H-1117 Budapest, Hungary
E-mail: derenyi@elte.hu

DI CRESCENZO ANTONIO
Dipartimento di Matematica e Informatica, Università di Salerno
Via Ponte don Melillo, 84084 Fisciano (SA), Italy
E-mail: adicrescenzo@unisa.it

DI MAIO VITO
Istituto di Cibernetica "E. Caianiello" del CNR
C/o Complesso Olivetti, Via Campi Flegrei, 34, 80078, Pozzuoli, Italy
E-mail: vdm@biocib.cib.na.cnr.it

DI NARDO ELVIRA
Dipartimento di Matematica e Informatica, Università degli Studi della Basilicata
Viale dell'Ateneo Lucano 1, I-85100, Potenza, Italy
E-mail: elvira.dinardo@unibas.it

DITLEVSEN SUSANNE
University of Copenhagen
Øster Farimagsgade 5, 1014 Copenhagen K, Denmark
E-mail: sudi@pubhealth.ku.dk

ÉRDI PÈTER
Center for Complex Systems Studies,
Kalamazoo College Kalamazoo, MI, U.S.A.
E-mail: perdi@kzoo.edu

ERRICO DANIELA
Dipartimento di Matematica e Applicazioni, Università di Napoli Federico II
Via Cintia, 80126 Napoli, Italy
E-mail: daniela_errico@virgilio.it

FALCONI MANUEL
Depto. De Matemáticas, Fac. De Ciencias, UNAM
C. Universitaria, México, D.F. 04510, México
E-mail: falconi@servidor.unam.mx

FERGOLA PAOLO
Dipartimento di Matematica e Applicazioni, Università di Napoli Federico II
Compl. MSA, Via Cintia, Napoli, Italy
E-mail: paolo.fergola@dma.unina.it

FESTA PAOLA
Dipartimento di Matematica e Applicazioni, Università di Napoli Federico II
Compl. MSA, Via Cintia, Napoli, Italy
E-mail: paola.festa@unina.it

FUCHS EINAT
Department of Zoology, Tel-Aviv University Ramat Aviv
Tel Aviv 69978, Israel
E-mail: fuchsein@post.tau.ac.il

GAMBA ANDREA
Politecnico di Torino
Corso Duca degli Abruzzi 24, 10129, Torino - Italy
E-mail: andrea.gamba@polito.it

GIARDINA IRENE
CNR-INFM , Department of Physics, University of Rome La Sapienza
P.le A. Moro 2, 00185 Roma, Italy
E-mail: irene.giardina@roma1.infn.it

GIGANTE GUIDO
Istituto Superiore di Sanità
Viale Regina Elena 299, 00161, Rome, Italy
E-mail: guido.gigante@iss.infn.it

GIORNO VIRGINIA
Università degli Studi di Salerno
Via Ponte Don Melillo, 84084, Fisciano (SA), Italy
E-mail: giorno@unisa.it

GIRAUDO MARIA TERESA
Department of Mathematics, University of Torino
Via Carlo Alberto 10, 10123 Torino, Italy
E-mail: mariateresa.giraudo@unito.it

GRÜNBAUM DANIEL
School of Oceanography, University of Washington of Washington
Seattle, WA 98195-357940 U.S.A.
E-mail: grunbaum@ocean.washington.edu

HÄNGGI PETER
University of Augsburg, Universitätsstraße 1, 86159
Augsburg, Germany
E-mail: hanggi@physik.uni-augsburg.de

HASTINGS ALAN
Department of Environmental Science and Policy, University of California
One Shields Avenue, Davis, CA 95616, U.S.A.
E-mail: amhastings@ucdavis.edu http:/two.ucdavis.edu/~me

HIDA TAKEYUKI
Nagoya University Hirabari Minami 2-903,
Tenpaku-ku, Nagoya 468-0020, Japan
E-mail: takeyuki@math.nagoya-u.ac.jp

HOLCOMBE MIKE
University of Sheffield
Regent Court - Portobello Street, Sheffield, S1 4DP, UK
E-mail: m.holcombe@dcs.shef.ac.uk

HÖPFNER REINHARD
Institut für Mathematik, Johannes Gutenberg Universität
Mainz Staudingerweg 9, D-55099 Mainz, Germany
E-mail: hoepfner@mathematik.uni-mainz.de

HOTANI HIROKAZU
Department of Molecular Biology, School of Science Nagoya University
Nagoya, 464-8602, Japan
E-mail: hotani@nano-nagoya.jst.go.jp

ISHII YOSHIHARU
Soft Nano-machine Project, CTRST, JST Osaka University
1-3 Yamadaoka, Suita, Osaka, 565-0871, Japan
E-mail: ishii@phys1.med.osaka-u.ac.jp

IWAKI MITSUHIRO
Graduate School of Frontier Biosciences, Osaka University
7th Floor, Nanobiology building, 1-3 Yamadaoka, Suita, Osaka, Japan
E-mail: iwaki@phys1.med.osaka-u.ac.jp

IZHIKEVICH EUGENE M.
The Neurosciences Institute
San Diego CA, 92121, U.S.A.
E-mail: Eugene.Izhikevich@nsi.edu

KOHNO TAKASHI
Complexity Modelling Project, ERATO, JST and The University of Tokyo
Komaba Open Laboratory, The University of Tokyo, 4-6-1 Komaba, Megro-ku, Tokyo, 153-8505, Japan
E-mail: kohno@sat.t.u-tokyo.ac.jp

KOMORI TOMOTAKA
Graduate School of Frontier Biosciences,
Osaka University 1-3 Yamadaoka, Suita, Osaka 565-0871, Japan
E-mail: komotomo@phys1.med.osaka-u.ac.jp

KOSTAL LUBOMIR
Institute of Physiology, Academy of Sciences of the Czech Republic
Videnska 1083, 142 20 Prague 4, Czech Republic
E-mail: kostal@biomed.cas.cz

LANSKY PETR
Institute of Physiology, Academy of Sciences of the Czech Republic
Vídenská 1083, 14220 Praha 4, Czech Republic
E-mail: lansky@biomed.cas.cz

LARA-APARICIO MIGUEL
Universidad Nacional Autónoma de México, Facultad de Ciencias, Departamento de Matemáticas
Circuito Exterior S/N, Ciudad Universitaria, México
E-mail: aparicio@servidor.unam.mx

LEVIN SIMON A.
Princeton University
106 Guyot Hall, 08544 Princeton, NJ, U.S.A.
E-mail: slevin@princeton.edu

LITVAK- HINENZON ANNA
Biomathematics Unit, Faculty of Life Sciences, Tel Aviv University
Ramat-Aviv, Tel Aviv, Israel
E-mail: annal@post.tau.ac.il

LONGOBARDI MARIA
Dipartimento di Matematica e Applicazioni, Università di Napoli Federico II
Via Cintia, 80126 Napoli, Italy
E-mail: maria.longobardi@unina.it

LONGTIN ANDRÉ
Department of Physics and Center for Neural Dynamics, University of Ottawa
150 Louis Pasteur, Ottawa, ON, Canada K1N 6N5
E-mail: alongtin@uottawa.ca

MAGNASCO MARCELO O.
Rockefeller University
1230 York Avenue, New York NY 10021, U.S.A.
E-mail: magnasco@rockefeller.edu

MALCHOW HORST
Institute of Environmental Systems Research
University of Osnabrück, 49069 Osnabrück, Germany
E-mail: malchow@uos.de

MARTINUCCI BARBARA
Dipartimento di Matematica e Informatica, Università di Salerno
Via Ponte don Melillo, 84084 Fisciano (SA), Italy
E-mail: bmartinucci@unisa.it

MESZÉNA GEZA
Department of Biological Physics, Eötvös University
Pázmány 1A, H-1117, Budapest, Hungary
E-mail: Geza.Meszena@elte.hu

MIMURA MASAYASU
Meiji Advanced Center for Mathematical Sciences, Department of Mathematics, Meiji University
1-1 Higashimita, Tama-ku, Kawasaki, Japan 214-8571
E-mail: mimura@math.meiji.ac.jp

MIURA SHIGEHIRO
Division of Bioengineering, Graduate School of Engineering Science, Osaka University
Toyonaka Osaka 560-8531 Japan
E-mail: miura@bpe.es.osaka-u.ac.jp

MOEHLIS JEFF
Department of Mechanical Engineering
University of California, Santa Barbara, CA, 93106, U.S.A.
E-mail: moehlis@engineering.ucsb.edu

MOHTASHEMI MOJDEH
MITRE/MIT
Cambridge, MA
E-mail: mojdeh@mit.edu

MONTAGNINI LEONE
Istituto di Cibernetica "E. Caianiello" Napoli, Italy
Via di Monteverde 57, 00151 - Roma, Italy
E-mail: leonemontagnini@katamail.com

MORALE DANIELA
Dipartimento di Matematica
Via Saldini 50, 20133 Milano, Italy
E-mail: daniela.morale@mat.unimi.it

MORENO-DIAZ ARMINDA
School of Computer Science. Universidad Politécnica de Madrid
Campus de Montegancedo, s/n. 28660. Madrid, Spain
E-mail: amoreno@fi.upm.es

MOSS FRANK
Physics Department, University of Missouri - St. Louis
One University Boulevard - St. Louis, MO 63121-4400 U.S.A. 314-516-5000
E-mail: mossf@umsl.edu

NAKAGAKI TOSHIYUKI
Hokkaido University
SOUSEI 5-302 N21 W10 Sapporo, JAPAN
E-mail: nakagaki@es.hokudai.ac.jp

NAKAOKA SHINJI
Aihara Complexity Modelling Project, ERATO, JST, The University of Tokyo
Komaba Open Laboratory, The University of Tokyo, 4-6-1, Komaba, Megro-ku Tokyo, 153-8505, Japan
E-mail: snakaoka@aihara.jst.go.jp

NALDI GIOVANNI
Politecnico di Torino
Corso Duca degli Abruzzi 24, 10129, Torino - Italy
E-mail: giovanni.naldi@mat.unimi.it

NOBILE AMELIA G.
Dipartimento di Matematica e Informatica, Università di Salerno
Via Ponte don Melillo, 84084 Fisciano (SA), Italy
E-mail: nobile@unisa.it

NOMURA TAISHIN
Division of Bioengineering, Graduate School of Engineering Science, Osaka University
The Center for Advanced Medical Engineering and Informatics, Osaka University
E-mail: taishin@bpe.es.osaka-u.ac.jp

OLSON DONALD B.
RSMAS/MPO University of Miami
4600 Rickenbacker Cswy
E-mail: dolson@rsmas.miami.edu

PELLEREY FRANCO
Dipartimento di Matematica, Politecnico di Torino,
Corso Duca degli Abruzzi, 24, I-10129 Torino, Italy
E-mail: franco.pellerey@polito.it

PIROZZI ENRICA
Dipartimento di Matematica e Applicazioni, Università di Napoli
Via Cintia, 80126 Napoli, Italy
E-mail: enrica.pirozzi@unina.it

POKORA ONDREJ
Department of Mathematics and Statistics, Faculty of Science, Masaryk University
Janáckovo námestí 2a, 60200 Brno, Czech Republic
E-mail: pokora@math.muni.cz

POSTNOVA SVETLANA
Institute of Physiology, University of Marburg
Deutschhausstr. 2, D-35037 Marburg, Germany
E-mail: postnova@staff.uni-marburg.de

PUGLIESE ANDREA
Dipartimento di Matematica, Università di Trento
Via Sommarive, 14 - I-38050 Povo (TN), Italy
E-mail: andrea.pugliese@unitn.it

RICCIARDI LUIGI MARIA
Dipartimento di Matematica e Applicazioni, Università di Napoli
Via Cintia, 80126 Napoli, Italy
E-mail: luigi.ricciardi@unina.it

RICCIARDI TONIA
Dipartimento di Matematica e Applicazioni, Università di Napoli
Via Cintia, 80126 Napoli, Italy
E-mail: tonia.ricciardi@unina.it

SACERDOTE LAURA
Dipartimento di Matematica, Università degli Studi di Torino
Via Carlo Alberto, 10, 10123 Torino
E-mail: laura.sacerdote@unito.it

SALERNO MARIO
Department of Physics "E.R. Caianiello", University of Salerno
Via S. Allende, 1 -84081 Baronissi (SA), Italy
E-mail: salerno@sa.infn.it

SATO MASAKO
Faculty of Liberal Arts and Sciences, Osaka Prefecture University
1-1 Gakuen-cho, Sakai, Osaka 599-8531, Japan
E-mail: sato@las.osakafu-u.ac.jp

SATO SHUNSUKE
Aino University
4-5-4, Higashi-Ohda 567-0012, Ibaraki, Osaka, Japan
E-mail: s-sato@pt-u.aino.ac.jp

SCACCHI SIMONE
Dipartimento di Matematica, Università di Pavia
Via Ferrata, 1 27100 Pavia
E-mail: simone.scacchi@unipv.it

SCALIA TOMBA GIANPAOLO
Dept. of Mathematics, University of Rome Tor Vergata
Via Ricerca Scientifica 1, 00196 Roma, Italy
E-mail: scaliato@mat.uniroma2.it

SCARPETTA SILVIA
Dept. Physics "E.R.Caianiello" University of Salerno, INFM and INFN gruppo coll. Salerno
Via S. Allende Baronissi (Salerno) Italy
E-mail: `silvia@sa.infn.it`

SCHIERWAGEN ANDREAS
University of Leipzig, Institute of Computer Science Intelligent Systems Department, Computational Neuroscience
Augustuspl. 10, 04109 Leipzig, Germany
E-mail: `schierwa@informatik.uni-leipzig.de`

SCHIMANSKY-GEIER LUTZ
Humboldt University at Berlin
Newtonstrasse 15, 12489 Berlin, Germany
E-mail: `alsg@physik.hu-berlin.de`

SENO HIROMI
Department of Mathematical and Life Sciences, Graduate School of Science, Hiroshima University
Kagamiyama 1-3-1, Higashi-hiroshima, Hiroshima 739-8526, Japan
E-mail: `seno@math.sci.hiroshima-u.ac.jp`

SHIGESADA NANAKO
Faculty of Culture and Information Science, Doshisha University
Kyotanabe 610-0394, Japan
E-mail: `nshigesa@mail.doshisha.ac.jp`

SHIMOKAWA TETSUYA
Graduate School of Frontier Biosciences, Osaka University
1-3, Yamadaoka, Suita, Osaka 565-0871, Japan
E-mail: `simokawa@phys1.med.osaka-u.ac.jp`

SIROVICH ROBERTA
Department of Mathematics, University of Torino
Via Carlo Alberto 10, 10123 Torino, Italy
E-mail: `roberta.sirovich@unito.it`

SMITH E. CHARLES
Biomath Program, North Carolina State University
Dept. Statistics, Box 8203, NCSU, Raleigh, NC 27695-8203, U.S.A.
E-mail: `bmasmith@ncsu.edu`

SZOLLOSI J. GERGELY
Department of Biological Physics, Eötvös University
Pázmány P. stny. 1A, H-1117 Budapest, Hungary
E-mail: `ssolo@angel.elte.hu`

TAKAGI SEIJI
Research Institute for Electronic Science, Hokkaido University
Sapporo 060-0812, Japan
E-mail: takagi@es.hokudai.ac.jp

TALKNER PETER
Universität Augsburg
Universitätsstrasse 1, 86135 Augsburg, Germany
E-mail: peter.talkner@physik.uni-augsburg.de

TERMINI SETTIMO
Istituto di Cibernetica "E. Caianiello" del CNR
C/o Complesso Olivetti, Via Campi Flegrei, 34, 80078, Pozzuoli, Italy
E-mail: s.termini@cib.na.cnr.it

TERO ATSUSHI
Hokkaido University
SOUSEI 2-306 N21 W10 Sapporo, JAPAN
E-mail: tero@topology.coe.hokudai.ac.jp

THIEULLEN MICHÈLE
Laboratoire de Probabilités et Modèles Aléatoires, Univ. Paris 6
4 Place Jussieu, 75252 Paris cedex 05
E-mail: mth@ccr.jussieu.fr

TUCKWELL HENRY
Max Planck Institute for Mathematics in the Sciences
Inselstr. 22 Leipizig 04103 Germany
E-mail: tuckwell@mis.mpg.de

YANAGIDA TOSHIO
Graduate School of Frontier Biosciences, Osaka University
1-3 Yamadaoka, Suita, Osaka, 565-0871, Japan
E-mail: yanagida@phys1.med.osaka-u.ac.jp

YOSHIOKA MASAHIKO
Dept. Physics "E.R.Caianiello" University of Salerno
Via S. Allende Baronissi (Salerno) Italy
E-mail: yoshioka@sa.infn.it

VISENTIN FRANCESCA
Dipartimento di Matematica e Applicazioni, Università di Napoli
Via Cintia, 80126 Napoli, Italy
E-mail: francesca.visentin@unina.it

Author Index